Marine Microbiomes

This book examines the diversity, ecological roles, and biotechnological potential of marine microorganisms. The introductory chapters provide the vast biodiversity of marine archaebacteria, actinobacteria, cyanobacteria, and algal and fungal ecosystems. The subsequent chapters examine the contribution of these microbiomes to biogeochemical cycles and the establishment of symbiotic relationships within marine ecosystems. The text also highlights the biotechnological insights gained from studying marine microbes, including the discovery, characterization, and utilization of novel marine enzymes and the burgeoning bioeconomy centered on marine microbial products. Further, this book explores the mechanisms behind microorganism-mediated metal solubilization, the degradation of marine micropollutants, and the breakdown of petroleum hydrocarbons. The chapters also cover the realms of microbial bioprospecting for functional foods, novel medicinal compounds, cosmeceuticals, and the development of microbial biosensors. Toward the end, *Marine Microbiomes: Exploring Ecology, Biotechnology, and Environmental Impact* reviews human activities such as ocean acidification, rising temperatures, heavy metal pollution, and the introduction of microplastics and pathogens affecting marine microbial communities. This book caters to researchers and students in marine biology, microbiology, environmental science, and biotechnology.

Key features

- Provides an in-depth examination of the vast diversity of marine microorganisms, including archaebacteria, actinobacteria, cyanobacteria, algae, fungi, and viruses
- Examines the ecological roles and biotechnological potential of these microorganisms, emphasizing the richness and complexity of life in marine extreme habitats
- Explores the crucial role played by marine microbiomes in biogeochemical cycles
- Covers microorganism-mediated metal solubilization, bioleaching, the degradation of micropollutants, and the breakdown of petroleum hydrocarbons
- Discusses the potential of marine microorganisms as sources for novel applications in functional foods, medicines, cosmeceuticals, and biosensors

Bioprospecting and Sustainable Development Goals
Series Editor: Neeta Raj Sharma

Marine Microbiomes: Exploring Ecology, Biotechnology, and Environmental Impact
Manish Thakur, Tejinder Kaur, Surajit Pathak (ISBN 9781032834818)

For more information about this series, please visit https://www.routledge.com/Bioprospecting-and-Sustainable-Development-Goals/book-series/BSDG

Marine Microbiomes
Exploring Ecology, Biotechnology, and Environmental Impact

Edited by
Manish Thakur, Tejinder Kaur, and Surajit Pathak

CRC Press
Taylor & Francis Group
Boca Raton London New York

CRC Press is an imprint of the
Taylor & Francis Group, an **informa** business

Designed cover image: Shutterstock

First edition published 2026
by CRC Press
2385 NW Executive Center Drive, Suite 320, Boca Raton FL 33431

and by CRC Press
4 Park Square, Milton Park, Abingdon, Oxon, OX14 4RN

CRC Press is an imprint of Taylor & Francis Group, LLC

ISBN: 9781032834818 (hbk)
ISBN: 9781032834825 (pbk)
ISBN: 9781003509530 (ebk)

DOI: 10.1201/9781003509530

Typeset in Times
by codeMantra

Contents

About the Editors

Dr Manish Thakur, who earned a PhD in microbiology, is an expert in the field, focusing on microbiology and the identification of bioactive components. His vast teaching and research experience has led to significant contributions, documented through various publications in esteemed journals. He has published more than 25 articles in peer-reviewed international journals and books. His research excellence has been acknowledged with numerous awards and certificates. Additionally, he has engaged with the scientific community by presenting his research at both national and international conferences, sharing his insights with a broader audience.

Dr Tejinder Kaur, a distinguished academician and researcher, has been the coordinator of the Department of Zoology at DAV University, Jalandhar, India since 2013. She completed her PhD at Panjab University, Chandigarh, in 2011, demonstrating her dedication to academic achievement. Dr. Kaur's significant impact in her field is reflected in her substantial publication portfolio, which includes over 37 papers in both national and international journals. Her research interests are diverse, covering immunoprophylaxis, chemotherapy, immunochemotherapy, immunodiagnostics of *Leishmania donovani*, the developmental genetics of *Drosophila* and *Tribolium*, as well as antimicrobial resistance.

Dr Surajit Pathak, PhD, has been a professor at the Chettinad Academy of Research and Education, Chennai, India since 2015. He has also worked at Linkoping University, Linkoping, Sweden as a principal research scientist (2022–2022). He earned his PhD from India in 2007 and completed his postdoctoral training (2007–2015) at the University of Alabama, USA; the University of Padova, Italy; and the University of Linkoping, Sweden. His current research focuses on "biomarker discoveries for early detection of colorectal cancer." He has published more than 100 research articles in peer-reviewed international journals of repute. He received various research grants, including core international and national research grants from SERB-CRG (DST, Govt. of India), DST Indo-Italy Bilateral Research Projects, and other international funding agencies. He has various active international and national research collaborations. He is the editorial board member of various renowned international high-impact journals. He is a member of various professional research bodies in India, including the Indian Association of Cancer Research, the Indian Science Congress, and the Aesthetic Multispecialty Society (AMS).

1 The Enigmatic Depths
Exploring Marine Extreme Habitats

Pratikshya Das[1], *Manish Thakur*[2], *and Tejinder Kaur*[1]
[1]Department of Zoology, DAV University, Jalandhar, Punjab, India
[2]Department of Microbiology, School of Bioengineering and Biosciences, Lovely Professional University, Phagwara, Punjab, India

1.1 INTRODUCTION

Marine extreme habitats encompass a variety of environments characterized by extreme physical or chemical conditions. High pressure is a fundamental characteristic, with pressure increasing approximately 1 atm for every 10 m of depth, reaching a staggering 1,000 atm in the deepest ocean trenches (Kolb, 2024). Most deep-sea ecosystems hover between 2°C and 4°C (35.6°F and 39.2°F) (Rohling et al., 2022), except around hydrothermal vents where the superheated water can surpass 400°C (752°F) (Dahlhoff et al., 1991). This great pressure is matched with low temperatures. Another important consideration is the lack of sunlight beyond 200 m as the light is insufficient for photosynthesis, which results in ecosystems dependent on other energy sources such as chemosynthesis (Jannasch, 1985).

1.2 CLASSIFICATION OF MARINE EXTREME HABITATS

1.2.1 By Physical Characteristics

Hydrothermal Vents: Found along mid-ocean ridges and subduction zones, these vents spew superheated, mineral-rich water into the ocean depths. Temperatures here can exceed 400°C (752°F). These vents create unique ecosystems where chemosynthetic microorganisms convert chemical energy from vent minerals into organic matter, forming the base of food webs that support organisms like giant tube worms, vent crabs, and specialized fish (Fisher et al., 2007).

Cold Seeps: Located along continental margins and deep-sea sediments, cold seeps are areas where hydrocarbons like methane and hydrogen sulfide seep from the seabed. Despite moderate temperatures, these seeps create chemically rich oases that support diverse communities of mussels, clams, and tube worms, which depend on chemosynthetic bacteria to convert the seeping gases into energy (Suess, 2014).

Abyssal Plains: These vast, flat regions cover much of the deep ocean floor at depths between 3,000 and 6,000 m. Characterized by extreme pressures, near-freezing temperatures, and total darkness, abyssal plains are inhabited by a diversity of organisms adapted to the nutrient-poor environment, including echinoderms, polychaete worms, and various benthic species (Cormier & Sloan, 2018).

Deep-Sea Trenches: Representing the ocean's deepest parts, such as the Mariana Trench, these trenches are characterized by immense pressures over a thousand times that at sea level and near-freezing temperatures (Liu et al., 2022). These habitats are home to extremophiles—organisms uniquely adapted to the crushing pressures and limited food availability. Notable residents include certain species of fish, amphipods, and microorganisms (Bartlett, 2009).

1.2.2 By Chemical Characteristics

Methane Hydrates: Found in deep-sea sediments, these crystalline structures trap methane within a lattice of water ice, creating unique ecological niches. Microbial communities that metabolize methane thrive here, supporting specialized food webs and impacting global carbon cycles (Gornitz & Fung, 1994).

Hypersaline Brine Pools: These highly saline bodies of water are often found on the ocean floor, associated with evaporite deposits. The extreme salinity supports specialized microbial life forms adapted to high-salinity conditions, such as halophilic bacteria and archaea (Oren, 2016).

1.3 ROLE IN GLOBAL BIOGEOCHEMICAL CYCLES

Marine extreme habitats are integral to the global biogeochemical cycles, acting as hotspots for carbon sequestration, facilitating sulfur and methane cycling through microbial processes, and enhancing nutrient availability in the ocean. These functions underscore the ecological significance of these habitats, highlighting their importance in maintaining the balance of Earth's biogeochemical systems and offering insights into the resilience and adaptability of life under extreme conditions.

DOI: 10.1201/9781003509530-1

1.3.1 Carbon Cycle

In the deep-sea environments and other extreme marine habitats, carbon sequestration and cycling occur through unique mechanisms. Hydrothermal vents and cold seeps, for example, contribute significantly to the global carbon cycle (Wu et al., 2022). Chemosynthetic bacteria, which thrive in these habitats, utilize carbon dioxide or methane as a carbon source, converting these compounds into organic matter through chemosynthesis (Minic & Thongbam, 2011). This process not only supports the local ecosystem's food web but also impacts global carbon dynamics. In methane hydrate regions, microorganisms metabolize methane, a potent greenhouse gas, and either convert it to biomass or oxidize it into carbon dioxide, which is then sequestered in deep-sea sediments (Orcutt et al., 2020). Thus, these habitats act as sinks for greenhouse gases, mitigating their release into the atmosphere and playing a crucial role in the long-term storage of carbon.

1.3.2 Sulfur and Methane Cycles

The sulfur and methane cycles are intricately linked to microbial activities in marine extreme habitats. In hydrothermal vent systems, sulfur-oxidizing bacteria play a pivotal role in converting toxic hydrogen sulfide, emitted from the Earth's crust, into elemental sulfur or sulfate (Zeng et al., 2021). These bacteria form the base of the vent ecosystems, providing essential nutrients to other organisms through symbiotic relationships or as primary producers. Similarly, methane seeps host methanotrophic archaea and bacteria that oxidize methane into carbon dioxide and water, integrating methane into broader marine and global biogeochemical cycles (Paul et al., 2017). These microbial processes not only detoxify the local environment but also contribute to the cycling of sulfur and methane on a global scale, influencing both the availability of these elements and their interaction with other biogeochemical processes.

1.3.3 Nutrient Cycling

Extreme marine habitats also significantly affect nutrient cycling, particularly for nitrogen and phosphorus. Chemosynthetic communities in hydrothermal vents and cold seeps engage in nitrogen fixation, converting atmospheric nitrogen into forms usable by living organisms, thus contributing to the overall nitrogen budget of the ocean (Enrich-Prast et al., 2014). Additionally, these habitats facilitate the release and recycling of phosphorus from deep-sea sediments through the decomposition of organic matter and the activity of specialized microorganisms. The interaction between microbes and the extreme physical conditions in these habitats often result in the formation of nutrient-rich zones that support diverse biological communities, even in otherwise nutrient-poor environments like abyssal plains and deep-sea trenches (Hiraoka et al., 2020).

These dynamic processes, operating under extreme conditions, underscore the profound influence of marine extreme habitats on Earth's elemental balances, serving as critical regulators and sentinels of our planet's biochemical equilibrium.

1.4 CHEMOSYNTHETIC PRIMARY PRODUCTION

In the hidden depths of our oceans, far from the reach of sunlight, life flourishes by tapping into the chemical riches of its surroundings through a process known as chemosynthesis (Chen et al., 2023). This form of primary production is crucial in sustaining the diverse and vibrant ecosystems found in marine extreme habitats.

1.4.1 Mechanism of Energy Production

In these dark and remote ecosystems, certain microorganisms, such as sulfur-oxidizing and methane-oxidizing bacteria, derive their energy by oxidizing inorganic molecules. At hydrothermal vents, for instance, sulfur-oxidizing bacteria utilize hydrogen sulfide, a chemical emitted from the seafloor (Zeng et al., 2021). They oxidize hydrogen sulfide into sulfur or sulfate, releasing energy that they use to convert carbon dioxide into organic compounds. A prime example of their significance is seen in their symbiotic relationship with the giant tube worm *Riftia pachyptila* (Hinzke et al., 2019). These tube worms lack a digestive system and rely on their internal bacteria to convert hydrogen sulfide from vent emissions into nutrients, supporting their survival in this otherwise hostile environment (Klose et al., 2016).

In a similar vein, methane-oxidizing bacteria, or methanotrophs, found in cold seeps, use methane as their energy source (Kalyuzhnaya et al., 2019). These bacteria oxidize methane, combining it with oxygen to produce carbon dioxide, water, and energy. This energy supports the synthesis of organic matter, forming the basis of complex food webs that sustain a variety of life forms from microorganisms to larger symbiotic organisms. By oxidizing methane, these bacteria transform this potent greenhouse gas into carbon dioxide and biomass, forming the foundation of unique ecological communities that include clams, mussels, and other specialized organisms (Guerrero-Cruz et al., 2021). This process not only supports local food webs but also contributes to the regulation of methane, influencing both marine ecosystems and global climate processes.

Together, these bacteria highlight the adaptability and ingenuity of life in the face of extreme conditions. Chemosynthesis exemplifies nature's capacity to exploit diverse energy sources, fostering rich ecosystems in places where traditional forms of life would struggle to exist.

1.5 GEOLOGICAL AND CHEMICAL FEATURES

The enigmatic depths of marine extreme habitats are defined not only by their biological wonders but also by their unique geological and chemical characteristics. These features shape the environments and provide the foundation for the extraordinary life forms that inhabit them. Understanding these geological and chemical aspects is crucial for comprehending how these habitats function and sustain life under such harsh conditions.

1.5.1 Formation and Structure

Marine extreme habitats owe their existence to a variety of geological processes that create environments starkly different from the rest of the ocean. Hydrothermal vents, for instance, are formed along mid-ocean ridges where tectonic plates diverge, allowing seawater to seep into the Earth's crust (Price & Giovannelli, 2017). As this water percolates down, it becomes superheated by underlying magma and is then expelled back into the ocean through vents, carrying dissolved minerals and chemicals. This process creates towering structures known as black smokers and white smokers, characterized by their mineral-laden plumes. Cold seeps, another form of extreme habitat, are found along continental margins and sediment-covered slopes (Levin, 2005). Here, hydrocarbons like methane and hydrogen sulfide seep slowly from underlying reservoirs or decaying organic matter buried in sediments. These seeps can form striking features such as methane hydrates—ice-like formations where methane molecules are trapped within water crystals. Abyssal plains, which constitute vast expanses of the ocean floor, are shaped by sediment deposition from above and tectonic activity below (Chakraborty, 2022). These plains are remarkably flat but are punctuated by occasional seamounts, trenches, and abyssal hills, adding to the complexity of these deep-sea landscapes. Deep-sea trenches represent the deepest parts of the ocean, formed by the subduction of one tectonic plate beneath another (Du et al., 2021). These trenches plunge several kilometers below the ocean surface, creating environments of extreme pressure and low temperatures.

1.5.2 Chemical Features

The chemical milieu of marine extreme habitats is dominated by high concentrations of elements and compounds that are rare or toxic in typical oceanic environments. Hydrothermal vents, for example, emit fluids rich in hydrogen sulfide, methane, and metals such as iron, manganese, and zinc (Meier et al., 2019). These chemicals precipitate upon contact with cold seawater, forming mineral-rich deposits around the vents. In cold seeps, methane and hydrogen sulfide are the principal components, diffusing slowly from the seafloor and forming chemosynthetic ecosystems (Zhang et al., 2023). Methane hydrates, found beneath sediments, are composed of methane molecules encased in ice, presenting a unique chemical structure that is stable under high-pressure and low-temperature conditions (Reichle, 2023). Abyssal plains, though generally lacking the chemical richness of vents or seeps, can accumulate organic matter and detritus that slowly decomposes, releasing nutrients and gases into the surrounding waters (Romero et al., 2021).

1.6 CONTRIBUTION TO BIODIVERSITY, NOVEL ADAPTATIONS, AND TROPHIC INTERACTIONS

Marine extreme habitats, ranging from the hydrothermal vents to the abyssal plains, are not only extraordinary for their harsh environmental conditions but also for their vibrant biodiversity and the intricate web of life they support. These ecosystems are teeming with species that have developed unique adaptations, both novel and specialized, enabling them to thrive under the extreme conditions found in these remote parts of the ocean. Understanding the biodiversity, adaptations, and trophic interactions in these habitats reveals how life can flourish against seemingly insurmountable odds.

1.6.1 Biodiversity: Unique and Endemic Species

Marine extreme habitats are a treasure trove of unique species, many of which are found nowhere else on Earth. These organisms have evolved to exploit the specific conditions of their habitats, leading to highly specialized and sometimes bizarre life forms. Hydrothermal vents, for example, are bustling with life, hosting species such as the giant tube worm *R. pachyptila*, which harbors symbiotic bacteria that utilize hydrogen sulfide from vent fluids to produce organic molecules through chemosynthesis (De Oliveira et al., 2022). The vent crabs, *Bythograea thermydron* (Jinks et al., 2002) and blind vent shrimp, *Rimicaris exoculata* (Zbinden & Cambon, 2020) have also adapted to the superheated, chemically rich environment around vent openings, showcasing the extreme specialization that these species undergo to survive. Cold seeps, characterized by steady emissions of hydrocarbons, support an equally fascinating array of species. These include seep mussels of *Bathymodiolus* spp. (Roeselers & Newton, 2012) and seep clams of *Calyptogena* spp. (Cavanaugh et al., 2006) which rely on symbiotic bacteria capable of oxidizing methane or hydrogen sulfide. These relationships enable the host organisms to thrive in the otherwise nutrient-poor environments of cold seeps, transforming the seepage of methane and sulfides into a rich source of sustenance. The abyssal plains, although seemingly barren, are home to diverse fauna such as deep-sea fishes, including the anglerfish, *Melanocetus johnsonii*, which employs bioluminescence to lure prey (Baker et al., 2019), and scavenging creatures like the giant isopod called *Bathynomus giganteus* (Lowry & Dempsey, 2006). In the deepest ocean trenches, species such as the hadal amphipod, *Hirondellea gigas* (Kobayashi et al., 2012), and the snailfish, *Pseudoliparis swirei*, have adapted to the crushing pressures and near-freezing temperatures of these remote depths, highlighting the extreme diversity of adaptations in these environments (Wang et al., 2019).

1.6.2 Adaptations: Surviving Under Extreme Conditions

1.6.2.1 Physiological Adaptations

Species in marine extreme habitats exhibit a variety of physiological adaptations that allow them to withstand high pressures, low temperatures, and other extreme conditions. Deep-sea organisms often possess pressure-resistant proteins and cell membrane structures that maintain their functionality despite the intense pressures (Marinebio Conservation Society, 2024). These adaptations ensure that enzymes and other cellular machinery remain effective, preserving metabolic processes even under conditions that would be fatal to surface-dwelling organisms.

To cope with the cold temperatures of deep-sea environments, many species produce antifreeze proteins that prevent ice crystal formation within their cells and maintain efficient biochemical pathways at low temperatures (Kar & Bhunia, 2015). Conversely, hydrothermal vent organisms are adapted to endure the high temperatures and rapid thermal fluctuations around vent openings. Their enzymes are heat-stable, enabling metabolic functions to persist in these otherwise inhospitable conditions (Minic et al., 2006).

1.6.2.2 Behavioral Adaptations

Behavioral strategies also play a crucial role in the survival of organisms in these extreme environments. Deep-sea fishes, such as the gulper eel, *Eurypharynx pelecanoides*, display unique feeding behaviors like expanding their mouth and stomach to consume prey much larger than themselves—a critical adaptation in a habitat where food is sparse (Nielsen et al., 1989). Hydrothermal vent organisms like the Pompeii worm, *Alvinella pompejana*, exhibit behaviors such as clustering around vent openings to capitalize on the warmth and nutrient-rich effluents while avoiding lethal temperatures (Bir et al., 2020). At cold seeps, species position themselves strategically to maximize their exposure to seeping hydrocarbons, facilitating the uptake of nutrients from their symbiotic bacteria and optimizing their survival strategy (Osman et al., 2023).

1.6.2.3 Morphological Adaptations

Morphological adaptations in marine extreme habitats are equally varied and remarkable. Many deep-sea creatures, such as the anglerfish, use bioluminescence for communication, prey attraction, or camouflage in the perpetual darkness (Razvy, 2024). This adaptation allows them to navigate and hunt in an environment devoid of light, turning the darkness to their advantage. Structurally, many deep-sea fishes have evolved elongated, soft bodies and reduced skeletal structures, which help them withstand the immense pressures of their environment without damage (Li et al., 2023). Tube worms at hydrothermal vents possess specialized plumes that extract sulfide from the vent water, providing it to their symbiotic bacteria housed within their bodies (McMullin et al., 2007). Cold seep organisms, like seep mussels, have developed specialized gills that harbor their symbiotic bacteria, enhancing their ability to oxidize methane or hydrogen sulfide (Duperron et al., 2005). These structural modifications optimize the interaction between the host and the symbiont, ensuring efficient nutrient exchange and survival in their harsh environment.

1.6.3 Trophic Interactions: Food Webs and Predator-Prey Dynamics

The food webs in marine extreme habitats are uniquely structured around chemosynthesis rather than photosynthesis, unlike the sunlit zones of the ocean. At hydrothermal vents, the base of the food web is formed by chemosynthetic bacteria that utilize chemicals like hydrogen sulfide to produce organic matter (Moriarty, 2010). These bacteria support a variety of primary consumers, including tube worms, vent mussels, and clams. Predators such as vent crabs and vent octopuses, *Vulcanoctopus hydrothermalis*, occupy higher trophic levels, preying on the primary consumers and smaller scavengers (Voight, 2000). The trophic dynamics in these habitats involve complex interactions, with organisms exploiting both the chemosynthetic production and detrital inputs from surface waters. In cold seep ecosystems, methane-oxidizing bacteria form the foundation of the food web. Symbiotic relationships are common, with many seep inhabitants relying on bacterial partners for nutrition. Predators in these environments, such as fishes and crustaceans, feed on the seep-associated fauna, creating a dynamic food web driven by the availability of methane and other hydrocarbons (Lin et al., 2020). On the abyssal plains, food webs are driven by the sparse organic matter that settles from the surface. Detritivores and scavengers, such as sea cucumbers and giant isopods, play key roles in consuming and recycling this material (Jeffreys et al., 2011). Predators include deep-sea fishes and cephalopods that have evolved to hunt in the darkness and withstand the pressures of the deep ocean (Drazen & Sutton, 2017).

The unique biodiversity, novel adaptations, and intricate trophic interactions found in marine extreme habitats underscore the remarkable resilience and ingenuity of life. These ecosystems reveal how species can not only survive but also thrive under conditions that would be lethal to most organisms. Their adaptations and interactions provide valuable insights into the potential for life in similar extreme environments beyond Earth, highlighting the broader implications of studying these extraordinary and diverse marine communities.

1.7 HUMAN IMPACT AND CONSERVATION

The human footprint has reached even the most remote and extreme marine habitats, raising significant concerns about the future of these unique ecosystems. Deep-sea mining, pollution, and climate change are among the major threats impacting these environments. Conservation efforts face numerous challenges, necessitating comprehensive strategies and robust international policies to protect these invaluable habitats.

1.7.1 Impact: Deep-Sea Mining, Pollution, and Climate Change

1.7.1.1 Deep-Sea Mining

As technological advancements make the extraction of minerals from the ocean floor increasingly feasible, deep-sea mining has emerged as a significant threat to marine extreme habitats (Sharma, 2015). The pursuit of resources like polymetallic nodules, seafloor massive sulfides, and cobalt-rich ferromanganese crusts targets areas rich in biodiversity, such as hydrothermal vents and abyssal plains (Miller et al., 2018). Mining activities disturb the seafloor, destroying habitats, displacing species, and potentially causing long-term ecological damage. For instance, the removal of polymetallic nodules disrupts the slow-growing communities of sponges and corals that rely on these substrates (Chin & Hari, 2020). The plumes of sediment generated by mining operations can also smother organisms over a wide area, impacting food webs and nutrient cycles (Christiansen et al., 2020).

1.7.1.2 Pollution

Pollution poses a multifaceted threat to marine extreme environments. Deep-sea ecosystems are vulnerable to the accumulation of pollutants, including heavy metals, plastics, and chemical contaminants, which can be transported from coastal regions by ocean currents (Thushari & Senevirathna, 2020). Plastics, in particular, have been found at the deepest oceanic trenches, such as the Mariana Trench, highlighting the pervasive nature of marine pollution (Jamieson et al., 2019). These pollutants can have toxic effects on deep-sea species, disrupting biological processes and altering community structures. Additionally, oil spills and leaks from offshore drilling activities can introduce hydrocarbons into these environments, affecting organisms reliant on specific chemical balances (National Research Council, 1985).

1.7.1.3 Climate Change

Climate change exerts profound impacts on extreme marine habitats through ocean warming, acidification, and deoxygenation (Sweetman et al., 2017). Rising ocean temperatures can alter the distribution of species adapted to cold, stable environments, pushing some species beyond their thermal tolerance limits and potentially leading to local extinctions. Ocean acidification, resulting from increased atmospheric CO_2 levels, affects the carbonate chemistry of seawater, posing a threat to organisms that rely on calcium carbonate structures, such as deep-sea corals and some shell-forming species (Leung et al., 2022). Deoxygenation, the reduction of oxygen levels in seawater, can exacerbate the stress on deep-sea communities, particularly in regions like oxygen minimum zones (OMZs), which may expand under future climate scenarios (Kirchman, 2019).

1.7.2 Conservation: Strategies and Challenges in Protecting These Habitats

1.7.2.1 Strategies

Effective conservation of marine extreme habitats requires a multifaceted approach that combines scientific research, technological innovation, and policy measures. Marine protected areas (MPAs) have proven effective in safeguarding certain regions, such as hydrothermal vent fields and cold seep ecosystems, by restricting activities like mining and bottom trawling (Kriegl et al., 2021). Establishing networks of MPAs can enhance connectivity between protected areas, allowing species to migrate and maintain genetic diversity (Assis et al., 2021). Technological advancements in remote sensing and deep-sea exploration play a crucial role in monitoring these habitats. Autonomous underwater vehicles (AUVs) and remotely operated vehicles (ROVs) enable detailed mapping and assessment of deep-sea ecosystems, providing data critical for informed management and conservation (McLean et al., 2020). Genetic and ecological studies can identify key species and ecosystem functions, informing conservation priorities and strategies.

1.7.2.2 Challenges

Conserving marine extreme habitats is fraught with challenges. The remoteness and inaccessibility of these environments complicate monitoring and enforcement of conservation measures. Additionally, the lack of comprehensive baseline data on many deep-sea ecosystems hinders the assessment of human impacts and the effectiveness of conservation interventions. The high cost of deep-sea exploration and the need for specialized equipment further constrain conservation efforts.

1.7.3 Policies: Overview of International Agreements and Policies

1.7.3.1 International Agreements

Several international frameworks aim to protect marine biodiversity, including the habitats in extreme environments. The United Nations Convention on the Law of the Sea (UNCLOS) establishes a legal framework for the use and conservation of ocean resources, including provisions for the protection of the marine environment from pollution and over-exploitation (Hoagland et al., 2001). The International Seabed Authority (ISA) regulates deep-sea mining activities in areas beyond national jurisdiction, requiring environmental impact assessments and management plans to minimize ecological damage (Durden et al., 2018). The Convention on Biological Diversity (CBD) promotes the conservation of marine biodiversity through the establishment of

MPAs and the implementation of the Ecosystem Approach, which integrates ecological, social, and economic considerations in marine management (Enright & Boteler, 2020). Regional agreements, such as the OSPAR Convention in the North-East Atlantic, also contribute to the protection of deep-sea ecosystems by coordinating efforts among neighboring countries (Alarcon et al., 2023).

1.7.3.2 National Policies

National governments play a vital role in implementing conservation measures for marine extreme habitats within their jurisdictions. Policies often include the creation of MPAs, regulations on deep-sea mining and fishing, and initiatives to reduce marine pollution. For example, countries like Australia (Tan & Fischer, 2022) and New Zealand (Ellis et al., 2017) have established extensive networks of MPAs that encompass deep-sea environments, while nations with interests in deep-sea mining are developing regulations to balance resource extraction with environmental protection (Mendoza Alarcón, 2022).

Addressing the human impacts on marine extreme habitats and implementing effective conservation strategies are critical for preserving these unique ecosystems. A combination of international cooperation, national policies, and technological innovations is essential to mitigate the threats posed by deep-sea mining, pollution, and climate change. Ensuring the protection of these habitats not only safeguards their intrinsic biodiversity but also enhances our understanding of the ocean's role in global ecological and biogeochemical processes.

1.8 TECHNOLOGICAL ADVANCES IN EXPLORATION

The exploration of marine extreme habitats has been revolutionized by technological advancements, enabling scientists to study these remote and challenging environments with unprecedented precision and depth. Cutting-edge technologies such as submersibles, ROVs, deep-sea probes, and innovative sampling techniques have opened new frontiers in our understanding of these enigmatic ecosystems.

1.8.1 Exploration Technologies: Submersibles, ROVs, and Deep-Sea Probes

1.8.1.1 Submersibles

Manned submersibles, like Alvin (Rona, 1999) and Shinkai 6500 (Komuku et al., 2007), have played a pivotal role in the exploration of deep-sea environments. These craft can carry scientists directly to the ocean floor, allowing for real-time observations and the collection of samples from hydrothermal vents, cold seeps, and deep-sea trenches. Submersibles are equipped with high-resolution cameras, manipulator arms, and sample storage compartments, enabling the study of deep-sea geology, chemistry, and biology under direct human supervision.

1.8.1.2 Remotely Operated Vehicles

ROVs, such as Jason and Hercules, are unmanned, tethered devices controlled from surface vessels (Ballard et al., 2000). They have transformed deep-sea research by providing access to depths that are beyond the reach of manned submersibles (Spanier et al., 1994). ROVs are outfitted with sophisticated tools, including high-definition video cameras, robotic arms for collecting biological and geological samples, and a suite of sensors to measure environmental parameters like temperature, pressure, and chemical composition. Their ability to operate in harsh and confined environments makes them ideal for exploring complex habitats such as vent fields and methane hydrate deposits.

1.8.1.3 Deep-Sea Probes

AUVs and other deep-sea probes, such as the Hugin (Wynn et al., 2014) and Abyss (Linke & Lackschewitz, 2016), are untethered systems that can operate independently of surface vessels. These devices are programmed to follow predefined paths, conducting detailed surveys of the seafloor and water column. AUVs can map large areas with sonar and optical imaging systems, collect samples with integrated samplers, and monitor environmental conditions with onboard sensors. Their autonomy and versatility allow them to cover extensive areas and access regions that are difficult for tethered systems to reach (Fun Sang Cepeda et al., 2023).

1.8.2 Sampling Techniques: Collection Methods and Data Acquisition

1.8.2.1 Sample Collection

The collection of samples from extreme marine habitats involves specialized techniques to preserve the integrity of the specimens under high-pressure, low-temperature conditions. ROVs and submersibles use manipulator arms to gather biological specimens, sediment cores, and rock samples, which are then stored in pressurized containers to prevent decompression damage. These samples provide critical insights into the physiology, genetics, and biochemistry of deep-sea organisms, as well as the geochemistry of the seafloor.

1.8.2.2 Data Acquisition

Modern exploration tools are equipped with advanced data acquisition systems that record a wide range of environmental parameters in situ. Sensors measure temperature, salinity, pH, and chemical concentrations, while cameras and sonar systems capture high-resolution images and topographical maps. These data are crucial for understanding the physical and chemical conditions of extreme habitats and for identifying biological hotspots. Additionally, real-time data transmission from ROVs and AUVs enables scientists to analyze findings and make decisions about exploration and sampling strategies immediately.

1.8.3 Innovations: Recent Technological Advancements in Exploration

1.8.3.1 High-Resolution Imaging

Advancements in imaging technology have dramatically improved our ability to visualize deep-sea environments. High-definition cameras and laser scanning systems on ROVs and AUVs now provide detailed, three-dimensional maps of the seafloor, revealing fine-scale geological features and biological communities (Huvenne et al., 2018). These innovations have enabled the discovery of new vent fields, cold seeps, and unique biological structures, enhancing our understanding of the spatial organization of these habitats.

1.8.3.2 Genomic and Biogeochemical Sensors

Recent developments in in situ genomic and biogeochemical sensors have allowed for the direct analysis of microbial communities and chemical processes in deep-sea environments. Devices such as Environmental Sample Processors (ESPs) can filter water samples, extract DNA, and perform genetic analyses on site, providing insights into microbial diversity and metabolic pathways without the need to bring samples to the surface (Knapik et al., 2020). Biogeochemical sensors can detect and quantify trace gases, nutrients, and isotopes, helping to unravel the complex chemical interactions that sustain life in these extreme environments (Chai et al., 2020).

1.8.3.3 Autonomous Swarms and Networks

Innovations in autonomous swarming technology and sensor networks are expanding the scope of deep-sea exploration. Swarms of small, autonomous robots can coordinate their movements to cover large areas, conduct synchronized sampling, and relay data to surface vessels or satellite networks (Jaffe et al., 2017). These systems enhance the efficiency of exploration missions, providing comprehensive datasets on the physical, chemical, and biological characteristics of marine extreme habitats.

Technological advances have transformed our ability to explore and understand marine extreme habitats, enabling detailed investigations of these remote and challenging ecosystems. From manned submersibles and ROVs to autonomous probes and advanced sensors, these tools have provided unprecedented insights into the geology, chemistry, and biology of the deep ocean. As exploration technologies continue to evolve, they will play a crucial role in uncovering the mysteries of these unique environments, informing conservation efforts and expanding our knowledge of life's resilience in extreme conditions.

1.9 FUTURE PERSPECTIVE: UNLEASHING THE POTENTIAL OF DEEP-SEA BIOACTIVE COMPOUNDS

As the technological capabilities for exploring marine extreme habitats advance, so does the opportunity to discover and explore deep-sea bioactive compounds that can mark a transformative era in biotechnology and pharmaceutical research. These environments are rich sources of novel compounds that promise to drive significant innovations in drug development and environmental solutions (Karthikeyan et al., 2022). The following future directions highlight the critical areas of focus to fully leverage the potential of deep-sea bioactive compounds.

1.9.1 Harnessing Novel Compounds for Drug Development

Deep-sea organisms, particularly fungi and bacteria, are prolific producers of secondary metabolites with unique structural properties. The recent identification of 184 new compounds from deep-sea fungi, including polyketides, terpenoids, and nitrogen-containing compounds, underscores the potential of these microorganisms as sources of new drug leads (Wang et al., 2023). These compounds exhibit diverse biological activities, including anti-infective, cytotoxic, and enzyme-inhibitory properties, positioning them as promising candidates for the development of novel therapeutics.

To capitalize on these findings, future research should focus on:

Activating Cryptic Biosynthetic Pathways: Many deep-sea microorganisms harbor silent biosynthetic gene clusters that remain unexpressed under standard laboratory conditions (Chiang et al., 2011). Advanced genetic and molecular techniques, such as heterologous expression and pathway engineering, can be employed to activate these cryptic pathways, revealing new metabolites and expanding the chemical diversity available for drug discovery (Xu et al., 2022).

Streamlining Isolation and Screening: The isolation of novel microbial strains from deep-sea environments remains a challenging task (Zhang et al., 2018). Employing low-nutrient media that mimic deep-sea conditions can improve the isolation of unique microorganisms (Connon & Giovannoni, 2002). Additionally, advanced dereplication methods using mass spectrometry (MS) and nuclear magnetic resonance (NMR) can expedite the identification and characterization of bioactive compounds (Gaudêncio et al., 2023).

Exploring Bioactivity Profiles: Detailed bioactivity profiling of newly discovered compounds is essential for identifying their therapeutic potential (Barba-Ostria et al., 2022). This includes assessing their efficacy against various disease targets, such as cancer cells and infectious pathogens, to guide the development of specific drug candidates.

1.9.2 Innovations in Marine Biotechnology

Marine organisms are not only valuable for drug discovery but also hold promise for environmental biotechnology. Their adaptations to extreme conditions make them ideal candidates for applications in water purification and bioremediation. Future research should focus on:

Developing Biocatalysts for Environmental Solutions: The enzymes and metabolic pathways of deep-sea microorganisms can be harnessed to degrade pollutants and contaminants in marine and terrestrial environments (Bala et al., 2022). These biocatalysts can be integrated into systems for the bioremediation of oil spills, heavy metal contamination, and other environmental hazards, leveraging their resistance to harsh conditions.

Bioprospecting for Sustainable Industrial Processes: Deep-sea microorganisms offer a reservoir of biocatalysts for industrial processes that require resistance to extreme conditions, such as high pressure, temperature, or salinity (Gomes & Steiner, 2004). Their application in the production of sustainable chemicals and materials could revolutionize various industrial sectors, from pharmaceuticals to biofuels.

1.10 CONCLUSION

Marine extreme habitats, with their extraordinary biodiversity and novel adaptations, provide a window into the resilience and adaptability of life under the most severe conditions. They challenge our understanding of biological limits and offer insights into the potential for life in extreme environments beyond Earth. These ecosystems are not isolated wonders but integral components of the global biosphere, contributing to biogeochemical cycles and ecological processes that have far-reaching implications for our planet. As technology advances and our exploration continues, the study of these habitats will undoubtedly yield further revelations, enhancing our understanding of life's tenacity and the interconnectedness of Earth's ecosystems. Protecting these unique environments is essential for preserving their ecological and scientific value, ensuring that they continue to inspire and inform future generations.

REFERENCES

Alarcon Blazquez, M., Van Der Veeren, R., Gacutan, J., & James, P. (2023). Compiling preliminary SEEA Ecosystem Accounts for the OSPAR regional sea: Experimental findings and lessons learned. *One Ecosystem*, 8, e108030. https://doi.org/10.3897/oneeco.8.e108030

Assis, J., Failler, P., Fragkopoulou, E., Abecasis, D., Touron-Gardic, G., Regalla, A., Sidina, E., Dinis, H., & Serrao, E. A. (2021). Potential biodiversity connectivity in the network of marine protected areas in western Africa. *Frontiers in Marine Science*, 8, 765053. https://doi.org/10.3389/fmars.2021.765053

Baker, L. J., Freed, L. L., Easson, C. G., Lopez, J. V., Fenolio, D., Sutton, T. T., Nyholm, S. V., & Hendry, T. A. (2019). Diverse deep-sea anglerfishes share a genetically reduced luminous symbiont that is acquired from the environment. *eLife*, 8, e47606. https://doi.org/10.7554/eLife.47606

Bala, S., Garg, D., Thirumalesh, B. V., Sharma, M., Sridhar, K., Inbaraj, B. S., & Tripathi, M. (2022). Recent strategies for bioremediation of emerging pollutants: A review for a green and sustainable environment. *Toxics*, 10(8), 484. https://doi.org/10.3390/toxics10080484

Ballard, R. D., McCann, A. M., Yoerger, D., Whitcomb, L., Mindell, D., Oleson, J., Singh, H., Foley, B., Adams, J., Piechota, D., & Giangrande, C. (2000). The discovery of ancient history in the deep sea using advanced deep submergence technology. *Deep Sea Research Part I: Oceanographic Research Papers*, 47(9), 1591–1620.

Barba-Ostria, C., Carrera-Pacheco, S. E., Gonzalez-Pastor, R., Heredia-Moya, J., Mayorga-Ramos, A., Rodríguez-Pólit, C., Zúñiga-Miranda, J., Arias-Almeida, B., & Guamán, L. P. (2022). Evaluation of biological activity of natural compounds: Current trends and methods. *Molecules*, 27(14), 4490. https://doi.org/10.3390/molecules27144490

Bartlett, D. H. (2009). Microbial life in the trenches. *Marine Technology Society Journal*, 43(5), 128–131. https://doi.org/10.4031/MTSJ.43.5.5

Bir, J., Ibrahim Khalil, S. M., & Golder, M. (2020). Adaptation in extreme underwater vent ecosystem: A case study on Pompeii worm (*Alvinella pompejana*). *International Journal of Fauna and Biological Studies*, 7, 25–32.

Cavanaugh, C. M., McKiness, Z. P., Newton, I. L. G., & Stewart, F. J. (2006). Marine chemosynthetic symbioses. In M. Dworkin, S. Falkow, E. Rosenberg, K.-H. Schleifer, & E. Stackebrandt (Eds.), *The Prokaryotes* (pp. 475–507). Springer. https://doi.org/10.1007/0-387-30741-9_18

Chai, F., Johnson, K. S., Claustre, H., Xing, X., Wang, Y., Boss, E., Riser, S., Fennel, K., Schofield, O., & Sutton, A. (2020). Monitoring ocean biogeochemistry with autonomous platforms. *Nature Reviews Earth & Environment*, 1(6), 315–326. https://doi.org/10.1038/s43017-020-0053-y

Chakraborty, S. K. (2022). Ocean ecosystem and its multidimensional eco-functionality and significance. In R. Brinkmann (Ed.), *The Palgrave Handbook of Global Sustainability* (pp. 1–45). Springer International Publishing. https://doi.org/10.1007/978-3-030-38948-2_37-1

Chen, H., Li, C., Qiu, J.-W., Bettencourt, R., Feng, D., Wang, Y., Jing, H., & Sun, J. (2023). Editorial: Deep-sea chemosynthetic ecosystems: Living in extreme environments. *Frontiers in Marine Science*, 10, 1087465. https://doi.org/10.3389/fmars.2023.1087465

Chiang, Y. M., Chang, S. L., Oakley, B. R., & Wang, C. C. (2011). Recent advances in awakening silent biosynthetic gene clusters and linking orphan clusters to natural products in microorganisms. *Current Opinion in Chemical Biology*, 15(1), 137–143.

Chin, A., & Hari, K. (2020). *Predicting the Impacts of Mining of Deep Sea Polymetallic Nodules in the Pacific Ocean: A Review of Scientific Literature* (p. 52). Deep Sea Mining Campaign and MiningWatch, Canada.

Christiansen, B., Denda, A., & Christiansen, S. (2020). Potential effects of deep seabed mining on pelagic and benthopelagic biota. *Marine Policy*, 114, 103442. https://doi.org/10.1016/j.marpol.2019.02.014

Connon, S. A., & Giovannoni, S. J. (2002). High-throughput methods for culturing microorganisms in very-low-nutrient media yield diverse new marine isolates. *Applied and Environmental Microbiology*, 68(8), 3878–3885. https://doi.org/10.1128/AEM.68.8.3878-3885.2002

Cormier, M.-H., & Sloan, H. (2018). Abyssal hills and abyssal plains. In A. Micallef, S. Krastel, & A. Savini (Eds.), *Submarine Geomorphology* (pp. 389–408). Springer International Publishing. https://doi.org/10.1007/978-3-319-57852-1_20

Dahlhoff, E., O'Brien, J., Somero, G. N., & Vetter, R. D. (1991). Temperature effects on mitochondria from hydrothermal vent invertebrates: Evidence for adaptation to elevated and variable habitat temperatures. *Physiological Zoology*, 64(6), 1490–1508. https://doi.org/10.1086/physzool.64.6.30158226

DeOliveira, A. L., Mitchell, J., Girguis, P., & Bright, M. (2022). Novel insights on obligate symbiont lifestyle and adaptation to chemosynthetic environment as revealed by the giant tubeworm genome. *Molecular Biology and Evolution*, 39(1), msab347. https://doi.org/10.1093/molbev/msab347

Drazen, J. C., & Sutton, T. T. (2017). Dining in the deep: The feeding ecology of deep-sea fishes. *Annual Review of Marine Science*, 9, 337–366.

Du, M., Peng, X., Zhang, H., Ye, C., Dasgupta, S., Li, J., Li, J., Liu, S., Xu, H., Chen, C., Jing, H., Xu, H., Liu, J., He, S., He, L., Cai, S., Chen, S., & Ta, K. (2021). Geology, environment, and life in the deepest part of the world's oceans. *The Innovation*, 2(2), 100109. https://doi.org/10.1016/j.xinn.2021.100109

Duperron, S., Nadalig, T., Caprais, J.-C., Sibuet, M., Fiala-Médioni, A., Amann, R., & Dubilier, N. (2005). Dual symbiosis in a *Bathymodiolus* sp. Mussel from a methane seep on the Gabon continental margin (Southeast Atlantic): 16s rRNA phylogeny and distribution of the symbionts in gills. *Applied and Environmental Microbiology*, 71(4), 1694–1700. https://doi.org/10.1128/AEM.71.4.1694-1700.2005

Durden, J. M., Lallier, L. E., Murphy, K., Jaeckel, A., Gjerde, K., & Jones, D. O. B. (2018). Environmental impact assessment process for deep-sea mining in "the Area." *Marine Policy*, 87, 194–202. https://doi.org/10.1016/j.marpol.2017.10.013

Ellis, J. I., Clark, M. R., Rouse, H. L., & Lamarche, G. (2017). Environmental management frameworks for offshore mining: The New Zealand approach. *Marine Policy*, 84, 178–192. https://doi.org/10.1016/j.marpol.2017.07.004

Enrich-Prast, A., Bastviken, D., Crill, P., Santoro, A. L., Signori, C. N., & Sanseverino, A. M. (2014). Chemosynthesis. *Reference Module in Earth Systems and Environmental Sciences*. Elsevier. https://doi.org/10.1016/B978-0-12-409548-9.09054-0

Enright, S. R., & Boteler, B. (2020). The ecosystem approach in international marine environmental law and governance. In T. G. O'Higgins, M. Lago, & T. H. DeWitt (Eds.), *Ecosystem-Based Management, Ecosystem Services and Aquatic Biodiversity* (pp. 333–352). Springer International Publishing. https://doi.org/10.1007/978-3-030-45843-0_17

Fisher, C., Takai, K., & Le Bris, N. (2007). Hydrothermal vent ecosystems. *Oceanography*, 20(1), 14–23. https://doi.org/10.5670/oceanog.2007.75

FunSang Cepeda, M., Freitas Machado, M. D. S., Sousa Barbosa, F. H., Santana Souza Moreira, D., Legaz Almansa, M. J., Lourenço De Souza, M. I., & Caprace, J.-D. (2023). Exploring autonomous and remotely operated vehicles in offshore structure inspections. *Journal of Marine Science and Engineering*, 11(11), 2172. https://doi.org/10.3390/jmse11112172

Gaudêncio, S. P., Bayram, E., Lukić Bilela, L., Cueto, M., Díaz-Marrero, A. R., Haznedaroglu, B. Z., Jimenez, C., Mandalakis, M., Pereira, F., Reyes, F., & Tasdemir, D. (2023). Advanced methods for natural products discovery: Bioactivity screening, dereplication, metabolomics profiling, genomic sequencing, databases and informatic tools, and structure elucidation. *Marine Drugs*, 21(5), 308. https://doi.org/10.3390/md21050308

Gomes, J., & Steiner, W. (2004). The biocatalytic potential of extremophiles and extremozymes. *Food Technology and Biotechnology*, 42(4), 223–225.

Gornitz, V., & Fung, I. (1994). Potential distribution of methane hydrates in the world's oceans. *Global Biogeochemical Cycles*, 8(3), 335–347. https://doi.org/10.1029/94GB00766

Guerrero-Cruz, S., Vaksmaa, A., Horn, M. A., Niemann, H., Pijuan, M., & Ho, A. (2021). Methanotrophs: Discoveries, environmental relevance, and a perspective on current and future applications. *Frontiers in Microbiology*, 12, 678057. https://doi.org/10.3389/fmicb.2021.678057

Hinzke, T., Kleiner, M., Breusing, C., Felbeck, H., Häsler, R., Sievert, S. M., Schlüter, R., Rosenstiel, P., Reusch, T. B. H., Schweder, T., & Markert, S. (2019). Host-microbe interactions in the chemosynthetic Riftia pachyptila symbiosis. *mBio*, 10(6), e02243–e02319. https://doi.org/10.1128/mBio.02243-19

Hiraoka, S., Hirai, M., Matsui, Y., Makabe, A., Minegishi, H., Tsuda, M., Juliarni, Rastelli, E., Danovaro, R., Corinaldesi, C., Kitahashi, T., Tasumi, E., Nishizawa, M., Takai, K., Nomaki, H., & Nunoura, T. (2020). Microbial community and geochemical analyses of trans-trench sediments for understanding the roles of hadal environments. *The ISME Journal*, 14(3), 740–756. https://doi.org/10.1038/s41396-019-0564-z

Hoagland, P., Jacoby, J., & Schumacher, M. E. (2001). Law of the sea. In J. Kirk Cochran, H. J. Bokuniewicz, & P. L. Yager (Eds.), *Encyclopedia of Ocean Sciences* (pp. 1481–1492). Elsevier. https://doi.org/10.1006/rwos.2001.0415

Huvenne, V. A. I., Robert, K., Marsh, L., Lo Iacono, C., Le Bas, T., & Wynn, R. B. (2018). Rovs and auvs. In A. Micallef, S. Krastel, & A. Savini (Eds.), *Submarine Geomorphology* (pp. 93–108). Springer International Publishing. https://doi.org/10.1007/978-3-319-57852-1_7

Jaffe, J. S., Franks, P. J. S., Roberts, P. L. D., Mirza, D., Schurgers, C., Kastner, R., & Boch, A. (2017). A swarm of autonomous miniature underwater robot drifters for exploring submesoscale ocean dynamics. *Nature Communications*, 8(1), 14189. https://doi.org/10.1038/ncomms14189

Jamieson, A. J., Brooks, L. S. R., Reid, W. D. K., Piertney, S. B., Narayanaswamy, B. E., & Linley, T. D. (2019). Microplastics and synthetic particles ingested by deep-sea amphipods in six of the deepest marine ecosystems on Earth. *Royal Society Open Science*, 6(2), 180667. https://doi.org/10.1098/rsos.180667

Jannasch, H. W. (1985). The chemosynthetic support of life and the microbial diversity at deep-sea hydrothermal vents. *Proceedings of the Royal Society of London. Series B. Biological Sciences*, 225(1240), 277–297. https://doi.org/10.1098/rspb.1985.0062

Jeffreys, R. M., Lavaleye, M. S., Bergman, M. J., Duineveld, G. C., & Witbaard, R. (2011). Do abyssal scavengers use phytodetritus as a food resource? Video and biochemical evidence from the Atlantic and Mediterranean. *Deep Sea Research Part I: Oceanographic Research Papers*, 58(4), 415–428.

Jinks, R. N., Markley, T. L., Taylor, E. E., Perovich, G., Dittel, A. I., Epifanio, C. E., & Cronin, T. W. (2002). Adaptive visual metamorphosis in a deep-sea hydrothermal vent crab. *Nature*, 420(6911), 68–70. https://doi.org/10.1038/nature01144

Kalyuzhnaya, M. G., Gomez, O. A., & Murrell, J. C. (2019). The methane-oxidizing bacteria (Methanotrophs). In T. J. McGenity (Ed.), *Taxonomy, Genomics and Ecophysiology of Hydrocarbon-Degrading Microbes* (pp. 1–34). Springer International Publishing. https://doi.org/10.1007/978-3-319-60053-6_10-1

Kar, R. K., & Bhunia, A. (2015). Biophysical and biochemical aspects of antifreeze proteins: Using computational tools to extract atomistic information. *Progress in Biophysics and Molecular Biology*, 119(2), 194–204. https://doi.org/10.1016/j.pbiomolbio.2015.09.001

Karthikeyan, A., Joseph, A., & Nair, B. G. (2022). Promising bioactive compounds from the marine environment and their potential effects on various diseases. *Journal of Genetic Engineering and Biotechnology*, 20(1), 14. https://doi.org/10.1186/s43141-021-00290-4

Kirchman, D. L. (2019). Deoxygenation of the open oceans and its consequences. In P. Maurice (Ed.), *Encyclopedia of Water* (1st ed., pp. 1–12). Wiley. https://doi.org/10.1002/9781119300762.wsts0052

Klose, J., Aistleitner, K., Horn, M., Krenn, L., Dirsch, V., Zehl, M., & Bright, M. (2016). Trophosome of the deep-sea tubeworm Riftia pachyptila inhibits bacterial growth. PLoS One, 11(1), e0146446. https://doi.org/10.1371/journal.pone.0146446

Knapik, K., Bagi, A., Krolicka, A., & Baussant, T. (2020). Metatranscriptomic analysis of oil-exposed seawater bacterial communities archived by an environmental sample processor (ESP). *Microorganisms*, 8(5), 744. https://doi.org/10.3390/microorganisms8050744

Kobayashi, H., Hatada, Y., Tsubouchi, T., Nagahama, T., & Takami, H. (2012). The hadal amphipod hirondellea gigas possessing a unique cellulase for digesting wooden debris buried in the deepest seafloor. PLoS One, 7(8), e42727. https://doi.org/10.1371/journal.pone.0042727

Komuku, T., Matsumoto, K., Imai, Y., Sakurai, T., & Ito, K. (2007). 1000 dives by the shinkai 6500 in 18 years. *2007 Symposium on Underwater Technology and Workshop on Scientific Use of Submarine Cables and Related Technologies*, Tokyo, Japan, pp. 21–27. https://doi.org/10.1109/UT.2007.370833

Kriegl, M., Elías Ilosvay, X. E., Von Dorrien, C., & Oesterwind, D. (2021). Marine protected areas: At the crossroads of nature conservation and fisheries management. *Frontiers in Marine Science*, 8, 676264. https://doi.org/10.3389/fmars.2021.676264

Leung, J. Y. S., Zhang, S., & Connell, S. D. (2022). Is ocean acidification really a threat to marine calcifiers? A systematic review and meta-analysis of 980+ studies spanning two decades. *Small*, 18(35), 2107407. https://doi.org/10.1002/smll.202107407

Levin, L. A. (2005). Ecology of cold seep sediments: Interactions of fauna with flow, chemistry and microbes. *Oceanography and Marine Biology*, 43, 1–46.

Li, G., Wong, T.-W., Shih, B., Guo, C., Wang, L., Liu, J., Wang, T., Liu, X., Yan, J., Wu, B., Yu, F., Chen, Y., Liang, Y., Xue, Y., Wang, C., He, S., Wen, L., Tolley, M. T., Zhang, A.-M., Laschi, C., & Li, T. (2023). Bioinspired soft robots for deep-sea exploration. *Nature Communications*, 14(1), 7097. https://doi.org/10.1038/s41467-023-42882-3

Lin, Z.-Y., Chen, H.-W., & Lin, H.-J. (2020). Trophic model of a deep-sea ecosystem with methane seeps in the South China Sea. *Deep Sea Research Part I: Oceanographic Research Papers*, 159, 103251. https://doi.org/10.1016/j.dsr.2020.103251

Linke, P., & Lackschewitz, K. (2016). Autonomous underwater vehicle „ABYSS". *Journal of Large-Scale Research Facilities JLSRF*, 2, A79–A79. https://doi.org/10.17815/jlsrf-2-149

Liu, Y., Zhang, Z., Ji, M., Hu, A., Wang, J., Jing, H., Liu, K., Xiao, X., & Zhao, W. (2022). Comparison of prokaryotes between Mount Everest and the Mariana Trench. *Microbiome*, 10(1), 215. https://doi.org/10.1186/s40168-022-01403-y

Lowry, J. K., & Dempsey, K. (2006). The giant deep-sea scavenger genus *Bathynomus* (Crustacea, Isopoda, Cirolanidae) in the Indo-West Pacific. *Tropical Deep-Sea Benthos*, 24, 163–192.

McLean, D. L., Parsons, M. J. G., Gates, A. R., Benfield, M. C., Bond, T., Booth, D. J., Bunce, M., Fowler, A. M., Harvey, E. S., Macreadie, P. I., Pattiaratchi, C. B., Rouse, S., Partridge, J. C., Thomson, P. G., Todd, V. L. G., & Jones, D. O. B. (2020). Enhancing the scientific value of industry remotely operated vehicles (ROVs) in our oceans. *Frontiers in Marine Science*, 7, 220. https://doi.org/10.3389/fmars.2020.00220

McMullin, E. R., Bergquist, D. C., & Fisher, C. R. (2007). Metazoans in extreme environments: Adaptations of hydrothermal vent and hydrocarbon seep fauna. *Gravitational and Space Research*, 13(2), 13–23.

Meier, D. V., Pjevac, P., Bach, W., Markert, S., Schweder, T., Jamieson, J., Petersen, S., Amann, R., & Meyerdierks, A. (2019). Microbial metal-sulfide oxidation in inactive hydrothermal vent chimneys suggested by metagenomic and metaproteomic analyses. *Environmental Microbiology*, 21(2), 682–701. https://doi.org/10.1111/1462-2920.14514

MendozaAlarcón, J. J. V. (2022). Environmentally balanced approaches to future deep seabed mining. Dissertations. World Maritime University.

Miller, K. A., Thompson, K. F., Johnston, P., & Santillo, D. (2018). An overview of seabed mining including the current state of development, environmental impacts, and knowledge gaps. *Frontiers in Marine Science*, 4, 418. https://doi.org/10.3389/fmars.2017.00418

Minic, Z., & Thongbam, P. D. (2011). The biological deep sea hydrothermal vent as a model to study carbon dioxide capturing enzymes. *Marine Drugs*, 9(5), 719–738. https://doi.org/10.3390/md9050719

Minic, Z., Serre, V., & Hervé, G. (2006). Adaptation des organismes aux conditions extrêmes des sources hydrothermales marines profondes. *Comptes Rendus Biologies*, 329(7), 527–540. https://doi.org/10.1016/j.crvi.2006.02.001

Moriarty, P. (2010). Chemotrophy along seafloor hydrothermal vents. *Microbewiki*. October 26, 2010.

NationalResearch Council, Division on Engineering, Physical Sciences, Commission on Physical Sciences, Mathematics, Applications, & Steering Committee for the Petroleum in the Marine Environment Update. (1985). *Oil in the Sea: Inputs, Fates, and Effects* (Vol. 1). National Academies Press.

Nielsen, J. G., Bertelsen, E., & Jespersen, Å. (1989). The biology of Eurypharynx pelecanoides (Pisces, eurypharyngidae). *Acta Zoologica*, 70(3), 187–197. https://doi.org/10.1111/j.1463-6395.1989.tb01069.x

Orcutt, B. N., Bradley, J. A., Brazelton, W. J., Estes, E. R., Goordial, J. M., Huber, J. A., Jones, R. M., Mahmoudi, N., Marlow, J. J., Murdock, S., & Pachiadaki, M. (2020). Impacts of deep-sea mining on microbial ecosystem services. *Limnology and Oceanography*, 65(7), 1489–1510. https://doi.org/10.1002/lno.11403

Oren, A. (2016). Life in hypersaline environments. In C. J. Hurst (Ed.), *Their World: A Diversity of Microbial Environments* (Vol. 1, pp. 301–339). Springer International Publishing. https://doi.org/10.1007/978-3-319-28071-4_8

Osman, E. O., Vohsen, S. A., Girard, F., Cruz, R., Glickman, O., Bullock, L. M., Anderson, K. E., Weinnig, A. M., Cordes, E. E., Fisher, C. R., & Baums, I. B. (2023). Capacity of deep-sea corals to obtain nutrition from cold seeps aligned with microbiome reorganization. *Global Change Biology*, 29(1), 189–205. https://doi.org/10.1111/gcb.16447

Paul, B. G., Ding, H., Bagby, S. C., Kellermann, M. Y., Redmond, M. C., Andersen, G. L., & Valentine, D. L. (2017). Methane-oxidizing bacteria shunt carbon to microbial mats at a marine hydrocarbon seep. *Frontiers in Microbiology*, 8. https://doi.org/10.3389/fmicb.2017.00186

Price, R. E., & Giovannelli, D. (2017). A review of the geochemistry and microbiology of marine shallow-water hydrothermal vents. *Reference Module in Earth Systems and Environmental Sciences*. Elsevier. https://doi.org/10.1016/B978-0-12-409548-9.09523-3

Kolb, J. (2024). Project for sea. Retrieved June 14, 2024, from https://www.pmel.noaa.gov/eoi/nemo1998/education/pressure.html#:~:text=If%20you%20are%20at%20sea,the%20pressure%20at%20sea%20level.

Razvy, M. A. (2024, January 9). The fascinating reproduction of anglerfish: A deep-sea mystery unveiled. *International Network for Natural Sciences | Research Journal*, 24(1), 219–224. Retrieved, from https://innspub.net/the-fascinating-reproduction-of-anglerfish-a-deep-sea-mystery-unveiled/.

Reichle, D. E. (2023). Climate and climate models. In D. E. Reichle (Ed.), *The Global Carbon Cycle and Climate Change* (pp. 389–452). Elsevier. https://doi.org/10.1016/B978-0-443-18775-9.00010-3

Roeselers, G., & Newton, I. L. G. (2012). On the evolutionary ecology of symbioses between chemosynthetic bacteria and bivalves. *Applied Microbiology and Biotechnology*, 94(1), 1–10. https://doi.org/10.1007/s00253-011-3819-9

Rohling, E. J., Foster, G. L., Gernon, T. M., Grant, K. M., Heslop, D., Hibbert, F. D., Roberts, A. P., & Yu, J. (2022). Comparison and synthesis of sea-level and deep-sea temperature variations over the past 40 million years. *Reviews of Geophysics*, 60(4), e2022RG000775. https://doi.org/10.1029/2022RG000775

Romero-Romero, S., Miller, E. C., Black, J. A., Popp, B. N., & Drazen, J. C. (2021). Abyssal deposit feeders are secondary consumers of detritus and rely on nutrition derived from microbial communities in their guts. *Scientific Reports*, 11(1), 12594. https://doi.org/10.1038/s41598-021-91927-4

Rona, P. A. (1999). Deep-diving manned research submersibles. *Marine Technology Society Journal*, 33(4), 13–25. https://doi.org/10.4031/MTSJ.33.4.3

Sharma, R. (2015). Environmental issues of deep-sea mining. *Procedia Earth and Planetary Science*, 11, 204–211. https://doi.org/10.1016/j.proeps.2015.06.026

Spanier, E., Cobb, J. S., & Clancy, M. (1994). Impacts of Remotely Operated Vehicles (ROVs) on the behavior of marine animals: An example using American lobsters. *Marine Ecology Progress Series*, 104(3), 257–266. https://www.jstor.org/stable/24842188

Suess, E. (2014). Marine cold seeps and their manifestations: Geological control, biogeochemical criteria and environmental conditions. *International Journal of Earth Sciences*, 103(7), 1889–1916. https://doi.org/10.1007/s00531-014-1010-0

Sweetman, A. K., Thurber, A. R., Smith, C. R., Levin, L. A., Mora, C., Wei, C.-L., Gooday, A. J., Jones, D. O. B., Rex, M., Yasuhara, M., Ingels, J., Ruhl, H. A., Frieder, C. A., Danovaro, R., Würzberg, L., Baco, A., Grupe, B. M., Pasulka, A., Meyer, K. S., Dunlop, K., Henry, L.-A., & Roberts, J. M. (2017). Major impacts of climate change on deep-sea benthic ecosystems. *Elementa: Science of the Anthropocene*, 5, 4. https://doi.org/10.1525/elementa.203

Tan, J. S. D., & Fischer, A. M. (2022). Suggestions for marine protected area management in Australia: A review of temperature trends and management plans. *Regional Environmental Change*, 22(3), 92. https://doi.org/10.1007/s10113-022-01949-5

MarinebioConservation Society. (2024). The deep sea. Retrieved June 17, 2024, from https://www.marinebio.org/oceans/deep-sea/

Thushari, G. G. N., & Senevirathna, J. D. M. (2020). Plastic pollution in the marine environment. *Heliyon*, 6(8), e04709. https://doi.org/10.1016/j.heliyon.2020.e04709

Voight, J. R. (2000). A review of predators and predation at deep-sea hydrothermal vents. *Cahiers de Biologie Marine*, 41(2), 155–166.

Wang, K., Shen, Y., Yang, Y., Gan, X., Liu, G., Hu, K., Li, Y., Gao, Z., Zhu, L., Yan, G., He, L., Shan, X., Yang, L., Lu, S., Zeng, H., Pan, X., Liu, C., Yuan, Y., Feng, C., Xu, W., Zhu, C., Xiao, W., Dong, Y., Wang, W., Qiu, Q., & He, S. (2019). Morphology and genome of a snailfish from the Mariana Trench provide insights into deep-sea adaptation. *Nature Ecology & Evolution*, 3(5), 823–833. https://doi.org/10.1038/s41559-019-0864-8

Wang, Z., Qader, M., Wang, Y., Kong, F., Wang, Q., & Wang, C. (2023). Progress in the discovery of new bioactive substances from deep-sea associated fungi during 2020–2022. *Frontiers in Marine Science*, 10, 1232891. https://doi.org/10.3389/fmars.2023.1232891

Wu, N., Xu, C., Li, A., Cao, H., Chen, Y., Zhang, X., Geng, W., Zhai, B., Li, Q., & Sun, Z. (2022). Oceanic carbon cycle in a symbiotic zone between hydrothermal vents and cold seeps in the Okinawa Trough. *Geosystems and Geoenvironment*, 1(3), 100059. https://doi.org/10.1016/j.geogeo.2022.100059

Wynn, R. B., Huvenne, V. A. I., Le Bas, T. P., Murton, B. J., Connelly, D. P., Bett, B. J., Ruhl, H. A., Morris, K. J., Peakall, J., Parsons, D. R., Sumner, E. J., Darby, S. E., Dorrell, R. M., & Hunt, J. E. (2014). Autonomous Underwater Vehicles (AUVs): Their past, present and future contributions to the advancement of marine geoscience. *Marine Geology*, 352, 451–468. https://doi.org/10.1016/j.margeo.2014.03.012

Xu, Y., Du, X., Yu, X., Jiang, Q., Zheng, K., Xu, J., & Wang, P. (2022). Recent advances in the heterologous expression of biosynthetic gene clusters for marine natural products. *Marine Drugs*, 20(6), 341. https://doi.org/10.3390/md20060341

Zbinden, M., & Cambon-Bonavita, M. (2020). Rimicaris exoculata: Biology and ecology of a shrimp from deep-sea hydrothermal vents associated with ectosymbiotic bacteria. *Marine Ecology Progress Series*, 652, 187–222. https://doi.org/10.3354/meps13467

Zeng, X., Alain, K., & Shao, Z. (2021). Microorganisms from deep-sea hydrothermal vents. *Marine Life Science & Technology*, 3(2), 204–230. https://doi.org/10.1007/s42995-020-00086-4

Zhang, X., Luan, Z., Du, Z., Xi, S., Li, L., Lian, C., Zhang, J., Yin, Z., Ma, L., & Zhang, X. (2023). Gas hydrates in shallow sediments as capacitors for cold seep ecosystems: Insights from in-situ experiments. *Earth and Planetary Science Letters*, 624, 118469. https://doi.org/10.1016/j.epsl.2023.118469

Zhang, Z., Wu, Y., & Zhang, X.-H. (2018). Cultivation of microbes from the deep-sea environments. *Deep Sea Research Part II: Topical Studies in Oceanography*, 155, 34–43. https://doi.org/10.1016/j.dsr2.2017.07.008

2 Marine Archaebacteria, Actinobacteria, and Cyanobacteria

Smile Sharma, Khushbu, and Mahavir Joshi
University Institute of Biotechnology, Chandigarh University, Mohali, India

2.1 INTRODUCTION

Archaea constitute a separate domain of unicellular microorganisms, which encompass a separate branch and differ from both bacteria and eukaryotes. Chemically, archaea differ from bacteria with respect to cell wall composition, membrane lipids, and ribosomal RNA. Because many extreme environments are actually much more favorable to them than they will always find, some archaea are extremophiles, which allows them to inhabit many less extreme habitats such as those found in marine environments. Among these marine habitats are the deep-sea hydrothermal vents, where archaea withstand comparatively high temperatures and pressures, and hypersaline environments like salt lakes. This includes large populations in open oceans and major roles in nutrient cycling while contributing to the degradation of organic matter in marine sediments. Several marine planktonic archaea possess the potential of one of the largest organisms on Earth; they represent an important component of life on our planet that could influence the carbon and nitrogen cycles (DeLong, 2021; Marlow et al., 2014).

Actinobacteria are diverse and ecologically important phylum of Gram-positive bacteria, which is characterized by a high guanine and cytosine (G+C) content within its DNA. In both terrestrial and aquatic ecosystems, including marine environments, actinobacteria are found as ubiquitous organisms. Within these ecosystems, they take on an important role as decomposers and offer nutrient recycling. Some, for example, members of the genus *Streptomyces*, also produce antibiotics and contribute to cycling within the marine carbon cycle. Any actinobacteria display a lifestyle of mycelium, which involves branched growth forms and spore formation. Actinobacteria had been traditionally considered to be either transitional species between fungi and bacteria because many produce mycelium and reproduce via sporulation, which was thought to bridge the gap between them in morphological similarities. However, in modern classification schemes, they are firmly positioned as an independent phylum within the bacterial domain (Barka et al., 2016; Berry et al., 2023).

Cyanobacteria, or blue-green algae, are an important group of photosynthetic bacteria, a major member of marine environments. They are a major producer of oxygen and a carbon-fixing component using photosynthesis. Marine cyanobacteria are more than three million years in age and were reef builders. They are a dominant component of epiphytic, epilithic, and endolithic assemblages as well as microbial mats within coral reef communities (Charpy et al., 2012; Sánchez-Baracaldo, 2015). A number of populations of cyanobacteria are significant contributors to marine ecosystems. *Trichodesmium* is one dominant marine nitrogen-fixing organism, accounting for some 50% of total natural nitrogen fixation. *Prochlorococcus* and *Synechococcus* are found in most oceans and dominate most phytoplankton populations and make significant contributions to primary productivity (Charpy et al., 2012). Specific cyanobacteria, such as *Richelia* and *Calothrix*, have symbiosis with diatoms (Sánchez-Baracaldo, 2015). These microorganisms also engage in the nitrogen cycle of the biosphere, reducing nitrogen gas with nitrogenase enzymes to ammonia. Some cyanobacteria have developed specialized cell structures called heterocysts, which contain active nitrogenase and can only function in an oxygen-free environment during photosynthesis (El-Seedi et al., 2023). Besides being able to fix nitrogen, cyanobacteria perform the processes of calcification and decalcification (Charpy et al., 2012).

2.2 MARINE ARCHAEBACTERIA

Marine ecosystems are shaped by the intricate interactions of diverse microbial communities, including archaebacteria, actinobacteria, and cyanobacteria, each playing distinct yet interconnected roles. Archaebacteria, often overlooked, contribute to the decomposition of organic matter in seafloor sediments and can be one of the most abundant groups of organisms, significantly influencing both the carbon and nitrogen cycles.

2.2.1 Classification of Marine

Archaebacteria marine archaea belong to the domain archaea, distinct from bacteria and eukarya. Within archaea, there are several phyla commonly found in marine environments. Marine archaea, classified into four major groups (MG-I–MG-IV), serve as major players in the oceanic biogeochemical cycles. Still, their ecological functions and metabolic diversity have only recently begun to show clarity through advanced genomic tools.

DOI: 10.1201/9781003509530-2

Marine Group I (Nitrososphaerota) does dominate ammonia oxidation in global oceans, while Marine Groups II, III, and IV, all belonging to the Euryarchaeota, play different ecological roles along different ocean depths. Marine Group II (MG-II) is predominant in the surface ocean and the photic zone and significantly partakes in organic matter degradation and carbon cycling. Likely heterotrophic, MG-II uses dissolved organic compounds such as lipids and proteins for energy, and some members may even be photoheterotrophic, using rhodopsins to tap into light energy (Orsi et al., 2015). Marine Group III and Marine Group IV *Euryarchaeota* remain less understood than other marine archaeal groups, due to the scarcity of representatives within water column samples. Generally, MG-III archaea are considered low-abundance members of deep mesopelagic and bathypelagic communities. However, MG-III can be relatively abundant in certain locations, such as the deep Arctic Ocean and the oxygen minimum zone in the Eastern tropical South Pacific. A distinct clade of MG-III has also been found in the euphotic zone of the Mediterranean Sea, suggesting more diversity within this group than previously accepted. MG-III may perform anaerobic respiration. Novel MG-III genome sequence bins have been identified from the photic zone, containing numerous photolyase and rhodopsin genes, as well as genes for peptide and lipid uptake and degradation, suggesting a photoheterotrophic lifestyle (Haro-Moreno et al., 2017; Santoro et al., 2018). The first MG-IV clone was identified from depth of 3,000 m in the Antarctic polar front. Other MG-IV sequences originating from deep waters in the North Atlantic branch off at the base of the haloarchaea. Beyond these early observations, not much else is known about the distribution of the group (Santoro et al., 2018).

Marine microorganisms such as archaea, actinobacteria, and cyanobacteria perform important functions in ocean biogeochemical cycling. Marine archaea are dominant in nitrogen cycling and methane generation, which in turn affects the chemistry of oceans and the atmosphere. Actinobacteria carry out the decomposition of complex organic matter, leading to carbon cycling and the release of bioactive compounds vital to marine ecosystems. Cyanobacteria, being primary producers, are responsible for carbon fixation via oxygenic photosynthesis and are also significant nitrogen-fixing organisms sustaining marine food webs. Cyanobacteria are also involved in symbiotic associations that augment nutrient supply and ecosystem stability. In concert, these microorganisms maintain sea life through nutrient cycling control and carbon fixation, ultimately regulating world climate trends (Figure 2.1) (Dai et al., 2020; Pajares & Ramos, 2019; Marco, 2024; Voss et al., 2013).

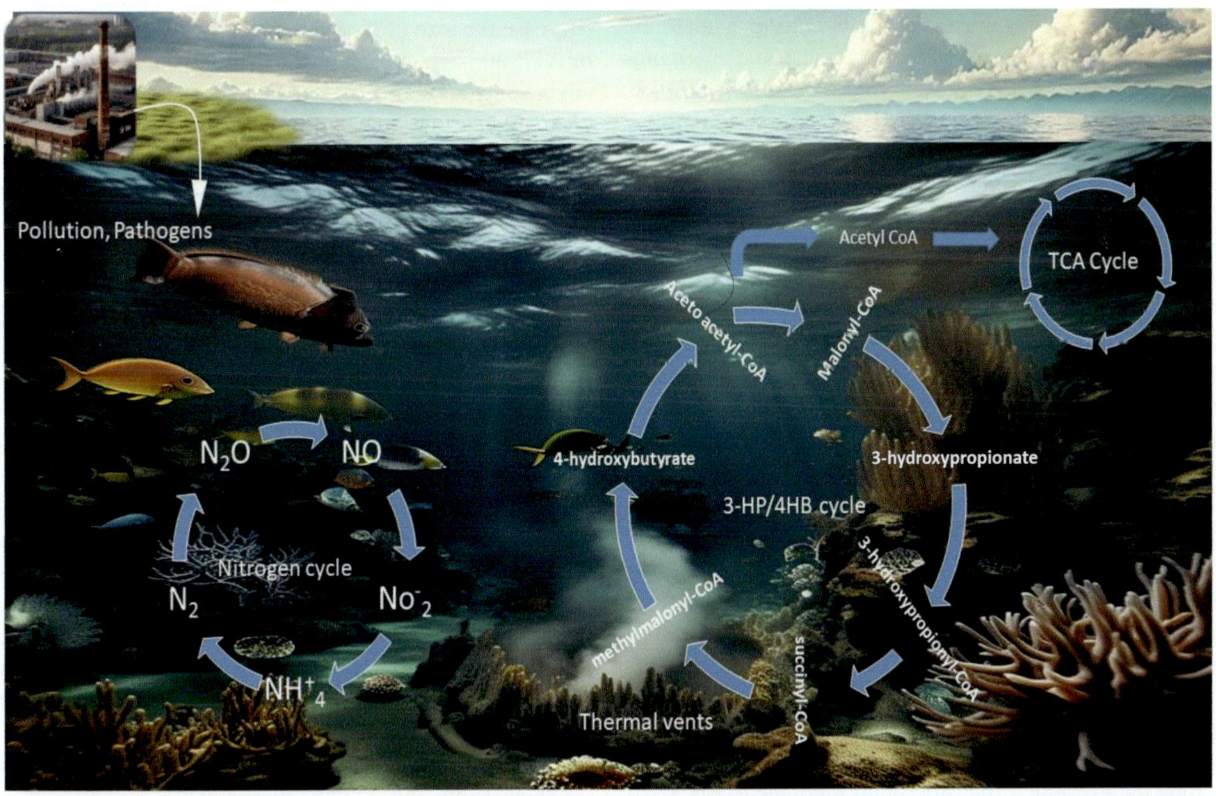

FIGURE 2.1 Various biochemical cycles occurring in the marine ecosystem, including the nitrogen cycle, 3-hydroxypropionate/4-hydroxybutyrate (3-HP/4-HB) cycle, and the TCA cycle. These cycles play a crucial role in regulating ocean chemistry, supporting marine life, and influencing global climate systems. Microbes, particularly bacteria and archaea, are the key drivers of these cycles.

2.3 DIVERGENT NICHES OF MARINE ARCHAEA

Archaebacteria can be basically grouped into four: methanogen, extreme halophiles, sulfate-reducing archaea, and extreme thermophiles. In marine environments, these microorganisms are typically found in specific niches, such as anaerobic sediments of shallow and deep-sea regions (free-living and symbiotic methanogens), hot springs and hydrothermal vents (methanogens, sulfate reducers, and extreme thermophiles), and highly saline, landlocked seas where halophiles flourish.

2.3.1 Methanogens

These anaerobic archaea exist in different marine locales such as deep-sea sediments, hydrothermal vents, and organic-rich mud volcanoes. Methanogens use substrates like CO_2, acetate, and methylated compounds to produce methane, often in association with other microbial communities. Their metabolic activities contribute to methane emissions and influence nutrient cycling and geochemical transformations within oceanic ecosystems.

In anaerobic marine environments, including the Gulf of Cádiz mud volcanoes, archaea are an important component of methane cycling; *Methanococcoides* methanogens are said to drive methanogenesis primarily by utilizing methyl substances like methylamine as methane substrates. In direct contrast, both anaerobic methanotrophic archaea (ANME), including *Methanosarcinales*, *Methanococcoides*, and ANME-1, oxidize methane through a reverse methanogenesis pathway, often in symbiosis with sulfate-reducing bacteria or metal oxides. Though metagenomic data corroborate this metabolism, an absence of wild-type isolates for direct biochemical verification presents a gap. The marine methanogens *Methanosarcina acetivorans* C2A and *Methanococcoides orientis* sp. nov. are reported to conserve energy via respiration, reducing ferrihydrite or humic acid while producing acetate and/or formate as the end product. Archaeal diversity differs with respect to the various mud volcano sites, since it reflects the geochemical heterogeneity; however, bacterial diversity is higher in comparison (Webster et al., 2023; Yan et al., 2023). Marine archaebacteria are essential for methane cycling within deep marine sediments, which may host ~26% of Earth's microbial biomass and constitute the largest methane reservoir. Within the Guaymas Basin (Gulf of California), methanogens were found in very low abundance, contributing to the microbial methane production, in particular at sites where thermal alteration had occurred in the past. Analyses of isotope data revealed that microbial methane exists at all depths studied, though its relative contribution diminishes with depth. Methylotrophic methanogenesis is an energetically superior pathway, generating more energy than hydrogenotrophic and acetoclastic processes. While the detection of *mcrA* genes was lowly present overall, a diversity of methyltransferase genes (*mtaB*, *mttB*) indicates potential methylotrophic activity. Microbial methanogenesis was likely sustained over geological timescales by small methanogen populations with thermal sediment alteration affecting methanogenic mechanisms (Bojanova et al., 2023).

In the South China Sea, fine-grained magnetite occurs in abundant amounts in iron-rich sediments, possibly under microbial influence. Dominating the turbidites are methanogenic archaea (*Methanosarcina* and *Methanocella*) and methanotrophic anaerobes that are associated with pyrite-rich layers. Formation appears linked to sulfide-free, iron-rich conditions, while goethite is a viable source for reactive iron. These findings illustrate the role of archaebacteria in methane metabolism and sediment magnetization in dynamic seep environments (Lin et al., 2024).

2.3.2 Extreme Halophiles

Halophilic archaea, or haloarchaea, are remarkable for their ability to survive demanding conditions such as high salinity, desiccation, and radiation. Some studies have shown that these microorganisms can live for desiccation and radiation, sometimes extending to millions of years. Some of the mechanisms include the production of protective materials, i.e., halomucin, and cells surviving in fluid inclusions of halite. In addition, haloarchaeans are able to decrease the size of their cells, decrease their consumption of ATP, and become adapted to limited nutrient availability environments (Stan-Lotter & Fendrihan, 2015). Within this context, a recent study focused on the molecular acclimation of *Halobacterium salinarum* housed in the halite brine inclusions. The proteomic analyses indicated that after two months, these cells exhibited protein expression similar to those in stationary-phase liquid cultures but significantly downregulated for ribosomal proteins. Such data point to a strategic reduction in protein synthesis machinery that would aid survival in the confined halite environment (Favreau et al., 2023). Extremely halophilic archaea, including Haloarchaea, Nanohaloarchaeota, Methanonatronarchaeia, and Halarchaeoplasmatales, thrive in hypersaline settings by maintaining osmotic balance. However, the evolutionary adaptations that made salt tolerance possible remain obscure, at least in view of some conflicting phylogeny from multiple lineages. Recent phylogenomic studies drawing upon enhanced taxon sampling and the most recent methodologies to correct proteomic compositional biases provide a more resolved phylogeny of these archaea. Newly identified uncultured lineages, Afararchaeaceae and Asbonarchaeaceae, help clarify the evolutionary placement of Haloarchaea and Nanohaloarchaeota. Metagenome-assembled genomes (MAGs) from the hypersaline aquatic systems of the Danakil Depression (Ethiopia) have further contributed to these insights. Phylogenomic analyses suggest at least four independent adaptations to extreme halophily in archaea, primarily facilitated by gene duplication and horizontal gene transfer. Dissemination of critical genes, such as those encoding potassium transporters, played a major role in the evolution of salt tolerance among these extremophiles (B. A. Baker et al., 2024). Halophilic microorganisms can persist in salt crystal brine inclusions, but this remains unclear as to how. The recent advancement of the halite surface sterilization has managed to isolate DNA as well as cells on the one hand, while facing some problems with getting rid of contaminants or maintaining the gene expression while facing "OMICS" studies. A recent proteomic analysis of *H. salinarum* NRC-1 residing within halite brine inclusions showed similarities to stationary-phase cultures, downregulation of motility-related proteins, and ribosomal proteins. Unique transporters indicated modified interactions with the brine microenvironment (Favreau et al., 2023). Four new species of *Natrinema* were isolated from marine and salt mine environments in China: *N. caseinilyticum, N. gelatinilyticum, N. marinum, and N. zhouii.* Molecular and genomic analyses showed their close clustering with *Natrinema* but attempting genome-related indexes, ANI, isDDH, AAI, constructing values that dropped below species boundary thresholds. They have all exhibited distinct phenotypes while possessing common compositions of their major polar lipids: PA, PG, PGP-Me, S-DGD-1, and S2-DGD, thereby distinguishing them from related species (Hu et al., 2023).

2.3.3 Sulfate-Reducing Archaea

Sulfate-reducing microorganisms (SRMs) are critical to the sulfur and carbon cycles of anoxic marine sediments as they are abundantly present in these environments, including fermentation byproducts like volatile fatty acids (VFAs) and hydrogen. The interactions of SRMs with other community species as well as with other microbial communities remain poorly understood. A recent study by Liang et al. has shown the ecological significance of SRMs among other ways by their activity in affecting microbial networks and structure of the microbial community; of particular interest is the degree to which SRM-induced changes in pH affect key bacteria such as *Marinilabiliales* (Bacteroidota), functioning in organic matter recycling (Wasmund, 2023). Research on sulfur metabolism in archaea has uncovered novel enzymes and pathways that enhance the understanding. There are diverse dissimilatory sulfur metabolisms within the Crenarchaeota and Euryarchaeota phyla flaunting aerobic elemental sulfur oxidation, anaerobic sulfur reduction and sulfate reduction, and organic sulfur respiration. For example, in assimilatory sulfur metabolism, novel processes include tRNA-dependent cysteine biosynthesis and distinct Fe–S cluster biogenesis. All archaea have eukaryotic-like ubiquitin systems contributing to protein degradation and sulfur cofactor biosynthesis. Special adaptation pathways also involve coenzyme M and coenzyme B crucial for methanogenesis (Gugliandolo & Maugeri, 2019; Liu et al., 2012). Only four genera of sulfate-reducing archaea have been cultured so far: *Archaeoglobus*, *Caldivirga*, *Thermocladium*, and *Vulcanisaeta*. Notably, these archaea are isolated in extreme niches, including oil deposits and hot springs (Demin et al., 2024). The three novel Helarchaeota MAGs in sulfate-rich sediments off Costa Rica coexist with sulfate-reducing bacteria (*Desulfobacteria*). Three MCR operon variants in these archaea suggest an anaerobic role in alkane oxidation stemming from thermogenic processes They expand the diversity within Helarchaeota while highlighting hydrocarbons as potential key energy sources in the deep marine biosphere (Zhao & Biddle, 2021).

2.3.4 Thermophilic Archaea

Archaea, particularly those thriving in extreme thermal environments, possess a remarkable suite of thermostable proteins essential for survival. Several proteins are involved in chromatin organization and DNA protection. Examples include Alba, Cren7, MC1, and histones found in various archaeal species including *Sulfolobus*, *Thermoplasma*, and marine group II archaea. These proteins are important for maintaining genome integrity and preventing DNA denaturation during high temperature. A yeast reverse gyrase is another crucial enzyme introducing positive supercoils to DNA and increasing its thermal stability even more. Chaperonins like GroEL/GroES homologues are also critical in assisting the protein-folding process and inhibiting aggregation under heat stress. There are fluorescent proteins which enable real-time visualization of cellular processes in these extreme environments. Thermostable enzymes like Rubisco are foundational for primary metabolism such as carbon fixation, enabling these organisms to grow and thrive in extreme environments. All these proteins and enzymes put a spotlight on the awesome adaptations enabling archaea to invade and thrive in some of Earth's unwelcoming environments, opening up much ground for biotechnological applications (Taguchi & Koike-Takeshita, 2023; Zhang et al., 2020). Thermophilic and hyperthermophilic archaeal enzymes exhibit distinctive structural and functional properties by virtue of which they retain stability even under extreme temperatures and pressures. They have a strong hydrophobic core and extensive electrostatic interactions, allowing them to remain active at very high temperatures. This loss of aerobic function is present in such hyperthermophilic archaea as the crenarchaeal genera *Archaeoglobus*, *Pyrodictium*, *Thermodiscus*, *Staphylothermus*, *Hyperthermus*, *Methanopyrus*, and *Pyrococcus*. If it is assumed that optimum growth temperatures reach from 75°C to 105°C, the maximum temperature of growth could attain values of even 113°C (*Pyrobaculum*) and potentially as high as 110°C in the case of *Pyrodictium occultum*. These are so well adapted to the high temperatures that no growth occurs below 80°C (Andrade & Garcia, 1999; Reed et al., 2013). A number of both thermophilic and extremely acidophilic Archaea possess tetraether lipids, which span the cell membrane and form monolayers instead of bilayers. These tetraether lipids are highly stable because of their restricted hydrocarbon chain mobility and the presence of isoprenoid methyl groups that inhibit water penetration. Tetraether membranes retain low proton permeability at even elevated temperatures, which is particularly crucial to sustaining a viable proton motive force central to ATP synthesis (Albers et al., 2000).

2.4 MAJOR ECOLOGICAL ROLES

Archaea, though fewer than bacteria in numbers, form an extremely important component of oceanic biomass, especially in deep waters, thus performing key ecological functions in marine ecosystems. They participate in carbon fixation, nitrogen cycling, and organic compound turnover.

2.4.1 Carbon and Nitrogen Cycling

Thaumarchaeota from the Marine Group I are the most abundant deep-sea microbes crucial for global carbon and nitrogen cycling through carbon dioxide fixation coupled with ammonium/ammonia oxidation that releases nitrous oxide as a byproduct (B. J. Baker et al., 2012; Könneke et al., 2014; Swan et al., 2014). Marine Group II generally dominates surface waters and can exhibit heterotrophic or photoheterotrophic lifestyles (Xia et al., 2017). The physiology and metabolism of Marine Groups III and IV are not well understood owing to their dominant habitat, the ocean's interior, and relatively low abundance. In subsurface and anoxic sediments, ANME, deep-sea archaeal group, and miscellaneous Crenarchaeota can comprise a substantial fraction of prokaryotic standing stocks, significantly contributing to biogeochemical cycles and global ecosystem functioning. Marine archaea are significant players in greenhouse gas cycling (Könneke et al., 2014; Xia et al., 2017).

2.4.2 Organic Matter Remineralization

In remineralization, organic matter is broken down by microorganisms into relatively simpler inorganic substances. Euryarchaeota from Marine Group II are likely to be important for organic matter remineralization in coastal marine environments. Different populations occur in winter and summer and can take part in the degradation of recalcitrant organic compounds and peptides (Orellana et al., 2019). Marine Thaumarchaeota, particularly isolates of *Nitrosopumilus*, mainly oxidize nitrogen from ammonium and urea rather than directly utilizing dissolved organic nitrogen (DON) compounds. Although oxidation of polyamine-N has been noted, it probably requires initial heterotrophic remineralization. In co-cultures with heterotrophic bacteria, inconsistent putrescine oxidation occurred, suggesting indirect metabolic interactions. Additionally, axenic cultures oxidized labeled putrescine during ammonia metabolism, suggesting co-metabolism or reactivity with metabolic byproducts ((Damashek et al., 2021). The effects of glacial-interglacial cycles regulate the distribution of such archaea, though they are extremely important for sedimentary organic carbon cycling. Archaeal communities in the South China Sea varied with changes in terrestrial input, sedimentation rates, and organic carbon levels. During the Last Glacial Period, increased terrestrial input led to a rise in ammonia-oxidizing Nitrososphaerales (SCG) archaea, while Thermoprofundales abundance correlated with total organic carbon, suggesting their role in deep-sea carbon processing. Interactions among benthic archaea, such as Bathyarchaeota and Thermoprofundales, highlight their importance in organic matter degradation, linking microbial dynamics to global climate changes (Lai et al., 2020). Meanwhile, recent studies have indicated that marine archaea play a pivotal role in the degradation of dissolved organic matter. Marine bacteria and archaea transform the optical properties of lignin, a complex organic polymer, in the open ocean. The portions of the lignin macromolecule may serve as a carbon source for certain archaea, thus becoming integral in the circulation of organic carbon in marine surroundings. In anoxic marine sediments, certain archaea, among which are ANME and members of the Bathyarchaeota phylum, degraded detrital proteins to enhance the processes involved in carbon cycling. Identified as uncultured archaea that can break down proteins in those environments, the implication is of an unrecognized role in protein remineralization. Thus, archaea enhanced organic matter decay under anoxic conditions and exerted a notable influence on methane production and consumption. The members of the ANME-2d clade assimilate both methane and dissolved inorganic carbon (DIC), indicating their roles in carbon cycling processes (Kurth et al., 2019; Weber et al., 2017; Yang et al., 2020). The role of prokaryotic communities in regulating the efficiency of the Southern Ocean biological carbon pump is critical (Dithugoe et al., 2023). Bacteria degrade complex organic carbon whereas archaea interfere with nitrogen metabolism, including nitrite and ammonia oxidation. Both groups contribute to chemoautotrophy and the production of recalcitrant dissolved organic carbon (RDOC), which furthers the sequestration of particulate organic matter (POM) (Dithugoe et al., 2023).

2.5 MARINE ACTINOBACTERIA

These microbes are fascinating and inhabit various marine environments. Evidence shows that they can tolerate a wide range of conditions. Description: Gram-positive, aerobic, nonmotile, thus characteristic of high DNA guanosine and cytosine base content. These filamentous bacteria almost look like fungi. Actinobacteria are a remarkable source of biologically active natural compounds, and over half of these are produced by members of the group concerned. In recent years, growing attention has been given to the "rare" actinobacteria from extreme environments, including marine environments, on the premise that the physiology unique to these extreme environments could induce the production of diverse new types of natural compounds. Hence, the marine environment has attracted considerable attention as a promising source of novel bioactive natural products. Though these natural compounds will often need some structural modifications and diversification to strengthen their biological or chemical properties for use in drugs and other applications, synthetic biology tools, metabolic engineering, and chemical synthesis platforms can aid further in these modifications (Jagannathan et al., 2021; Jose & Jha, 2017).

Marine actinobacteria, characterized by filamentous and branching growth forms, form intricate networks of hyphae-like fungal mycelia. This structural adaptation proves to be critical for their colonization of various marine substrates, including sediments and organic particles. During the past decade, these microorganisms' ecological relevance and biotechnological potential have been given attention in increasing measure by researches. They can synthesize antibacterial, antifungal, and anticancer bioactive compounds that have a wide range of uses. Studies dealing with marine systems have started to give increasing attention to the potential they offer with respect to unique reservoirs of rare actinobacteria, believed to be remarkable sources of novel bioactive compounds. This lacuna points toward the necessity of exploring the understudied and unstudied localities, for example, deep-sea sediments and unique marine niches, favoring the survival of these microbes (Subramani & Sipkema, 2019).

These microbes show unique physiological and biochemical adaptations rendering them survivable in saline conditions. Many are halotolerant and/or halophilic and can thus endure conditions of high salt concentration (Gunde-Cimerman et al., 2018). For instance, the marine species *Salinispora* has undergone notable evolutionary adaptations, allowing them to thrive in high-salinity habitats. *Salinispora* has no paragon Nqr (NADH:quinone oxidoreductase) system, unlike several marine bacteria. Instead, two uncompleted copies of NDH-1-generating machines expressed in *Salinispora* genomes compose the most critical electron transport unit. One such copy assembles a membrane portion of complex I: sodium or proton pumping leads to the accentuation of ionic motive forces. The other carries *mrp* genes that probably encode a sodium antiporter. This Mrp system, being different from already known active homologs, helps maintain low concentrations of sodium inside the cells, preventing toxicity and accommodating for pH regulation in seawater, which usually has a pH of 8.3 (Penn & Jensen, 2012).

2.6 ACTINOBACTERIA HABITATS

Marine actinobacteria are prevalent in various marine habitats ranging from shallow coastal waters to deep-sea sediments, where they perform important functions during nutrient cycling, organic matter decomposition, and symbiotic interactions. With the ability to tolerate extreme ranges of salinity, temperature, pressure, and nutrient conditions, these organisms occupy extreme and dynamic niches in marine ecosystems (Subramani & Sipkema, 2019). The primary marine habitats of actinobacteria are discussed below along with their ecological importance and examples.

2.6.1 Deep-Sea Environments

Deep-sea sediments, representing over 70% of the Earth's surface, host various actinobacterial communities that are adapted to pressure, low temperatures, and scarcity of organic input. Metabarcoding studies conducted in a culture-independent manner suggested that, unlike Atlantic abyssal sediments that exhibited greater diversity, actinobacteria are abundant in Arctic deep-sea sediments. For example, sediments sampled from the Mohn's treasure of the Arctic Mid-Ocean Ridge and the area of Macaronesia of the Atlantic presented isolates from the genera *Streptomyces*, *Rhodococcus*, and *Microbacterium*, signifying their metabolic versatility. The actinobacteria of these habitats complete the recycling of nutrients by breaking down organic matter, a process essential for the survival of deep-sea ecosystems. It has also been shown, for example, that the antibiotic, anticancer, and anti-inflammatory potentialities have arisen against multiresistant pathogens from *Streptomyces* strains that got isolated from the Arctic sediments. Moreover, their unexploited genetic diversity has been fostered by the conjunction of culture dependent and culture-independent techniques that position deep-sea sediments as reservoirs of bioproductive strains with outstanding biotechnological importance (Ribeiro et al., 2023; Siro et al., 2022).

2.6.2 Coral Reef Ecosystems

Coral reefs are generally extra stressful in thermally stressed parts like the Arabian Gulf. Actinobacteria occurs inside tissue and mucus layers. In corals such as *Coscinaraea columna*, *Platygyra daedalea*, and *Porites harrisoni*, distinct distributions of 19 genera, including *Kocuria*, *Brevibacterium*, and *Streptomyces*, were observed to be between tissue and mucus. For example, corals, *Kocuria* and *Brevibacterium*, dominate the tissues, while mucus hosts the majority of the *Streptomyces* and *Micromonospora*. These adaptations certainly explained their niche-specific community distributions. Though their symbiotic interaction played a role in antimicrobial production, that actinobacteria may alert the function of corals when already under thermal stress applied. The results revealed 82 isolates from Gulf corals having significant inhibitory effects against Gram-positive and Gram-negative pathogens, with the strongest spectrum of activity belonging to *Streptomyces*. Such antimicrobial activity could alleviate the pathogenicity of corals, allowing host survival experiences in warmer ocean waters. The spatiotemporal changes in community composition reflect further on the constantly changing interactions between the host and microbe, triggered by the effects of changing environments; it reflects the temporal and spatial changes in community composition triggered by changing environmental factors (Mahmoud & Kalendar, 2016).

2.6.3 Mangrove Swamps and Coastal Sediments

Mangrove ecosystems, along locations like Egypt's Red Sea coast, support actinobacterial populations with salinity-tolerant or anaerobic conditions. The rhizosphere soils of *Streptomyces* and *Micromonospora* hold potent antimicrobial activity among *Streptomyces parvulus* strains toward *Staphylococcus aureus* and its drug-resistant versions. It is noted that these communities are richest in organic sediments, where the function of degradation of organic carbon and development of nitrogen occurs. Mangrove-derived actinobacteria exhibit strong antibacterial and antioxidant properties, and show potential in anticancer applications. Ethyl acetate extracts from *Streptomyces* sp. isolated from Red Sea mangroves showed MIC values of less than 7.8 μg/mL against pathogenic bacteria, comparable with the clinical antibiotic ciprofloxacin. Such results placed mangroves in an unexplored collection of pharmacologically active strains, with a possibility for utility against antibiotic resistance (Hamed et al., 2021; Ye et al., 2023).

2.6.4 Marine Sediments in Area or Intertidal Condition

These intertidal environments with periodic wetting and drying periods are suitable for actinobacteria to stand up to such extreme fluctuations. Sandy sediments from Ilha do Mel, Brazil, produced *Streptomyces* and *Micromonospora* strains that had bioactive compound-producing capabilities under a varied salinity and nutrient regime. These isolates express enzymatic activity directed at the degradation of complex polymers found in intertidal organic debris, namely cellulose and chitin. The antifungal and antibacterial potentials of fungi and actinobacteria isolated from intertidal sediments were seen to be synergistic since the crude extracts inhibited the growth of pathogens such as *Candida albicans* and *Escherichia coli*. From Paranaguá Bay, novel strains of *Streptomyces* were isolated, backing a further assertion that many such biosphere areas have yet to owe biotechnological potential (Porsani et al., 2013). A study of intertidal sediments of Diu Island in the Arabian Sea reported the isolation of 148 cultivable actinobacteria, which were characterized into operational taxonomic units based on their 16S rRNA gene sequences. These bacteria showed considerable biosynthetic potential toward the production of secondary metabolites for genes coding for polyketide synthases and nonribosomal peptide synthetases. This biosynthetic potential is further associated with the antibacterial activities observed by the laboratory isolates (Jose & Jha, 2017).

2.6.5 Open Ocean and Marine Surface Waters

Marine actinobacteria are major microbiota in open oceans and marine surface waters and play a pivotal role in ecological and biogeochemical processes. These bacteria contribute to the microbiome in the pelagic zone by organic matter degradation, nutrient cycling, and acting as producers of bioactive compounds (Sarkar & Suthindhiran, 2022). Marine actinobacteria, such as *Candidatus actinomarina*, are abundant in the surface waters, and their abundances are also concerned with the organic matter supplied by the phytoplankton blooms and particulate detritus. Metagenomic studies have revealed their adaptation to oligotrophic conditions, wherein they utilize dissolved organic carbon. The bloom of actinobacteria during the algal bloom was indicative of an enhancement in the decomposition of the organic matter released by the phytoplankton. Genomic analysis of *C. actinomarina* indicates the possession of protease and glycoside hydrolases genes, indicating the viability in degrading POM (Ghai et al., 2013).

2.7 ECOLOGICAL SIGNIFICANCE OF MARINE ACTINOBACTERIA

These microbes constitute a critical component of oceanic microbial communities. They are a prominent contributor to biogeochemical cycling and ecosystem functioning. These Gram-positive bacteria exist in almost all marine environments, including open oceans, coastal waters, and deep-sea sediments. The ability of actinobacteria to thrive in oligotrophic waters is determined by their metabolic versatility, allowing them to degrade complex organic matter into forms that can be transferred in food webs. They also assist in the cycling of sulfur and carbon, affecting ocean chemistry and even atmospheric processes. Due to their diverse functional roles, marine actinobacteria play a critical role in sustaining marine ecosystem health (Ayala-Muñoz et al., 2022; Bhatt, 2021).

2.7.1 Contribution to the Carbon Cycle

They play an important role in the carbon cycle. They collectively mediate organic matter decomposition and its consequent carbon sequestration in various marine ecosystem types. The process involves the hydrolysis of complex organic polymers (e.g., chitin, cellulose, and lignin) to bioavailable nutrients underlying marine food webs. For example, during seasonal phytoplankton blooms, particle-attached (PA) actinobacteria have shown the ability to metabolize organic polymers while utilizing monomers as the blooms dissipate. Contributions of actinobacteria during seasonal phytoplankton blooms involve breaking down complex organic polymers into bioavailable monomers, completing nutrient recycling. They effectively utilize their activity, particularly with organic substrates, as an influence on microbial dynamics in food webs. In marine ecosystems such as the Baltic Sea, the actinobacteria exhibited a seasonal succession from polymer degradation (dominated in growing populations by Bacteroidetes) to the utilization of monomers, and thus supporting biogeochemical processes. Understanding their impact is imperative for the prediction of microbial ecosystem responses to environmental changes (Martínez-García et al., 2022).

2.7.2 Degradation of Lignin

Lignin is an aromatic polymer that imparts structural rigidity to plant cell walls. One of the most chemically resistant organic materials in nature, its degradation cannot be overemphasized for organic carbon recycling in marine environments. Lignin-degrading enzymes produced by marine actinobacteria such as *Streptomyces*, *Rhodococcus*, and *Amycolatopsis* include lignin peroxidases, manganese peroxidases, and laccases. These enzymes break down lignin into smaller aromatic compounds and organic acids (Grgas et al., 2023). For example, the marine actinobacteria strains isolated using vanillic acid, a compound derived from lignin, as the sole carbon and energy source have been characterized. The isolates were predominantly assigned to *Arthrobacter*. The *Arthrobacter* indicated that they could metabolize various lignin-derived compounds that include 4-hydroxybenzoic acid, ferulic acid, syringic acid, and benzoic acid. The strain exhibited protocatechuate 3,4-dioxygenase and 4,5-dioxygenase activities which facilitate the cleavage of the aromatic ring and correspond to genetic sequences characteristic of lignin-degrading actinobacteria.

2.7.3 Contribution to the Nitrogen Cycle

They contribute substantially to the nitrogen cycle through nitrogen fixation, nitrification, and denitrification processes across various oceanic zones. These are processes crucial to maintaining nitrogen balance to efficiently sustain marine ecosystems. The nitrogen fixed by these diazotrophic actinobacteria in the euphotic zone converts atmospheric nitrogen (N_2) into ammonia (NH_3) released for primary producers such as phytoplankton. This is particularly important in nutrient-poor regions where nitrogen limits productivity. Among the actinobacteria found in coastal ecosystems, *Micromonospora* increase nitrogen availability (Nafis et al., 2019; Pajares & Ramos, 2019; Trujillo et al., 2015).

2.7.4 Symbiotic Relationships

These microbes are recognized to be associated with several marine organisms, which include sponges, corals, and echinoderms, as well as vertebrate hosts, such as the pufferfish. Their associations may witness the production of unique secondary metabolites serving as chemical defenses against predation and microbial infections. Ecological niches of marine actinomycetes include free-floating planktonic forms, biofilm, and communities that dwell among sediments. Those prove particularly diverse, with certain species being described at as great a depth as 500 m, including representative genera such as *Streptomyces*, *Micromonospora*, and *Actinomyces*. These sedimentary actinobacteria are influenced by environmental parameters like temperature, pH, salinity, and organic carbon content, which varies across locations and depths (Chen et al., 2021; Kamjam et al., 2017; Subramani & Sipkema, 2019). They play critical ecological roles in nutrient cycling and in shaping metabolic pathways of their hosts. For example, they are responsible for nitrogen and carbon recycling by hosts like sponges and corals. Diverse microbiota may also constitute up to 35% of sponge biomass. Such microbial communities generate bioactive compounds to protect the sponge against predation and disease, and to provide some extent of ecological stability. Actinomycetes in corals also produce bioactive metabolites that promote their survival by enhancing stress tolerance (Brinkmann et al., 2017).

Marine-associated bacterial communities show astonishing potential in producing bioactive molecules bearing biomedical applications. Research has shown that a diverse assemblage of actinobacterial life forms may be associated with many of the marine invertebrates through traditional culturing techniques and metagenomic approaches. Such actinobacteria belong largely to genera of *Streptomyces*, *Nocardiopsis*, *Kocuria*, *Salinospora*, *Nocardia*, *Rhodococcus*, *Nonomuraea*, *Actinokinespora*, and *Saccharopolyspora*. These remain mainstream as contributors to the natural synthesis of secondary metabolites, including antibiotics, anticancer agents, and other bioactive compounds that nurture the host's defense mechanisms and ecological resilience. Such data provide evidence that bacteria associated with marine invertebrates are a promising source to discover novel natural products with therapeutic potentials (Majithiya & Gohel, 2022).

2.8 APPLICATION OF ACTINOBACTERIA

2.8.1 Drug Discovery and Pharmaceuticals

Many genera of actinomycetes include *Streptomyces*, *Micromonospora*, *Actinomadura*, *Amycolatopsis*, *Streptoverticillium*, *Actinoplanes*, *Nocardia*, *Saccharopolyspora*, *Streptosporangium*, *Streptoalloteichus*, *Dactylosporangium* and *Frankia* are getting recognized in producing numerous types of antimicrobial compounds. These bioactive metabolites are essential for pharmaceutical industries. Actinomycetes produce numerous major classes of antibiotics including β-lactams, aminoglycosides, lipopeptides, glycopeptides, ansamycins, anthracyclines, nucleosides, peptides, polyenes, polyethers, tetracyclines, and macrolides (Alharbi, 2016).

In 2003, salinosporamide A with anticancer properties was isolated from *Salinispora tropica*. Chemically related to the known proteasome inhibitor omuralide, this compound was later found to be a more potent 20S proteasome inhibitor (1.3 nM) than the previously identified metabolite. Also, salinosporamide A was orally efficacious against human tumors in mouse models (Fenical & Jensen, 2006). An obligate marine *Streptomyces* strain that yielded marinopyrroles A and B, these actinobacteria are highly halogenated, axially chiral metabolites that feature a rare bispyrrole structure. Notably, the marinopyrroles exhibited potent antibiotic activity against methicillin-resistant *S. aureus* (MRSA) (Hughes et al., 2008). The strain SCSIO 01127 of *Streptomyces*, isolated from sediment of South China sea and identified through 16S rDNA analysis, produced two new spirotetronate antibiotics, lobophorins E(1) and F(2), along with its known analogs lobophorins A(3) and B(4). Lobophorin F(2) showed an improved antibacterial activity against both *S. aureus* and *Enterococcus faecalis* and it was found to be more cytotoxic than lobophorin B(4) against SF-268, MCF-7, and NCI-H460 cancer cell lines (Niu et al., 2011), *Micromonospora palythoicola* sp. nov. and *Streptomyces poriticola* sp. nov. *S. poriticola* demonstrated antimicrobial activity and selective cytotoxicity against MCF-7 breast cancer cells, while *M. palythoicola* exhibited antimicrobial effects against *Streptococcus mutans* and *Kocuria rhizophila* (Kanchanasin et al., 2024). A study conducted between 2017 and 2021 on marine actinobacteria yielded 77 novel alkaloids, predominantly indoles, diketopiperazines, glutarimides, indolizidines, and pyrroles. While microbial extract analysis has remained the primary mode of discovery, genome mining provides a very interesting alternative that minimizes compound rediscovery while fast tracking the discovery of novel drugs. The synthetic biology integrates it and increases this capability highlighting actinobacteria as an important source of bioactive compounds for pharmaceutical development (De Rop et al., 2022) (Table 2.1).

2.8.2 Bioremediation and Environmental Applications

Marine actinobacteria play a crucial role in bioremediation, exhibiting unique metabolic adaptations for hydrocarbon degradation. A study by Fernandes et al. (2024) identified 21 actinobacteria, primarily from *Rhodococcus*, *Gordonia*, and *Pseudonocardia*, capable of degrading oil-derived pollutants, with *R. ruber* completely breaking down 100 mg/L of fluorene. *Streptomyces gougerotti* and *Micromonospora matsumotoense* demonstrated potential for biodegradable plastic

TABLE 2.1
Marine Actinobacteria and Their Bioactive Compounds

Strain of Actinobacteria	Compound Produced	Isolated from Marine Habitat	Biological Activity	Reference
Micromonospora palythoicola sp. nov.	Not specified	Marine invertebrates	Antimicrobial and anticancer	Kanchanasin et al. (2024)
Streptomyces poriticola sp. nov.	Not specified	Marine invertebrates	Antimicrobial and anticancer	Kanchanasin et al. (2024)
Streptomyces sp. RKBH-B7	Guanahanolide A	Marine invertebrates	Anticancer	Marchbank et al. (2020)
Streptomyces koyangensis SCSIO 5802	Glycoabyssomicin A	Deep-sea	Antimicrobial	Zhu et al. (2024)
Streptomyces sp. SCSIO 40068	Kebanmycins	Mangrove rhizosphere	Antimicrobial and cytotoxic	Zhao et al. (2024)
Actinomadura glauciflava NA03286	Maduraflavacins A-E	Marine sediment	Antimicrobial	Zou et al. (2024)
Nocardiopsis sp. HDN154086	*p*-Terphenyl derivatives, namely, nocarterphenyls D–H	Marine sediment	Antibacterial activity against MRSA	Chang et al. (2021)
		Deep sea		
Dermacoccus abyssi MT1.1^T	New phenazine	Mariana Trench sediment	No bioactive test reported	Abdel-Mageed et al. (2020)
Nonomuraea sp. MM565M-173N2	Sealutomicin A	Deep-sea sediment	Antibacterial against carbapenem-resistant Enterobacteriaceae	Igarashi et al. (2021)
Streptomyces sp. *SY2111*	Streptonaphthyridine A	Mariana Trench sediment	Anti-proliferation human glioma U87MG and U251	Qin et al. (2023)
Micromonospora ferruginea	Quinocycline B (“kosinostatin”) and isoquinocycline B	Deep-Sea Sponge	Antimicrobial	Back et al. (2021, 2023)
Salinispora spp.	Salinosporamide A	Deep-sea sediments	Anticancer (proteasome inhibitor)	Russo et al. (2015)

production. The integration of physicochemical and biological methods can enhance environmental decontamination, emphasizing the untapped potential of marine actinobacteria (Fernandes et al., 2024).

Rhodococcus sp. NCIM 5126, isolated from a chronically oil-polluted coastal region near Mumbai Harbour, plays a significant role in bioremediation. This strain can degrade the aliphatic and aromatic fractions of crude oil, achieving degradation rates of 72%, 60%, and 35% for Bombay High, Assam, and Gujarat crude oils, respectively. It thrives under optimized conditions: 70 mM nitrogen as urea, 0.1 mM phosphorus as K_2HPO_4, pH 8.0, at 30°C, and 150 rpm for 72 hours. As a facultative halophile, it can grow in both seawater and distilled water, tolerating up to 1.7 M NaCl. This versatility makes it valuable for cleaning hydrocarbon-contaminated marine and freshwater ecosystems, particularly in tropical regions where natural temperatures enhance degradation processes. Supplying nitrogen to these ecosystems could accelerate the cleanup of such contaminants (Sharma & Pant, 2001).

These microorganisms have adapted to living in extreme conditions, making them prime candidates for bioremediation activities in various marine ecosystems, such as coastal waters, deep-sea sediments, and sites contaminated by oil. They are instrumental in bioremediation because they utilize their metabolic capacities to degrade pollutants and help restore the contaminated marine environments. Because of this capability to degrade hydrocarbons, heavy metals, plastics, and toxic compounds, they are most valuable for addressing marine pollution and oil spills (Aliko et al., 2022; Rathore et al., 2021). In the hydrothermal region of the Southern Mid-Atlantic Ridge, six consortia were isolated which were capable of degrading oil in room temperature enrichment experiments. The primary degradative bacteria belonged to the genera *Nitratireductor*, *Pseudonocardia*, *Brevundimonas*, and *Acinetobacter*. These consortia exhibited a diversity of hydrocarbon-degrading bacteria, noticeably abundant in sediments preserved at 4°C in proximity to hydrothermal vents. It shows that there is an extensive variety of oil-degrading microorganisms in the deep sea environment, which lays a strong base for improving strategies for oil bioremediation (Ma et al., 2021).

2.8.3 Enzyme Production for Industrial Use

Marine actinobacteria are a rich source of enzymes for use in various industries that function under harsh conditions, such as extreme salinity, temperature, and pressure. These enzymes are characterized by their thermal stability and activity, which render them quite useful for a range of biotechnological applications, including food processing, pharmaceuticals, detergent formulations, and biofuel production; for instance, the marine actinobacteria produce multienzyme cellulases that have high pH stability and thermal tolerance, which find application in biofuel production. Their enzymes are also used in the textile and food industries for their versatility and effectiveness. These peculiar conditions in marine environments become a habit where enzyme activity develops with unique properties from that of enzymes produced from terrestrial sources (González et al., 2020; Samuel et al., 2022). A recent bioactivity study on marine enzymes showed *Streptomyces rochei* produced a remarkably larger hydrolytic zone of 24 mm on starch agar plates compared to other isolates. In the Plackett-Burman experimental design of optimized cultivation conditions for a number of *Streptomyces* species, among them *S. griseorubens*, *S. rochei*, and *S. parvus*, the specific activities of α-amylases significantly increased. With further optimization through the Box-Behnken design, *S. rochei* achieved a particular activity up by a significant 12.22-fold. The maximum activity corresponded with pH 6 and 65°C, demonstrating thermal stability at 70°C for 40 minutes and showing an optimal salt concentration of 1 M. Presence of cationic metal ions, such as Cu^{2+}, Zn^{2+}, and Fe^{2+}, showed significant enhancement of the enzyme activity. Increased synthesis of α-amylase by *S. rochei* thus becomes a rightful avenue of commercial development (Al-Agamy et al., 2021). The marine natural products, MNPs, provided a good source of bioactive compounds for drug discovery. Advances in sequencing have revealed the existence of many biosynthetic gene clusters in marine microorganisms; many of these, however, exhibit poor expression under standard laboratory conditions. Moreover, a sizable number of these microbes are non-cultivable or genetically intractable. Heterologous expression systems allow a promising strategy to activating silent BGCs and hence optimizing metabolite production, thus aiding in the discovery of novel marine-derived compounds (Hassan & Jin, 2024). The increasing problem of pollution in the environment affects various species throughout the globe, thus requiring innovative solutions for the remediation work. Conventional chemical and physical methods are expensive and have an adverse effect on the environment; thus, there is a gradual drift toward the bioremediation option. This method of employing microorganisms enables the degradation of pollutants: more ecologically sound and cheaper. Advances in genetics have seen the rise of "genetically engineered microbes," selected for better degradation capabilities. By incorporating traits into bacteria, fungi, and algae, GEMs can degrade complex pollutants like oil spills, halobenzoates, and volatile organic compounds in a matter of hours. With the aid of these engineered microorganisms, a quick adaptation to new pollutants is feasible and allows rapid degradation in contributing to a cleaner environment. Genetic engineering increases the efficacy of bioremediation, thus benefiting environmental health and the well-being of humankind. Marine actinobacteria promise a vast potentiality within marine biotechnology via contributing to biopolymers, biofuels, marine bioproducts, and bioenergy development. Adaptation of these microorganisms toward survival in extreme marine climate provides the offspring with possibilities to utilize novel means of bioactive compounds in many sectors of industries (Patel et al., 2023).

2.8.4 Key Applications in Marine Biotechnology

Marine actinobacteria are a valuable source of biopolymers, biofuels, and cosmetics. They produce some species of marine-derived biopolymers, which are biodegradable plastics and bio-based polymers that provide a sustainable alternative to petroleum-based plastics. These eco-friendly materials can be explored for sustainable packaging and biomedical applications (Sharma et al., 2022). These microbes also facilitate the production of biofuels by being involved in biomass conversion to bioethanol and biodiesel, providing renewable energy alternatives to fossil fuels (Das et al., 2023). The potential to produce antioxidants, UV-protective compounds, and anti-aging compounds makes marine microorganism valuable in the cosmetic and skincare sector (Ghosh et al., 2023). These compounds are involved in conferring anti-inflammatory and anti-aging activity that may be useful for cosmeceuticals. Current research is focused on harnessing marine actinobacteria for the production of biofuels by constructing efficient microbial routes for the production of bioethanol and biodiesel. Scientists are exploring marine-derived biomaterials such as biopolymer-based films and coatings for biomedical and sustainable packaging applications (Singh et al., 2023). Aside from this, UV-protective and anti-aging compounds from marine actinobacteria are being explored for potential inclusion in natural sunscreens and skin-rejuvenating treatments (Yadav et al., 2023). With the growing interest in sustainable biotechnology, marine actinobacteria are opening up a novel alternative for renewable energy, biodegradable materials, and bioproducts for health. Metabolic versatility offered to them is a thrilling attraction to put more of the spotlight into research and commercialization opportunities across industries.

2.9 MARINE CYANOBACTERIA

Marine cyanobacteria is one of the most ancient photosynthetic bacterial groups, which have profoundly impacted the Earth's biosphere for more than 3.5 billion years. The ancient oxygenic photosynthetic contributors were key players in the Great Oxidation Event, which transformed the Earth's atmosphere in a way that paved the way for the evolution of aerobic life forms. Presently, these organisms are thriving in a variety of marine ecosystems, serving as crucial primary producers and nitrogen fixers in global biogeochemical cycles. *Prochlorococcus* and *Synechococcus* are the principal contributors to oceanic carbon fixation via primary production. They are found symbiotically with corals, sponges, and mollusks, providing vital nutrients essential for their survival. Additionally, these cyanobacteria are producers of a diverse array of bioactive compounds with antimicrobial, anticancer, and anti-inflammatory activities. Recent studies emphasize the importance of marine cyanobacteria to elucidate ocean productivity, microbial ecology, and the effects of climate change on the distribution and functioning of these organisms. Through the fixation of carbon dioxide via various trophic levels of photosynthesis, cyanobacteria are pivotal as global regulators of carbon cycles and climate change mitigation (Cesar-Ribeiro et al., 2023).

2.9.1 Diversity of Marine Cyanobacteria

These microorganisms are highly diverse, with extensive morphological, physiological, and phylogenetic differences. This diversity is also expressed in the ecophysiological specialists of the microorganism, which have evolved to occupy specific environmental niches in varied marine ecosystems, ranging from surface waters to benthic mats. Advanced molecular methods have greatly refined our knowledge on the cyanobacterial diversity revealing species and strains complexly able to inhabit distinct ecological niches. The Caribbean has been characterized by studies that reported 76 genera and 119 species, exemplifying the regional unexplored biodiversity potential (Vargas et al., 2023). Contrarily, genetic investigations such as 16S rRNA sequencing serve to further merge morphological and genetic perspectives in the taxonomic and phylogenetic placement of cyanobacterial isolates, which conversely emphasizes the necessity of merging morphologic and genetic means of diversity study. From another perspective, the diversity of marine cyanobacteria is also revealed by the variation in pigment types such as those found in *Synechococcus*, which have profound effects on oceanic biological productivity. Furthermore, the chemical diversity within filamentous benthic cyanobacteria echoes the high prospects for discovering novel compounds of biotechnological origin (Table 2.2) (Chagas et al., 2024; Faldu et al., 2014; Grébert et al., 2022).

2.10 ECOLOGICAL ROLES OF MARINE *CYANOBACTERIA*

Marine cyanobacteria form one of the core components of marine ecosystems, providing key functions in primary production, nutrient cycling, and food-web functioning. Marine *cyanobacteria* are minute organisms that thrive in diverse environments ranging from the sunlit upper ocean to the deep sea due to their significant influence on global biogeochemical cycles. As primary producers, these microorganisms convert sunlight into organic matter that constitutes the basis of the marine food chain. Besides, the fixation of nitrogen further enhances nutrient availability to consequent marine organisms. The marine *cyanobacteria* provide immense ecological services, not only in carbon and nitrogen cycles but also in sustaining marine ecosystems in terms of resilience and productivity, indeed becoming a key concern for wholesome understanding of oceanic health and sustainability (El-Seedi et al., 2023; Nweze, 2009).

TABLE 2.2
Key Genera of Marine Cyanobacteria and Their Characteristics

Genus	Ecological Role	Distribution	Notable Features	References
Prochlorococcus	Dominant primary producer in oligotrophic waters	Tropical and subtropical open ocean	Possesses unique chlorophyll b-containing pigments	Biller et al. (2015) and Laurenceau et al. (2020)
Synechococcus	Significant primary producer in coastal and open ocean	Widespread, tolerates a wider range of conditions	Possesses phycobilins for light harvesting, photosynthetic cell factories	Scanlan and West (2006)
Trichodesmium	Major nitrogen fixer in tropical and subtropical waters	Tropical and subtropical open ocean	Forms blooms, impacting nutrient availability	Ani et al. (2023) and Capone et al. (1997)
Cyanothece	Nitrogen fixer, tolerates high salinity	Coastal lagoons, hypersaline environments	Possesses a circadian rhythm for nitrogen fixation	du Plooy et al. (2015), Liu et al. (2025), and Wu et al. (2025)
Lyngbya	Benthic primary producer	Coastal areas, coral reefs	Produces potent toxins (e.g., lyngbyatoxin A)	Ahern et al. (2007), and Gärtner et al. (2021), and Lima (2023)
Nodularia	Nitrogen fixer, forms blooms in brackish waters	Baltic Sea, other brackish estuaries	Produces the potent hepatotoxin nodularin	Lopes and Vasconcelos (2011), and Mazur-Marzec et al. (2024)

2.10.1 Photosynthetic Adaptations

Marine *cyanobacteria* are oxygenic phototrophs using chlorophyll a and phycobiliproteins for the capturing of light energy and converting CO_2 into organic matter (Assunçao et al., 2022; Saini et al., 2018). When rehydrated, desiccated cyanobacterial mats that dominate the arid zones of Earth would resume photosynthesis practically instantaneously. During the rehydration of marine mats from Exmouth Gulf, Australia, chlorophyll a concentrations increased two to five times within 15 minutes of rehydrating. This more likely happened due to resynthesis metabolism from reserves of stored precursors. Photosynthesis resumed rapidly, followed by migration of cyanobacteria to the mat surface over time intervals of between 15 minutes and 48 hours, providing an additional impetus for photosynthesis. This mechanism of rapid recovery occurs in both marine and terrestrial systems, but with mat-specific variations tied to environmental conditions (Chennu et al., 2015).

Cyanobacteria like *Fischerella thermalis* can switch from using visible light to far-red light by incorporating chlorophyll f into their photosystem I complexes, which allows them to perform photosynthesis under weak filtered sunlight, such as beneath pond surfaces or leaf litter (Gisriel et al., 2020). *Cyanobacteria* use phycobilisomes, that is, the light-harvesting antennae comprising of phycobiliproteins (e.g., phycoerythrin, phycocyanin), as well as linker proteins to enhance light absorption under low-light conditions and engage in photoprotective quenching under high-light intensities in order to protect themselves from ROS damage. These cyanobacteria dissipate excess energy absorbed through non-photochemical quenching mechanisms when exposed to high light intensities, protecting their photosynthetic apparatus from photo-damage, particularly for those exposed to sunlight in shallow coastal waters (Veaudor et al., 2020). *Cyanobacteria* have evolved to exhibit reduced genome size and streamlined photosynthetic processes to bring efficiency in specific ecological niches such as low-nutrient or low-light environments (Noell et al., 2023).

2.10.2 Nutrients Acquisition and Cycling

Nutrients, especially nitrogen and phosphorous, are often limiting factors for primary production in marine ecosystems. Marine *cyanobacteria* exhibit diverse strategies for nutrient acquisition. In the resurrection of the marine oscillatorium mats of Exmouth Gulf in Australia, chlorophyll a concentrations increased, doubling and tripling in about 15 minutes of rehydration. This was expected because there was a chance of resynthesis at the metabolic level from reserved stock of precursors (El-Seedi et al., 2023). Non-heterocystous *cyanobacteria*, while usually referred to as non-diazotrophic, can acquire nitrogen from the dissolved inorganic and organic sources of nitrogen suspended in the water column (Tee et al., 2020). Uptake of phosphorus is the last mode and one of great importance. Marine *cyanobacteria* use many phosphate transporters and alkaline phosphatases to access different phosphorus sources and efficiently use phosphorus in phosphate-limited environments. In addition, some cyanobacteria employ mixotrophy, combining photosynthetic carbon fixation with the uptake of dissolved organic carbon, giving them an advantage under nutrient-poor conditions (El-Seedi et al, 2023; Mutalipassi et al., 2021).

2.10.3 Stress Responses and Environmental Adaptations

Marine environments are faced with multiple stressors, including fluctuating salinity and temperature, UV radiation, oxidative stress, and grazing pressure. To deal with these stresses, marine cyanobacteria have developed elaborate responses. Osmotic stress due to salinity changes is a major concern, with cyanobacteria making use of several osmolytes, such as sucrose and glucosylglycerol, to keep the cellular turgor (Rachedi et al., 2020). The influence of extremes of temperature on the production of different forms of heat shock proteins and cold shock proteins protects cellular proteins from undergoing denaturation or aggregates (Yadav et al., 2022). UV exposure may cause DNA and photosynthetic machinery damage; hence, cyanobacteria produce UV-absorbing compounds such as mycosporine-like amino acids and scytonemin for photoprotection and DNA repair procedures to alleviate UV damage (Singh et al., 2023). Antioxidant enzymes such as superoxide dismutase and catalase play a critical role in detoxifying reactive oxygen species generated under oxidative stress. Morphological adaptation like mucilage production and formation of colonies can afford protection against grazing and viral infection (Cassier-Chauvat et al., 2021; Jin et al., 2020).

2.11 CONCLUSION AND FUTURE ASPECTS

The development of marine *archaea*, actinobacteria, and cyanobacteria highlights their important ecological roles and interactions in marine ecosystems. Marine cyanobacteria like *Prochlorococcus* and *Synechococcus* are primary producers contributing significantly to global carbon fixation forming the base of the marine food web. Thus, they contribute to the carbon cycle via photosynthesis, sequestering CO_2, particularly in the oligotrophic waters. Also, *cyanobacteria* like *Trichodesmium* fix atmospheric nitrogen, thereby contributing bioavailable nitrogen in nitrogen limited environments more conducive to marine productivity and nutrient cycling. Marine actinobacteria decompose organic matter, releasing their nutrients which are recycled back in the ecosystem enhancing nutrient availability for other life like phytoplankton. Marine archaea, in particular ammonia-oxidizing archaea, thrive under extreme conditions and play a significant role in nitrification and biogeochemical cycles. Their metabolic flexibility enables them to adapt to a wide range of marine environments. Furthermore, cyanobacteria are responsible for harmful algal blooms that release toxins that can disrupt ecosystems and pose risks to marine life and human health. These blooms also alter microbial community structure through stimulation of heterotrophic bacterial activity. The interactions between cyanobacteria and other marine organisms provide a basis on which nutrient exchange occurs, thus allowing stability of the ecosystem. This understanding illuminates future predictions on changes to marine ecosystems, especially under climate change influence. The studies of marine archaea, actinobacteria, and cyanobacteria, with their enormous potential, can advance scientific answers to some of the more ruthless global challenges that society faces. These microorganisms are implicated in the biogeochemical cycles that drive major forms of primary production and interactions in ecological systems, making them among the most prime targets for work in this future area until their biotechnological application is fully understood. Marine microbes thrive in extremely harsh environments like deep-sea hydrothermal vents, polar ice caps, hypersaline lagoons, and deep anoxic basins.

REFERENCES

Abdel-Mageed, W. M., Juhasz, B., Lehri, B., Alqahtani, A. S., Nouioui, I., Pech-Puch, D., Tabudravu, J. N., Goodfellow, M., Rodríguez, J., Jaspars, M., & Ebel, R. (2020). Whole genome sequence of Dermacoccus abyssi MT1.1 isolated from the Challenger Deep of the Mariana Trench reveals phenazine biosynthesis and environmental adaptation gene clusters. *Marine Drugs*, *18*(3), 131.

Ahern, K. S., Ahern, C. R., Savige, G. M., & Udy, J. W. (2007). Mapping the distribution, biomass and tissue nutrient levels of a marine benthic cyanobacteria bloom (Lyngbya majuscula). *Marine and Freshwater Research*, *58*(10), 883–904.

Al-Agamy, M. H., Alhuzani, M. R., Kelany, M. S., & Hamed, M. M. (2021). Production and partial characterization of α-amylase enzyme from marine actinomycetes. *BioMed Research International*, 2021, Article 5289848.

Albers, S.-V., de Vossenberg, J. L., Driessen, A. J., & Konings, W. N. (2000). Adaptations of the archaeal cell membrane to heat stress. *Front Biosci*, *5*(1), 813.

Alharbi, N. S. (2016). Novel bioactive molecules from marine actinomycetes. *Bioscience Biotechnology Research Asia*, *13*(4), 1905–1927.

Aliko, V., et al. (2022). Biodiversity and biological applications of marine actinomycetes isolated from marine sediments of Abu-Qir Bay, Egypt. *Journal of Genetic Engineering and Biotechnology*, 21(1), Article 69.

Andrade, H. A., & Garcia, C. A. E. (1999). Skipjack tuna fishery in relation to sea surface temperature off the southern Brazilian coast. *Fisheries Oceanography*, *8*(4), 245–254.

Ani, A., Kumar, S., & Matsuda, K. (2023). An ANI gap within bacterial species that advances the definitions of microbial species and strain boundaries. *mBio*, 14(1), e02696–23.

Ayala-Muñoz, D., Burgos, W. D., Sánchez-España, J., Falagán, C., Couradeau, E., & Macalady, J. L. (2022). Novel microorganisms contribute to biosulfidogenesis in the deep layer of an acidic pit lake. Frontiers in Bioengineering and Biotechnology, 10, Article 867321.

Assunção, J. P., Bragança, A., & Dahis, R. (2022). The effects of transportation infrastructure on deforestation in the Amazon: A general equilibrium approach. *The World Bank Policy Research Working Paper Series*.

Back, L., Wang, Q., & Chen, X. (2021). The impact of the COVID-19 pandemic on marine pollution research: A bibliometric analysis. *Frontiers in Marine Science*, *8*, 720.

Baker, B. A., Gutiérrez-Preciado, A., del Rio, A., McCarthy, C. G. P., López-García, P., Huerta-Cepas, J., Susko, E., Roger, A. J., Eme, L., & Moreira, D. (2024). Expanded phylogeny of extremely halophilic archaea shows multiple independent adaptations to hypersaline environments. *Nature Microbiology*, *9*(4), 964–975.

Baker, B. J., Lesniewski, R. A., & Dick, G. J. (2012). Genome-enabled transcriptomics reveals archaeal populations that drive nitrification in a deep-sea hydrothermal plume. *The ISME Journal*, *6*(12), 2269–2279.

Barka, E. A., Vatsa, P., Sanchez, L., Gaveau-Vaillant, N., Jacquard, C., Klenk, H.-P., Clément, C., Ouhdouch, Y., & van Wezel, G. P. (2016). Taxonomy, physiology, and natural products of Actinobacteria. *Microbiology and Molecular Biology Reviews*, *80*(1), 1–43.

Berry, O., Briand, E., Bagot, A., Chaigné, M., Meslet-Cladière, L., Wang, J., Grovel, O., Jansen, J. J., Ruiz, N., Du Pont, T. R., & others. (2023). Deciphering interactions between the marine dinoflagellate Prorocentrum lima and the fungus Aspergillus pseudoglaucus. *Environmental Microbiology*, *25*(2), 250–267.

Biller, S. J., Berube, P. M., Lindell, D., & Chisholm, S. W. (2015). Prochlorococcus: the structure and function of collective diversity. *Nature Reviews Microbiology*, *13*(1), 13–27.

Bhatt, P., et al. (2021). Bioprospection and secondary metabolites profiling of marine actinomycetes from Devbagh and Tilmati beaches, Arabian Sea. *Saudi Journal of Biological Sciences*, *28*(11), 6400-6408.

Bojanova, D. P., De Anda, V. Y., Haghnegahdar, M. A., Teske, A. P., Ash, J. L., Young, E. D., Baker, B. J., LaRowe, D. E., & Amend, J. P. (2023). Well-hidden methanogenesis in deep, organic-rich sediments of Guaymas Basin. *The ISME Journal*, *17*(11), 1828–1838.

Brinkmann, H., Besch, G., Sader, J. E., & Lalk, M. (2017). Antimicrobial-producing actinobacteria isolated from marine sponges: Diversity and biotechnological potential. *Frontiers in Microbiology*, 8, Article 1106.

Capone, D. G., Zehr, J., Paerl, H., Bergman, B., & Carpenter, E. J. (1997). Trichodesmium: A globally significant marine cyanobacterium. *Science*, *276*(5316), 1221–1229.

Cassier-Chauvat, C., Blanc-Garin, V., & Chauvat, F. (2021). Genetic, genomics, and responses to stresses in cyanobacteria: Biotechnological implications. *Genes*, *12*(4), 500.

Cesar-Ribeiro, C., Leite, R., & Santini, A. (2023). Marine biodiversity discovery: the metrics of new species descriptions. *Frontiers in Marine Science*, *10*, 929989.

Charpy, L., Casareto, B. E., Langlade, M.-J., & Suzuki, Y. (2012). Cyanobacteria in coral reef ecosystems: a review. *Journal of Marine Sciences*, *2012*(1), 259571.

Chang, Y.-C., Huang, Y.-C., & Chang, R.-C. (2021). Editorial: Advances in marine environmental protection: Challenges, solutions and perspectives. *Frontiers in Marine Science*, *12*, 1593179.

Chen, J., Xu, L., Zhou, Y., & Han, B. (2021). Natural products from actinomycetes associated with marine organisms. *Marine Drugs*, *19*(11), 629.

Chennu, S., Firth, N. C., Menon, D. K., & Bekinschtein, T. A. (2015). Brain Connectivity Dissociates Responsiveness from Drug Exposure during Propofol-Induced Transitions of Consciousness. *PLoS Computational Biology*, *10*(7), e1004669.

Damashek, J., Okotie-Oyekan, A. O., Gifford, S. M., Vorobev, A., Moran, M. A., & Hollibaugh, J. T. (2021). Transcriptional activity differentiates families of Marine Group II Euryarchaeota in the coastal ocean. *ISME Communications*, *1*(1), 5.

Das, S., Deka, N., & Deka, A. (2023). Marine actinomycetes. Journal of King Saud University - Science.

DeLong, E. F. (2021). Exploring marine planktonic archaea: then and now. *Frontiers in Microbiology*, *11*, 616086.

Demin, K. A., Prazdnova, E. V, Minkina, T. M., & Gorovtsov, A. V. (2024). Sulfate-reducing bacteria unearthed: ecological functions of the diverse prokaryotic group in terrestrial environments. *Applied and Environmental Microbiology*, *90*(4), e01390--23.

De Rop, A., Ramírez-León, D., Sánchez-Murillo, S., Quintero-Bucheri, K., Benavides, M., Morales, M., & Cuervo-Cruz, M. (2022). Marine actinobacteria metabolites: Unlocking new treatments for acne vulgaris. *Frontiers in Microbiology*, 15, Article 1501951.

Dithugoe, C. D., Bezuidt, O. K. I., Cavan, E. L., Froneman, W. P., Thomalla, S. J., & Makhalanyane, T. P. (2023). Bacteria and archaea regulate particulate organic matter export in suspended and sinking marine particle fractions. *Msphere*, *8*(3), e00420--22.

El-Seedi, H. R., El-Mallah, M. F., Yosri, N., Alajlani, M., Zhao, C., Mehmood, M. A., Du, M., Ullah, H., Daglia, M., Guo, Z., & others. (2023). Review of marine cyanobacteria and the aspects related to their roles: chemical, biological properties, nitrogen fixation and climate change. *Marine Drugs*, *21*(8), 439.

Favreau, C., Tribondeau, A., Marugan, M., Guyot, F., Alpha-Bazin, B., Marie, A., Puppo, R., Dufour, T., Huguet, A., Zirah, S., & others. (2023). Molecular acclimation of Halobacterium salinarum to halite

Faldu, N., Patel, S., Vishwakarma, N., Singh, A., Patel, K., & Pandhi, N. (2014). Genetic diversity of marine and fresh water cyanobacteria from the Gujarat state of India. *Advances in Bioscience and Biotechnology*, *5*(2), 86-95.brine inclusions. *Frontiers in Microbiology*, *13*, 1075274.

Fenical, W., & Jensen, P. R. (2006). Developing a new resource for drug discovery: marine actinomycete bacteria. *Nature Chemical Biology*, *2*(12), 666–673.

Fernandes, R., Hemmati, B. W., Dijkstra, K., & Khajeh, S. (2024). Marine actinomycetes: A hidden treasure trove for antibacterial discovery. *Frontiers in Marine Science*, 12, Article 1558320.

Gärtner, A., Drews, M., Nedelcu, A. M., Muler, M. N., & Schiel, D. R. (2021). The Importance of Marine Research Infrastructures in Capturing Extreme Events and Changes. *Frontiers in Marine Science*, 8, 626668.

Ghai, R., Mizuno, C. M., Picazo, A., Camacho, A., & Rodriguez-Valera, F. (2013). Metagenomics uncovers a new group of low GC and ultra-small marine Actinobacteria. *Scientific Reports*, *3*(1), 2471.

Gisriel, C. J., Bryant, D. A., Brudvig, G. W., & Cardona, T. (2020). The structure of Photosystem I acclimated to far-red light illuminates an ecologically important acclimation mechanism in photosynthesis. *Science Advances*, *6*(7), eaay6415.

González, V., Sarmiento-Vizcaíno, A., Braña, A. F., Palacios, J. J., Otero, L., Fernández, J., Molina, A., Kulik, A., Vázquez, F., & Acuña, J. L. (2020). Bioprospecting marine actinomycetes for antileishmanial and antitrypanosomal agents. *Marine Drugs*, *18*(8), 403.

Ghosh, S., Hussein, M. K., El Badan, D., El Sayed, N., Sayed, A., & Mousa, M. (2023). Biodiversity and biological applications of marine actinomycetes—Abu-Qir Bay, Mediterranean Sea, Egypt. *Journal of Genetic Engineering and Biotechnology*, 21, Article 150

Grébert, T., Garczarek, L., Daubin, V., Humily, F., Marie, D., Ratin, M., Devailly, A., Farrant, G. K., Mary, I., Mella-Flores, D., Tanguy, G., Labadie, K., Wincker, P., Kehoe, D. M., & Partensky, F. (2022). Diversity and evolution of pigment types in marine Synechococcus cyanobacteria. *Genome Biology and Evolution*, *14*(4), evac035.

Grgas, D., Hegazy, G. E., & Abdelmohsen, U. R. (2023). Biodiversity and biological applications of marine actinomycetes. *Journal of Genetic Engineering and Biotechnology*, 21(1), Article 69.

Gugliandolo, C., & Maugeri, T. L. (2019). Phylogenetic diversity of archaea in shallow hydrothermal vents of Eolian Islands, Italy. *Diversity*, *11*(9), 156.

Gunde-Cimerman, N., Plemenitaš, A., & Oren, A. (2018). Strategies of adaptation of microorganisms of the three domains of life to high salt concentrations. *FEMS Microbiology Reviews*, *42*(3), 353–375.

Hamed, M. M., Abdrabo, M. A. A., & Youssif, A. M. (2021). *Biosurfactant production by marine actinomycetes isolates Streptomyces althioticus RG3 and Streptomyces californicus RG8 as promising sources of antimicrobial and antifouling effects.*

Haro-Moreno, J. M., Rodriguez-Valera, F., López-García P, P., Moreira, D., & Martin-Cuadrado, A.-B. (2017). New insights into marine group III Euryarchaeota, from dark to light. *The ISME Journal*, *11*(5), 1102–1117.

Hassan, S. S. U., & Jin, H. (2024). Marine actinomycetes: A hidden treasure trove for antibacterial discovery. *Frontiers in Marine Science*, 12, Article 1558320.

Hu, Y., Zhang, J., Jiang, Z., Li, Y., & Li, S. (2023). Influence of the oceanic crust structure on marine magnetic anomalies: Review and forward modelling. *Geological Journal*, *58*(3), 1069–1082.

Hughes, C. C., Fenical, W., & Jensen, P. R. (2008). The marinopyrroles, antibiotics of an unprecedented structure class from a marine Streptomyces species. *Organic Letters*, *10*(4), 629–631.

Igarashi, H., Lan, K. W., Chang, J., Wang, J., & Tsz, C. (2021). Effects of climate change on the abundance and distribution of bigeye tuna cohorts in the western and central Pacific Ocean. *Frontiers in Marine Science*, *8*, Article 614594.

Jagannathan, S. V, Manemann, E. M., Rowe, S. E., Callender, M. C., & Soto, W. (2021). Marine actinomycetes, new sources of biotechnological products. *Marine Drugs*, *19*(7), 365.

Jin, P., Hutchins, D. A., & Gao, K. (2020). The impacts of ocean acidification on marine food quality and its potential food chain consequences. *Frontiers in Marine Science*, *7*, 543979.

Jose, P. A., & Jha, B. (2017). Intertidal marine sediment harbours Actinobacteria with promising bioactive and biosynthetic potential. *Scientific Reports*, *7*(1), 10041.

Kamjam, M., Sivalingam, P., Deng, Z., & Hong, K. (2017). Biological significance of marine actinobacteria of East Coast of Andhra Pradesh, India: Diversity and antimicrobial potential. *Frontiers in Microbiology*, 8, Article 1201.

Kanchanasin, P., Pan, C., Hassan, S. S. U., Ishaq, M., Yan, S., & Jin, H. (2024). Discovery of two new actinobacteria, Micromonospora sp. S2-005T and Streptomyces sp. C6-003T, isolated from marine invertebrates of Thailand with antimicrobial and anticancer activities. *Frontiers in Marine Science*, 12, Article 1558320.

Könneke, M., Schubert, D. M., Brown, P. C., Hügler, M., Standfest, S., Schwander, T., von Borzyskowski, L., Erb, T. J., Stahl, D. A., & Berg, I. A. (2014). Ammonia-oxidizing archaea use the most energy-efficient aerobic pathway for CO2 fixation. *Proceedings of the National Academy of Sciences*, *111*(22), 8239–8244.

Kurth, J. M., Smit, N. T., Berger, S., Schouten, S., Jetten, M. S. M., & Welte, C. U. (2019). Anaerobic methanotrophic archaea of the ANME-2d clade feature lipid composition that differs from other ANME archaea. *FEMS Microbiology Ecology*, *95*(7), fiz082.

Lai, D., Hedlund, B. P., Xie, W., Liu, J., Phelps, T. J., Zhang, C., & Wang, P. (2020). Impact of terrestrial input on deep-sea benthic archaeal community structure in South China Sea sediments. *Frontiers in Microbiology*, *11*, 572017.

Laurenceau, J.-P., Rivera, L. M., & Kleiman, E. M. (2020). The interpersonal process model of intimacy, burden, and psychological distress in caregiving relationships. *Journal of Social and Personal Relationships*, 37(9), 2555-2575.

Lima, F. P. (2023). A review of artificial intelligence in marine science. *Frontiers in Earth Science*, 11, Article 1090185.

Lin, Z., Lin, H., Strauss, H., Roberts, A. P., Niu, M., Fang, Y., Yang, T., Sun, X., Birgel, D., & Peckmann, J. (2024). Methanogenic archaea as catalysts for magnetite formation in iron-rich marine sediments. *Journal of Geophysical Research: Solid Earth*, *129*(6), e2023JB028312.

Liu, Y., Lei, Y., Zhang, X., Gao, Y., Xiao, Y., & Peng, H. (2012). Identification and phylogenetic characterization of a new subfamily of $α$-amylase enzymes from marine microorganisms. *Marine Biotechnology*, *14*(3), 253–260.

Liu, Y., Ren, Y., Li, K., Tan, Y., Liu, H., Ke, Z., et al. (2025). Effects of monsoon-driven currents on copepods in the Northeast South China Sea. *Journal of Geographical Sciences*.

Lopes, V. R., & Vasconcelos, V. M. (2011). Planktonic and benthic cyanobacteria of European brackish waters: A perspective on estuaries and brackish seas. *Estuarine, Coastal and Shelf Science*, *95*(2), 221–234.

Ma, J., et al. (2021). Marine actinomycetes as prolific sources of novel antibiotics: A review of discovery, taxonomy, and biosynthetic potential. *Frontiers in Microbiology*, 12, Article 672400.

Mahmoud, H. M., & Kalendar, A. A. (2016). Coral-associated actinobacteria: diversity, abundance, and biotechnological potentials. *Frontiers in Microbiology*, *7*, 204.

Marlow, J. J., Steele, J. A., Case, D. H., Connon, S. A., Levin, L. A., & Orphan, V. J. (2014). Microbial abundance and diversity patterns associated with sediments and carbonates from the methane seep environments of Hydrate Ridge, OR. *Frontiers in Marine Science*, *1*, 44.

Martínez-García, S., Bunse, C., Pontiller, J., & Gasol, J. M. (2022). Seasonal dynamics in carbon cycling of marine bacterioplankton in the Baltic Sea. *Frontiers in Microbiology*, 13, Article 896540

Marchbank, T., Playford, R. J., & Cattell, M. (2020). Marked variability in bioactivity between commercially available bovine colostrum for human use: Implications for clinical trials. *PLoS ONE*, *15*(6), e0234719.

Majithiya, V. R., & Gohel, S. D. (2022). Isolation and characterization of marine actinobacteria

Maya-Ramírez, C. E., Saih, A., Méndez Tenorio, A., Wong Baeza, C., & Chagas-Ribeiro, C. (2024). Exploring marine natural compounds: Innovative therapeutic candidates against Chagas disease through virtual screening and molecular dynamics. *ACS Omega*. Advance online publication.associated with seaweeds. *Journal of the Marine Biological Association of India*, *64*(1), 34–40.

Mazur-Marzec, H., Anders, F., & Spoof, L. (2024). Biodiversity of microorganisms in the Baltic Sea: The power of novel methods in the identification of marine microbes. *FEMS Microbiology Reviews*, 48(5), fuae024.

Mutalipassi, M., Riccio, G., Mazzella, V., Galasso, C., Somma, E., Chiarore, A., et al. (2021). Symbioses of Cyanobacteria in marine environments: Ecological insights and biotechnological perspectives. *Marine Drugs*, *19*(4), 227.

Nafis, A., Simões, M., Gaspar, H., & Amaro, H. (2019). Actinobacteria isolated from Laminaria ochroleuca: A source of new bioactive compounds. *Frontiers in Microbiology*, 10, Article 683.

Niu, S.-B., Sun, K., & Zhang, D. (2011). Marine rare actinomycetes: New sources of bioactive compounds. *Journal of Microbiology and Biotechnology*, *21*(1), 1–12.

Noell, P. J., Sills, R. B., Benzerga, A. A., & Boyce, B. L. (2023). Void nucleation during ductile rupture of metals: A review. *Progress in Materials Science*, *135*, 101085.

Nweze, E. I. (2009). Virulence properties of diarrheagenic *Escherichia coli* and etiology of diarrhea in infants, young children and other age groups in Southeast Nigeria. *American-Eurasian Journal of Scientific Research*, *4*(3), 173–179.

Ocasio, C. A., & Scanlan, T. S. (2006). Design and characterization of a thyroid hormone receptor-alpha (TRalpha)-specific agonist. *ACS Chemical Biology*, *1*(9), 585–593.

Orellana, L. H., Francis, T. Ben, Krüger, K., Teeling, H., Müller, M.-C., Fuchs, B. M., Konstantinidis, K. T., & Amann, R. I. (2019). Niche differentiation among annually recurrent coastal Marine Group II Euryarchaeota. *The ISME Journal*, *13*(12), 3024–3036.

Orsi, W. D., Smith, J. M., Wilcox, H. M., Swalwell, J. E., Carini, P., Worden, A. Z., & Santoro, A. E. (2015). Ecophysiology of uncultivated marine euryarchaea is linked to particulate organic matter. *The ISME Journal*, *9*(8), 1747–1763.

Patel, N., Chauhan, M. K., Nair, S., & Verma, A. K. (2023). Isolation and identification of pigments from marine actinomycetes and their antibacterial, antioxidant, and toxicological properties. *Journal of Aquatic Biology & Fisheries*, *9*(1), 45–57.

Penn, K., & Jensen, P. R. (2012). Comparative genomics reveals evidence of marine adaptation in Salinispora species. *BMC Genomics*, *13*(1), 86.

Porsani, M. V, Amatuzzi, R. F., Oliveira, B. H., Baratto, L. C., Dalitz, C. A., Bozza, A., Marangoni, P. R., Dalzoto, P. R., Kolm, H. E., & Pimentel, I. C. (2013). Antimicrobial potential of fungi and actinobacteria isolated from sandy sediments of intertidal regions. *International Journal of Pharmaceutical, Chemical \& Biological Sciences*, *3*(3).

Qin, H., Chen, Y., Wang, C., Liu, J., & Li, X. (2023). Lightweight marine biological target detection algorithm based on YOLOv5. *Frontiers in Marine Science*, *10*, 1219155.

Rachedi, R., Foglino, M., & Latifi, A. (2020). Stress signaling in cyanobacteria: A mechanistic overview. *Life*, *10*(12), 312.

Rathore, A., Prajapati, B., & Dube, B. (2021). Cultivation and characterization of deep-sea actinomycetes: Novel strains with biodegradation capabilities and biotechnological potential. *Frontiers in Microbiology*, 12, Article 656789.

Reed, A. J., Morris, J. P., Linse, K., & Thatje, S. (2013). Plasticity in shell morphology and growth among deep-sea protobranch bivalves of the genus Yoldiella (Yoldiidae) from contrasting Southern Ocean regions. *Deep Sea Research Part I: Oceanographic Research Papers*, *81*, 14–24.

Ribeiro, I., Antunes, J. T., Alexandrino, D. A. M., Tomasino, M. P., Almeida, E., Hilário, A., Urbatzka, R., Leão, P. N., Mucha, A. P., & Carvalho, M. F. (2023). Actinobacteria from Arctic and Atlantic deep-sea sediments—Biodiversity and bioactive potential. *Frontiers in Microbiology*, *14*, 1158441.

Russo, T., Maiello, G., Talarico, L., Baillie, C., Colosimo, G., DAndrea, L., Maio, F. D., Fiorentino, F., Franceschini, S., Garofalo, G., Scancella, D., Cataudella, S., & Mariani, S. (2015). All is fish that comes to the net: metabarcoding for rapid fisheries catch assessment. *Ecological Applications*, *31*(2), e2273.

Saini, V., Datt, M., Singh, J., & Singh, N. (2018). An updated checklist of bee diversity from Arunachal Pradesh, India, with notes on their biology and conservation. *Journal of Asia-Pacific Biodiversity, 11*(1), 50-58.

Samuel, S. A., Mane, M., Rathod, K., Suthar, H., & Mandal, S. K. (2022). Bioprospection and secondary metabolites profiling of marine actinomycetes isolated from Devbagh and Tilmati beaches, India. *Saudi Journal of Biological Sciences*, *29*(1), 573-585.

Sánchez-Baracaldo, P. (2015). Origin of marine planktonic cyanobacteria. *Scientific Reports*, *5*(1), 17418.

Santoro, G., Vrontis, D., Thrassou, A., & Dezi, L. (2018). The Internet of Things: Building a knowledge management system for open innovation and knowledge management capacity. *Technological Forecasting and Social Change*, *136*, 347–354.

Sarkar, G., & Suthindhiran, K. (2022). Diversity and biotechnological potential of marine actinomycetes from India. *Indian Journal of Microbiology*, *62*(4), 475–493.

Sharma, S. L., & Pant, A. (2001). Crude oil degradation by a marine actinomycete Rhodococcus sp. *Indian Journal of Marine Sciences*, *30*(3), 146-150.

Singh, A., Laha, S., Banerjee, A., Sharma, P., & Thamban, M (2023). Climate sensitivity of the summer runoff of two glacierised Himalayan catchments with contrasting climate. *Hydrology and Earth System Sciences*, *27*(2), 627-645.

Siro, G., Donald, L., & Pipite, A. (2022). The diversity of deep-sea actinobacteria and their natural products: an epitome of curiosity and drug discovery. *Diversity*, *15*(1), 30.

Stan-Lotter, H., & Fendrihan, S. (2015). Halophilic archaea: life with desiccation, radiation and oligotrophy over geological times. *Life*, *5*(3), 1487–1496.

Subramani, R., & Sipkema, D. (2019). Marine rare actinomycetes: a promising source of structurally diverse and unique novel natural products. *Marine Drugs*, *17*(5), 249.

Swan, B. K., Chaffin, M. D., Martinez-Garcia, M., Morrison, H. G., Field, E. K., Poulton, N. J., Masland, E. D. P., Harris, C. C., Sczyrba, A., Chain, P. S. G., & others. (2014). Genomic and metabolic diversity of Marine Group I Thaumarchaeota in the mesopelagic of two subtropical gyres. *PloS One*, *9*(4), e95380.

Taguchi, H., & Koike-Takeshita, A. (2023). In vivo client proteins of the chaperonin GroEL-GroES provide insight into the role of chaperones in protein evolution. *Frontiers in Molecular Biosciences*, *10*, 1091677.

Tee, M. L., Tee, C. A., Anlacan, J. P., Aligam, K. J. G., Reyes, P. W. C., Kuruchittham, V., & Ho, R. (2020). Psychological impact of COVID-19 pandemic in the Philippines. *Frontiers in Psychology*, *11*, Article 566898.

Trujillo, M. E., Meier-Kolthoff, J. P., Suzuki, K., Rohde, M., & Goodfellow, M. (2015). Taxonomy, physiology, and natural products of Actinobacteria. *Microbiology and Molecular Biology Reviews*, *79*(1), 1-43.

Vargas, C. A., Gonzalez, R. E., & Riascos, J. M. (2023). Editorial: Technological advances for measuring planktonic communities in aquatic ecosystems. *Frontiers in Marine Science*, 10, Article 1100998.

Veaudor, T., Blanc-Garin, V., Chenebault, C., Diaz-Santos, E., Sassi, J.-F., Cassier-Chauvat, C., & Chauvat, F. (2020). Recent advances in the photoautotrophic metabolism of cyanobacteria: Biotechnological implications. *Life*, *10*(5), 71.

Wasmund, K. (2023). Deciphering community interactions of sulfate-reducing microorganisms in complex microbial communities of marine sediments. *Mbio*, *14*(4), e00513--23.

Weber, M., Makarow, D., Unger, B., Mortier, N., De Wilde, B., van Eekert, M., Schuman, E., Tosin, M., Pognani, M., Innocenti, F. D., & others. (2017). Assessing marine biodegradability of plastic—towards an environmentally relevant international standard test scheme. *Proceedings of the International Conference on Microplastic Pollution in the Mediterranean Sea*, 189–193.

Webster, J., Nyadjro, E., Kaltenberger, G., Toft, T., Collazo, L., Lau, Y., Wang, Z., Boyer, T., Mickle, P., & Larsen, K. (2023). Tracking Marine Microplastic Monitoring Through the NOAA NCEI Global Marine Microplastic Concentrations ESRI ArcGIS Online Experience Builder Web Application. *OCEANS 2023-MTS/IEEE US Gulf Coast*, 1–5.

Wu, J., Sala, J., Giglio, D., Capotondi, A., Sukianto, T., & Kuusela, M. (2025). Leading dynamical processes of global marine heatwaves. *Earth System Science Data Discussions*.

Xia, Y., Lü, C., Hou, N., Xin, Y., Liu, J., Liu, H., & Xun, L. (2017). Sulfide production and oxidation by heterotrophic bacteria under aerobic conditions. *The ISME Journal*, *11*(12), 2754–2766.

Yadav, A. K., Singh, A., Dhiman, M., et al. (2022). Extractive text summarization using deep learning approach. *International Journal of Information Technology*, 14, 2407-2415.

Yadav, R., Lee, S. W., Kim, S., & Kim, Y. H. (2023). Marine mudflat actinomycetes as a novel natural products source. *Frontiers in Marine Science*, 10, Article 1297446.

Yan, Y., Zhang, J., Wang, Y., Tao, Y., Xu, Y., & Gu, W. (2023). Spatiotemporal distribution characteristics of sea ice disasters in the Northern China Sea from 2001 to 2020. *Ocean \& Coastal Management*, *246*, 106889.

Yang, T., Jiang, Z., Sun, R., Cheng, N., & Feng, H. (2020). Maritime search and rescue based on group mobile computing for unmanned aerial vehicles and unmanned surface vehicles. *IEEE Transactions on Industrial Informatics*, *16*(12), 7700–7708.

Ye, C., Tang, Y., An, D., Wang, F., Zheng, Y., & Van Esch, B. P. M. (2023). Investigation on stall characteristics of marine centrifugal pump considering transition effect. *Ocean Engineering*, *280*, 114823.

Zou, X., Chen, H., Wang, Y., & Liu, D. (2024). Assessment of carrying capacity of marine fishery resources in China using remote sensing and surplus production models. *Marine Policy*, *152*, 105734.

Zhang, Y., He, S., Bonneil, É., & Simpson, B. K. (2020). Generation of antioxidative peptides from Atlantic sea cucumber using alcalase versus trypsin: In vitro activity, de novo sequencing, and in silico docking for in vivo function prediction. *Food Chemistry*, *306*, 125581.

Zhao, R., & Biddle, J. F. (2021). Helarchaeota and co-occurring sulfate-reducing bacteria in subseafloor sediments from the Costa Rica Margin. *ISME Communications*, *1*(1), 25.

Zhu, S., Loo, Y. T., Veerapathiran, S., Loo, T. Y. J., Tran, B. N., Teh, C., Zhong, J., Matsudaira, P., Saunders, T. E., & Wohland, T. (2024). Receptor binding and tortuosity explain morphogen local-to-global diffusion coefficient transition. *Biophysical Journal*. Advance online publication.

3 Marine Algal and Fungal Ecosystems

Harleen Kaur and Ashish Sharma
Department of Botany and Environment Science, DAV University, Jalandhar, India

3.1 INTRODUCTION

Coastal and marine areas can extend as far as 100 km inland and reach out to the continental shelf, encompassing oceanic regions with depths of up to 50 m. These habitats support a variety of unique ecosystems, including coastal wetlands, estuaries, coral reefs, oyster reefs, sandy beaches, dunes, seagrass beds (Barbier 2017). Many unique marine habitats are highly valued for their cultural significance and various other benefits. People may find satisfaction simply in knowing that these ecosystems exist or in the assurance that they will be preserved for future generations to enjoy as these values hold significant importance for indigenous communities in rural regions, as their cultural, heritage, and traditional knowledge are deeply connected to the natural world (Barbier 2017). As coastal areas become more densely populated and the global population continues to grow rapidly, our reliance on marine resources has never been greater (Crain et al. 2009).

Marine algae/macroalgae are autotrophic organisms that flourish in coastal regions because of their remarkable adaptability. Based on the types of pigments present, they are broadly categorized into three groups Chlorophyta, Phaeophyta, and Rhodophyta (Yu et al. 2014). The composition, biomass, and various other characteristics of algal communities are analyzed in assessments of aquatic ecosystems to identify potential threats to drinking water, fisheries, and recreational water activities (Stevenson et al. 2010). Algae are chlorophyllous organisms (Bold & Wynne 1985) that exist either as single cells, form colonies, or develop into multicellular organisms, sometimes working together as simple tissues. They are found in nearly every environment on Earth, including oceans, rivers, lakes, soil, and even on walls. Algae also live as symbionts within animals and plants. Essentially, they can thrive in any location that provides light for photosynthesis (El Gamal 2010).

Over the past four decades, a vast array of compounds have been extracted from marine organisms that possess promising biological activities (El Gamal 2010). The ocean is inhabited by over 80% of the earth's species (Jha and Zi-Rong 2004), which includes more than 150,000 seaweed species found in intertidal and tropical regions. This diversity makes the ocean a crucial source of naturally occurring products (Falcão 2006). To adapt to a diverse environments, macroalgae generate numerous biologically active compounds and metabolites (Hultberg et al. 2013). Macroalgae are among the extensively studied as well as utilized marine resource (Chew et al. 2017), leading to significant research on the bioactivities of their various components.

Fungi are key players in terrestrial ecosystems, flourishing in mutualism with flora and fauna, playing an important role in nutrient cycling across various environments. Wherever there is moisture and an available carbon source in the terrestrial biosphere, fungi are likely to be found (Gladfelter et al. 2019). In marine fungi, numerous alkaloids, terpenoids, polyketides, peptides, and other compounds with notable bioactive properties have been discovered (Xu et al. 2019). Sponges, which host a variety of microbes, are particularly valuable as they hold significant pharmaceutical potential (Sajeevan et al. 2020). Interest in fungal metabolites has significantly increased, especially following the discovery of active compounds produced by them (Hasan et al. 2015; Imhoff 2016). It is now considered as a vital source of secondary metabolites (SMs), out of which several possess distinct biological properties. These metabolites are often produced by several multi-functional enzymes, such as polyketide synthases (Gallo et al. 2013). In light of the above discussion, the current chapter highlights the diverse potential of marine algal and fungal ecosystems in producing numerous bioactive compounds. These compounds offer a wide array of applications and help reduce reliance on artificially produced substances.

3.2 MARINE ALGAE

Macroalgae, or seaweeds, are large, macroscopic organisms found naturally or cultivated along extensive low-tide shores. Marine algae thrive in nutrient-dense oceans and are especially plentiful in coastal regions, particularly in countries with vast shorelines like Japan, Malaysia, Thailand, Singapore, United States, India, Australia, etc. It has been estimated that worldwide, approximately 8 million tons fresh seaweeds are extracted annually. In 2008, global production of macroalgae reached 15.5 million tons of fresh weight, with 93% classified as commercially valuable. According to an estimate, total annual market value is around six billion US$, with consumable food products accounting for five billion US$ of this total (Kılınç et al. 2013; FAO 2014).

Marine algae are among the earliest forms of plant life, with a history spanning three billion years. These are found in diverse environments, ranging from polar to tropical zones, and from nutrient-loaded coastal waters to nutrient-poor ocean areas. These are characterized by their

DOI: 10.1201/9781003509530-3

ability to perform photosynthesis, are unified mainly by the absence of structures typical of higher plants, as well as by the universal presence of chlorophyll a pigment. They vary greatly in their size, ranging from unicellular microalgae to large species like *Macrocystis*, a type of macroalgae that can grow over 30 m in length (Qin et al. 2012).

Macroalgae contain very less dry matter ranging from 10% to 15% (Roesijadi et al. 2010). Dehydrated macroalgae are composed of a significant amount of carbohydrates, around 60%. The primary sources of these carbohydrates are the cell wall components of the macroalgae (Chen et al. 2015). Various compounds such as carrageenan, mannan, agar, ulvan, mannitol, laminarin, fucoidan, alginate, fucose and uronic acid (Jung et al. 2013) make macroalgae different from microalgae. Marine algae are essential in animal aquaculture diets and serve as a significant source of valuable compounds for the chemical and healthcare industries. These compounds include oils (such as triglycerides), polysaccharides (like algin and agar), pigments (such as phycobiliproteins and carotenoids), and potent new pharmaceuticals (Apt & Behrens 1999; Chisti 2007; Hanzhi et al. 2011).

In recent decades, an investigation of marine organism derived bioactive compounds has resulted in exploration of significant number of extracts with pharmaceutical and industrial uses. Throughout history, algal members have been employed as food source and for their various products, such as alginate and carrageenan. Recent advances in technology have resulted in discovery of antibacterial compounds within the SMs produced by different algal groups, including Phaeophyceae, Rhodophyceae, Chlorophyceae, Chrysophyceae, and Bacillariophyceae (Ördög et al. 2004; Venugopal 2008; Cox et al. 2010; Abu-Ghannam & Rajauria 2013; Rizzo et al. 2016).

In previous years, various research reports had been published on the health benefits and applications of seaweeds, as well as the polysaccharides produced by algae (Gomez d'Ayala et al. 2008; Costa et al. 2010; Laurienzo 2010; Kraan 2012; Albuquerque et al. 2013; Fedorov et al. 2013; Wang et al. 2014). In 2013 alone, more than a thousand marine-derived pharmacologically active compounds were identified globally, showing potential effectiveness against bacteria, viruses, fungi, and several other conditions (Shannon and Abu-Ghannam 2016). Additionally, with low lignin content of up to 0.03 g/kg of dry matter, macroalgae are well-suited for downstream processes like anaerobic digestion or fermentation without the need for expensive pretreatment (Horn 2000). The various benefits provided by marine algae are discussed in the sections that follow.

3.2.1 Algal Allelopathy

Organisms that can produce and release allelopathic compounds might gain a competitive edge in their environment, making allelopathy a potential self-regulation strategy within algal communities (Hay 2009). Not only does allelopathy regulate succession, but it also serves as an active mechanism for maintaining species diversity, particularly in confined environments like benthic ecosystems (Pezzolesi et al. 2021). Allelopathy, observed in both terrestrial and marine environments, plays a role in generating potentially antibacterial compounds. Allelopathy is associated with several ecological functions, such as signaling within populations, deterring feeding or antagonism, and repelling grazers (Pezzolesi et al. 2021). Marine organisms produce various allelopathic compounds, including alkaloids, fatty acids, low-molecular-weight (LMW) peptides, phenols, and derivatives like aldehydes. Recent studies have demonstrated that certain compounds produced by algae serve as chemical defenses, effectively deterring various natural threats, including competitors, pathogenic bacteria, etc. (Amsler and Fairhead 2005; Pereira and Da Gama 2008).

Macroalgal allelopathy involves the impact of substances produced by macroalgae on other target organisms (Harlin and Rice 1987). These effects can impact the origin, growth, health, or population dynamics of both source and recipient organisms (Alvarez-Hernández et al. 2019; Harlin and Rice 1987). Many algae release allelochemicals into their surroundings, having either beneficial/harmful effects on competing organisms, ultimately providing a competitive advantage to the algae (Shannon and Abu-Ghannam 2016). Macroalgal compounds can both inhibit and stimulate microbial growth, potentially causing rapid and adverse effects on organisms near the algal thalli (Morrow et al. 2011). The allelopathic interactions of macroalgae involve a complex mechanism that relies on generation of allelopathic compounds and their successful transfer to adjacent organisms (Lewis 1986).

A significant body of knowledge has been developed related to the allelopathic effects of macroalgae, such as *Ulva* on other microalgae such as *Heterosigma akashiwo*, *Alexandrium tamarense*, *Prorocentrum micans*, *Chaetoceros gracilis* (Nan et al. 2008; Wang et al. 2007; Tang and Gobler 2011). Dopamine, an allelopathic compound produced by *Ulvaria obscura* (a green alga), has been found to inhibit the development of neighboring aquatic animals (Van Alstyne et al. 2014; Tocher and Craigie 1966). Tichocarpols A and Tichocarpols B, extracted from *Tichocarpus crinitus* (a red alga), demonstrate antifeeding properties against *Strongylocentrotus intermedius* (a sea urchin) (Ishii et al. 2004). Research by Wium-Andersen et al. (1982, 1983) stated that *Chara globularis*, a freshwater alga, can inhibit phytoplanktonic communities through the release of two sulfur-based compounds: dithiolane and trithiane. On tropical reefs, macroalgae frequently compete intensely with corals and other benthic organisms, particularly in regions with low herbivory and elevated nutrient levels (Fong & Paul 2011; Rasher & Hay 2010), manifesting through a range of physical and chemical mechanisms (Chadwick & Morrow 2011).

3.2.2 Bioactive Compounds Derived from Marine Algae

Typically, algae comprises three main compounds viz. proteins, carbohydrates, and lipids. In addition to this, biorefinery processes can extract other valuable substances from algae, including antioxidants, vitamins, natural colorants, pharmaceuticals, biofertilizers, etc. (Trivedi et al. 2015). Marine macroalgae are a significant resource of dietary fiber, vitamins, proteins, and minerals. Recent advancements in the marine bioprocessing industry have led to the use of marine food and their by-products as valuable foods (Sudhakar et al. 2019). Also several SMs with beneficial activities are produced by macroalgae (Table 3.1), some of which are distinct from those found in terrestrial plants (Milchakova 2011). Algae produce bioactive compounds that are rich in antioxidants and exhibit notable pharmacological properties. The diverse morphological and physiological characteristics of algae contribute to the synthesis of these valuable metabolites, which hold considerable commercial potential (Kumar et al. 2019).

Algal based polyphenols, particularly phlorotannins found in brown algae, are renowned for their exceptional antioxidant properties. In addition to phlorotannins, various other algae-derived compounds such as carotenoids, sterols, peptides within algal proteins, and mycosporine-like amino acids also display varying degrees of antioxidant activity (Plaza et al. 2008). Macroalgal tissue extracts have diverse pharmacological applications. To date, approximately 83 species globally have been shown to have pharmacological efficacy. These extracts are used for various purposes, including antiviral activities (e.g., against colds and influenza), antibacterial effects (e.g., against tuberculosis), and antifungal properties (Negreanu-Pirjol et al. 2022).

According to König et al. (2000) 17 out of 39 marine natural products were derived from species belonging to genus *Laurencia*. Eight β-chamigrane-type sesquiterpenes were extracted from the dichloromethanolic extract of *Laurencia scoparia*, and their effectiveness was demonstrated against parasitic stage of *Nippostrongylus brasiliensis* (Davyt et al. 2001). Ioannou et al. (2011) extracted 17 dolabellane diterpenes, including eight newly identified compounds, from brown *Dilophus spiralis* (a member of brown alga), sourced from Greece (Guiry and Guiry 2018). Researchers investigating *Sargassum sagamianum* (a brown algae) from Japan identified two known compounds, sargaquinoic acid and

TABLE 3.1
List of Several Active Compounds Produced by Different Algal Members and the Bioactivity Associated with Compounds

S.No.	Algal Member	Bioactive Compounds	Bioactivity	References
1	*Cladophora glomerata, Chara fragilis, Ulva flexuosa*	Carotenoids and phenols	Antioxidant activity	Fabrowska et al. (2016)
2	*Cladophora glomerata, Polysiphoniucoides, Ulva flexuosa, Ulva clathrata*	Auxins, cytokinins, polyphenols, microelements as well as macroelements	Growth promotion in plants	Michalak et al. (2016)
3	*Gracilaria mammillaris*	Polyphenols, carotenes	Antioxidant activity	Ospina et al. (2017)
4	*Saccharina japonica, Sargassum horneri*	Fucoxanthin, fatty acids, polyphenols	Anti-oxidative, anti-microbial, and anti-hypertensive activity	Periaswamy Sivagnanam et al. (2015)
5	*Pelvetia babingtonii*	Fucosterol	α-Glucosidase inhibitory activity	Ohta et al. (2002)
6	*Ecklonia stolonifera*	Polyphenol	In vitro α-glucosidase inhibitory activity as well as inhibition of increased plasma glucose level	Gouveia et al. (2007) and Iwai (2008)
7	*Laminaria japonica*	Butyl-isobutyl-phthalate	Hypoglycemic effect	Liu et al. (2011)
8	*Sargassum yezoense*	Sargahydroquinoic acid, sargaquinoic acid	Lipid and glucose metabolism	Adeghate et al. (2011) and Kim et al. (2012)
9	*Porphyra* genus	Catechol, hesperidin, rutin	Anti-inflammatory property	Herrero et al. (2013)
10	*Alaria esculenta, Ascophyllum nodosum, Cystoseira tamariscifolia, Ecklonia arborea, Ecklonia bicyclis, Ecklonia cava, Ecklonia stolonifera, Hizikia fusiformis, Ishige foliacea, Palmaria palmate, Ulva lactuca*	Phlorotannins	Anticancerous, anti-inflammatory, anti-microbial, anti-proliferative, anti-allergic, antiaging activity	Li et al. (2011), Balboa et al. (2013), Lee and Jeon (2013), Ngo and Kim (2013), and Sanjeewa et al. (2016)
11	*Dunaliella salina*	β-Carotene	Anticancerous, anti-oxidant activity	Talero et al. (2015)

sargachromenol, along with two newly discovered derivatives of sargaquinoic acid: 15′-hydroxysargaquinolide and a compound resulting from its oxidative cleavage (Falkenberg et al. 2019). Afolayan et al. (2008) reported the isolation of several compounds such as sargaquinoic acid, sargaquinol, fucoxanthin from *Sargassum heterophyllum* collected in South Africa. These compounds have been noted for their strong antioxidant properties.

Sulfated polysaccharides from algae, along with carotenoids, sterols, peptides, mycosporine-like amino acids all display different degrees of antioxidative properties (Plaza et al. 2008). These compounds have diverse uses such as anti-viral properties (effective against influenza, colds, etc.), others as anti-bacterial agents (such as those effective against tuberculosis), and some have anti-fungal properties. Polyphenols are a key class of SMs found in marine macroalgae along with other terrestrial plants, though the chemical structures of polyphenols from these two sources differ significantly (Lee & Jeon 2013). Several studies have focused on green algae from the *Ulva* genus, commonly known as sea lettuces, in animal experiments. The biomass of *Ulva* species contains a high protein content, positioning it as a promising protein source for its use in animal feed (Nielsen et al. 2012). Red macroalgae species, such as those belonging to *Porphyra* genus like *Porphyra dioica*, *Porphyra purpurea*, *Porphyra yezoensis*, and have also been investigated as protein-rich sources in fish diets. Brown macroalgae, known for their high productivity, are readily available in various regions; however, they are less suitable for utilization in animal diets owing to their chemical. However, incorporating extracts from brown seaweed like *Laminaria digitata*, which are rich in laminarin and fucoidan, has been found to improve pork quality, prolong shelf life, and decrease lipid oxidation in muscle tissue (Moroney et al. 2012).

3.2.3 Role of Macroalgae in Biofuel Production

Biofuels are derived from biological sources and can be utilized in industrial processes and in vehicles. They are divided into first, second, and third generations based on the type of biomass used (Sharmila et al. 2021). Macroalgae/seaweeds, are abundant in phycocolloids, including alginate, carrageenan, laminari, etc. These compounds are not only edible but also possess a range of commercial applications at industrial level. Due to the complexity of breaking down phycocolloids, specialized pretreatment processes are required for biofuel production (Sudhakar et al. 2019).

In previous years, keen interest had grown in the research and commercial use of microalgae. Their quick growth rate in comparison to terrestrial plant species makes them a promising renewable energy source (Tredici 2010). Biofuels can be produced from macroalgae (Table 3.2) using various methods such as anaerobic digestion, fermentation, liquefaction, enabling the production of several kinds of fuels such as bioethanol, biodiesel, etc. (Chen et al. 2015).

TABLE 3.2
List of Macroalgal Members Utilized for Production of Biofuels

S.No.	Macroalgae Name	Biofuels	References
1	*Caulerpa peltata*	Bio-diesel	Tamilarasan and Sahadevan (2014)
2	*Chaetomorpha gracilis*	Liquid biofuel	Sánchez-Borroto et al. (2018)
3	*Enteromorpha* sp.	Bio-diesel	Suganya et al. (2013)
4	*Gelidium amansii*	Bio-ethanol	Park et al. (2012)
5	*Gelidium elegans*	Bio-ethanol	Yanagisawa et al. (2013)
6	*Kappaphycus alvarezii*	Bio-ethanol	Khambhaty et al. (2012) and Ingle et al. (2018)
7	*Laminaria hyperborea*	Bio-ethanol	Horn et al. (2000)
8	*Macrocystis pyrifera*	Bio-methane	Hanssen et al. (1987)
9	*Saccharina latissima*	Bio-methane	Vivekanand et al. (2012)
10	*Saccharina* sp.	Bio-ethanol	Horn et al. (2000)
11	*Sargassum fulvellum*	Bio-ethanol	Aizawa et al. (2007)
12	*Ulva lactuca*	Bio-methane	Bruhn et al. (2011)
13	*Ulva* sp.	Bio-methane	Adams et al. (2009) and Costa et al. (2012)

Hydrogen production from macroalgae is a notable focus in renewable energy research. Recent studies have identified *Laminaria japonica* and *Gelidium amansii*, a brown and red alga, respectively, as promising sources for bio-hydrogen production via anaerobic fermentation (Park et al. 2011). Sugars found in seaweeds, including glucose, cellulose, and galactose, can be efficiently employed for production of biofuel via fermentation processes (Sudhakar et al. 2017). Many algal species accumulate a significant amount of lipids, approximately 50% of their dry weight, as storage materials. These lipids are chemically similar to those found in oilseeds making algae a potential source for the production of bio-diesel (Griffiths and Harrison 2009). As reported by Osman et al. (2020), bio-diesel produced by *Ulva intestinalis*, resulted in the production of 32.3 mg/g dry weight. Similarly, Sharmila et al. (2012) documented that bio-diesel produced from *Gracilaria corticata* and *Chaetomorpha antennina* generated yields of 2 mL/10 g and 2.4 mL/10 g algal biomass, respectively.

3.3 MARINE FUNGI

Fungi represent a crucial component of the Earth's microbial diversity (Hawksworth 2001; O'Brien et al. 2005; Hibbett et al. 2007; Mueller et al. 2007). Estimates suggest that the number of fungal species ranges between 1.5 and 1.6 million (Hawksworth 1991, 2001). Marine fungi are classified as either "true fungi" or "stramenopila fungi" (Damare et al. 2011). To date, around 530 species of obligate marine fungi have been identified, primarily found on decomposing lignocellulosic materials along coasts, with some also discovered on decaying wood in deep-sea environments (Jones et al. 2009). The largest marine fungal family is Halosphaeriaceae, consisting of 59 genera having 142 species. Among these, genera *Candida*, *Aspergillus*, and *Penicillium* are the most diverse comprising of 47, 64, and 39 species, respectively (Jones et al. 2015). Marine fungi have been extracted from algae, corals, sponges, tunicates, mollusks, plants, wood, and fishes. Their ecology, diversity, and phylogeny have been extensively studied (Arifeen & Liu 2018). In marine ecosystems, fungi are typically associated with organic matter like driftwood, living organisms, and seabed sediments. They play a vital role in nutrient cycling, particularly in deep-sea sediments (Baltar et al. 2021).

Marine fungi account for less than 1% of all known fungal species, making them a relatively underexplored group. Recent research suggests that over 10,000 species of marine fungi belongs to the phyla Ascomycota and Basidiomycota. Due to their enzyme producing ability and production of antimicrobial compounds, fungi have numerous medical applications. Studies indicate that fungi generate more antimicrobial and anticancerous substances than those reported for bacteria (Noman et al. 2021). The distinct physico-chemical conditions of the marine environment lead to the production of a diverse range of molecules by nearly every class of marine life, each with unique structural properties (Jin et al. 2016).

Marine-derived fungi have been frequently referenced in natural product chemistry, with Jensen (2002) being notable contributors to this field. Pang et al. (2016) explored the differences between "marine fungi" and "marine-derived fungi," offering a clearer and more nuanced definition. Fungi-based SMs display a variety of properties such as anticancerous, anti-microbial, anti-viral, anti-oxidant, and anti-inflammatory properties (Tarman 2020; Queirós et al. 2009). Although culture-dependent techniques remain dominant in this research, the majority of recovered fungi belongs to Ascomycota and Basidiomycota, where over 90% of all known species are marine fungal ones (Jones et al. 2019). Penicillin, the pioneering antibiotic sourced from nature, was discovered by Sir Alexander Fleming in 1928 and is produced by fungi named *Penicillium notatum* (Houbraken et al. 2011). In 1945, the first marine-derived antibiotics, cephalosporins, were identified. These antibiotics are obtained from *Acremonium chrysogenum* (Silber et al. 2016). Since then, the several marine fungi-based antibiotic compounds have surged significantly (Silber et al. 2016). Currently, over 3,500 SMs originating from marine fungi have been recorded, with many more likely awaiting discovery (Carroll et al. 2019).

3.3.1 Fungi: A Source of Bio-Active Compounds

Fungi isolated from deep-sea environments present a promising, yet largely underexplored, resource for biologically active compounds. This is likely because of the richness and diversity of their SMs, which hold significant potential for drug development and pharmacology (Xu et al. 2018). Marine fungi have emerged as a significant source of microbial compounds, owing to their remarkable chemical diversity and prominent biological activities. Of the 22,000 known bioactive microbial metabolites, approximately 38% are derived from fungi. However, with only about 5% of the world's fungal species identified, there exists vast potential for uncovering new bioactive SMs within this group (Schulz et al. 2008).

While considerable advances have been made in identifying valuable drugs from marine fungi, the pace of discovering new biologically active natural products from these organisms has been relatively slow. It wasn't until the late 1980s that researchers began to place more emphasis on fungi originating from marine environments (Jin et al. 2016). Blunt et al. (2015) pointed out that marine fungi often produce natural compounds with superior oral bioavailability and improved physicochemical properties. Marine natural products are gaining increasing importance in biomedical research and drug development (Molinski et al. 2009). In recent decades, there has been a noticeable trend of repeatedly isolating previously known compounds and eventual decline in discovery of marine macroorganism-based novel compounds (Debbab et al. 2010). Marine fungi continue to be a promising source of bioactive metabolites, showing potential as novel agents for treating cancer, bacterial infections, malaria, inflammation, and viral diseases (Table 3.3).

Bio-active compounds synthesized by marine fungi are garnering growing interest as promising candidates for medicinal applications and plant protection. In addition to their therapeutic potential, these compounds have been found to play a role in preserving ecological balance and facilitating host-microorganism interactions (Duarte et al. 2012). The SMs of marine fungi are notable for their unique and diverse chemical structures, making them promising compounds for the development of new pharmaceuticals (Brereton et al. 2007; Liberati et al. 2009). Hu et al. (2015) indicated that more than 2,225 compounds obtained from marine organisms have demonstrated anticancerous properties. Some of the most potent anticancer compounds found in marine organisms include polyphenols, polysaccharides, and alkaloids (Shin 2020; Zin et al. 2017). Several SMs from marine fungi exhibit anticancer effects by causing structural changes in cells, inducing DNA damage, and increasing the accumulation of drugs in lung cancer cells. These compounds also hinder cell proliferation, enhance the uptake of Doxorubicin (Dox), and promote apoptosis. Such compounds were isolated from *Neosartorya* spp. (Noman et al. 2021).

3.3.2 Industrial Applications of Marine Fungi

Marine fungi are often studied owing to their ability to produce several enzymes of industrial importance, such as amylase, chitinase, β-glucosidase, cellulases, alkaline

TABLE 3.3
List of Properties Exhibited by Marine Fungi

S.No.	Fungi Name	Properties	References
1	*Aspergillus niger*	Anticancerous activities	Vala et al. (2018)
2	*Aspergillus terreus*	Anticancerous activities	Paul and Tiwary (2020)
3	*Aspergillus tubingensis*	Anticancerous activities	Doriya and Kumar (2018)
4	*Arthrinium arundinis*	Cytotoxicities	Wang et al. (2015)
5	*Ascochyta salicorniae*	Anti-fungal property	Bhadury et al. (2006)
6	*Aspergillus* sp.	Anti-bacterial property	Nguyen et al. (2007)
7	*Aspergillus terreus*	Anti-viral property	Nong et al. (2014)
8	*Aspergillus ustus*	Cytotoxicity	Liu et al. (2009)
9	*Cosmospora* sp.	Antidiabetic	Seo et al. (2009)
10	*Curvularia* sp.	Cytotoxicity	Greve et al. (2008)
11	*Diaporthaceae* sp. PSU-SP2/4	Anti-bacterial property	Khamthong et al. (2014)
12	*Emericella unguis*	Anti-bacterial property	Bhadury et al. (2006)
13	*Exophiala* sp.	Anti-bacterial property	Zhang et al. (2008)
14	*Fusarium* sp.	Anticancerous property	El-Gendy et al. (2021)
15	*Nigrospora oryzae* SCSGAF 0111	Anti-fungal property	Dong et al. (2014)
16	*Nigrospora* sp.	MRSA	Trisuwan et al. (2009)
17	*Paraphaeosphaeria* sp., *Penicillium montanense*	Anti-fungal property	Bhadury et al. (2006)
18	*Sarocladium strictum*	Anticancerous activities	Fan et al. (2020)
19	*Scytalidium* sp.	Anti-viral property	Imhoff et al. (2011)
20	*Spicellum roseum*	Cytotoxicity	Kralj et al. (2007)
21	*Stachybotrys* sp. RF-7260	Anti-viral property	Yasuhara-Bell and Lu (2010)
22	*Talaromyces pinophilus*	Anticancerous activities	Krishnapura and Belur (2016)
23	*Trichoderma virens*	Anti-bacterial property	Bhadury et al. (2006)
24	*Trichoderma viride*	Anticancerous activities	Jenila and Gnanadoss (2018)
25	*Trichosporon asahii*	Anticancerous activities	Baskar and Sree (2020)
26	*Zopfiella latipes*	Anti-fungal property	Bhadury et al. (2006)

protease, lactase, etc. all of which have considerable economic significance (Velmurugan and Lee 2012). Enzymes like proteases, laccases, amylases, xylanases, and cellulases derived from marine fungi have diverse industrial applications such as textiles, paper, leather, pulp, biofuels, medicine, pharmaceuticals, food, and beverages. Additionally, they are important for environmental purposes, animal feed production, and research (Arifeen and Liu 2018). Marine fungi produces several enzymes such as lignin peroxidases, xylanases, laccases, which break down complex substances like industrial pollutants (Atalla et al. 2010).

Enzymes, for example α-amylase, acetylcholinesterase, carboxymethyl cellulase, protease, lipase, phosphatase, are essential for several industries, such as food and beverages, textiles, leather processing, wastewater treatment, starch processing, paper production, pharmaceuticals, etc. (Mohapatra et al. 2003; Wang 2006). Among these, proteases are particularly valuable for detergent formulations and hold significant commercial importance, while both lipases and proteases are utilized in dairy industry (Parte et al. 2017). Studies on marine fungi-mediated degradation of plastics suggest that polyurethanes are particularly vulnerable to fungal decomposition (Jones and Le Campion-Alsumard 1970), with species such as *Aspergillus flavus*, *Aspergillus terreus*, *Aspergillus niger*, *Aspergillus fumigatus*, and various *Penicillium* species, extracted from seawater, identified as effective in breaking down polyethylene (Alshehrei 2017).

Several enzymes have been extracted from marine fungi for application in biotechnological processes. These fungi can produce enzymes that are resistant to salt, pressure, heavy metals, and heat (Velmurugan and Lee 2012). Raghukumar et al. (1994) discovered that fungi sourced from mangrove environments are capable of producing laccase, xylanase, and cellulase. Marine-derived ascomycetes and basidiomycetes generate laccases, which can effectively decolorize and modify pollutants found in textile wastewater (Verma et al. 2010). In addition to their ecological functions, marine fungi are essential for the breakdown and decomposition of organic materials.

Marine fungi are recognized for producing active cellulases and laccases, as well as specific glycoside hydrolases (GHs) that are adapted to marine environments, particularly when agricultural plant waste or algal polysaccharides are present in growth medium (Balabanova et al. 2018). Additionally, a strain of *Kodamaea ohmeri* isolated from the gut of a marine fish produces phytase, an enzyme used as an additive in animal feed to improve the nutritional value of grains and oilseeds by hydrolyzing phytic acid to release inorganic phosphate (Li et al. 2009).

In the paper industry, enzyme xylanases are utilized for improvement of lignin dissolution and to decrease reliance on chlorine compounds during the bleaching process, leading to better pulp properties in addition to reduced pollution. Marine fungi are recognized as a significant source of these enzymes. In a study, 84% of the marine ascomycete strains tested exhibited xylanolytic activity (Bucher et al. 2004), while all samples obtained from Antarctic sponges showed effectiveness in this area (Del-Cid et al. 2014).

3.4 CONCLUSION

Marine fungal and algal ecosystems serve as invaluable sources of biologically active compounds with significant potential. These compounds, which arise from the unique and often extreme conditions of marine environments, exhibit a diverse range of biological activities, such as antimicrobial activity, antiviral property, anticancerous and anti-inflammatory effects. Additionally, marine fungi and marine algae-derived substances have numerous industrial applications. However, utilization of such marine organism-based compounds is still in the early stages, leaving much to be uncovered regarding their full potential. Integrating marine biotechnology with conservation efforts will be essential to balance usage and preservation. Marine fungi and algae represent a largely untapped reservoir of bioactive compounds that could have extensive applications. Ongoing research in this area could lead to the discovery of new therapeutic agents and bio-based products, contributing to health, industry, and environmental sustainability.

REFERENCES

Abu-Ghannam, N., & Rajauria, G. (2013). Antimicrobial activity of compounds isolated from algae. In *Functional Ingredients from Algae for Foods and Nutraceuticals*, Dominguez, H., Ed., Woodhead Publishing Ltd., pp. 287–306.

Adams, J. M., Gallagher, J. A., & Donnison, I. S. (2009). Fermentation study on *Saccharina latissima* for bioethanol production considering variable pre-treatments. *Journal of Applied Phycology*, *21*, 569–574.

Adeghate, E., Adem, A., Hasan, M. Y., Tekes, K., & Kalasz, H. (2011). Suppl 2: medicinal chemistry and actions of dual and Pan PPAR modulators. *The Open Medicinal Chemistry Journal*, *5*, 93.

Afolayan, A. F., Bolton, J. J., Lategan, C. A., Smith, P. J., & Beukes, D. R. (2008). Fucoxanthin, tetraprenylated toluquinone and toluhydroquinone metabolites from *Sargassum heterophyllum* inhibit the in vitro growth of the malaria parasite *Plasmodium falciparum*. *Zeitschrift für Naturforschung C*, *63*(11–12), 848–852.

Aizawa, M., Asaoka, K., Atsumi, M., & Sakou, T. (2007, September). Seaweed bioethanol production in Japan—The ocean sunrise project. In *OCEANS 2007*, Vancouver, British Columbia, Canada (pp. 1–5). IEEE.

Albuquerque, I. R. L., Cordeiro, S. L., Gomes, D. L., Dreyfuss, J. L., Filgueira, L. G. A., Leite, E. L., Nader, H. B., & Rocha, H. A. O. (2013). Evaluation of anti-nociceptive and anti-inflammatory activities of a heterofucan from *Dictyota menstrualis*. *Marine Drugs*, *11*(8), 2722–2740.

Alshehrei, F. 2017. Biodegradation of low-density polyethylene by fungi isolated from Red sea water. *International Journal of Current Microbiology and Applied Sciences*, *6*, 1703–1709. doi: 10.20546/ijcmas.2017.602.204.

Alvarez-Hernández, S., Lozano-Ramírez, C., & Rodríguez-Palacio, M. (2019). Influence of the habitat on marine macroalgae toxicity. *Annual Research & Review in Biology*, *33*(1), 1–9.

Amsler, C. D., & Fairhead, V. A. (2005). Defensive and sensory chemical ecology of brown algae. *Advances in Botanical Research*, *43*, 1–91.

Apt, K. E., & Behrens, P. W. (1999). Commercial developments in microalgal biotechnology. *Journal of Phycology*, *35*(2), 215–226.

Arifeen, M. Z. U., & Liu, C. H. (2018). Novel enzymes isolated from marine-derived fungi and its potential applications. *Journal of Biotechnology and Bioengineering*, *1*, 1–11.

Atalla, M. M., Zeinab, H. K., Eman, R. H., Amani, A. Y., & Abeer, A. A. E. A. (2010). Screening of some marine-derived fungal isolates for lignin degrading enzymes (LDEs) production. *Agriculture and Biology Journal of North America*, *1*(4), 591–9.

Balabanova, L., Slepchenko, L., Son, O., & Tekutyeva, L. (2018). Biotechnology potential of marine fungi degrading plant and algae polymeric substrates. *Frontiers in Microbiology*, *9*, 1527.

Balboa, E. M., Conde, E., Moure, A., Falqué, E., & Domínguez, H. (2013). In vitro antioxidant properties of crude extracts and compounds from brown algae. *Food Chemistry*, *138*(2–3), 1764–1785.

Baltar, F., Zhao, Z., & Herndl, G. J. (2021). Potential and expression of carbohydrate utilization by marine fungi in the global ocean. *Microbiome*, *9*(1), 106.

Barbier, E. B. (2017). Marine ecosystem services. *Current Biology*, *27*(11), R507–R510.

Baskar, G., & Sree, N. S. (2020). Synthesis, characterization and anticancer activity of β-cyclodextrin-*Asparaginase* nanobiocomposite on prostate and lymphoma cancer cells. *Journal of Drug Delivery Science and Technology*, *55*, 101417.

Bhadury, P., Mohammad, B. T., & Wright, P. C. (2006). The current status of natural products from marine fungi and their potential as anti-infective agents. *Journal of Industrial Microbiology and Biotechnology*, *33*(5), 325.

Blunt, J.W., Copp, B.R., Munro, M.H.G., Northcote, P.T., & Prinsep, M.R. (2011). Marine natural products. *Natural Product Reports, 28*, 196–268.

Bold, H. C., &Wynne, M. J. (1985). *Introduction to the Algae Structure and Reproduction*, second ed., Prentice-Hall Inc., pp. 1–33.

Brereton, P., Kitchenham, B. A., Budgen, D., Turner, M., & Khalil, M. (2007). Lessons from applying the systematic literature review process within the software engineering domain. *Journal of Systems and Software*, *80*(4), 571–583.

Bruhn, A., Dahl, J., Nielsen, H. B., Nikolaisen, L., Rasmussen, M. B., Markager, S., Olesen, B., Arias, C., & Jensen, P. D. (2011). Bioenergy potential of Ulva lactuca: biomass yield, methane production and combustion. *Bioresource Technology*, *102*(3), 2595–2604.

Bucher, V. V. C., Hyde, K. D., Pointing, S. B., & Reddy, C. A. (2004). Production of wood decay enzymes, mass loss and lignin solubilization in wood by marine ascomycetes and their anamorphs. *Fungal Diversity*, *15*, 1–14.

Carroll, A. R., Copp, B. R., Davis, R. A., Keyzers, R. A., & Prinsep, M. R. (2019). Marine natural products. *Natural Product Reports*, *36*(1), 122–173.

Chadwick, N. E., & Morrow, K. M. (2011). Competition among sessile organisms on coral reefs. In *Coral Reefs: An Ecosystem in Transition*, Dubinsky, Z., & Stambler, N., Eds., Springer, pp. 347–371.

Chen, H., Zhou, D., Luo, G., Zhang, S., & Chen, J. (2015). Macroalgae for biofuels production: progress and perspectives. *Renewable and Sustainable Energy Reviews*, *47*, 427–437.

Chew, K. W., Yap, J. Y., Show, P. L., Suan, N. H., Juan, J. C., Ling, T. C., Lee, D. J., & Chang, J. S. (2017). Microalgae biorefinery: high value products perspectives. *Bioresource Technology*, 229, 53–62.

Chisti, Y. (2007). Biodiesel from microalgae. *Biotechnology Advances*, *25*(3), 294–306.

Costa, J. C., Gonçalves, P. R., Nobre, A., & Alves, M. M. (2012). Biomethanation potential of macroalgae *Ulva* spp. and *Gracilaria* spp. and in co-digestion with waste activated sludge. *Bioresource Technology*, *114*, 320–326.

Costa, L. S., Fidelis, G. P., Cordeiro, S. L., Oliveira, R. M., Sabry, D. A., Câmara, R. B. G., Nobre, L. T. D. B., Costa, M. S. S. P., Almeida-Lima, J., Farias, E. H. C., Leite, E. L., & Rocha, H. A. O. (2010). Biological activities of sulfated polysaccharides from tropical seaweeds. *Biomedicine & Pharmacotherapy*, *64*(1), 21–28.

Cox, S., Abu-Ghannam, N., & Gupta, S. (2010). An assessment of the antioxidant and antimicrobial activity of six species of edible Irish seaweeds. *International Food Research Journal 17*, 205–220.

Crain, C. M., Halpern, B. S., Beck, M. W., & Kappel, C. V. (2009). Understanding and managing human threats to the coastal marine environment. *Annals of the New York Academy of Sciences*, *1162*(1), 39–62.

Damare, S., Singh, P., & Raghukumar, S. (2011). Biotechnology of marine fungi. *Progress in Molecular and Subcellular Biology*, *53*, 277–297.

Davyt, D., Fernandez, R., Suescun, L., Mombrú, A. W., Saldaña, J., Domínguez, L., Coll, J., Fujii M.T., & Manta, E. (2001). New sesquiterpene derivatives from the red alga laurencia s coparia. Isolation, structure determination, and anthelmintic activity. *Journal of Natural Products*, *64*(12), 1552–1555.

Debbab, A., Aly, A. H., Lin, W. H., & Proksch, P. (2010). Bioactive compounds from marine bacteria and fungi. *Microbial Biotechnology*, *3*(5), 544–563.

Del-Cid, A., Ubilla, P., Ravanal, M. C., Medina, E., Vaca, I., Levicán, G., Eyzaguirre, J., & Chávez, R. (2014). Cold-active xylanase produced by fungi associated with Antarctic marine sponges. *Applied Biochemistry and Biotechnology*, *172*, 524–532.

Dong, J. J., Bao, J., Zhang, X. Y., Xu, X. Y., Nong, X. H., & Qi, S. H. (2014). Alkaloids and citrinins from marine-derived fungus *Nigrospora oryzae* SCSGAF 0111. *Tetrahedron Letters*, *55*(16), 2749–2753.

Doriya, K., & Kumar, D. S. (2018). Solid state fermentation of mixed substrate for L-asparaginase production using tray and in-house designed rotary bioreactor. *Biochemical Engineering Journal*, *138*, 188–196.

Duarte, K., Rocha-Santos, T. A., Freitas, A. C., & Duarte, A. C. (2012). Analytical techniques for discovery of bioactive compounds from marine fungi. *TrAC Trends in Analytical Chemistry*, *34*, 97–110.

El Gamal, A. A. (2010). Biological importance of marine algae. *Saudi Pharmaceutical Journal*, *18*(1), 1–25.

El-Gendy, M. M. A. A., Awad, M. F., El-Shenawy, F. S., & El-Bondkly, A. M. A. (2021). Production, purification, characterization, antioxidant and antiproliferative activities of extracellular L-asparaginase produced by *Fusarium equiseti* AHMF4. *Saudi Journal of Biological Sciences*, *28*(4), 2540–2548.

Fabrowska, J., Ibañez, E., Łęska, B., & Herrero, M. (2016). Supercritical fluid extraction as a tool to valorize underexploited freshwater green algae. *Algal Research*, *19*, 237–245.

Falcão, V. D. R. (2006). Molecular aspects of nitrate reductase from the marine macroalgae *Gracilaria tenuistipitata* (Rhodophyta): gene sequencing and study of messenger RNA expression. Doctoral dissertation, University of São Paulo.

Falkenberg, M., Nakano, E., Zambotti-Villela, L., Zatelli, G. A., Philippus, A. C., Imamura, K. B., Velasquez, A. M. A., Freitas, R. P., Tallarico, L. F., Colepicolo, P., & Graminha, M. A. (2019). Bioactive compounds against neglected diseases isolated from macroalgae: a review. *Journal of Applied Phycology*, *31*, 797–823.

Fan, B., Dewapriya, P., Li, F., Grauso, L., Blümel, M., Mangoni, A., & Tasdemir, D. (2020). Pyrenosetin D, a new pentacyclic decalinoyltetramic acid derivative from the algicolous fungus *Pyrenochaetopsis* sp. FVE–087. *Marine Drugs*, *18*(6), 281.

FAO. (2014). Fisheries and Aquaculture Information and Statistics Service, 16/03/2014.

Fedorov, S. N., Ermakova, S. P., Zvyagintseva, T. N., & Stonik, V. A. (2013). Anticancer and cancer preventive properties of marine polysaccharides: some results and prospects. *Marine Drugs*, *11*(12), 4876–4901.

Fong, P., & Paul, V. J. (2011). Coral reef algae. *Coral Reefs: An Ecosystem in Transition*, Dubinsky, Z., & Stambler, N., Eds., Springer, pp. 241–272.

Gallo, A., Ferrara, M., & Perrone, G. (2013). Phylogenetic study of polyketide synthases and nonribosomal peptide synthetases involved in the biosynthesis of mycotoxins. *Toxins*, *5*(4), 717–742.

Gladfelter, A. S., James, T. Y., & Amend, A. S. (2019). Marine fungi. *Current Biology*, *29*(6), R191–R195.

Gomez d'Ayala, G., Malinconico, M., & Laurienzo, P. (2008). Marine derived polysaccharides for biomedical applications: chemical modification approaches. *Molecules*, *13*(9), 2069–2106.

Gouveia, L., Nobre, B. P., Marcelo, F. M., Mrejen, S., Cardoso, M. T., Palavra, A. F., & Mendes, R. L. (2007). Functional food oil coloured by pigments extracted from microalgae with supercritical CO_2. *Food Chemistry*, *101*(2), 717–723.

Greve, H., Schupp, P. J., Eguereva, E., Kehraus, S., Kelter, G., Maier, A., Fiebig, H. H., & König, G. M. (2008). Apralactone A and a new stereochemical class of curvularins from the marine fungus *Curvularia* sp. *European Journal of Organic Chemistry*, *2008*(30), 5085. doi: 10.1002/ejoc.200800522.

Griffiths, M. J., & Harrison, S. T. (2009). Lipid productivity as a key characteristic for choosing algal species for biodiesel production. *Journal of Applied Phycology*, *21*, 493–507.

Guiry, M. D., & Guiry, G. M. (2018). *AlgaeBase*. World-wide electronic publication. National University of Ireland. https://www.algaebase.org. Accessed 20 February 2018.

Hanssen, J. F., Indergaard, M., Østgaard, K., Bævre, O. A., Pedersen, T. A., & Jensen, A. (1987). Anaerobic digestion of *Laminaria* spp. and *Ascophyllum nodosum* and application of end products. *Biomass*, *14*(1), 1–13.

Hanzhi, L., Song, Q., & Peng, J. (2011). Biotechnology of seaweeds: facing the coming decade. In *Handbook of Marine Macroalgae: Biotechnology and Applied Phycology*, Mitra, A., Ed., Wiley, pp. 424–430.

Harlin, M. M., & Rice, E. L. (1987). Allelochemistry in marine macroalgae. *Critical Reviews in Plant Sciences*, *5*(3), 237–249.

Hasan, S., Ansari, M. I., Ahmad, A., & Mishra, M. (2015). Major bioactive metabolites from marine fungi: a review. *Bioinformation*, *11*(4), 176.

Hawksworth, D. L. (1991). The fungal dimension of biodiversity: magnitude, significance, and conservation. *Mycological Research*, *95*(6), 641–655.

Hawksworth, D. L. (2001). The magnitude of fungal diversity: the 15 million species estimate revisited. *Mycological Research*, *105*(12), 1422–1432.

Hay, M. E. (2009). Marine chemical ecology: chemical signals and cues structure marine populations, communities, and ecosystems. *Annual Review of Marine Science*, *1*(1), 193–212.

Herrero, M., Mendiola, J. A., Plaza, M., & Ibañez, E. (2013). Screening for bioactive compounds from algae. In *Advanced Biofuels and Bioproducts*, Lee, J. W., Ed., Springer, pp. 833–872.

Hibbett, D. S., Binder, M., Bischoff, J. F., Blackwell, M., Cannon, P. F., Eriksson, O. E., Huhndorf, S., James, T., Kirk, P. M., Lücking, R., Lumbsch, H. T., Lutzoni, F., Matheny, P. B., McLaughlin, D. J., Powell, M. J., Redhead, S., Schoch, C. L., Spatafora, J. W., Stalpers, J. A., Vilgalys, R., & Zhang, N. (2007). A higher-level phylogenetic classification of the Fungi. *Mycological Research*, *111*(5), 509–547.

Horn, S. J. (2000). Bioenergy from brown seaweeds. Fakultet for naturvitenskap og teknologi.

Horn, S. J., Aasen, I. M., & Østgaard, K. (2000). Ethanol production from seaweed extract. *Journal of Industrial Microbiology and Biotechnology*, *25*(5), 249–254.

Houbraken, J., Frisvad, J. C., & Samson, R. A. (2011). Fleming's penicillin producing strain is not *Penicillium chrysogenum* but *P. rubens*. *IMA Fungus*, *2*, 87–95.

Hu, Y., Chen, J., Hu, G., Yu, J., Zhu, X., Lin, Y., Chen, S., & Yuan, J. (2015). Statistical research on the bioactivity of new marine natural products discovered during the 28 years from 1985 to 2012. *Marine Drugs*, *13*(1), 202–221.

Hultberg, M., Carlsson, A. S., & Gustafsson, S. (2013). Treatment of drainage solution from hydroponic greenhouse production with microalgae. *Bioresource Technology*, *136*, 401–406.

Imhoff, J. F. (2016). Natural products from marine fungi—still an underrepresented resource. *Marine Drugs*, *14*(1), 19.

Imhoff, J. F., Labes, A., & Wiese, J. (2011). Bio-mining the microbial treasures of the ocean: new natural products. *Biotechnology Advances*, *29*(5), 468–482.

Ingle, K., Vitkin, E., Robin, A., Yakhini, Z., Mishori, D., & Golberg, A. (2018). Macroalgae biorefinery from *Kappaphycus alvarezii*: conversion modeling and performance prediction for India and Philippines as examples. *BioEnergy Research*, *11*, 22–32.

Ioannou, E., Quesada, A., Rahman, M. M., Gibbons, S., Vagias, C., & Roussis, V. (2011). Dolabellanes with antibacterial activity from the brown alga Dilophus spiralis. *Journal of Natural Products*, *74*(2), 213–222.

Ishii, T., Okino, T., Suzuki, M., & Machiguchi, Y. (2004). Tichocarpols A and B, two novel phenylpropanoids with feeding-deterrent activity from the red alga *Tichocarpus crinitus*. *Journal of Natural Products*, *67*(10), 1764–1766.

Iwai, K. (2008). Antidiabetic and antioxidant effects of polyphenols in brown alga *Ecklonia stolonifera* in genetically diabetic KK-A y mice. *Plant Foods for Human Nutrition*, *63*, 163–169.

Jenila, A. V., & Gnanadoss, J. J. (2018). Formulation of a suitable medium and its optimization for maximizing L-asparaginase production from endophytic fungi *Fusarium* sp. LCJ273. *Biosciences Biotechnology Research Asia*, *15*(4), 887–898.

Jensen, P. R. (2002). Secondary metabolites from marine fungi. In *Fungi in Marine Environments*, Hyde, K. D., Ed., Fungal Diversity Press, 293–315.

Jha, R. K., & Zi-Rong, X. (2004). Biomedical compounds from marine organisms. *Marine Drugs*, *2*(3), 123–146.

Jin, L., Quan, C., Hou, X., & Fan, S. (2016). Potential pharmacological resources: natural bioactive compounds from marine-derived fungi. *Marine Drugs*, *14*(4), 76.

Jones, E. B. G., Sakayaroj, J., Suetrong, S., Somrithipol, S., & Pang, K. L. (2009). Classification of marine *Ascomycota*, anamorphic taxa and *Basidiomycota*. *Fungal Diversity*, *35*(1), 187.

Jones, E. G., & Le Campion-Alsumard, T. (1970). The biodeterioration of polyurethane by marine fungi. *International Biodeterioration Bulletin*, *6*(3), 119–124.

Jones, E. G., Pang, K. L., Abdel-Wahab, M. A., Scholz, B., Hyde, K. D., Boekhout, T., Ebel, R., Rateb, M. E., Henderson, L., Sakayaroj, J., Suetrong, S., Dayarathne, M. C., Kumar, V., Raghukumar, S., Sridhar, K. R., Bahkali, A. H. A., Gleason, F. H., & Norphanphoun, C. (2019). An online resource for marine fungi. *Fungal Diversity*, *96*, 347–433.

Jones, E. G., Suetrong, S., Sakayaroj, J., Bahkali, A. H., Abdel-Wahab, M. A., Boekhout, T., & Pang, K. L. (2015). Classification of marine ascomycota, basidiomycota, blastocladiomycota and chytridiomycota. *Fungal Diversity*, *73*, 1–72.

Jung, K. A., Lim, S. R., Kim, Y., & Park, J. M. (2013). Potentials of macroalgae as feedstocks for biorefinery. *Bioresource Technology*, *135*, 182–190.

Khambhaty, Y., Mody, K., Gandhi, M. R., Thampy, S., Maiti, P., Brahmbhatt, H., Eswaran, K., & Ghosh, P. K. (2012). *Kappaphycus alvarezii* as a source of bioethanol. *Bioresource Technology*, *103*(1), 180–185.

Khamthong, N., Rukachaisirikul, V., Phongpaichit, S., Preedanon, S., & Sakayaroj, J. (2014). An antibacterial cytochalasin derivative from the marine-derived fungus *Diaporthaceae* sp. PSU-SP2/4. *Phytochemistry Letters*, *10*, 5–9.

Kılınç, B., Cirik, S., Turan, G., Tekogul, H., & Koru, E. (2013). Seaweeds for food and industrial applications. In *Food Industry*, Muzzalupo, I., Ed., IntechOpen, pp. 736–748.

Kim, K. J., Yoon, K. Y., & Lee, B. Y. (2012). Fucoidan regulate blood glucose homeostasis in C57BL/KSJ m+/+ db and C57BL/KSJ db/db mice. *Fitoterapia*, *83*(6), 1105–1109.

König, G. M., Wright, A. D., & Franzblau, S. G. (2000). Assessment of antimycobacterial activity of a series of mainly marine derived natural products. *Planta Medica*, *66*(04), 337–342.

Kraan, S. (2012). Algal polysaccharides, novel applications and outlook. In *Carbohydrates-Comprehensive Studies on Glycobiology and Glycotechnology*, Chang, C.-F., Ed., IntechOpen, pp. 489–532.

Kralj, A., Kehraus, S., Krick, A., van Echten-Deckert, G., & König, G. M. (2007). Two new depsipeptides from the marine fungus *Spicellum roseum*. *Planta Medica*, *73*(04), 366–371.

Krishnapura, P. R., & Belur, P. D. (2016). Partial purification and characterization of L-asparaginase from an endophytic *Talaromyces pinophilus* isolated from the rhizomes of *Curcuma amada*. *Journal of Molecular Catalysis B: Enzymatic*, *124*, 83–91.

Kumar, B. R., Deviram, G., Mathimani, T., Duc, P. A., & Pugazhendhi, A. (2019). Microalgae as rich source of polyunsaturated fatty acids. *Biocatalysis and Agricultural Biotechnology*, *17*, 583–588.

Laurienzo, P. (2010). Marine polysaccharides in pharmaceutical applications: an overview. *Marine Drugs*, *8*(9), 2435–2465.
Lee, S. H., & Jeon, Y. J. (2013). Anti-diabetic effects of brown algae derived phlorotannins, marine polyphenols through diverse mechanisms. *Fitoterapia*, *86*, 129–136.
Lewis Jr, W. M. (1986). Evolutionary interpretations of allelochemical interactions in phytoplankton algae. *The American Naturalist*, *127*(2), 184–194.
Li, X., Liu, Z., Chi, Z., Li, J., & Wang, X. (2009). Molecular cloning, characterization, and expression of the phytase gene from marine yeast *Kodamaea ohmeri* BG3. *Mycological Research*, *113*(1), 24–32.
Li, Y. X., Wijesekara, I., Li, Y., & Kim, S. K. (2011). Phlorotannins as bioactive agents from brown algae. *Process Biochemistry*, *46*(12), 2219–2224.
Liberati, A., Altman, D. G., Tetzlaff, J., Mulrow, C., Gøtzsche, P. C., Ioannidis, J. P., Clarke, M., Devereaux, P. J., Kleijnen, J., & Moher, D. (2009). The PRISMA statement for reporting systematic reviews and meta-analyses of studies that evaluate health care interventions: explanation and elaboration. *Annals of Internal Medicine*, *151*(4), W65.
Liu, H., Edrada-Ebel, R., Ebel, R., Wang, Y., Schulz, B., Draeger, S., Muller, W.E.G., Wray, V., Lin, W., & Proksch, P. (2009). Drimane sesquiterpenoids from the fungus *Aspergillus ustus* isolated from the marine sponge *Suberites domuncula*. *Journal of Natural Products*, *72*(9), 1585–1588.
Liu, M., Zhang, W., Qiu, L., & Lin, X. (2011). Synthesis of butyl-isobutyl-phthalate and its interaction with α-glucosidase in vitro. *The Journal of Biochemistry*, *149*(1), 27–33.
Michalak, I., Górka, B., Wieczorek, P. P., Rój, E., Lipok, J., Łęska, B., Messyasz, B., Wilk, R., Schroeder, G., Dobrzynska-Inger, A., & Chojnacka, K. (2016). Supercritical fluid extraction of algae enhances levels of biologically active compounds promoting plant growth. *European Journal of Phycology*, *51*(3), 243–252.
Milchakova, N. A. (2011). *Marine Plants of the Black Sea. An Illustrated Field Guide*, Digit Print Press.
Mohapatra, B. R., Bapuji, M., & Sree, A. J. A. B. (2003). Production of industrial enzymes (amylase, carboxymethylcellulase and protease) by bacteria isolated from marine sedentary organisms. *Acta Biotechnologica*, *23*(1), 75–84.
Molinski, T. F., Dalisay, D. S., Lievens, S. L., & Saludes, J. P. (2009). Drug development from marine natural products. *Nature Reviews Drug Discovery*, *8*(1), 69–85.
Moroney, N. C., O'Grady, M. N., O'Doherty, J. V., & Kerry, J. P. (2012). Addition of seaweed (*Laminaria digitata*) extracts containing laminarin and fucoidan to porcine diets: influence on the quality and shelf-life of fresh pork. *Meat Science*, *92*(4), 423–429.
Morrow, K. M., Paul, V. J., Liles, M. R., & Chadwick, N. E. (2011). Allelochemicals produced by Caribbean macroalgae and cyanobacteria have species-specific effects on reef coral microorganisms. *Coral Reefs*, *30*, 309–320.
Mueller, G. M., Schmit, J. P., Leacock, P. R., Buyck, B., Cifuentes, J., Desjardin, D. E., Halling, R. E., Hjortstam, K., Iturriaga, T., Larsson, K., Lodge, D. J., May, T. W., Minter, D., Rajchenberg, M., Redhead, S. A., Ryvarden, L., Trappe, J. M., Watling, R., & Wu, Q. (2007). Global diversity and distribution of macrofungi. *Biodiversity and Conservation*, *16*, 37–48.
Nan, C., Zhang, H., Lin, S., Zhao, G., & Liu, X. (2008). Allelopathic effects of *Ulva lactuca* on selected species of harmful bloom-forming microalgae in laboratory cultures. *Aquatic Botany*, *89*(1), 9–15.
Negreanu-Pirjol, B. S., Negreanu-Pirjol, T., Popoviciu, D. R., Anton, R. E., & Prelipcean, A. M. (2022). Marine bioactive compounds derived from macroalgae as new potential players in drug delivery systems: a review. *Pharmaceutics*, *14*(9), 1781.
Ngo, D. H., & Kim, S. K. (2013). Sulfated polysaccharides as bioactive agents from marine algae. *International Journal of Biological Macromolecules*, *62*, 70–75.
Nguyen, H. P., Zhang, D., Lee, U., Kang, J. S., Choi, H. D., & Son, B. W. (2007). Dehydroxychlorofusarielin B, an antibacterial polyoxygenated decalin derivative from the marine-derived fungus *Aspergillus* sp. *Journal of Natural Products*, *70*(7), 1188–1190.
Nielsen, M. M., Bruhn, A., Rasmussen, M. B., Olesen, B., Larsen, M. M., & Møller, H. B. (2012). Cultivation of *Ulva lactuca* with manure for simultaneous bioremediation and biomass production. *Journal of Applied Phycology*, *24*, 449–458.
Noman, E., Al-Shaibani, M. M., Bakhrebah, M. A., Almoheer, R., Al-Sahari, M., Al-Gheethi, A., Mohamed, R. M. S. R., Almulaiky, Y. Q. & Abdulaal, W. H. (2021). Potential of anti-cancer activity of secondary metabolic products from marine fungi. *Journal of Fungi*, *7*(6), 436.
Nong, X. H., Wang, Y. F., Zhang, X. Y., Zhou, M. P., Xu, X. Y., & Qi, S. H. (2014). Territrem and butyrolactone derivatives from a marine-derived fungus *Aspergillus terreus*. *Marine Drugs*, *12*(12), 6113–6124.
O'Brien, H. E., Parrent, J. L., Jackson, J. A., Moncalvo, J. M., & Vilgalys, R. (2005). Fungal community analysis by large-scale sequencing of environmental samples. *Applied and Environmental Microbiology*, *71*(9), 5544–5550.
Ohta, T., Sasaki, S., Oohori, T., Yoshikawa, S., & Kurihara, H. (2002). α-Glucosidase inhibitory activity of a 70% methanol extract from Ezoishige (*Pelvetia babingtonii* de Toni) and its effect on the elevation of blood glucose level in rats. *Bioscience, Biotechnology, and Biochemistry*, *66*(7), 1552–1554.
Ördög, V., Stirk, W. A., Lenobel, R., Bancířová, M., Strnad, M., Van Staden, J., Szigeti, J. & Németh, L. (2004). Screening microalgae for some potentially useful agricultural and pharmaceutical secondary metabolites. *Journal of Applied Phycology*, *16*, 309–314.
Osman, M. E., Abo-Shady, A. M., Elshobary, M. E., Abd El-Ghafar, M. O., & Abomohra, A. E. F. (2020). Screening of seaweeds for sustainable biofuel recovery through sequential biodiesel and bioethanol production. *Environmental Science and Pollution Research*, *27*, 32481–32493.
Ospina, M., Castro-Vargas, H. I., & Parada-Alfonso, F. (2017). Antioxidant capacity of Colombian seaweeds: 1. Extracts obtained from *Gracilaria mammillaris* by means of supercritical fluid extraction. *The Journal of Supercritical Fluids*, *128*, 314–322.
Pang, K. L., Overy, D. P., Jones, E. G., da Luz Calado, M., Burgaud, G., Walker, A. K., Johnson, A. J., Kerr, R. G., Cha, H. J., & Bills, G. F. (2016). "Marine fungi" and "marine-derived fungi" in natural product chemistry research: toward a new consensual definition. *Fungal Biology Reviews*, *30*(4), 163–175.
Park, J. H., Hong, J. Y., Jang, H. C., Oh, S. G., Kim, S. H., Yoon, J. J., & Kim, Y. J. (2012). Use of *Gelidium amansii* as a promising resource for bioethanol: a practical approach for continuous dilute-acid hydrolysis and fermentation. *Bioresource Technology*, *108*, 83–88.
Park, J. H., Yoon, J. J., Park, H. D., Kim, Y. J., Lim, D. J., & Kim, S. H. (2011). Feasibility of biohydrogen production from *Gelidium amansii*. *International Journal of Hydrogen Energy*, *36*(21), 13997–14003.

Parte, S., Sirisha, V. L., & D'Souza, J. S. (2017). Biotechnological applications of marine enzymes from algae, bacteria, fungi, and sponges. In *Advances in Food and Nutrition Research*, Toldra, F., Ed., Academic Press, Vol. 80, pp. 75–106.

Paul, V., & Tiwary, B. N. (2020). An investigation on the acrylamide mitigation potential of L-asparaginase from *Aspergillus terreus* BVC strain. *Biocatalysis and Agricultural Biotechnology*, *27*, 101677.

Pereira, R. C., & Da Gama, B. A. P. (2008). Macroalgal chemical defenses and their roles in structuring tropical marine communities. In *Algal Chemical Ecology*, Amsler, C. D., Ed., Springer, Berlin, Heidelberg, pp. 25–55.

Periaswamy Sivagnanam, S., Yin, S., Choi, J. H., Park, Y. B., Woo, H. C., & Chun, B. S. (2015). Biological properties of fucoxanthin in oil recovered from two brown seaweeds using supercritical CO_2 extraction. *Marine Drugs*, *13*(6), 3422–3442.

Pezzolesi, L., Accoroni, S., Rindi, F., Samorì, C., Totti, C., & Pistocchi, R. (2021). Survey of the allelopathic potential of Mediterranean macroalgae: production of long-chain polyunsaturated aldehydes (PUAs). *Phytochemistry*, *189*, 112826.

Plaza, M., Cifuentes, A., & Ibáñez, E. (2008). In the search of new functional food ingredients from algae. *Trends in Food Science & Technology*, *19*(1), 31–39.

Qin, S., Lin, H., & Jiang, P. (2012). Advances in genetic engineering of marine algae. *Biotechnology Advances*, *30*(6), 1602–1613.

Queirós, B., Barreira, J. C., Sarmento, A. C., & Ferreira, I. C. (2009). In search of synergistic effects in antioxidant capacity of combined edible mushrooms. *International Journal of Food Sciences and Nutrition*, *60*(sup6), 160–172.

Raghukumar, C., Raghukumar, S., Chinnaraj, A., Chandramohan, D., D'souza, T. M., & Reddy, C. A. (1994). Laccase and other lignocellulose modifying enzymes of marine fungi isolated from the coast of India. *Botanica Marina*, *37*(6), 515–524.

Rasher, D. B., & Hay, M. E. (2010). Chemically rich seaweeds poison corals when not controlled by herbivores. *Proceedings of the National Academy of Sciences of United States of America*, *107*(21), 9683–9688.

Rizzo, L., Fraschetti, S., Alifano, P., Tredici, M. S., & Stabili, L. (2016). Association of Vibrio community with the Atlantic Mediterranean invasive alga *Caulerpa cylindracea*. *Journal of Experimental Marine Biology and Ecology*, *475*, 129–136.

Roesijadi, G., Jones, S. B., Snowden-Swan, L. J., & Zhu, Y. (2010). Macroalgae as a biomass feedstock: a preliminary analysis. No. PNNL-19944. Pacific Northwest National Laboratory, Richland, WA.

Sajeevan, T. P., Umar, M. D., Dhaneesha, M., Joseph, R., Ravinesh, R., & Sajeevan, T. P. (2020). Endophytic fungi isolated from the marine sponges as a source of potential bioactive compounds. *Journal of Aquatic Biology & Fisheries*, *8*, 58–66.

Sánchez-Borroto, Y., Lapuerta, M., Melo-Espinosa, E. A., Bolonio, D., Tobío-Perez, I., & Piloto-Rodríguez, R. (2018). Green-filamentous macroalgae *Chaetomorpha cf. gracilis* from Cuban wetlands as a feedstock to produce alternative fuel: a physicochemical characterization. *Energy Sources, Part A: Recovery, Utilization, and Environmental Effects*, *40*(10), 1279–1289.

Sanjeewa, K. K. A., Kim, E. A., Son, K. T., & Jeon, Y. J. (2016). Bioactive properties and potentials cosmeceutical applications of phlorotannins isolated from brown seaweeds: a review. *Journal of Photochemistry and Photobiology B: Biology*, *162*, 100–105.

Schulz, B., Draeger, S., dela Cruz, T. E., Rheinheimer, J., Siems, K., Loesgen, S., Bitzer, J., Schloerke, O., Zeeck, A., Kock, I., Hidayat, H., Dai, J., & Krohn, K. (2008). *Screening Strategies for Obtaining Novel, Biologically Active, Fungal Secondary Metabolites from Marine Habitats*, Walter de Gruyter.

Seo, C., Sohn, J. H., Oh, H., Kim, B. Y., & Ahn, J. S. (2009). Isolation of the protein tyrosine phosphatase 1B inhibitory metabolite from the marine-derived fungus *Cosmospora* sp. SF-5060. *Bioorganic & Medicinal Chemistry Letters*, *19*(21), 6095–6097.

Shannon, E., & Abu-Ghannam, N. (2016). Antibacterial derivatives of marine algae: an overview of pharmacological mechanisms and applications. *Marine Drugs*, *14*(4), 81.

Sharmila, S., Jeyanthi Rebecca, L., & Das, M. P. (2012). Production of biodiesel from *Chaetomorpha antennina* and *Gracilaria corticata*. *Journal of Chemical and Pharmaceutical Research*, *4*(11), 4870–4874.

Sharmila, V. D., Kumar, D., Pugazhendi, A., Bajhaiya, A. K., & Gugulothu, P. (2021). Biofuel production from Macroalgae: present scenario and future scope. *Bioengineered*, *12*(2), 9216.

Shin, H. J. (2020). Natural products from marine fungi. *Marine Drugs*, *18*(5), 230.

Silber, J., Kramer, A., Labes, A., & Tasdemir, D. (2016). From discovery to production: biotechnology of marine fungi for the production of new antibiotics. *Marine Drugs*, *14*(7), 137.

Stevenson, R. J., Pan, Y. D., & van Dam, H. (2010). Assessing environmental conditions in rivers and streams with diatoms, In *CABDirect2. The Diatoms: Applications for the Environmental and Earth Sciences*, pp. 57–85.

Sudhakar, M. P., Jegatheesan, A., Poonam, C., Perumal, K., & Arunkumar, K. (2017). Biosaccharification and ethanol production from spent seaweed biomass using marine bacteria and yeast. *Renewable Energy*, *105*, 133–139.

Sudhakar, M. P., Kumar, B. R., Mathimani, T., & Arunkumar, K. (2019). A review on bioenergy and bioactive compounds from microalgae and macroalgae-sustainable energy perspective. *Journal of Cleaner Production*, *228*, 1320–1333.

Suganya, T., Gandhi, N. N., & Renganathan, S. (2013). Production of algal biodiesel from marine macroalgae Enteromorpha compressa by two step process: optimization and kinetic study. *Bioresource Technology*, *128*, 392–400.

Talero, E., García-Mauriño, S., Ávila-Román, J., Rodríguez-Luna, A., Alcaide, A., & Motilva, V. (2015). Bioactive compounds isolated from microalgae in chronic inflammation and cancer. *Marine Drugs*, *13*(10), 6152–6209.

Tamilarasan, S., & Sahadevan, R. (2014). Ultrasonic assisted acid base transesterification of algal oil from marine macroalgae *Caulerpa peltata*: optimization and characterization studies. *Fuel*, *128*, 347–355.

Tang, Y. Z., & Gobler, C. J. (2011). The green macroalga, *Ulva lactuca*, inhibits the growth of seven common harmful algal bloom species via allelopathy. *Harmful Algae*, *10*(5), 480–488.

Tarman, K. (2020). Marine fungi as a source of natural products. *Encyclopedia of Marine Biotechnology*, *4*, 2147–2160.

Tocher, R. D., & Craigie, J. S. (1966). Enzymes of marine algae: II. Isolation and identification of 3-hydroxytyramine as the phenolase substrate in *Monostroma fuscum*. *Canadian Journal of Botany*, *44*(5), 605–608.

Tredici, M. R. (2010). Photobiology of microalgae mass cultures: understanding the tools for the next green revolution. *Biofuels*, *1*(1), 143–162.

Trisuwan, K., Rukachaisirikul, V., Sukpondma, Y., Phongpaichit, S., Preedanon, S., & Sakayaroj, J. (2009). Lactone derivatives from the marine-derived fungus *Penicillium* sp. PSU-F44. *Chemical and Pharmaceutical Bulletin*, *57*(10), 1100–1102.

Trivedi, J., Aila, M., Bangwal, D. P., Kaul, S., & Garg, M. O. (2015). Algae based biorefinery—how to make sense?. *Renewable and Sustainable Energy Reviews*, *47*, 295–307.

Vala, A. K., Sachaniya, B., Dudhagara, D., Panseriya, H. Z., Gosai, H., Rawal, R., & Dave, B. P. (2018). Characterization of L-asparaginase from marine-derived *Aspergillus niger* AKV-MKBU, its antiproliferative activity and bench scale production using industrial waste. *International Journal of Biological Macromolecules*, *108*, 41–46.

Van Alstyne, K. L., Harvey, E. L., & Cataldo, M. (2014). Effects of dopamine, a compound released by the green-tide macroalga *Ulvaria obscura* (Chlorophyta), on marine algae and invertebrate larvae and juveniles. *Phycologia*, *53*(2), 195–202.

Velmurugan, N., & Lee, Y. S. (2012). Enzymes from marine fungi: current research and future prospects, In *Marine Fungi and Fungal-like Organisms (Marine and Freshwater Botany)*, Jones, E. B. G., & Pang, K.-P., Eds., Walter de Gruyter GmbH, pp. 441–474.

Ventura, M. R., & Castañón, J. I. R. (1998). The nutritive value of seaweed (*Ulva lactuca*) for goats. *Small Ruminant Research*, *29*(3), 325–327.

Venugopal, V. (2008). *Marine Products for Healthcare: Functional and Bioactive Nutraceutical Compounds from the Ocean*. CRC Press.

Verma, A. K., Raghukumar, C., Verma, P., Shouche, Y. S., & Naik, C. G. (2010). Four marine-derived fungi for bioremediation of raw textile mill effluents. *Biodegradation*, *21*, 217–233.

Vivekanand, V., Eijsink, V. G., & Horn, S. J. (2012). Biogas production from the brown seaweed *Saccharina latissima*: thermal pretreatment and codigestion with wheat straw. *Journal of Applied Phycology*, *24*, 1295–1301.

Wang, G. (2006). Diversity and biotechnological potential of the sponge-associated microbial consortia. *Journal of Industrial Microbiology and Biotechnology*, *33*(7), 545.

Wang, J., Wei, X., Qin, X., Lin, X., Zhou, X., Liao, S., Yang, B., Liu, J., Tu, Z., & Liu, Y. (2015). Arthpyrones A–C, pyridone alkaloids from a sponge-derived fungus *Arthrinium arundinis* ZSDS1-F3. *Organic Letters*, *17*(3), 656–659.

Wang, L., Wang, X., Wu, H., & Liu, R. (2014). Overview on biological activities and molecular characteristics of sulfated polysaccharides from marine green algae in recent years. *Marine Drugs*, *12*(9), 4984–5020.

Wang, R., Xiao, H., Wang, Y., Zhou, W., & Tang, X. (2007). Effects of three macroalgae, *Ulva linza* (Chlorophyta), *Corallina pilulifera* (Rhodophyta) and *Sargassum thunbergii* (Phaeophyta) on the growth of the red tide microalga *Prorocentrum donghaiense* under laboratory conditions. *Journal of Sea Research*, *58*(3), 189–197.

Witvrouw, M., & De Clercq, E. (1997). Sulfated polysaccharides extracted from sea algae as potential antiviral drugs. *General Pharmacology: The Vascular System*, *29*(4), 497–511.

Wium-Andersen, S., Anthoni, U., & Houen, G. (1983). Elemental sulphur, a possible allelopathic compound from *Ceratophyllum demersum*. *Phytochemistry*, *22*(11), 2613.

Wium-Andersen, S., Anthoni, U., Christophersen, C., & Houen, G. (1982). Allelopathic effects on phytoplankton by substances isolated from aquatic macrophytes (Charales). *Oikos*, *39*, 187–190.

Xu, J. L., Liu, H. X., Chen, Y. C., Tan, H. B., Guo, H., Xu, L. Q., Li, S. N., Huang, Z. L., Li, H. H., Gao, X. X., & Zhang, W. M. (2018). Highly substituted benzophenone aldehydes and eremophilane derivatives from the deep-sea derived fungus *Phomopsis lithocarpus* FS508. *Marine Drugs*, *16*(9), 329.

Xu, J., Yi, M., Ding, L., & He, S. (2019). A review of anti-inflammatory compounds from marine fungi, 2000–2018. *Marine Drugs*, *17*(11), 636.

Yanagisawa, M., Kawai, S., & Murata, K. (2013). Strategies for the production of high concentrations of bioethanol from seaweeds: production of high concentrations of bioethanol from seaweeds. *Bioengineered*, *4*(4), 224–235.

Yasuhara-Bell, J., & Lu, Y. (2010). Marine compounds and their antiviral activities. *Antiviral Research*, *86*(3), 231–240.

Yu, K. X., Jantan, I., Ahmad, R., & Wong, C. L. (2014). The major bioactive components of seaweeds and their mosquitocidal potential. *Parasitology Research*, *113*, 3121–3141.

Zhang, D., Yang, X., Kang, J. S., Choi, H. D., & Son, B. W. (2008). Chlorohydroaspyrones A and B, antibacterial aspyrone derivatives from the marine-derived fungus *Exophiala* sp. *Journal of Natural Products*, *71*(8), 1458–1460.

Zin, S. R. M., Kassim, N. M., Alshawsh, M. A., Hashim, N. E., & Mohamed, Z. (2017). Biological activities of *Anastatica hierochuntica* L.: a systematic review. *Biomedicine & Pharmacotherapy*, *91*, 611–620.

4 The Mysterious World of Oceanic Viruses

Priya Yadav, Harshvardhan Singh Salathia, and Manish Thakur
School of Bioengineering and Biosciences, Lovely Professional University, Phagwara, India

4.1 INTRODUCTION

The planet Earth is covered with more than 70% of its surface with water, accounting for over 90% of its crust volume. The oceans are complex and consist of a wide range of life forms that can survive in conditions of extreme pressure, salinity, and temperature (Steward et al., 2013). Marine microorganisms make up a total biomass of around 70% or even up to 90% in oceans. Together, they form the marine microbiome. Recent research shows that marine viruses have total biomass of 0.03 Gt of carbon (Jian et al., 2021). While it is believed that most marine viruses are bacteriophages, the diversity and abundance of marine eukaryotic viruses is also increasing. These viruses can be found in various sizes, ranging from 20 nm to 1.2 µm and vary widely in genetic composition, with some having only a few genes while others contain thousands of genes encoding over a thousand proteins (Zayed et al., 2022). The study of marine viruses continues to unveil new insights into their biological activities and evolution.

Viruses are present everywhere in the environment and are found in various ecosystems, including the deep oceans, marines, polar regions, hydrothermal vents, and other extreme environment condition (Dominguez-Huerta et al., 2022). Some viruses are released from the ocean in atmosphere as they are believed to be aerosol which can transmit through air. These viruses play a major role in controlling the toxication by infecting the photosynthetic algae, as these algae form algal blooms which leads to the destruction of marine life as it covers the surface of ocean preventing sunlight and oxygen from reaching the organisms underneath. These algal blooms also play a major role in maintaining the ecological balance of the planet, serving as hotspots of primary production that fuel the entire marine food web and contribute to carbon export into the deep sea (Almutairi et al., 2025). However, these blooms can be terminated by viral infections, causing the lysis of phytoplankton cells and releasing intracellular content into the seawater, which then becomes available for bacterial respiration and growth. On the other hand, infected cells can form large aggregates, causing phytoplankton to sink faster and driving the sequestration of carbon to the deep ocean.

The extensive variety and abundance of viruses in the marine environment were first observed in the 1980s when microscopic counts of virus particles showed millions of viruses within just a milliliter of seawater. Earlier in the 1970s, it was discovered that marine bacteria are diverse and abundant, with initial reports enumerating more than one million bacteria per milliliter in marine environments (An et al., 2024). This revelation of marine bacterial abundance further emphasized the critical importance of microbial processes in the global oceans.

The first virus found to infect a marine eukaryotic organism was reported in the 1970s in a unicellular alga—*Micromonas pusilla* (a prasinophyte, family Mamiellaceae)—that is a primary producer in some ocean waters. This discovery suggested that viruses could play significant roles in influencing the ecology of important marine primary producers. In the early 1980s, studies of viruses infecting freshwater Chlorella algal symbionts led to the taxonomic classification of the Phycodnaviridae, a family of algal viruses. Within the Phycodnaviridae, the *phaeoviruses* that infect marine filamentous brown algae of the *Ectocarpale*s were further identified. As methodologies for investigating viruses in the marine environment advance, our understanding of the roles and significance of viral infection expands, including the influence of viruses on host population abundance and distribution, evolution, and horizontal gene transfer (HGT) (Danovaro et al., 2011).

4.2 MARINE VIRUSES AND THEIR HOST

According to Breitbart (2012), viral abundances are closely linked to their host and are generally more abundant in phytoplankton and Cyanobacteria than in other microorganisms. Previous large-scale spatial studies of viral distribution in the Pacific and South Atlantic Oceans, using flow cytometry, revealed high viral abundance in the water surface column up to 200 m in tropical and subtropical regions, but lower levels in Antarctic waters. A virus "hot-spot" was identified in the mid-range area of the North Pacific, supporting the theory that the large-scale distribution patterns of viruses are influenced by host distributions and physical processes. Brum and Sullivan (2015) utilized cultivation-independent methods such as metagenomics to show that the Southern Ocean is dominated by lysogenic viruses, which are referred to as "seasonal time bombs" because they can transition to a lytic cycle as the production of their bacterial hosts increases. Virome data collected by Roux et al. (2019) identified virally encoded AMGs

DOI: 10.1201/9781003509530-4

(Auxiliary Metabolic Genes) in surface waters that appear to be involved in nitrogen and sulfur cycling.

The Pacific Ocean Virome (POV) is a highly significantly curated dataset of double-stranded DNA (dsDNA) viromes that includes 32 quantitatively representative viromes collected from various depths and seasons in transects spanning coastal waters to open-ocean waters in the Pacific Ocean (Luo et al., 2022). The POV study revealed a decrease in virus richness from deeper to surface waters, from winter to summer, and in surface layers with distance from the shore. Additionally, data from the TARA ocean expedition were utilized to create a global virome map of dsDNA viruses sampled in both surface and deep-ocean waters, along with genomic data from the Malaspina 2010–2011 (Zeng et al., 2025) Circumnavigation Expedition, which assessed pelagic processes across the Indian, Pacific, and Atlantic Oceans. The researchers discovered that 38 out of 867 viral clusters were locally or globally abundant, representing nearly 50% of the viral population in any given Global Ocean Virome (GOV) (Al-Shayeb et al., 2020). These extensive datasets are crucial for enhancing our understanding of marine viruses on a global scale and are instrumental in uncovering uncultivated novel viruses and developing essential marine ecosystem models.

4.3 THE ROLE OF VIRUSES IN THE OCEAN

Viruses are inalienable part of the ocean's ecosystem, influencing various biological and chemical processes. Here are some key points about their role:

1. **Microbial Control:** The ocean is filled with tiny organisms, many of which are single-celled organisms. These organisms, such as bacteria, microalgae, and protozoa, come in a great diversity regarding their shapes and sizes; there are also viruses present in the sea, which require a host to survive and replicate (Vincent & Vardi, 2023). Viruses consist of protein shell with genetic material (either DNA or RNA) typically measuring between 20 and 200 nm in size. Viruses due to their size are the most abundant organisms in the ocean. A mere milliliter of coastal water from the ocean's surface can harbor up to ten million viruses (Jin et al., 2024). The virus count diminishes the farther we move offshore and into deeper waters. In the open sea, it is estimated that there are approximately 100,000 viruses per milliliter at a depth of 4,000 m (Rohwer & Thurber, 2009). Around 20% of ocean microbes are believed to be killed by viruses daily. When these viruses replicate and infect bacterial and algal cells, the cells release their contents into the water, adding to the dissolved organic and inorganic nutrients in the ocean. The substances released are recycled by smaller organisms for growth (Khotimchenko & Shchelkanov, 2024). These smaller organisms are then consumed by larger creatures like zooplankton and fish, which in turn are prey for sharks, whales, and humans. This process allows microscopic viruses to recycle the essential materials needed for life to thrive. Given that the ocean spans over 70% of Earth's surface, their impact greatly shapes life across much of our planet (Suttle, 2007).
2. **Nutrient Cycling:** When viruses lyse their host cells, they release organic matter and nutrients into the environment. This process is known as viral shunt and it facilitates recycling of nutrients and sustains the microbial loop, which is crucial for the nutrient dynamics in the ocean. This viral shunt process prevents marine microbial POM (particulate organic matter) from moving up the trophic levels by converting it into DOM (dissolved organic matter), this DOM is readily used by different micoorganisms (Tong & Xu, 2024). This mechanism enhances the efficiency of the biological pump by retaining essential nutrients in the surface waters. Additionally, the viral shunt plays a significant role in the biogeochemical cycles, influencing the carbon cycle and the overall productivity of marine ecosystems (Tran & Anantharaman, 2021).
3. **Genetic Exchange:** Viruses play a crucial role in genetic exchange among marine microorganisms through mechanisms like transduction, where they transfer genetic material between cells. This process increases genetic diversity and facilitates the evolution of new traits, enhancing the adaptability of microbial communities (Fogg, 2025). By influencing HGT, viruses help spread advantageous genes, impacting nutrient and carbon cycling in the ocean. This genetic exchange drives the evolution and ecological dynamics of marine ecosystems, contributing to their resilience and functionality.
4. **Increased Genetic Diversity:** Genetic exchange introduces new genetic material into microbial populations, increasing genetic diversity (Sousa et al., 2021). This diversity is crucial for the adaptability and resilience of microbial communities in the face of environmental changes.
5. **Evolution of New Traits**: The integration of new genes can lead to the evolution of new traits, such as antibiotic resistance, metabolic capabilities, and virulence factors (Rocha & Bikard, 2022). This can have significant implications for the ecological roles and interactions of marine microorganisms.
6. **Horizontal Gene Transfer (HGT):** HGT facilitated by viruses allows for the rapid spread of advantageous genes across different species and genera (Sousa et al., 2023). This can lead to the emergence of new microbial strains with enhanced capabilities, such as improved nutrient utilization or resistance to environmental stressors.

7. **Carbon Cycling:** During infection, viruses lyse the phytoplankton which results in the release of cellular contents into the surrounding water which also includes the organic carbon. This process is known as the viral shunt which diverts carbon from the traditional food web into DOM. The DOM released by lysed cells becomes available for uptake by bacteria and other microorganisms, acting as fuel for the microbial metabolism (Braga et al., 2022). This microbial loop helps recycle nutrients and organic matter within the euphotic zone which makes it a critical component of the ocean's carbon cycle. The viral shunt also plays a major role in the isolation of carbon in the deep ocean. By breaking down phytoplankton cells, viruses reduce the size and sinking rate of particulate organic carbon (POC) (Middelboe & Brussaard, 2017). Smaller particles are less likely to sink rapidly and are more prone to being remineralized in the upper ocean, where they can be respired back into carbon dioxide. This process limits the amount of carbon that reaches the deep ocean and is sequestered for long periods (Tranvik et al., 2009). Overall, this viral lysis of phytoplankton plays a significant role in regulating the distribution of carbon in marine ecosystems.
8. **Reducing Algal Bloom:** Viruses play a significant role in regulating harmful algal blooms (HABs), which can have severe ecological and economic consequences. These blooms, often caused by species like dinoflagellates and cyanobacteria, can produce toxins harmful to marine life and humans, disrupt ecosystems, and lead to substantial economic losses in fisheries and tourism (Tong & Xu, 2024). When viruses infect algal cells, they initiate a lytic cycle, leading to the destruction of the host cells. This process releases cellular contents, including organic matter and nutrients, back into the environment. By lysing algal cells, viruses effectively reduce the population density of the algae, preventing the bloom from reaching levels that could cause significant harm (Vincent et al., 2021). The impact of viral lysis on algal blooms is multifaceted. Firstly, it helps control the size and duration of the bloom, ensuring that it does not become too large or persist for too long. This regulation is crucial because extended blooms can reduce the oxygen levels in the water, killing the marine life (Kuhlisch et al., 2021). Secondly, the release of organic matter and nutrients from lysed cells can stimulate the growth of other microorganisms, including bacteria, which can further influence the dynamics of the bloom and the surrounding ecosystem.

4.4 METHODS TO STUDY VIRUSES

Studying marine viruses involves a variety of methods which gives unique insight into the abundance, diversity, and ecological roles (Figure 4.1):

4.4.1 Microscopy Techniques

- TEM uses a beam of electrons directed through an ultra-thin sample to construct a picture of the internal structure of the virus. It affords a resolution high enough to permit study of minute details such as the structure of nucleic acids and viral proteins.
- SEM is presented as an electron beam-operated microscoping attempt that produces detailed images showing the surface of a virus only. Examining surface natural features and morphological properties of a virus is another successful use of SEM.

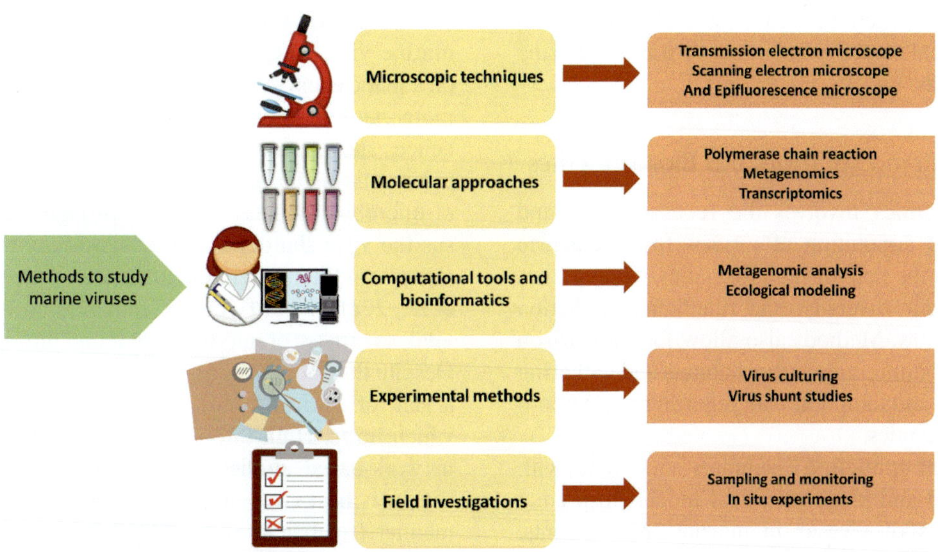

FIGURE 4.1 Different methods to study marine viruses.

Epifluorescence microscopy uses fluorescent dyes binding selectively to viral nucleic acids; thus, under specific wavelengths of light, viruses can be seen in a microscope. It has become one of the most commonly used methods for counting viruses in water samples and studying their spatial distribution, particularly in ecological studies focused upon virus abundance and transmission(Turzynski et al., 2021).

4.4.2 Molecular Approaches

- **Polymerase Chain Reaction (PCR):**
  Traditional PCR detects small quantities of viral RNA or DNA in environmental samples by amplifying specific sequences. Scientists can detect and count single viruses by having primers complementary to viral sequences.
  Quantitative PCR (qPCR), or real-time PCR, amplifies and measures viral DNA in real time, allowing accurate measurement of viral abundance in samples. The quantity of fluorescence generated during the reaction is in proportion to the quantity of DNA present, providing quantitative information on viral load.
- **Metagenomics** is sequencing all the viruses within a sample. The method does not involve virus cultivation; thus, it can be applied in the research of marine viruses that are recalcitrant to culture. Metagenomic sequencing provides extensive information on viral diversity, uncovering the presence of known and unknown viruses and allowing the understanding of viruses' functional roles in the environment through the detection of genes involved in significant processes.
- **Transcriptomics** is the study of RNA transcripts to detect viral genes expressed in the environment. RNA sequencing allows scientists to detect which genes are turned on or off in response to environmental cues. This helps in a better understanding of viruses' functional roles and interactions with host animals and the impact of viruses on marine ecosystems.

4.4.3 Computational Tools and Bioinformatics

- **Metagenomics** involves the reconstruction and functional annotation of viruses from sequence data with the help of bioinformatics. Virus detection, genetic diversity, and functional prediction would follow. Methods also allow for comparison of viral communities across habitats and gathering of spatial and temporal data regarding viral population dynamics.
- **Ecological modeling** integrates viral data with ecological and biogeochemical models to predict the effects of viruses in marine environments. Models can predict how viral infections can alter nutrient cycling, microbial community structure, and ecosystem health. The incorporation of viral data into the models gives scientists a reasonable understanding of the role viruses play in the biogeochemical cycles in the ocean.

4.4.4 Experimental Methods

- **Virus Cultivation:** Scientists study virus-host interaction, life cycle, infection process, and host influence in the laboratory setting.
- **Viral Shunt Studies:** Viral shunt, which explains how viruses affect nutrient cycling in oceanic ecosystems by lysis of host cells and release of organic matter, presents an integrated perspective of their effect on energy and matter flux.

4.4.5 Field Investigations

- Regular water sampling across different marine habitats enables one to monitor viral abundance and diversity over time to enhance understanding of the reaction of viral populations to environmental change.
- Ocean experiments can enable one to study viral dynamics in their natural environment.

These diverse methodologies collectively contribute to a comprehensive understanding of marine viruses, including their diversity, ecological roles, and impacts on marine ecosystems.

4.5 FUTURE APPROACHES AND CHALLENGES

The impact of climate change on host-virus dynamics is crucial, as it will likely alter species habitats and affect the location and magnitude of viral infections in the ocean (Mugimba et al., 2021). Long-term studies using specific viral biomarkers are essential for monitoring changes, such as tracking melting ice as a new reservoir for ancient marine viruses. Quantifying viral infection and its impact on aquatic ecosystems is a significant challenge that requires more natural environment measurements. Although evidence shows low infection levels in cyanobacteria or eukaryotic algae in the ocean, the ecological significance of microbial biomass recycling through the viral shunt versus the viral shuttle, and their sensitivity to environmental stress conditions, remain unanswered questions (Dittami et al., 2021). Identifying the authentic hosts of microbes is vital, as viruses of microbes are found throughout the ocean. Despite the abundance of genomic information, there is still a lack of in-depth understanding of which viruses infect which hosts. High-throughput single-cell technologies and network-based studies hold great potential for discovering new host-virus interactions in a culture-independent manner (Lauber & Seitz, 2022) (Table 4.1). Exploring the

TABLE 4.1
Discovery of Key Human and Marine-Associated Viruses Relevant to Host–Virus Dynamics Studies

Virus	Year of Discovery	Reference
Coxsackievirus (members of the enterovirus group belonging to Picornaviridae)	1947	Dalldorf and Sickles (1948)
Hepatitis viruses	1950	Bradley (1992)
Norwalk-like viruses	1968	Gerba et al. (1996)
Rotavirus (belonging to Reoviridae)	1973	Gerba et al. (1996)
Enteroviruses	1979	Gerba et al. (1979)
Adenoviruses	1998	Pina et al. (1998)
HAV	1998	Pina et al. (1998)
SRSV group I	1998	Green et al. (1998)
Reoviruses	2001	Aulicino et al. (2001)

mechanisms of viral resistance in phytoplankton populations is important, as diverse populations can regrow after viral-induced lysis. However, the molecular mechanisms underlying this resistance are still unknown, and none of the known microbial defense systems have been characterized in the marine environment (Zhang et al., 2022). Developing single-cell approaches can provide novel insights into gene function and the physiology of virocells, as well as help elucidate the complex diversity of viral life cycles. Single-cell methods can also help identify the source of resistance when it emerges in a rare subpopulation and enable mapping and quantifying different states and outcomes of infection.

4.6 MARINE VIRUSES: IMPACT OF CLIMATE CHANGE

The influx of atmospheric and oceanic CO_2 is significantly changing key abiotic conditions in marine environments, including temperature patterns, circulation, stratification, and ocean pH (Zhang et al., 2021b). Temperature, salinity, mixing patterns, and nutrient availability are examples of physical and chemical characteristics that are critical in determining how viruses behave in marine environments. According to regional research, the abundance of virioplankton gradually declines from seas that are productively coastal to those that are oligotrophic offshore (Piedade et al., 2024). Similar to this, two factors influence the distribution of benthic viruses: the flux of vertical particles and the formation of parasitic primary, indicating a possible connection between trophic status and viriobenthos. Furthermore, the productivity of water bodies is positively correlated with viral abundance and the role viruses play in prokaryotic mortality; environments with significant eutrophication have the largest proportion of infected cells (Lemieux et al., 2022). This demonstrates the strong beneficial relationships that exist between bacterial respiration and viral generation. The abundance of marine viruses is dependent on the availability of their hosts. Variations in the population size, metabolic activity, and replication rate of prokaryotic hosts will influence viral abundance. While correlation analyses have established that viral abundance in marine ecosystems is closely linked to the abundance of potential host cells, our understanding of the factors regulating viral abundance over extensive spatial scales (thousands of kilometers) remains limited. Wilhelm et al. (2003) observed that the significant variability in viral abundance in surface waters along a ~5,000 km latitudinal transect in the southeastern Pacific Ocean may be influenced by variations in sunlight regimes (which affect viral infectivity) and oceanographic features (which impact viral transport and residence time).

Ongoing climate change has been projected to increase the sea-surface temperatures, with the most significant rises in oceans (around 4°C–6°C). These temperature changes significantly impact microbial communities in ocean. Although the data vary, there is observation of increase in prokaryotic heterotrophic production with temperature up to about 15°C (Khotimchenko & Shchelkanov, 2024). This increase is limited by a corresponding rise in respiration, which reduces growth efficiency by about 2.5%/1°C increase in temperature. As viral replication and life cycles are closely tied to host metabolism, temperature increases are likely to affect virus-cell interactions. For instance, in prokaryotes, higher growth rates lead to shorter lytic cycles and larger burst sizes (Buscaglia et al., 2024), resulting in increased virus production. Although evidence is limited, natural systems have shown that higher production rates correlate with larger burst sizes. Evidence has suggested that there will be a significant impact on the viral population due to the rise in sea-surface temperatures. During one temperature effect in temperate-open oceans, a slight temperature increase led to a doubling of viral abundance (Locke et al., 2022). This indicated that virioplankton distribution is more or less influenced by more complex factors than temperature alone. While during one observation there was detection of a positive relationships between viral abundance and temperature, a generalized global analysis shows a general decrease in viral population with increase in temperature. The patterns of viriobenthos abundance differed from those of water-column data, but the limited data set prevents firm conclusions (Li et al., 2021). The inverse relationship between virioplankton abundance and temperature suggests that latitudinal changes, affecting sunlight regimes and trophic characteristics, can cascade to influence host growth rates and viral infectivity (Rahlff et al., 2023). Multivariate analyses from the Atlantic Ocean indicated that temperature significantly controls virioplankton distribution in the Sargasso Sea but not in the northeastern Atlantic. Additionally, recent experimental studies showed that short-term temperature changes do not directly affect viral abundance.

4.7 POTENTIAL APPLICATIONS OF MARINE VIRUSES

Marine viruses have different applications in different sectors (Figure 4.2).

1. **The pharmaceutical industry** investigates marine viruses since they possess unique genetics and structure. They serve to influence several focal points:
   - **Antiviral Drug Development:** The proteins and enzymes from marine viruses carry unique properties that can lead to the creation of novel antiviral drugs. Such agents may perhaps narrow their target to specific viral infections, providing new cures for diseases that resist current treatment modalities (Sukmarini, 2022).
   - **Exploration of the Antibacterial Agents:** The enzymes and metabolites of marine viruses allow for antibacterials to be produced. These have shown some promise against drug-resistant bacteria, thus to treat MLS diseases (Sansone et al., 2020).
   - **Investigating Anticarcinogenic Drug:** Although chemists believe it may be useful in treating malignant tumors, marine viruses may possess all of those desirable chemical properties. Some can inhibit cancer cell growth while others may induce apoptosis (programmed cell death) and are thereby seen as promising potential anticancer agents.
   - **Drug Delivery:** Marine viruses can be developed that deliver drugs specifically to various cells. This strategized approach has the potential to increase the effectiveness of treatment while decreasing side effects by making sure drugs attain the target cells, causing no harm to healthy tissues (Santos et al., 2020).
   - **Investigation of Immunomodulatory Agents:** Marine viruses possess the ability to enhance immune responses or to suppress inflammation (Al-Khayri et al., 2022). These features provide an opportunity to develop therapies for autoimmune disorders and inflammatory diseases.
2. **Environmental Remediation:** The enrichment of aquatic systems with nutrients, particularly phosphorus (P), can lead to eutrophication, a widespread issue affecting water quality. Effluent wastewater containing phosphorus must be treated to remove it before being returned to the environment (Aliko et al., 2022). Current technologies for reducing phosphorus levels in wastewater include chemical precipitation, physical, and biological methods. One biological process, enhanced biological phosphorus removal (EBPR), relies on microorganisms that accumulate large amounts of phosphorus intracellularly. However, this process can deteriorate under certain conditions. An alternative approach for phosphorus remediation may be found in marine phages, including cyanophages, which are abundant and diverse components of marine viral communities (Ru et al., 2023). Recent studies have discovered previously unrecognized genes within marine phages that are involved in key cellular processes. For example, some cyanophages have been found to contain a gene encoding a high-affinity phosphate-binding protein (pstS), which enables them to improve phosphorus uptake in cyanobacteria. This ability to enhance host phosphorus acquisition in phosphorus-limited

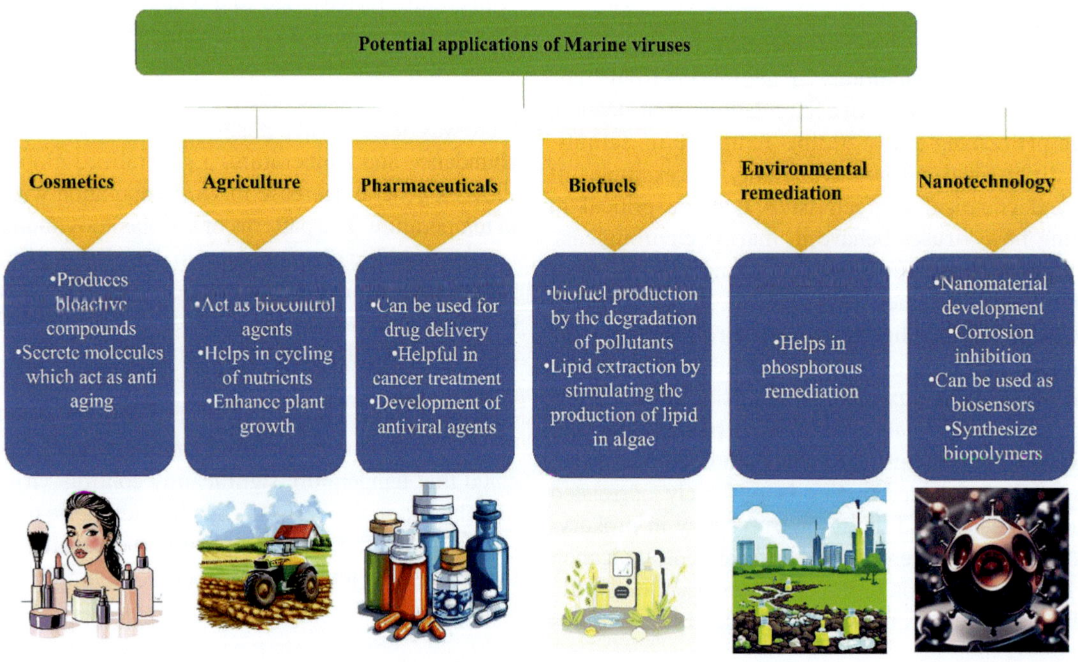

FIGURE 4.2 Potential applications of marine viruses.

environments could provide significant advantages to these phages. The high affinity of this protein for phosphorus suggests potential applications in environmental biotechnology.

3. **Agriculture:**
    - **Soil Health and Nutrient Cycling:** Marine viruses, especially bacteriophages (viruses that infect bacteria), play a significant role in nutrient cycling. When bacteriophages infect and lyse bacterial cells, they release organic matter and nutrients back into the environment. This process, known as the "viral shunt," helps maintain the balance of microbial communities and enhances soil fertility by making nutrients more available to plants. This can be particularly beneficial in agricultural soils where maintaining nutrient levels is crucial for crop productivity.
    - **Biocontrol Agents:** Marine viruses can be harnessed as biocontrol agents to target specific bacterial pathogens that affect crops. For example, phages can be used to control bacterial diseases in plants, reducing the need for chemical pesticides. This approach is environmentally friendly and can help mitigate the development of antibiotic resistance in pathogens (Varanda et al., 2021). By using phages that are specific to harmful bacteria, farmers can protect their crops without harming beneficial microbes in the soil. The use of phages for biocontrol purposes is increasingly being embraced across various sectors such as agriculture and food production. A few noteworthy instances include:
    - **Controlling Plant Pathogens:** Phages are utilized to combat bacterial pathogens that harm crops. For instance, they have been effectively employed to manage bacterial spot disease in tomatoes and peppers caused by *Xanthomonas* species. By doing so, phages aid in minimizing crop losses and reducing the reliance on chemical pesticides.
    - **Food Safety:** Within the food industry, phages play a crucial role in enhancing food safety by targeting foodborne pathogens. For example, they are employed to control *Listeria monocytogenes* in ready-to-eat foods such as deli meats and cheeses. This helps in lowering the risk of foodborne illnesses and elevating the safety standards of food products.
    - **Aquaculture:** Phages find application in aquaculture for the control of bacterial infections in fish and shellfish. They have been utilized, for instance, to manage infections caused by *Vibrio species* in shrimp farming (Hashiro & Yasueda, 2022). This contributes to the maintenance of healthy aquaculture systems and reduces reliance on antibiotics.
    - **Post-Harvest Protection:** Phages can be deployed to safeguard harvested fruits and vegetables from bacterial spoilage. The application of phages on the surface of produce can extend the shelf life and mitigate post-harvest losses, particularly beneficial for perishable items like leafy greens and berries (Song et al., 2023).
    - **Soil Health:** Introduction of phages into the soil can help combat specific bacterial pathogens affecting plant roots, thereby supporting a healthy soil microbiome and bolstering plant growth. For instance, phages have been utilized to control *Ralstonia solanacearum*, the causative agent of bacterial wilt in crops like potatoes and tomatoes.
    - **Bioremediation:** Phages are being explored for their potential in bioremediation, where they aid in degrading harmful bacteria in contaminated soils and water. Such an application holds promise for improving environmental health and promoting sustainable agricultural practices.
    - **Enhancing Plant Growth:** Healthy microbial communities are essential for plant growth and resilience. Marine viruses can indirectly promote plant growth by regulating the populations of soil bacteria. When viruses lyse bacterial cells, they release nutrients that can be taken up by plants. Additionally, by controlling harmful bacterial populations, viruses can help maintain a healthy microbial balance, which supports plant health and growth. This can lead to more robust crops and higher yields.
    - **Research and Development:** Advances in molecular biology and DNA sequencing technologies have opened new avenues for studying marine viruses and their interactions with microbial communities. Researchers are exploring how marine viruses can be applied in agriculture to improve soil health, control plant diseases, and enhance crop productivity (Song et al., 2023). For instance, metagenomic studies allow scientists to identify and characterize viral communities in different environments, providing insights into their potential applications.
    - **Climate Change Considerations:** Climate change can impact the dynamics between marine viruses and their hosts. Changes in temperature, pH, and other environmental factors can influence viral activity and the composition of microbial communities. Understanding these dynamics is crucial for developing effective strategies to use marine viruses in agriculture, especially in the context

of a changing climate. Researchers are investigating how climate change might affect the efficacy of phage-based biocontrol methods and nutrient cycling processes.

Potential Challenges and Future Direction: While the potential benefits of using marine viruses in agriculture are promising, there are challenges to consider. These include ensuring the specificity of phages to target only harmful bacteria, avoiding unintended ecological impacts, and developing scalable methods for phage production and application. Ongoing research aims to address these challenges and unlock the full potential of marine viruses in sustainable agriculture.

4. **Biofuels:** Marine viruses may prove advantageous in the production of biofuels by interacting with algae. The subsequent aspects illustrate how these viruses can be made useful in the production of biofuels:
    - **Algal Biomass Production:** Some virus species infect algae and lead to the release of organic matter and vital nutrients into the oceanic ecosystem (Allen et al., 2022). This further promotes the setting of other algal communities with essential materials available, leading to the increase in total biomass—important in biofuel production.
    - **Lipid Extraction Enhancement:** Some marine viruses induce lipid accumulation in algae and hence are useful for biodiesel production. Researchers can increase algal lipid content by careful manipulation of viral infections, allowing a more effective and cheaper extraction process.
    - **Genetic Engineering:** Marine viruses can serve as vectors for genetic engineering to enhance the biofuel-producing capabilities of algae. By introducing specific genes into algal genomes, scientists can optimize metabolic pathways to increase the yield of biofuel precursors such as lipids and carbohydrates.
    - **Bioreactor Optimization:** In controlled environments like bioreactors, marine viruses can be employed to manage algal populations and prevent the overgrowth of undesirable species, thus maintaining a stable and productive algal culture—an essential aspect of consistent biofuel production.
    - **Bioremediation and Biofuel Production:** Marine viruses can also contribute to bioremediation by aiding in the degradation of pollutants in contaminated water bodies. The organic matter released during this process can be a substrate for algal growth, linking bioremediation efforts with biofuel production.
    - **Climate Change Mitigation:** By promoting the growth of algae, marine viruses can aid in carbon sequestration, as algae absorb carbon dioxide during photosynthesis. Their increased growth can help mitigate the effects of climate change, and the harvested algal biomass can be converted into biofuels, providing a renewable energy source.

    Challenges and Future Directions: While the potential of marine viruses in biofuel production is significant, there are challenges to address, such as ensuring the specificity of viral infections to target desired algal species, avoiding unintended ecological impacts, and developing scalable methods for viral application. Ongoing research aims to overcome these challenges and fully harness the potential of marine viruses in sustainable biofuel production.

5. **Cosmetics:** Because of their distinctive properties and prospective applications, marine viruses are becoming increasingly popular in the cosmetic industry.
    - **Bioactive Compounds:** Marine viruses can yield bioactive compounds with skincare benefits, exhibiting antioxidant, anti-inflammatory, and antimicrobial activities that are beneficial to the skin (Ding et al., 2022).
    - **Anti-Aging:** Certain marine viruses yield enzymes and proteins that help in the production of collagen and elastin, which are necessary for the elasticity of skin and wrinkle prevention.
    - **Natural Ingredients:** The shift toward cosmetics featuring natural and eco-friendly ingredients has generated interest related to marine viruses as a promising source of new ingredients, as natural ones are less irritating and more skin-compatible than synthetic chemicals (Zhang et al., 2021a).
    - **Biotechnology:** Recently developed biotechnology techniques have made it feasible to extract and purify together the biologically active compounds from marine viruses, which means that their applications may be very attractive in development as leverage for enhancing the action of creams, lotions, and other skincare products.
    - **Sustainability:** Using marine viruses is in line with the movement toward sustainable and eco-friendly products. Marine viruses can be easily harvested without compromising marine habitats, which makes them a sustainable source of bioactive ingredients.
    - **Research and Development:** Additional research must now be done to get a more thorough knowledge of the potential of marine viruses in cosmetics, and work continues

while new viruses are discovered and their characteristics are understood to yield ones useful for topical skincare. For example, in the past ten years, researchers showed that a virus called EhV and its genome contains a cluster of at least seven genes responsible for producing essential components of sphingolipid biosynthesis, leading to the production of ceramide. This gene cluster appears to be active in enriching cell membranes. Ceramides play a role in regulating important processes such as aging, cell differentiation, apoptosis, and cell cycle arrest. They are also crucial structural lipids in the skin, hair, and nails of both humans and animals, serving as a vital epidermal barrier against water loss. Because of their significance, there has been notable interest in these compounds as active ingredients in the cosmetic industry.

While the use of marine viruses in cosmetics is still emerging, it offers great promise for innovative and effective skincare products.

6. **Nanotechnology:** Marine viruses' diverse properties and interactions are drawing material sciences' attention. The discipline has diverse interesting applications and research domains:

   Nanotechnology:
   - **Nanomaterial Development:** Marine viruses have potential for nanomaterial development with diverse properties: nanowire and nanotube engineering for electronic, sensor, and energy storage devices.
   - **Virus-Templated Synthesis:** Marine viruses' morphology can be used as a template for nanomaterial synthesis and will be used in the preparation of other ordered and uniform nanostructures.

   Biocorrosion and Biofouling:
   - **Corrosion Inhibition:** Sea viruses may alter the biofilm on surfaces and, in turn, could affect the corrosion of materials in seawaters.
   - **Fouling Control:** Marine viruses can effect biofouling control, thus preserving the integrity and functions of marine vessels and structures.

   Biopolymers:
   - **Sustainable Materials:** Marine viruses can produce biopolymers that are green and biodegradable, enabling their use in packaging, medical devices, and other sustainable materials.

   Environmental Sensors:
   - **Biosensors:** Engineered marine viruses can work on the detection of environmental pollutants, thus enabling real-time monitoring of water quality for early warnings on contamination.

   Bioremediation:
   - **Pollutant Degradation:** Marine viruses can help bioremediation by assisting in cleaning up oil spills and other contaminants by infecting and lysing bacteria that break down toxic chemicals. Though the exploration of marine viruses in material sciences is still in its early stages, the potential applications appear to be extensive and promising. Ongoing research efforts continue to reveal new ways of leveraging the capabilities of these diminutive yet powerful organisms.

4.8 CONCLUSION

In conclusion, marine viruses are fundamental to the ocean's ecological and biogeochemical processes. Far from being mere agents of disease, they play crucial roles in nutrient cycling, genetic exchange, and overall marine productivity. The viral shunt exemplifies how viruses recycle essential nutrients, enhancing the biological pump's efficiency and sustaining the microbial loop. This process is vital for maintaining marine life's balance and supporting intricate food webs from microorganisms to large marine animals. Understanding marine viruses' distribution patterns and interactions with environmental factors is essential. Stable environments like the Sargasso Sea show more predictable viral distributions compared to dynamic systems like the North Atlantic, highlighting the need for extensive research across different oceanic realms. As climate change continues to impact sea-surface temperatures and other oceanographic conditions, the role of viruses in marine ecosystems may become even more critical. Their ability to mediate genetic exchange and adapt to changing environments could influence the resilience and evolution of marine microbial communities. Overall, insights from studying marine viruses not only enhance our understanding of oceanic ecosystems but also open up potential applications in biotechnology. By leveraging the unique properties of these viruses, we can explore innovative solutions for environmental and medical challenges. This conclusion encapsulates the document's key themes and findings, emphasizing the significance of marine viruses and their diverse roles in the ocean.

REFERENCES

Aliko, V., Multisanti, C. R., Turani, B., & Faggio, C. (2022). Get rid of marine pollution: Bioremediation an innovative, attractive, and successful cleaning strategy. *Sustainability*, *14*(18), 18. https://doi.org/10.3390/su141811784

Al-Khayri, J. M., Asghar, W., Khan, S., Akhtar, A., Ayub, H., Khalid, N., Alessa, F. M., Al-Mssallem, M. Q., Rezk, A. A.-S., & Shehata, W. F. (2022). Therapeutic potential of marine bioactive peptides against human immunodeficiency virus: Recent evidence, challenges, and future trends. *Marine Drugs*, *20*(8), 8. https://doi.org/10.3390/md20080477

Allen, M. J., Cicéron, F., & Monier, A. (2022). The potential of nature's unseen industrious heroes: Marine viruses. *The Biochemist*, *44*(2), 18–21. https://doi.org/10.1042/bio_2022_108

Almutairi, A., Suttle, C. A., & Gustavsen, J. A. (2025). The hypersaline northwestern Arabian Gulf contains a phylogenetically diverse and highly uneven community of viruses related to cyanophages and pelagiphages. *Aquatic Microbial Ecology*, *91*, 1–14. https://doi.org/10.3354/ame02014

Al-Shayeb, B., Sachdeva, R., Chen, L.-X., Ward, F., Munk, P., Devoto, A., Castelle, C. J., Olm, M. R., Bouma-Gregson, K., Amano, Y., He, C., Méheust, R., Brooks, B., Thomas, A., Lavy, A., Matheus-Carnevali, P., Sun, C., Goltsman, D. S. A., Borton, M. A., … Banfield, J. F. (2020). Clades of huge phages from across Earth's ecosystems. *Nature*, *578*(7795), 425–431. https://doi.org/10.1038/s41586-020-2007-4

An, L., Liu, X., Wang, J., Xu, J., Chen, X., Liu, X., Hu, B., Nie, Y., & Wu, X.-L. (2024). Global diversity and ecological functions of viruses inhabiting oil reservoirs. *Nature Communications*, *15*(1), 6789. https://doi.org/10.1038/s41467-024-51101-6

Aulicino, F. A., Orsini, P., Carere, M., & Mastrantonio, A. (2001). Bacteriological and virological quality of seawater bathing areas along the Tyrrhenian coast. *International Journal of Environmental Health Research*, *11*(1), 5–11. https://doi.org/10.1080/0960312002001546

Bradley, D. W. (1992). Hepatitis E: Epidemiology, aetiology and molecular biology. *Reviews in Medical Virology*, *2*(1), 19–28. https://doi.org/10.1002/rmv.1980020104

Braga, L. P. P., Orland, C., Emilson, E. J. S., Fitch, A. A., Osterholz, H., Dittmar, T., Basiliko, N., Mykytczuk, N. C. S., & Tanentzap, A. J. (2022). Viruses direct carbon cycling in lake sediments under global change. *Proceedings of the National Academy of Sciences of the Unites States of America*, *119*(41), e2202261119. https://doi.org/10.1073/pnas.2202261119

Breitbart, M. (2012). Marine viruses: Truth or dare. *Annual Review of Marine Science*, *4*(2012), 425–448. https://doi.org/10.1146/annurev-marine-120709-142805

Brum, J. R., & Sullivan, M. B. (2015). Rising to the challenge: Accelerated pace of discovery transforms marine virology. *Nature Reviews Microbiology*, *13*(3), 147–159. https://doi.org/10.1038/nrmicro3404

Buscaglia, M., Iriarte, J. L., Schulz, F., & Díez, B. (2024). Adaptation strategies of giant viruses to low-temperature marine ecosystems. *The ISME Journal*, *18*(1), wrae162. https://doi.org/10.1093/ismejo/wrae162

Dalldorf, G., & Sickles, G. M. (1948). An unidentified, filtrable agent isolated from the feces of children with paralysis. *Science*, *108*(2794), 61–62. https://doi.org/10.1126/science.108.2794.61

Danovaro, R., Corinaldesi, C., Dell'Anno, A., Fuhrman, J. A., Middelburg, J. J., Noble, R. T., & Suttle, C. A. (2011). Marine viruses and global climate change. *FEMS Microbiology Reviews*, *35*(6), 993–1034. https://doi.org/10.1111/j.1574-6976.2010.00258.x

Ding, J., Wu, B., & Chen, L. (2022). Application of marine microbial natural products in cosmetics. *Frontiers in Microbiology*, *13*. https://doi.org/10.3389/fmicb.2022.892505

Dittami, S. M., Arboleda, E., Auguet, J.-C., Bigalke, A., Briand, E., Cárdenas, P., Cardini, U., Decelle, J., Engelen, A. H., Eveillard, D., Gachon, C. M. M., Griffiths, S. M., Harder, T., Kayal, E., Kazamia, E., Lallier, F. H., Medina, M., Marzinelli, E. M., Morganti, T. M., … Not, F. (2021). A community perspective on the concept of marine holobionts: Current status, challenges, and future directions. *PeerJ*, *9*, e10911. https://doi.org/10.7717/peerj.10911

Dominguez-Huerta, G., Zayed, A. A., Wainaina, J. M., Guo, J., Tian, F., Pratama, A. A., Bolduc, B., Mohssen, M., Zablocki, O., Pelletier, E., Delage, E., Alberti, A., Aury, J.-M., Carradec, Q., da Silva, C., Labadie, K., Poulain, J., Tara Oceans Coordinators, Bowler, C., … Sullivan, M. B. (2022). Diversity and ecological footprint of Global Ocean RNA viruses. *Science*, *376*(6598), 1202–1208. https://doi.org/10.1126/science.abn6358

Fogg, P. C. M. (2025). Gene transfer agents: The ambiguous role of selfless viruses in genetic exchange and bacterial evolution. *Molecular Microbiology*, *123*(2), 124–131. https://doi.org/10.1111/mmi.15251

Gerba, C. P., Goyal, S. M., LaBelle, R. L., Cech, I., & Bodgan, G. F. (1979). Failure of indicator bacteria to reflect the occurrence of enteroviruses in marine waters. *American Journal of Public Health*, *69*(11), 1116–1119. https://doi.org/10.2105/ajph.69.11.1116

Gerba, C. P., Rose, J. B., & Haas, C. N. (1996). Sensitive populations: Who is at the greatest risk? *International Journal of Food Microbiology*, *30*(1–2), 113–123. https://doi.org/10.1016/0168-1605(96)00996-8

Green, J., Henshilwood, K., Gallimore, C. I., Brown, D. W. G., & Lees, D. N. (1998). A nested reverse transcriptase pcr assay for detection of small round-structured viruses in environmentally contaminated molluscan shellfish. *Applied and Environmental Microbiology*, *64*(3), 858–863. https://doi.org/10.1128/AEM.64.3.858-863.1998

Hashiro, S., & Yasueda, H. (2022). RNA interference-based pesticides and antiviral agents: Microbial overproduction systems for double-stranded RNA for applications in agriculture and aquaculture. *Applied Sciences*, *12*(6), 6. https://doi.org/10.3390/app12062954

Jian, H., Yi, Y., Wang, J., Hao, Y., Zhang, M., Wang, S., Meng, C., Zhang, Y., Jing, H., Wang, Y., & Xiao, X. (2021). Diversity and distribution of viruses inhabiting the deepest ocean on Earth. *The ISME Journal*, *15*(10), 3094–3110. https://doi.org/10.1038/s41396-021-00994-y

Jin, M., Gong, Y., Geslin, C., He, T., & Zhang, X. (2024). Editorial: The role of viruses in marine environments. *Frontiers in Microbiology*, *15*. https://doi.org/10.3389/fmicb.2024.1437050

Khotimchenko, Y. S., & Shchelkanov, M. Y. (2024). Viruses of the ocean: On the shores of the aqua incognita. Horizons of taxonomic diversity. *Russian Journal of Marine Biology*, *50*(1), 1–24. https://doi.org/10.1134/S106307402401005X

Kuhlisch, C., Schleyer, G., Shahaf, N., Vincent, F., Schatz, D., & Vardi, A. (2021). Viral infection of algal blooms leaves a unique metabolic footprint on the dissolved organic matter in the ocean. *Science Advances*, *7*(25), eabf4680. https://doi.org/10.1126/sciadv.abf4680

Lauber, C., & Seitz, S. (2022). Opportunities and challenges of data-driven virus discovery. *Biomolecules*, *12*(8), 8. https://doi.org/10.3390/biom12081073

Lemieux, A., Colby, G. A., Poulain, A. J., & Aris-Brosou, S. (2022). Viral spillover risk increases with climate change in High Arctic lake sediments. *Proceedings of the Royal Society B: Biological Sciences*, *289*(1985), 20221073. https://doi.org/10.1098/rspb.2022.1073

Li, Z., Pan, D., Wei, G., Pi, W., Zhang, C., Wang, J.-H., Peng, Y., Zhang, L., Wang, Y., Hubert, C. R. J., & Dong, X. (2021). Deep sea sediments associated with cold seeps are a subsurface reservoir of viral diversity. *The ISME Journal*, *15*(8), 2366–2378. https://doi.org/10.1038/s41396-021-00932-y

Locke, H., Bidle, K. D., Thamatrakoln, K., Johns, C. T., Bonachela, J. A., Ferrell, B. D., & Wommack, K. E. (2022). Chapter Two – Marine viruses and climate change: Virioplankton, the carbon cycle, and our future ocean. In M. J. Roossinck (Ed.), *Advances in Virus Research* (Vol. 114, pp. 67–146). Academic Press. https://doi.org/10.1016/bs.aivir.2022.09.001

Luo, E., Leu, A. O., Eppley, J. M., Karl, D. M., & DeLong, E. F. (2022). Diversity and origins of bacterial and archaeal viruses on sinking particles reaching the abyssal ocean. *The ISME Journal*, *16*(6), 1627–1635. https://doi.org/10.1038/s41396-022-01202-1

Middelboe, M., & Brussaard, C. P. D. (2017). Marine viruses: Key players in marine ecosystems. *Viruses*, *9*(10), 10. https://doi.org/10.3390/v9100302

Mugimba, K. K., Byarugaba, D. K., Mutoloki, S., Evensen, Ø., & Munang'andu, H. M. (2021). Challenges and solutions to viral diseases of finfish in marine aquaculture. *Pathogens*, *10*(6), 6. https://doi.org/10.3390/pathogens10060673

Piedade, G. J., Schön, M. E., Lood, C., Fofanov, M. V., Wesdorp, E. M., Biggs, T. E. G., Wu, L., Bolhuis, H., Fischer, M. G., Yutin, N., Dutilh, B. E., & Brussaard, C. P. D. (2024). Seasonal dynamics and diversity of Antarctic marine viruses reveal a novel viral seascape. *Nature Communications*, *15*(1), 9192. https://doi.org/10.1038/s41467-024-53317-y

Pina, S., Puig, M., Lucena, F., Jofre, J., & Girones, R. (1998). Viral pollution in the environment and in shellfish: Human adenovirus detection by PCR as an index of human viruses. *Applied and Environmental Microbiology*, *64*(9), 3376–3382. https://doi.org/10.1128/aem.64.9.3376-3382.1998

Rahlff, J., Esser, S. P., Plewka, J., Heinrichs, M. E., Soares, A., Scarchilli, C., Grigioni, P., Wex, H., Giebel, H.-A., & Probst, A. J. (2023). Marine viruses disperse bidirectionally along the natural water cycle. *Nature Communications*, *14*(1), 6354. https://doi.org/10.1038/s41467-023-42125-5

Rocha, E. P. C., & Bikard, D. (2022). Microbial defenses against mobile genetic elements and viruses: Who defends whom from what? *PLOS Biology*, *20*(1), e3001514. https://doi.org/10.1371/journal.pbio.3001514

Rohwer, F., & Thurber, R. V. (2009). Viruses manipulate the marine environment. *Nature*, *459*(7244), 207–212. https://doi.org/10.1038/nature08060

Roux, S., Adriaenssens, E. M., Dutilh, B. E., Koonin, E. V., Kropinski, A. M., Krupovic, M., Kuhn, J. H., Lavigne, R., Brister, J. R., Varsani, A., Amid, C., Aziz, R. K., Bordenstein, S. R., Bork, P., Breitbart, M., Cochrane, G. R., Daly, R. A., Desnues, C., Duhaime, M. B., … Eloe-Fadrosh, E. A. (2019). Minimum information about an uncultivated virus genome (MIUViG). *Nature Biotechnology*, *37*(1), 29–37. https://doi.org/10.1038/nbt.4306

Ru, J., Xue, J., Sun, J., Cova, L., & Deng, L. (2023). Unveiling the hidden role of aquatic viruses in hydrocarbon pollution bioremediation. *Journal of Hazardous Materials*, *459*, 132299. https://doi.org/10.1016/j.jhazmat.2023.132299

Sansone, C., Brunet, C., Noonan, D. M., & Albini, A. (2020). Marine algal antioxidants as potential vectors for controlling viral diseases. *Antioxidants*, *9*(5), 5. https://doi.org/10.3390/antiox9050392

Santos, J. D., Vitorino, I., Reyes, F., Vicente, F., & Lage, O. M. (2020). From ocean to medicine: Pharmaceutical applications of metabolites from marine bacteria. *Antibiotics*, *9*(8), 8. https://doi.org/10.3390/antibiotics9080455

Song, H., Liu, Y., & Wang, Q. (2023). Research progress of anti-plant virus agents based on marine natural products. *Advanced Agrochem*, *2*(1), 31–38. https://doi.org/10.1016/j.aac.2022.12.001

Sousa, J. A. M. de, Pfeifer, E., Touchon, M., & Rocha, E. P. C. (2021). Causes and consequences of bacteriophage diversification via genetic exchanges across lifestyles and bacterial taxa (p. 2020.04.14.041137). bioRxiv. https://doi.org/10.1101/2020.04.14.041137

Sousa, J. M. de, Lourenço, M., & Gordo, I. (2023). Horizontal gene transfer among host-associated microbes. *Cell Host & Microbe*, *31*(4), 513–527. https://doi.org/10.1016/j.chom.2023.03.017

Steward, G. F., Culley, A. I., Mueller, J. A., Wood-Charlson, E. M., Belcaid, M., & Poisson, G. (2013). Are we missing half of the viruses in the ocean? *The ISME Journal*, *7*(3), 672–679. https://doi.org/10.1038/ismej.2012.121

Sukmarini, L. (2022). Antiviral peptides (AVPs) of marine origin as propitious therapeutic drug candidates for the treatment of human viruses. *Molecules*, *27*(9), 9. https://doi.org/10.3390/molecules27092619

Suttle, C. A. (2007). Marine viruses—Major players in the global ecosystem. *Nature Reviews Microbiology*, *5*(10), 801–812. https://doi.org/10.1038/nrmicro1750

Tong, D., & Xu, J. (2024). Element cycling by environmental viruses. *National Science Review*, *11*(12), nwae459. https://doi.org/10.1093/nsr/nwae459

Tran, P. Q., & Anantharaman, K. (2021). Biogeochemistry goes viral: Towards a multifaceted approach to study viruses and biogeochemical cycling. *mSystems*, *6*(5). https://doi.org/10.1128/msystems.01138-21

Tranvik, L. J., Downing, J. A., Cotner, J. B., Loiselle, S. A., Striegl, R. G., Ballatore, T. J., Dillon, P., Finlay, K., Fortino, K., Knoll, L. B., Kortelainen, P. L., Kutser, T., Larsen, Soren., Laurion, I., Leech, D. M., McCallister, S. L., McKnight, D. M., Melack, J. M., Overholt, E., … Weyhenmeyer, G. A. (2009). Lakes and reservoirs as regulators of carbon cycling and climate. *Limnology and Oceanography*, *54*(6 part 2), 2298–2314. https://doi.org/10.4319/lo.2009.54.6_part_2.2298

Turzynski, V., Monsees, I., Moraru, C., & Probst, A. J. (2021). Imaging techniques for detecting prokaryotic viruses in environmental samples. *Viruses*, *13*(11), 11. https://doi.org/10.3390/v13112126

Varanda, C., Félix, M. do R., Campos, M. D., & Materatski, P. (2021). An overview of the application of viruses to biotechnology. *Viruses*, *13*(10), 10. https://doi.org/10.3390/v13102073

Vincent, F., Sheyn, U., Porat, Z., Schatz, D., & Vardi, A. (2021). Visualizing active viral infection reveals diverse cell fates in synchronized algal bloom demise. *Proceedings of the National Academy of Sciences of the United States of America*, *118*(11), e2021586118. https://doi.org/10.1073/pnas.2021586118

Vincent, F., & Vardi, A. (2023). Viral infection in the ocean—A journey across scales. *PLOS Biology*, *21*(1), e3001966. https://doi.org/10.1371/journal.pbio.3001966

Wilhelm, J., Pingoud, A., & Hahn, M. (2003). Real-time PCR-based method for the estimation of genome sizes. *Nucleic Acids Research*, *31*(10), e56. https://doi.org/10.1093/nar/gng056

Zayed, A. A., Wainaina, J. M., Dominguez-Huerta, G., Pelletier, E., Guo, J., Mohssen, M., Tian, F., Pratama, A. A., Bolduc, B., Zablocki, O., Cronin, D., Solden, L., Delage, E., Alberti, A., Aury, J.-M., Carradec, Q., da Silva, C., Labadie, K.,

Poulain, J., … Sullivan, M. B. (2022). Cryptic and abundant marine viruses at the evolutionary origins of Earth's RNA virome. *Science*, *376*(6589), 156–162. https://doi.org/10.1126/science.abm5847

Zeng, Y., Zhong, X., Chen, Y., Gou, M., Yu, K., & Tang, Y.-Q. (2025). Characteristics of phages and their interactions with hosts in anaerobic reactors. *Environmental Microbiology*, *27*(1), e70040. https://doi.org/10.1111/1462-2920.70040

Zhang, J., Zeng, R., & Labes, A. (2021a). Editorial: Marine microbial-derived molecules and their potential medical and cosmetic applications. *Frontiers in Microbiology*, *12*. https://doi.org/10.3389/fmicb.2021.706152

Zhang, Q.-Y., Ke, F., Gui, L., & Zhao, Z. (2022). Recent insights into aquatic viruses: Emerging and reemerging pathogens, molecular features, biological effects, and novel investigative approaches. *Water Biology and Security*, *1*(4), 100062. https://doi.org/10.1016/j.watbs.2022.100062

Zhang, R., Weinbauer, M. G., & Peduzzi, P. (2021b). Aquatic viruses and climate change. *Current Issues in Molecular Biology*, *41*(1), 1. https://doi.org/10.21775/cimb.041.357

5 Marine Microbiomes and Biogeochemical Cycles

Awadhesh Kumar Verma[1], *Aryan Misra*[1], *Pawanpreet Kaur*[1], *Karishma Vivek Kathpalia*[1], *Tanya Singh*[1], *Nisha Shankhwar*[2], *and Neeta Raj Sharma*[1]
[1]Department of Biotechnology & Bioinformatics, School of Bioengineering & Biosciences, Lovely Professional University, India
[2]Special Centre for Nanoscience, Jawaharlal Nehru University, New Delhi, India

5.1 INTRODUCTION TO MARINE MICROBIOMES

5.1.1 Overview of Marine Microbiomes

Marine microbiomes consist of complex communities of microorganisms, including bacteria, archaea, viruses, and eukaryotic microbes, that are integral to oceanic processes. They drive biogeochemical cycles, such as carbon and nitrogen cycling, and contribute to primary production and decomposition. These microbiomes are pivotal in maintaining the stability of marine ecosystems, influencing global climate dynamics and supporting biodiversity through their interactions with other marine organisms (Table 5.1) (Vanwonterghem & Webster, 2020).

5.1.1.1 The Diversity and Ubiquity of Marine Microbes

Marine microbiomes encompass an extensive array of microorganisms, including bacteria, archaea, viruses, and single-celled eukaryotes. These microbes are omnipresent across various marine habitats, from the surface waters to the deep-sea trenches. The microbiome is key to understanding and sustaining the services that ocean ecosystems provide. The marine microbiome—an ensemble of microscopic organisms that inhabit water columns, sediments, and aquatic organisms—contains members spanning in size from viruses of a few tens of nanometers to metazoans of several centimeters. Bacteria are generally less than 1–2 μm in size except for the largest bacterium *Epulopiscium fishelsoni*, a Gram-positive species 200–800 μm long, that lives in the gut of a Red Sea fish. In the marine environment, 90% of bacteria are Gram-negative with different characteristics and the Gram-negative cell wall is better adapted for survival in the marine environment (Goodwin & Lacoursière-Roussel, 2023). Marine bacteria, such as *Prochlorococcus* and *Synechococcus*, represent a significant portion of the global photosynthetic biomass. The ubiquity and diversity of marine microbes reflect their pivotal roles in biogeochemical processes and ecosystem functions. Advances in molecular techniques, including high-throughput sequencing, have revealed an unprecedented diversity of microbial life, underscoring their ecological significance (Figure 5.1) (Goodwin & Lacoursière-Roussel, 2023).

TABLE 5.1
Major Microbial Groups in Marine Microbiomes

Microbial Group	Main Functions	Typical Habitats	Key Roles in Marine Ecosystems	Examples	References
Bacteria	Decomposition, nutrient cycling, primary production	Surface waters, sediments, hydrothermal vents	Carbon and nitrogen cycling, organic matter degradation	Prochlorococcus, Synechococcus	(Berube et al., 2018)
Archaea	Methane production, ammonia oxidation	Deep-sea, hypersaline environments, sediments	Methane cycling, nitrogen transformations	Methanococcus, Nitrosopumilus	(Pfeifer et al., 2021)
Viruses	Viral lysis of microbes, gene transfer	Present in both open ocean and coastal marine environments	Regulation of microbial populations, gene exchange	Marine bacteriophages	(Focardi et al., 2020)
Eukaryotic Microbes	Primary production, symbiotic relationships	Phytoplankton, zooplankton, coral reefs	Photosynthesis, nutrient Cycling, food web dynamics	Diatoms, Dinoflagellates, Foraminifera	(Ollison et al., 2023)

DOI: 10.1201/9781003509530-5

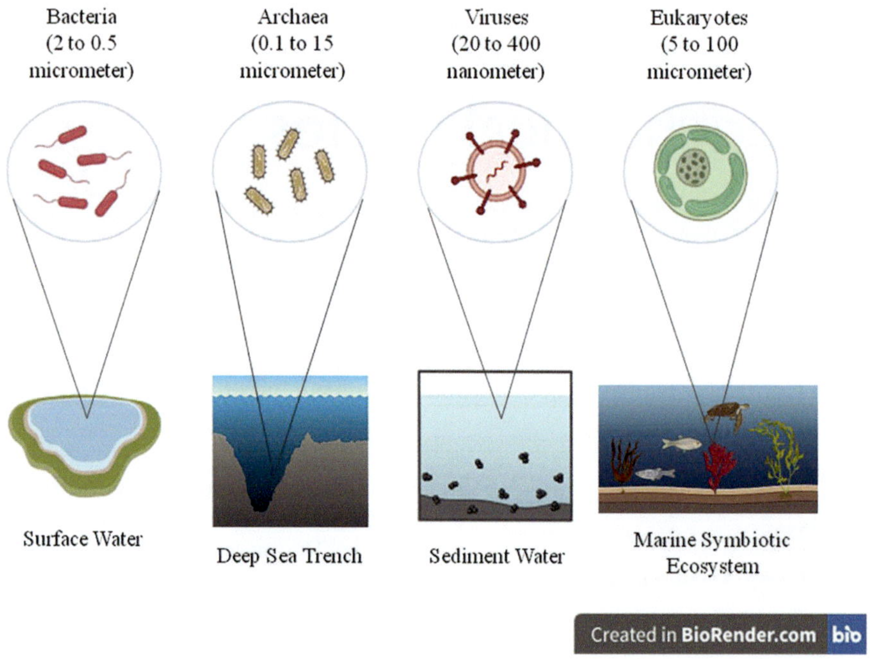

FIGURE 5.1 Marine microbiome diversity across habitats, illustrating microbial (bacteria, archaea, viruses, eukaryotes) variation in surface waters, deep-sea trenches, sediments, and symbiotic environments. (Self-developed using Biorender.com.)

5.1.1.2 Historical Perspective on Marine Microbiology

The field of marine microbiology has evolved from the early observations of microorganisms in seawater to the sophisticated genomic studies of contemporary research. Initially, the study of marine microbes was constrained by limited cultivation techniques and rudimentary microscopy. The discovery of marine bacteria and their roles in nutrient cycling marked the beginning of a broader understanding of marine microbial ecology (Adler & Dücker, 2018). The development of molecular biology techniques, such as PCR and metagenomics, has revolutionized the field, enabling the characterization of microbial communities without the need for cultivation and providing insights into their functional capacities and ecological interactions. The contributions of Certes and Regnard have been largely overlooked in the historical narratives of microbiology and marine science. However, a closer examination of their experiments is essential for understanding the origins and development of marine microbiology. Their work offers key insights into the scientific context of the late 19th century, a period when marine microbiology was taking shape as a distinct discipline (Adler & Dücker, 2018). This field emerged from the convergence of several scientific areas, including terrestrial microbial ecology, which studied microorganisms in soil environments, experimental physiology, which explored the biological functions of organisms, and the nascent field of deep-sea biology, which was just beginning to investigate life in the ocean's depths. By blending these disciplines, early marine microbiologists laid the groundwork for a new scientific field dedicated to understanding the microbial life that thrives in marine environments, from surface waters to the deep sea (Adler & Dücker, 2018).

5.1.2 Significance of Marine Microbiomes in Ocean Ecosystems

5.1.2.1 Microbial Roles in Marine Food Webs

Microbial food web organisms are at the base of the functioning of pelagic ecosystems and support the whole marine food web. They are very reactive to environmental changes and their interactions are modified in response to different productive periods such as phytoplankton bloom and non-bloom as well as contrasted climatic years (Trombetta et al., 2020). Marine microbes occupy fundamental positions in oceanic food webs, serving as primary producers, decomposers, and contributors to nutrient cycling. Phytoplankton, including cyanobacteria and microalgae, form the base of the marine food web, converting inorganic carbon into organic matter through photosynthesis. Bacteria and archaea decompose organic matter, recycling nutrients and influencing the availability of key elements such as nitrogen and phosphorus. This microbial activity supports the growth of higher trophic levels, including zooplankton, fish, and marine mammals, thereby sustaining oceanic productivity and biodiversity (Figure 5.2) (Glibert & Mitra, 2022).

5.1.2.2 Interactions with Other Marine Organisms

Marine microbes engage in complex interactions with other marine organisms, including symbiotic, parasitic, and mutualistic relationships. Symbiotic associations between

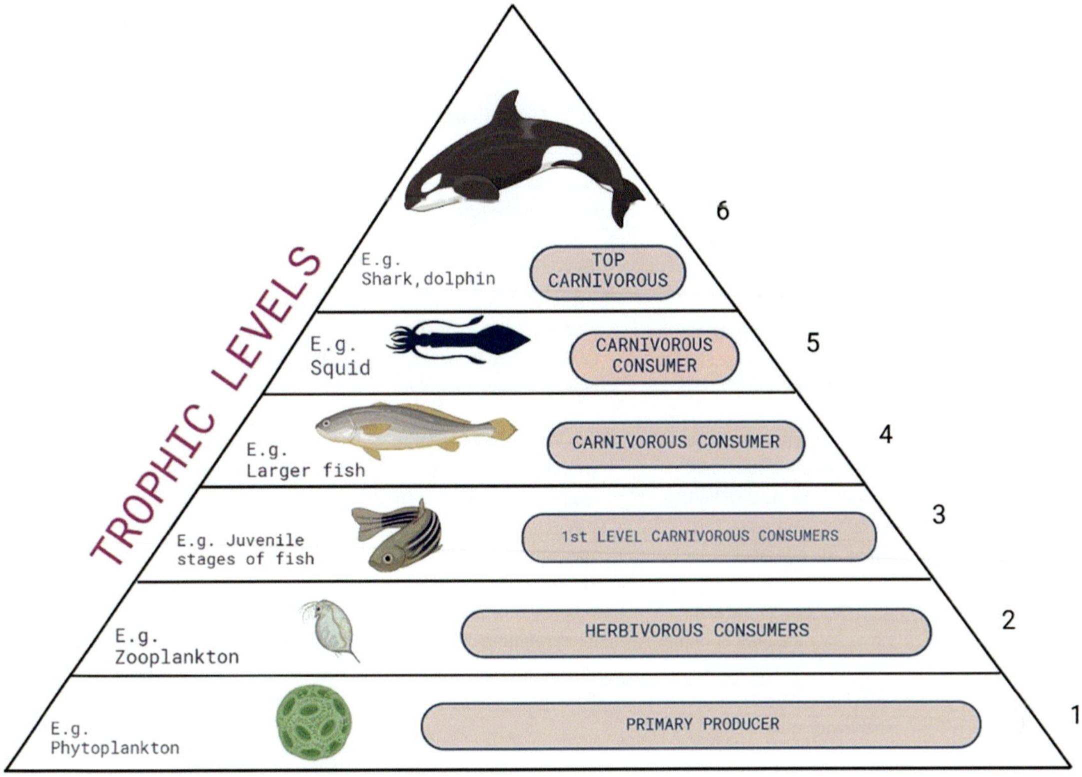

FIGURE 5.2 Marine food web with emphasis on microbial contributions: A diagrammatic food web showing the roles of marine microbes as primary producers, decomposers, and contributors to nutrient cycling. (Self-developed using Biorender.com.)

microbes and marine invertebrates, such as corals and sponges, are essential for nutrient acquisition and stress resistance. In contrast, pathogenic microbes can impact marine species health, influencing population dynamics and ecosystem stability. These interactions highlight the integral role of microbes in shaping marine ecosystems and their responses to environmental changes (Wilkins et al., 2019).

5.2 BIOGEOCHEMICAL CYCLES IN THE MARINE ENVIRONMENT

5.2.1 Key Biogeochemical Cycles

Key biogeochemical cycles in marine ecosystems include the carbon, nitrogen, and sulfur cycles, all of which are heavily mediated by marine microbiomes. Microbes facilitate the conversion of carbon dioxide into organic matter through photosynthesis, regulate nitrogen availability through processes like nitrogen fixation and denitrification, and drive sulfur transformations that impact ocean chemistry. These cycles are essential for maintaining the productivity and health of marine environments, influencing both local ecosystems and global climate patterns (Table 5.2) (Li et al., 2018).

5.2.1.1 The Carbon Cycle: Marine Perspectives

The marine carbon cycle is a critical component of the global carbon cycle, involving complex processes that regulate atmospheric CO_2 levels and climate. Marine microorganisms have a central place in the global carbon cycle as they function as a biological pump, sequestering anthropogenic carbon dioxide from the atmosphere in the deep ocean (York, 2018). Marine phytoplankton play a crucial role in carbon sequestration through photosynthesis, converting CO_2 into organic matter. This organic carbon can be transported to the deep ocean via the biological pump, where it contributes to long-term carbon storage. Microbial processes, including the degradation of organic matter and remineralization of carbon, also influence carbon fluxes and storage in marine systems (Figure 5.3a) (Nowicki et al., 2022).

5.2.1.2 Nitrogen Cycle in Ocean Ecosystems

The nitrogen cycle in marine environments is characterized by dynamic transformations of nitrogen compounds, including fixation, nitrification, denitrification, and ammonification. Marine microbes, particularly diazotrophic bacteria, are essential for nitrogen fixation, converting atmospheric N_2 into bioavailable forms. Nitrifying bacteria oxidize ammonia to nitrite and nitrate, while denitrifying bacteria reduce nitrates to nitrogen gas, completing the cycle. These processes are vital for maintaining nitrogen availability and supporting marine productivity (Voss et al., 2013). The nitrogen cycle on the Earth has evolved over three billion years through biogeochemical and microbial processes coupled via natural feedbacks that keep the nitrogen cycle of the oceans in approximate balance. However,

TABLE 5.2
Comparison of Traditional and Modern Techniques in Marine Microbiome Research

Technique	Method	Advantages	Limitations	Typical Applications	References
Traditional Techniques	Culturing	Isolates and studies individual microbial species; provides physiological and biochemical insights.	Limited to culturable microbes; may not represent the full community; time-consuming.	Identification and characterization of culturable microbes; studies of growth and behavior.	(Pajares & Ramos, 2019)
	Microscopy	Direct visualization of microbial morphology and abundance; relatively simple and cost-effective	Low resolution; often unable to differentiate between species or detect non-culturable microbes	Visual assessment of microbial populations and their morphology; qualitative analysis	(Mason et al., 2023)
Modern Techniques	High-throughput Sequencing	Provides comprehensive and high-resolution microbial profiles; can identify both culturable and non-culturable microbes	High cost; generates large volumes of data that require substantial computational resources and expertise	Community composition analysis; identification of microbial diversity and rare taxa	(Joint et al., 2010)
	Metagenomics	Allows for functional and taxonomic analysis of entire microbial communities; can detect novel and low-abundance microbes	Complex data analysis and interpretation; requires advanced bioinformatics skills	Functional genomics; assessment of microbial functions and interactions within ecosystems	(Patin & Goodwin, 2023)
	Meta transcriptomics	Provides insight into active gene expression and microbial functions in situ; captures dynamic processes	Expensive; data analysis is challenging due to complexity and volume	Understanding gene expression profiles; studying functional dynamics of microbial communities	(Ayon, 2023)
	Proteomics	Analyzes protein expression and functional activity of microbial communities; provides insight into functional proteins	High cost; requires specialized equipment and expertise	Analysis of protein functions and interactions; insights into microbial metabolic pathways	(Kennedy et al., 2010)

recent findings strongly suggest a significant imbalance of the oceans' nitrogen budget toward higher losses than inputs. Moreover, new processes have been identified, such as the anaerobic ammonium oxidation (anammox) processes the conversion of ammonia and nitrite to dinitrogen gas or the denitrification by eukaryotic species such as *Foraminifera* (Figure 5.3b) (Voss et al., 2013).

5.2.2 Microbial Contributions to Biogeochemical Cycles

5.2.2.1 Microbial-Mediated Carbon Sequestration

Microbes significantly contribute to carbon sequestration through various mechanisms, including the production of extracellular polymeric substances (EPS) and the formation of marine snow. EPS enhance the aggregation of organic particles, facilitating their sinking to the deep ocean. Additionally, microbial-mediated processes, such as methane oxidation and sulfate reduction, influence carbon cycling and sequestration in marine sediments. These processes are crucial for regulating atmospheric CO_2 levels and mitigating climate change (Arnosti, 2011). Inoculation of soil with living microbes or propagules has grown in interest and in application due to the modification/degradation of soil systems (including native microbial communities), and the need to maintain agricultural yields with fewer synthetic inputs. However, while beneficial microorganisms such as plant growth promoting bacteria (PGPB) and mycorrhizal fungi have been employed via inoculation, their utilization as an agronomic tool remains trivial in the context of large-scale commercial agriculture (Arnosti, 2011).

5.2.2.2 Nitrogen Fixation and Nitrification by Marine Microbes

Marine microbes are central to nitrogen cycling through processes such as nitrogen fixation and nitrification. Diazotrophic microbes, including cyanobacteria and certain archaea, fix atmospheric N_2 into ammonium, which is then utilized by other organisms or further processed by nitrifying bacteria. Nitrification, the conversion of ammonia to nitrite and nitrate, is essential for maintaining nitrogen availability and supporting primary production.

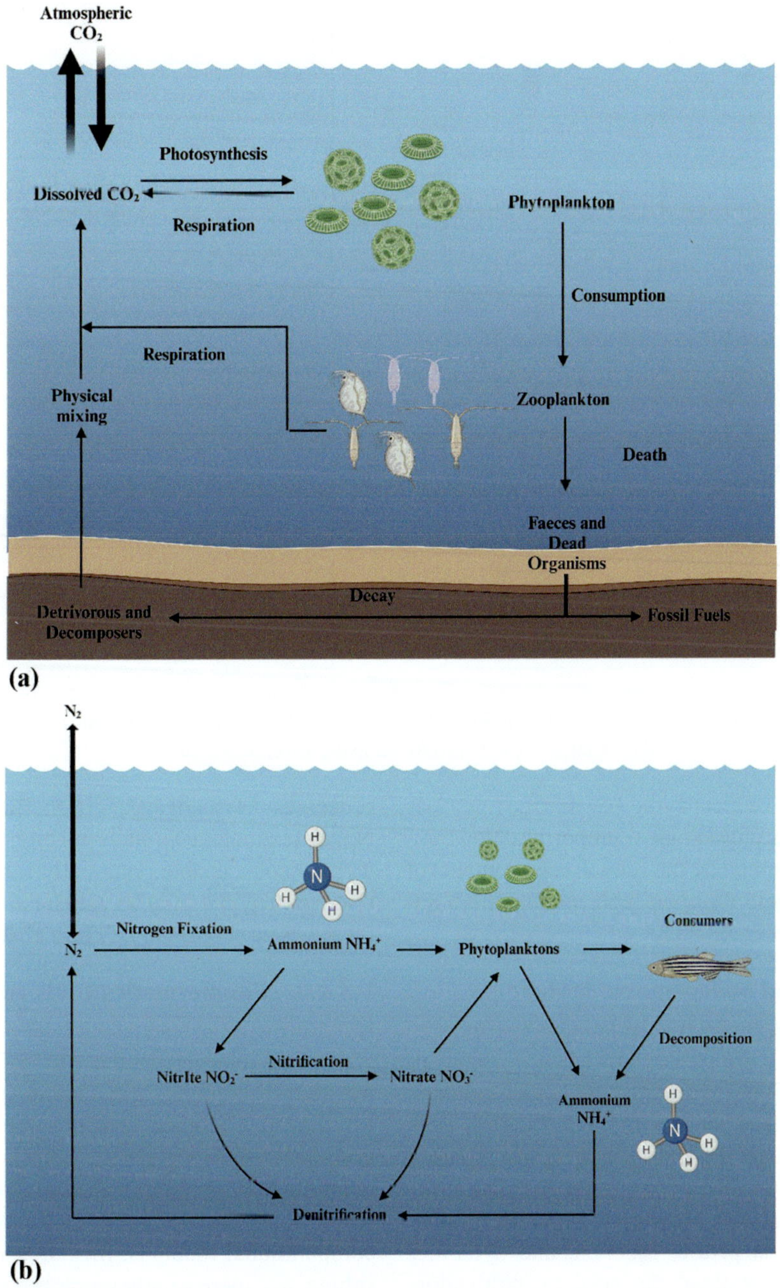

FIGURE 5.3 (a): The Marine Carbon Cycle: A flowchart illustrating the marine Carbon cycle, highlighting the role of marine microorganisms in carbon sequestration. (Self-developed using Biorender.com.) (b): The Marine Nitrogen Cycle: A flowchart illustrating the marine Nitrogen cycle, highlighting the role of marine microorganisms in nitrogen sequestration. (Self-developed using Biorender.com.)

These microbial processes ensure the sustainability of marine ecosystems and their resilience to environmental changes (Wang et al., 2022).

5.3 METHODOLOGIES FOR STUDYING MARINE MICROBIOMES

5.3.1 Traditional and Modern Techniques

Traditional techniques in marine microbiome research relied on culturing microorganisms, which captured only a fraction of microbial diversity (Mason et al., 2023; Pajares & Ramos, 2019). Modern techniques, such as high-throughput sequencing and metagenomics, now enable comprehensive analysis of microbial communities, allowing for a deeper understanding of their complexity, functions, and roles in marine ecosystems (Joint et al., 2010; Patin & Goodwin, 2023).

5.3.1.1 Cultivation-Based Approaches

Traditional cultivation-based techniques have provided foundational insights into marine microbiology by isolating and characterizing individual microbial species. Methods such as selective media and enrichment cultures

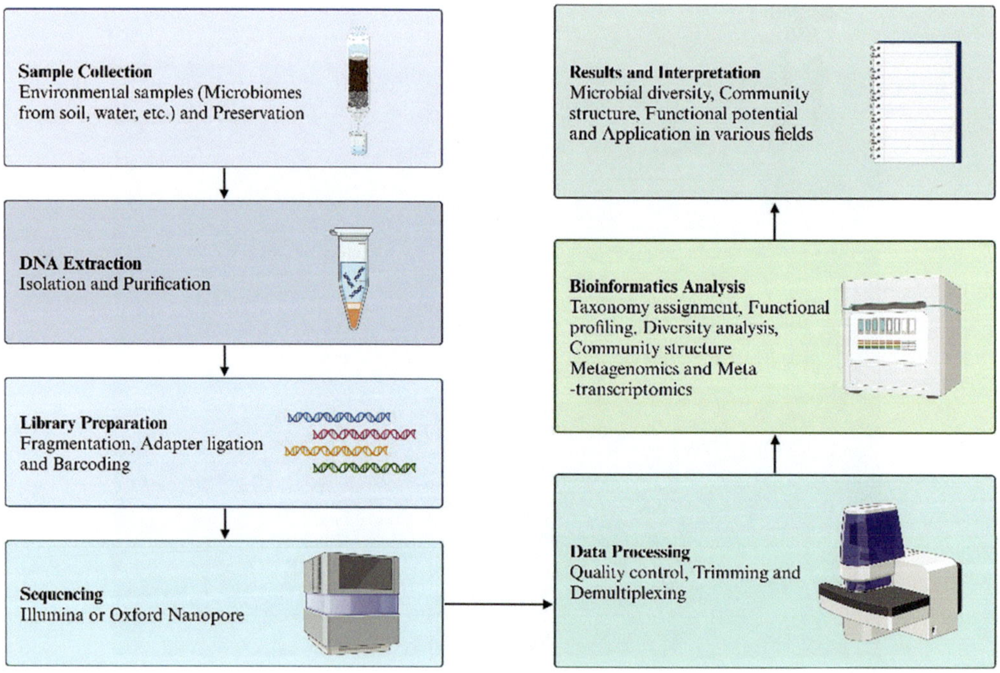

FIGURE 5.4 High-Throughput Sequencing in Marine Microbiome Research: A schematic that outlines the steps in high-throughput sequencing, from sample collection to data analysis. (Self-developed using Biorender.com.)

have been instrumental in identifying key marine microbes and understanding their physiological properties. However, these approaches are limited by the growth requirements of microorganisms and their ability to replicate under laboratory conditions (Bonnet et al., 2020).

5.3.1.2 Genomic and Metagenomic Techniques

Modern genomic and metagenomic techniques have revolutionized marine microbiome research by enabling the analysis of microbial communities directly from environmental samples. High-throughput sequencing technologies, such as 16S rRNA gene sequencing and shotgun metagenomics, provide comprehensive profiles of microbial diversity, functional genes, and metabolic pathways. These techniques offer insights into the structure, function, and dynamics of marine microbiomes without the need for cultivation, revealing novel microbial taxa and their ecological roles (Pérez-Cobas et al., 2020).

5.3.2 Advances in Marine Microbiome Research

5.3.2.1 High-Throughput Sequencing Technologies

High-throughput sequencing technologies, including Illumina and Oxford Nanopore platforms, have advanced the study of marine microbiomes by providing detailed and large-scale sequence data. These technologies enable the identification of microbial taxa, functional genes, and genomic features, facilitating a deeper understanding of microbial diversity and ecosystem functions. The integration of metagenomic, meta-transcriptomic, and metaproteomic data further enhances our comprehension of microbial interactions and their roles in biogeochemical processes (Armengaud, 2023; Bashiardes et al., 2016). Novel technologies, such as high-throughput sequencing technologies, enable potentially full-spectrum detection of trace pathogenic microorganisms in complex environmental matrices (Figure 5.4) (Zhao et al., 2023).

5.3.2.2 Bioinformatics Tools and Applications

Bioinformatics tools and applications are essential for analyzing and interpreting the vast amounts of data generated by high-throughput sequencing. Tools such as QIIME, Mothur, and Kraken facilitate the processing, classification, and visualization of sequence data, enabling the identification of microbial taxa and functional pathways (Galloway-Peña & Hanson, 2020). Advanced statistical and computational methods, including machine learning algorithms, are increasingly employed to uncover patterns and relationships within complex microbial datasets, providing insights into microbial community dynamics and their responses to environmental changes.

5.4 MARINE MICROBIOMES AND CLIMATE CHANGE

5.4.1 Impact of Climate Change on Marine Microbiomes

Climate change is reshaping marine microbiomes, the microscopic communities essential to ocean health. Key factors like rising sea temperatures and ocean acidification are altering microbial diversity and functions (Kroeker et al., 2013). These shifts can disrupt marine ecosystems and biogeochemical cycles, potentially affecting global climate regulation and marine biodiversity. Understanding these changes is crucial for predicting broader ecological impacts and developing strategies to mitigate them.

5.4.1.1 Ocean Acidification and Microbial Communities

Ocean acidification, resulting from increased atmospheric CO_2, has significant impacts on marine microbiomes. Changes in seawater pH affect microbial metabolic processes, community structure, and interactions with other marine organisms. Acidification can alter the abundance and activity of key microbial groups involved in nutrient cycling, potentially disrupting ecosystem functions and food web dynamics. The ocean absorbs excess atmospheric CO_2, which reacts with seawater to form carbonic acid, leading to ocean acidification. This process can disrupt biogeochemical cycles (Abirami et al., 2021). Understanding the responses of marine microbes to acidification is crucial for predicting the impacts of climate change on marine ecosystems (Figure 5.5).

5.4.1.2 Temperature Changes and Microbial Activity

Temperature changes associated with global warming influence microbial activity, diversity, and distribution in marine environments. Elevated temperatures can alter microbial metabolic rates, enzyme activities, and community compositions. These changes can impact biogeochemical processes, such as nutrient cycling and organic matter degradation, with implications for ocean productivity and ecosystem stability. Research into the effects of temperature on marine microbes is essential for understanding their role in climate change and developing strategies for adaptation and mitigation (Pogoda et al., 2019).

5.4.1.3 Responses of Marine Organisms to UVB Radiation

UVB radiation, a part of the ultraviolet light spectrum, can have significant effects on marine microbiota, including bacteria, phytoplankton, and other microorganisms. Here's a breakdown of how UVB radiation impacts these organisms:

1. **DNA Damage:** UVB radiation can cause direct DNA damage in marine microorganisms by inducing the formation of pyrimidine dimers and other photoproducts. This damage can lead to mutations, which might impair cellular functions or lead to cell death if repair mechanisms are insufficient (García-Corral et al., 2020).
2. **Photosynthesis Disruption:** In phytoplankton, UVB radiation can damage photosynthetic pigments and enzymes, disrupting photosynthesis. This can affect primary productivity in marine ecosystems, potentially leading to changes in nutrient dynamics and energy flow (Santos et al., 2012).
3. **Altered Community Structure:** UVB radiation can differentially affect various microbial species, potentially leading to shifts in community composition. UVB-resistant species may become more dominant, while sensitive ones decline, potentially affecting nutrient cycling and energy transfer (Shick et al., 1996).
4. **Impact on Symbiotic Relationships:** Many marine microorganisms, including coral-associated bacteria and algae, live in symbiotic relationships.

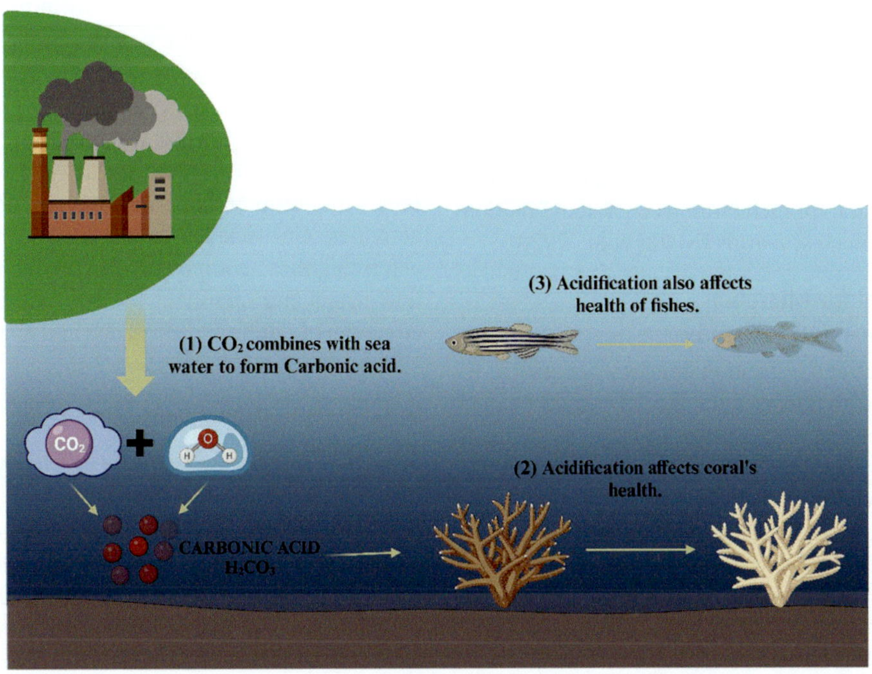

FIGURE 5.5 Impact of Ocean Acidification on Marine Microbial Communities: A comparative figure showing how ocean acidification affects marine microbial communities and their roles in biogeochemical cycles. (Self-developed using Biorender.com.)

UVB-induced stress can disrupt these relationships, leading to issues like coral bleaching or impaired nutrient exchange (Mohamed et al., 2023).

5. **Enhanced Genetic Variability:** The stress induced by UVB radiation might increase genetic variability within microbial populations due to higher mutation rates. This could potentially drive evolutionary changes over time (Wang et al., 2023).
6. **Biofilm Formation:** UVB can influence the formation and stability of microbial biofilms. For some microorganisms, UVB exposure might promote biofilm formation as a protective strategy, while for others, it could weaken biofilm structures (Fuentes et al., 2022).
7. **Interactions with Other Environmental Stressors:** UVB effects can be compounded by other environmental factors such as temperature, salinity, and nutrient availability. The combined stressors can lead to more complex impacts on microbial communities (Matallana-Surget et al., 2012).

Understanding these effects is crucial for assessing the health of marine ecosystems, especially as UVB radiation levels can be influenced by factors like ozone depletion and climate change.

5.4.2 Adaptive Responses of Microbiomes

5.4.2.1 Microbial Adaptation to Changing Environments

Marine microbes exhibit various adaptive strategies to cope with changing environmental conditions, including alterations in metabolic pathways, gene expression, and community structure. Adaptations may involve shifts in microbial community composition, enhanced tolerance to stressors, and the acquisition of novel functional traits. Studying these adaptive responses provides insights into microbial resilience and the potential for maintaining ecosystem functions under changing climate scenarios (Malhi et al., 2020).

5.4.2.2 Potential for Microbial Resilience and Mitigation

Microbial resilience and mitigation potential are critical for sustaining marine ecosystems in the face of climate change. Some marine microbes possess the capacity to adapt to environmental stressors and contribute to ecosystem stability through processes such as carbon sequestration and nutrient recycling. Exploring the mechanisms underlying microbial resilience and identifying key taxa with potential for mitigating climate impacts are essential for developing strategies to protect and restore marine ecosystems (Borchert et al., 2021).

5.5 APPLICATIONS AND FUTURE DIRECTIONS

The study of marine microbiomes has significant applications in environmental monitoring, bioremediation, and biotechnology. Advancements in genomic and bioinformatics tools are enabling more precise characterization of microbial communities and their functions. Future research aims to explore how these microbiomes can be harnessed to address climate change, restore degraded marine ecosystems, and develop sustainable marine resources. Understanding their resilience and adaptability will be crucial for mitigating the impacts of environmental changes and ensuring ocean health (Malla et al., 2018).

5.5.1 Biotechnological Applications

5.5.1.1 Marine Microbes in Biotechnology

Marine microbes have significant biotechnological potential, offering unique enzymes, metabolites, and bioactive compounds with applications in various industries. Marine-derived enzymes, such as cellulases and lipases, are utilized in biotechnology for processes including waste management, biofuel production, and pharmaceutical development. Marine microbiomes are a rich source of bioactive compounds. Biotechnology techniques are used to isolate and characterize novel antibiotics, anticancer agents, and other therapeutic molecules from marine microorganisms (Karthikeyan et al., 2022). The exploration of marine microbiomes for novel compounds and biocatalysts holds promise for advancing biotechnological innovations and addressing global challenges.

5.5.1.2 Potential for New Antibiotics and Bioactive Compounds

The marine environment is a rich source of novel antibiotics and bioactive compounds with potential therapeutic applications. Marine microorganisms, including bacteria and fungi, produce a diverse array of secondary metabolites with antimicrobial, anti-inflammatory, and anticancer activities (Gerwick & Fenner, 2013). Identifying and characterizing these compounds can lead to the development of new drugs and treatment options, addressing the growing concerns of antibiotic resistance and emerging diseases.

5.5.2 Future Research Directions

5.5.2.1 Emerging Frontiers in Marine Microbiome Research

Future research in marine microbiomes will focus on addressing knowledge gaps and exploring emerging frontiers in the field. Key areas of interest include the integration of multi-omic approaches to elucidate microbial functions, the study of microbial interactions and network dynamics, and the investigation of microbiome-ecosystem interactions under varying environmental conditions. Advances in technology and methodology will continue to drive discoveries and enhance our understanding of marine microbiomes (Sharma et al., 2022).

5.5.2.2 Integrating Microbial Insights into Ocean Management Policies

Integrating microbial insights into ocean management policies is essential for addressing the challenges of climate change, pollution, and biodiversity loss. Incorporating microbial data into marine conservation and management strategies can improve our understanding of ecosystem health and inform decision-making processes. Collaborative efforts between researchers, policymakers, and stakeholders will be crucial for developing effective policies and practices that promote the sustainability of marine ecosystems (Wang et al., 2024).

5.6 SUMMARY AND CONCLUSION

5.6.1 Summary of Key Findings

The study of marine microbiomes has illuminated their critical roles within ocean ecosystems, revealing their profound influence on biogeochemical cycles, marine food webs, and overall ecosystem health. Marine microbes, including bacteria, archaea, and viruses, exhibit extraordinary diversity and are integral to processes such as carbon sequestration and nitrogen cycling. Their ability to mediate complex biogeochemical cycles underscores their importance in regulating global climate and supporting marine productivity.

Historically, marine microbiology has transitioned from rudimentary observational techniques to sophisticated genomic and bioinformatics methods, providing unprecedented insights into microbial diversity and functionality. Advances in high-throughput sequencing and computational tools have enhanced our understanding of microbial community structures and their ecological interactions. These methodologies have uncovered the intricate relationships between marine microbes and their environments, shedding light on their responses to environmental stressors such as climate change.

The impact of climate change on marine microbiomes is profound, with ocean acidification and temperature changes influencing microbial activity, diversity, and ecosystem functions. Understanding microbial adaptation mechanisms and their potential for resilience is crucial for predicting future ecological shifts and developing strategies for mitigation. The potential of marine microbes in biotechnology, including the discovery of new antibiotics and bioactive compounds, highlights their value beyond ecological roles, offering opportunities for addressing global challenges.

5.6.2 Implications for Future Studies

The continued investigation of marine microbiomes highlights several critical areas for future research. Addressing gaps in our knowledge of microbial functions and interactions, especially in the context of environmental change, will deepen our understanding of marine ecosystems. Upcoming research should prioritize the integration of multi-omic approaches to clarify microbial community dynamics, the discovery of new microbial taxa and their roles in ecosystem processes, and the evaluation of how microbial shifts impact ocean management and conservation efforts. Moreover, incorporating microbial insights into ocean management policies is crucial for tackling the challenges of climate change, pollution, and biodiversity loss. Successful strategies to enhance the sustainability and resilience of marine ecosystems will require collaborative efforts among researchers, policymakers, and stakeholders.

5.6.3 Conclusion

Marine microbiomes are fundamental to the stability and function of ocean ecosystems, driving essential biogeochemical cycles and regulating marine food webs, while also playing a critical role in global climate processes. Their remarkable diversity and adaptability position them as key players in responding to environmental changes, including those caused by climate change. With the rapid advancements in genomic and bioinformatics tools, our understanding of these microbial communities has deepened, revealing their vast potential for addressing pressing challenges such as antibiotic resistance and climate mitigation. This chapter provides a fresh perspective on the importance of marine microbiomes, both in terms of ecosystem functions and their potential for biotechnological innovation. By incorporating the latest research and technologies, it offers a current understanding of microbial dynamics, emphasizing their role in tackling global issues such as climate change and antibiotic resistance. Readers will gain valuable insights into the resilience and adaptability of marine microbiomes, and their contributions to biogeochemical cycles, making this chapter a key resource for researchers, environmental managers, and policymakers working toward marine conservation and sustainable management. It also lays the groundwork for future studies by advocating for the integration of microbial data into ocean governance, offering a pathway for interdisciplinary research and practical applications in marine science. As research progresses, it becomes increasingly evident that marine microbiomes are not only essential for ecosystem stability but also offer innovative solutions to pressing global challenges like antibiotic resistance and climate change. Future research must focus on how microbial communities adapt to changing environmental conditions and how this knowledge can inform ocean management and conservation. Leveraging the potential of marine microbiomes will enable us to implement sustainable practices that support the long-term health and resilience of marine ecosystems while addressing critical global issues.

ABBREVIATIONS

CO_2	Carbon dioxide
DNA	Deoxyribonucleic Acid
N_2	Nitrogen
PCR	Polymerase Chain Reaction
PGPB	Plant Growth Promoting Bacteria
QIIME	Quantitative Insights into Microbial Ecology
rRNA	Ribosomal Ribonucleic Acid
UVB	Ultraviolet-B radiation

ACKNOWLEDGMENT

The authors sincerely acknowledge the financial support from the Department of Biotechnology (DBT), Government of India, under Project No. DBT/2022-23/July/N/2240, as part of the DBT-RA Program and School of Bioengineering and Biosciences, Lovely Professional University, Punjab, India.

REFERENCES

Abirami, B., Radhakrishnan, M., Kumaran, S., & Wilson, A. (2021). Impacts of global warming on marine microbial communities. *Science of the Total Environment*, *791*, 147905. https://doi.org/10.1016/j.scitotenv.2021.147905

Adler, A., & Dücker, E. (2018). When pasteurian science went to sea: The birth of marine microbiology. *Journal of the History of Biology*, *51*(1), 107–133. https://doi.org/10.1007/s10739-017-9477-8

Armengaud, J. (2023). Metaproteomics to understand how microbiota function: The crystal ball predicts a promising future. *Environmental Microbiology*, *25*(1), 115–125. https://doi.org/10.1111/1462-2920.16238

Arnosti, C. (2011). Microbial extracellular enzymes and the marine carbon cycle. *Annual Review of Marine Science*, *3*(1), 401–425. https://doi.org/10.1146/annurev-marine-120709-142731

Ayon, N. J. (2023). High-throughput screening of natural product and synthetic molecule libraries for antibacterial drug discovery. *Metabolites*, *13*(5), 625. https://doi.org/10.3390/metabo13050625

Bashiardes, S., Zilberman-Schapira, G., & Elinav, E. (2016). Use of metatranscriptomics in microbiome research. *Bioinformatics and Biology Insights*, *10*, BBI.S34610. https://doi.org/10.4137/BBI.S34610

Berube, P. M., Biller, S. J., Hackl, T., Hogle, S. L., Satinsky, B. M., Becker, J. W., Braakman, R., Collins, S. B., Kelly, L., Berta-Thompson, J., Coe, A., Bergauer, K., Bouman, H. A., Browning, T. J., De Corte, D., Hassler, C., Hulata, Y., Jacquot, J. E., Maas, E. W., … Chisholm, S. W. (2018). Single cell genomes of *Prochlorococcus*, *Synechococcus*, and sympatric microbes from diverse marine environments. *Scientific Data*, *5*(1), 180154. https://doi.org/10.1038/sdata.2018.154

Bonnet, M., Lagier, J. C., Raoult, D., & Khelaifia, S. (2020). Bacterial culture through selective and non-selective conditions: The evolution of culture media in clinical microbiology. *New Microbes and New Infections*, *34*, 100622. https://doi.org/10.1016/j.nmni.2019.100622

Borchert, E., Hammerschmidt, K., Hentschel, U., & Deines, P. (2021). Enhancing microbial pollutant degradation by integrating eco-evolutionary principles with environmental biotechnology. *Trends in Microbiology*, *29*(10), 908–918. https://doi.org/10.1016/j.tim.2021.03.002

Focardi, A., Ostrowski, M., Goossen, K., Brown, M. V., & Paulsen, I. (2020). Investigating the diversity of marine bacteriophage in contrasting water masses associated with the East Australian Current (EAC) system. *Viruses*, *12*(3), 317. https://doi.org/10.3390/v12030317

Fuentes, E., Pérez-Velón, D., & Prieto, B. (2022). Effects of changes in UV-B radiation levels on biofilm-forming organisms commonly found in cultural heritage surfaces. *Environmental Research*, *214*, 114061. https://doi.org/10.1016/j.envres.2022.114061

Galloway-Peña, J., & Hanson, B. (2020). Tools for analysis of the microbiome. *Digestive Diseases and Sciences*, *65*(3), 674–685. https://doi.org/10.1007/s10620-020-06091-y

García-Corral, L., Duarte, C., & Agusti, S. (2020). Impact of UV radiation on plankton net community production: Responses in Western Australian estuarine and coastal waters. *Marine Ecology Progress Series*, *651*, 45–56. https://doi.org/10.3354/meps13456

Gerwick, W. H., & Fenner, A. M. (2013). Drug discovery from marine microbes. *Microbial Ecology*, *65*(4), 800–806. https://doi.org/10.1007/s00248-012-0169-9

Glibert, P. M., & Mitra, A. (2022). From webs, loops, shunts, and pumps to microbial multitasking: Evolving concepts of marine microbial ecology, the mixoplankton paradigm, and implications for a future ocean. *Limnology and Oceanography*, *67*(3), 585–597. https://doi.org/10.1002/lno.12018

Goodwin, K. D., & Lacoursière-Roussel, A. (2023). Editorial: Marine microbiomes: Towards standard methods and best practices. *Frontiers in Microbiology*, *14*. https://doi.org/10.3389/fmicb.2023.1219958

Joint, I., Mühling, M., & Querellou, J. (2010). Culturing marine bacteria—An essential prerequisite for biodiscovery. *Microbial Biotechnology*, *3*(5), 564–575. https://doi.org/10.1111/j.1751-7915.2010.00188.x

Karthikeyan, A., Joseph, A., & Nair, B. G. (2022). Promising bioactive compounds from the marine environment and their potential effects on various diseases. *Journal of Genetic Engineering and Biotechnology*, *20*(1), 14. https://doi.org/10.1186/s43141-021-00290-4

Kennedy, J., Flemer, B., Jackson, S. A., Lejon, D. P. H., Morrissey, J. P., O'Gara, F., & Dobson, A. D. W. (2010). Marine metagenomics: New tools for the study and exploitation of marine microbial metabolism. *Marine Drugs*, *8*(3), 608–628. https://doi.org/10.3390/md8030608

Kroeker, K. J., Kordas, R. L., Crim, R., Hendriks, I. E., Ramajo, L., Singh, G. S., Duarte, C. M., & Gattuso, J. (2013). Impacts of ocean acidification on marine organisms: Quantifying sensitivities and interaction with warming. *Global Change Biology*, *19*(6), 1884–1896. https://doi.org/10.1111/gcb.12179

Li, Y., Tang, K., Zhang, L., Zhao, Z., Xie, X., Chen, C.-T. A., Wang, D., Jiao, N., & Zhang, Y. (2018). Coupled carbon, sulfur, and nitrogen cycles mediated by microorganisms in the water column of a shallow-water hydrothermal Ecosystem. *Frontiers in Microbiology*, *9*. https://doi.org/10.3389/fmicb.2018.02718

Malhi, Y., Franklin, J., Seddon, N., Solan, M., Turner, M. G., Field, C. B., & Knowlton, N. (2020). Climate change and ecosystems: Threats, opportunities and solutions. *Philosophical Transactions of the Royal Society B: Biological Sciences*, *375*(1794), 20190104. https://doi.org/10.1098/rstb.2019.0104

Malla, M. A., Dubey, A., Yadav, S., Kumar, A., Hashem, A., & Abd Allah, E. F. (2018). Understanding and designing the strategies for the microbe-mediated remediation of environmental contaminants using omics approaches. *Frontiers in Microbiology*, *9*. https://doi.org/10.3389/fmicb.2018.01132

Mason, A. R. G., Salomon, M. J., Lowe, A. J., & Cavagnaro, T. R. (2023). Microbial solutions to soil carbon sequestration. *Journal of Cleaner Production*, *417*, 137993. https://doi.org/10.1016/j.jclepro.2023.137993

Matallana-Surget, S., Villette, C., Intertaglia, L., Joux, F., Bourrain, M., & Lebaron, P. (2012). Response to UVB radiation and oxidative stress of marine bacteria isolated from South Pacific Ocean and Mediterranean Sea. *Journal of Photochemistry and Photobiology B: Biology*, *117*, 254–261. https://doi.org/10.1016/j.jphotobiol.2012.09.011

Mohamed, A. R., Ochsenkühn, M. A., Kazlak, A. M., Moustafa, A., & Amin, S. A. (2023). The coral microbiome: Towards an understanding of the molecular mechanisms of coral–microbiota interactions. *FEMS Microbiology Reviews*, *47*(2). https://doi.org/10.1093/femsre/fuad005

Nowicki, M., DeVries, T., & Siegel, D. A. (2022). Quantifying the carbon export and sequestration pathways of the ocean's biological carbon pump. *Global Biogeochemical Cycles*, *36*(3). https://doi.org/10.1029/2021GB007083

Ollison, G. A., Hu, S. K., Hopper, J. V., Stewart, B. P., Beatty, J. L., & Caron, D. A. (2023). Physiology governing diatom vs. dinoflagellate bloom and decline in coastal Santa Monica Bay. *Frontiers in Microbiology*, *14*. https://doi.org/10.3389/fmicb.2023.1287326

Pajares, S., & Ramos, R. (2019). Processes and microorganisms involved in the marine nitrogen cycle: Knowledge and gaps. *Frontiers in Marine Science*, *6*. https://doi.org/10.3389/fmars.2019.00739

Patin, N. V., & Goodwin, K. D. (2023). Capturing marine microbiomes and environmental DNA: A field sampling guide. *Frontiers in Microbiology*, *13*. https://doi.org/10.3389/fmicb.2022.1026596

Pérez-Cobas, A. E., Gomez-Valero, L., & Buchrieser, C. (2020). Metagenomic approaches in microbial ecology: An update on whole-genome and marker gene sequencing analyses. *Microbial Genomics*, *6*(8). https://doi.org/10.1099/mgen.0.000409

Pfeifer, K., Ergal, İ., Koller, M., Basen, M., Schuster, B., & Rittmann, S. K.-M. R. (2021). Archaea biotechnology. *Biotechnology Advances*, *47*, 107668. https://doi.org/10.1016/j.biotechadv.2020.107668

Pogoda, B., Brown, J., Hancock, B., Preston, J., Pouvreau, S., Kamermans, P., Sanderson, W., & von Nordheim, H. (2019). The Native Oyster Restoration Alliance (NORA) and the Berlin Oyster Recommendation: Bringing back a key ecosystem engineer by developing and supporting best practice in Europe. *Aquatic Living Resources*, *32*, 13. https://doi.org/10.1051/alr/2019012

Santos, A. L., Oliveira, V., Baptista, I., Henriques, I., Gomes, N. C. M., Almeida, A., Correia, A., & Cunha, A. (2012). Effects of UV-B radiation on the structural and physiological diversity of bacterioneuston and bacterioplankton. *Applied and Environmental Microbiology*, *78*(6), 2066–2069. https://doi.org/10.1128/AEM.06344-11

Sharma, P., Singh, S. P., Iqbal, H. M. N., & Tong, Y. W. (2022). Omics approaches in bioremediation of environmental contaminants: An integrated approach for environmental safety and sustainability. *Environmental Research*, *211*, 113102. https://doi.org/10.1016/j.envres.2022.113102

Shick, J. M., Lesser, M. P., & Jokiel, P. L. (1996). Effects of ultraviolet radiation on corals and other coral reef organisms. *Global Change Biology*, *2*(6), 527–545. https://doi.org/10.1111/j.1365-2486.1996.tb00065.x

Trombetta, T., Vidussi, F., Roques, C., Scotti, M., & Mostajir, B. (2020). Marine microbial food web networks during phytoplankton bloom and non-bloom periods: Warming favors smaller organism interactions and intensifies trophic cascade. *Frontiers in Microbiology*, *11*. https://doi.org/10.3389/fmicb.2020.502336

Vanwonterghem, I., & Webster, N. S. (2020). Coral reef microorganisms in a changing climate. *IScience*, *23*(4), 100972. https://doi.org/10.1016/j.isci.2020.100972

Voss, M., Bange, H. W., Dippner, J. W., Middelburg, J. J., Montoya, J. P., & Ward, B. (2013). The marine nitrogen cycle: Recent discoveries, uncertainties and the potential relevance of climate change. *Philosophical Transactions of the Royal Society B: Biological Sciences*, *368*(1621), 20130121. https://doi.org/10.1098/rstb.2013.0121

Wang, J., Yang, Z., Lu, P., Sun, Y., Xue, S., Tang, X., & Xiao, H. (2023). Effects of UV-B radiation on epiphytic bacterial communities on male and female *Sargassum thunbergii*. *Scientific Reports*, *13*(1), 3985. https://doi.org/10.1038/s41598-022-26494-3

Wang, S., Yang, Y., & Jing, J. (2022). A synthesis of viral contribution to marine nitrogen cycling. *Frontiers in Microbiology*, *13*. https://doi.org/10.3389/fmicb.2022.834581

Wang, Z., Wang, T., Zhang, X., Wang, J., Yang, Y., Sun, Y., Guo, X., Wu, Q., Nepovimova, E., Watson, A. E., & Kuca, K. (2024). Biodiversity conservation in the context of climate change: Facing challenges and management strategies. *Science of the Total Environment*, *937*, 173377. https://doi.org/10.1016/j.scitotenv.2024.173377

Wilkins, L. G. E., Leray, M., O'Dea, A., Yuen, B., Peixoto, R. S., Pereira, T. J., Bik, H. M., Coil, D. A., Duffy, J. E., Herre, E. A., Lessios, H. A., Lucey, N. M., Mejia, L. C., Rasher, D. B., Sharp, K. H., Sogin, E. M., Thacker, R. W., Vega Thurber, R., Wcislo, W. T., … Eisen, J. A. (2019). Host-associated microbiomes drive structure and function of marine ecosystems. *PLOS Biology*, *17*(11), e3000533. https://doi.org/10.1371/journal.pbio.3000533

York, A. (2018). Marine biogeochemical cycles in a changing world. *Nature Reviews Microbiology*, *16*(5), 259–259. https://doi.org/10.1038/nrmicro.2018.40

Zhao, Y., Huang, F., Wang, W., Gao, R., Fan, L., Wang, A., & Gao, S.-H. (2023). Application of high-throughput sequencing technologies and analytical tools for pathogen detection in urban water systems: Progress and future perspectives. *Science of the Total Environment*, *900*, 165867. https://doi.org/10.1016/j.scitotenv.2023.165867

6 Symbiotic Relationships in Marine Microbial Communities

Nisha Shankhwar[1], Awadhesh Kumar Verma[2], Kalakuntla Sriharshitha Rao[2], Vivekanand[2], Tanya Singh[2], Karishma Vivek Kathpalia[2], and Neeta Raj Sharma[2]
[1]Special Centre for Nanoscience, Jawaharlal Nehru University, New Delhi, India
[2]Department of Biotechnology and Bioinformatics, School of Bioengineering and Biosciences, Lovely Professional University, Jalandhar, India

6.1 INTRODUCTION TO SYMBIOSIS IN MARINE MICROBIOMES

6.1.1 Definition and Types of Symbiotic Relationships

Symbiosis, defined as the interaction between two or more different biological organisms living in close physical proximity, is a critical component of marine microbial ecology. This interaction encompasses three primary types: mutualism, commensalism, and parasitism, each characterized by distinct dynamics that influence microbial community structures and ecosystem processes (Srivastava et al., 2023). Mutualistic interactions are those in which both organisms involved derive benefits from the relationship. The symbiosis between corals and dinoflagellates is a quintessential example of mutualism. The dinoflagellates, residing within coral tissues, perform photosynthesis and provide essential nutrients to the coral, while the coral offers a protected environment and access to light for the dinoflagellates. This mutualistic relationship is crucial for the health of coral reefs, which are biodiversity hotspots and play significant roles in marine ecosystems (Li, 2024). Commensalism involves interactions where one organism benefits while the other is neither helped nor harmed. Commensal microorganisms may reside on or within larger host organisms, such as fish or invertebrates, without significantly affecting their host. This can lead to increased microbial diversity and influence the stability of microbial communities by providing niche spaces and resources. Parasitism, on the other hand, benefits one organism at the expense of another. Marine microbial parasites can impact their hosts' health and survival, which in turn can affect broader ecological interactions and community dynamics. Parasitic microbes in fish can influence fish population dynamics and predator-prey relationships within marine ecosystems (Åkesson et al., 2021).

6.1.2 Historical Overview of Symbiotic Research

The study of symbiosis in marine microbiomes has evolved from early observations of microbial interactions to sophisticated molecular and genomic analyses. Initial research focused on visualizing microbial associations using microscopy, while contemporary approaches utilize advanced techniques such as metagenomics, transcriptomics, and imaging technologies (Perreau & Moran, 2022). These advancements have provided deeper insights into the complexity and diversity of symbiotic relationships in marine environments, revealing previously unknown interactions and their ecological significance (Mutalipassi et al., 2021).

The historical background of symbiotic study highlights how multidisciplinary the discipline is, incorporating viewpoints from evolutionary biology, ecology, and microbiology. A more thorough knowledge of how symbiotic connections affect ecosystem dynamics and evolutionary processes has resulted from this combination. Furthermore, the attention has switched to seeing the collective genomes of host species and their symbionts as a single evolutionary unit as a result of the development of the hologenome theory (Raina et al., 2018).

6.2 TYPES OF SYMBIOTIC RELATIONSHIPS IN MARINE MICROBIAL COMMUNITIES

Marine microbial communities exhibit diverse symbiotic relationships that are critical to the functioning and stability of oceanic ecosystems. These interactions range from mutualistic, where both organisms benefit, to commensal and parasitic, which involve varying degrees of dependency and impact on the host (Figure 6.1 and Table 6.1). Each type of symbiotic relationship plays a unique role in influencing microbial diversity, nutrient cycling, and ecosystem resilience. Understanding these dynamics is essential for comprehending how marine microbes contribute to the broader ecological processes that sustain ocean life.

6.2.1 Mutualism

Mutualistic relationships in marine microbiomes involve reciprocal benefits between microorganisms and their hosts or other microbial partners. One notable example is the relationship between coral reefs and their associated dinoflagellates, *Symbiodinium* (Krediet et al., 2013). These photosynthetic microorganisms provide corals with

DOI: 10.1201/9781003509530-6

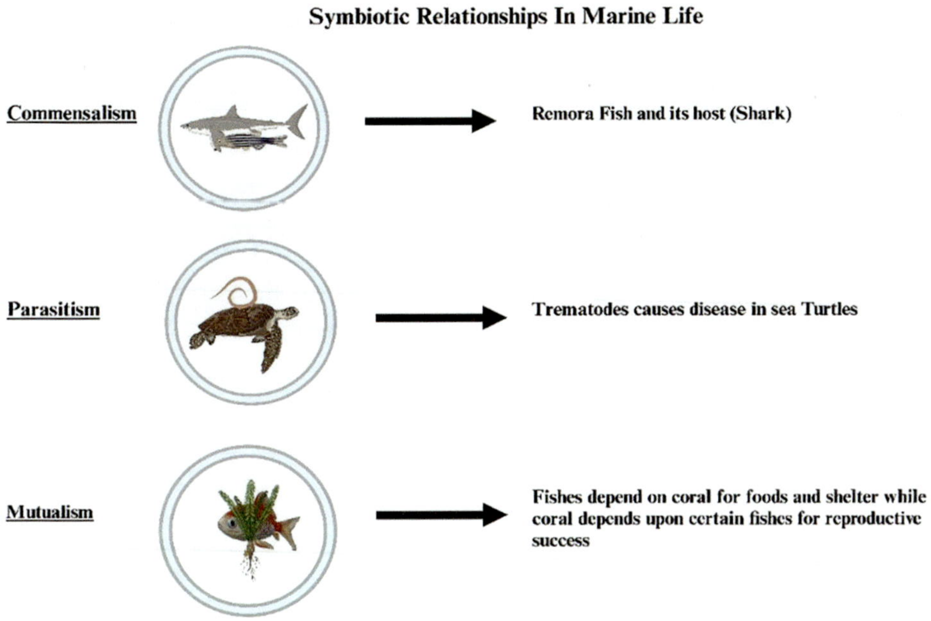

FIGURE 6.1 A visual diagram of the different types of symbiotic relationships (mutualism, commensalism, parasitism) using marine examples. (Self-developed using Biorender.com.)

TABLE 6.1
Types of Symbiotic Relationships and Their Ecological Roles

Type of Symbiosis	Definition	Ecological Impact	Example	Reference
Mutualism	Both organisms benefit.	Enhances nutrient cycling and productivity.	Coral reefs and *Symbiodinium*.	Jiang et al. (2021)
Commensalism	One organism benefits, the other is unaffected.	May support nutrient cycling and provide microbial stability.	Bacteria colonizing marine organisms without harm.	Iwasa-Arai et al. (2023)
Parasitism	One organism benefits at the expense of the other.	Influences microbial diversity and nutrient release.	Bacteriophages infecting bacteria.	Secor and Dandekar (2020)

essential nutrients through photosynthesis, while the coral host offers a protected environment and access to inorganic nutrients. This mutualism is vital for coral health and reef productivity. Another example includes nitrogen-fixing bacteria associated with marine plants such as seagrasses, where the bacteria provide fixed nitrogen in exchange for organic carbon from the plant (Mohamed et al., 2023). The Hawaiian bobtail squid (*Euprymna scolopes*) and the bioluminescent bacteria *Vibrio fischeri* have another interesting mutualistic interaction. In this association, the squid gives the bacteria a space in its light organ; in return, the bacteria release light that aids in the squid's ability to disguise itself from predators and increases its chances of survival. This interaction serves as an example of how mutualism may be extremely important to the ecology and behavior of marine creatures (Nyholm & McFall-Ngai, 2021).

6.2.2 Commensalism

Commensal relationships in marine microbiomes involve interactions where one organism benefits without significantly affecting the other. For instance, certain bacteria may colonize the surfaces of marine organisms such as fish or invertebrates without harming their hosts (Wilkins et al., 2019). These bacteria can gain access to nutrients and a stable environment, while the host organism remains largely unaffected. Commensal microbes can also contribute to the host's health by outcompeting pathogenic microbes or producing antimicrobial compounds (McLaren & Callahan, 2020).

In marine ecosystems, commensal relationships may be important for the cycling of nutrients. For instance, organic debris can be broken down by commensal bacteria found in marine environments, which releases nutrients back into the surrounding ecosystem. By keeping nutrients available for other creatures, this process not only helps the commensal bacteria by giving them food, but it also promotes the general health of the environment (Di Costanzo et al., 2023).

6.2.3 Parasitism

Parasitic relationships involve interactions where one organism benefits at the expense of the other. In marine

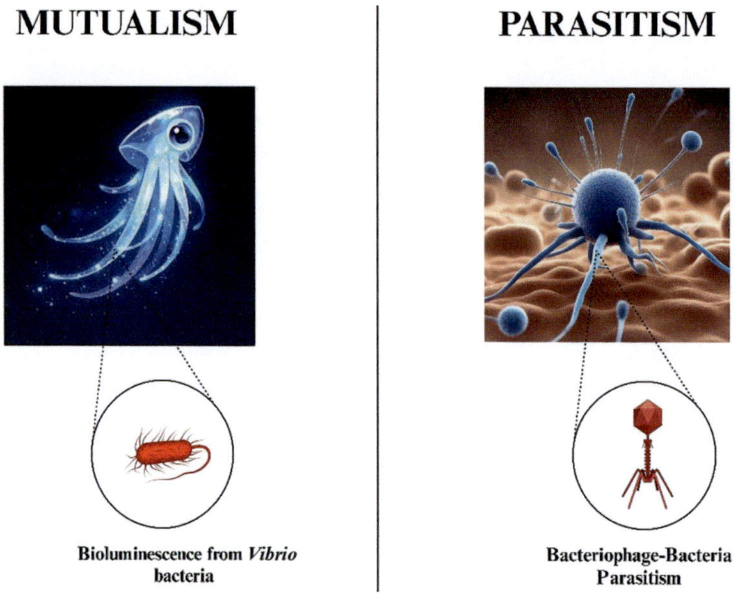

FIGURE 6.2 A split image showing the Squid-*Vibrio* mutualism on one side and the Bacteriophage-Bacteria parasitism on the other side. (Self-developed using Biorender.com.)

microbiomes, parasitic interactions can be observed between certain marine viruses and their microbial hosts (Drew et al., 2021). For example, bacteriophages infect and lyse bacterial cells, leading to the release of viral progeny and the death of the bacterial host. This parasitism can influence microbial community dynamics and nutrient cycling, with implications for ecosystem health and stability (Naureen et al., 2020).

Both the parasite and the host may undergo evolutionary changes as a result of parasitic partnerships. For example, the ongoing pressure from parasitic infections might encourage the evolution of more potent parasite methods while also causing microbial hosts to develop resistance mechanisms. This dynamic coevolution emphasizes the intricate relationship between parasitism and ecological balance by influencing the resilience of marine ecosystems overall and microbial population patterns (Figure 6.2) (Ebert & Fields, 2020).

6.3 FUNCTIONAL IMPLICATIONS OF SYMBIOTIC RELATIONSHIPS

6.3.1 Nutrient Cycling and Ecosystem Function

Symbiotic relationships significantly impact nutrient cycling and ecosystem functions in marine environments. Mutualistic interactions, such as those between marine plants and their associated nitrogen-fixing bacteria, contribute to nutrient availability and primary productivity (Fiore et al., 2010). Commensal and parasitic interactions can also influence nutrient dynamics by affecting microbial community structures and functions. For example, the lysis of bacterial cells by bacteriophages can release nutrients that are then utilized by other microorganisms, thereby influencing nutrient availability and cycling (Figure 6.3) (Alkhalil, 2023).

Marine habitat creation and preservation may involve symbiotic partnerships. For instance, the development and growth of coral reefs, which offer resources and shelter to a wide variety of marine animals, depend on the mutualistic interaction between coral polyps and their algae symbionts, or zooxanthellae. These vital habitats may deteriorate as a result of the loss of these symbiotic interactions, as demonstrated by coral bleaching events, which can have a domino impact on the ecosystem as a whole (Sheppard et al., 2012).

6.3.2 Microbial Diversity and Community Dynamics

Symbiotic relationships play a crucial role in shaping microbial diversity and community dynamics. Mutualistic interactions can enhance microbial diversity by fostering the growth of specific microbial taxa that are adapted to the symbiotic environment (Wiesmann et al., 2023). Conversely, parasitic interactions can lead to shifts in microbial community composition by selectively reducing the abundance of certain microbial groups. Understanding these dynamics is essential for elucidating the factors that regulate microbial community structures and functions in marine ecosystems (Bang et al., 2018).

By allowing several microbial species to survive in the same environment, commensal relationships can support the preservation of microbial diversity. Commensal microorganisms can avoid direct competition and coexist in a stable way by occupying different niches or making use of different resources, maintaining the diversity of the microbial community as a whole (Kern et al., 2021).

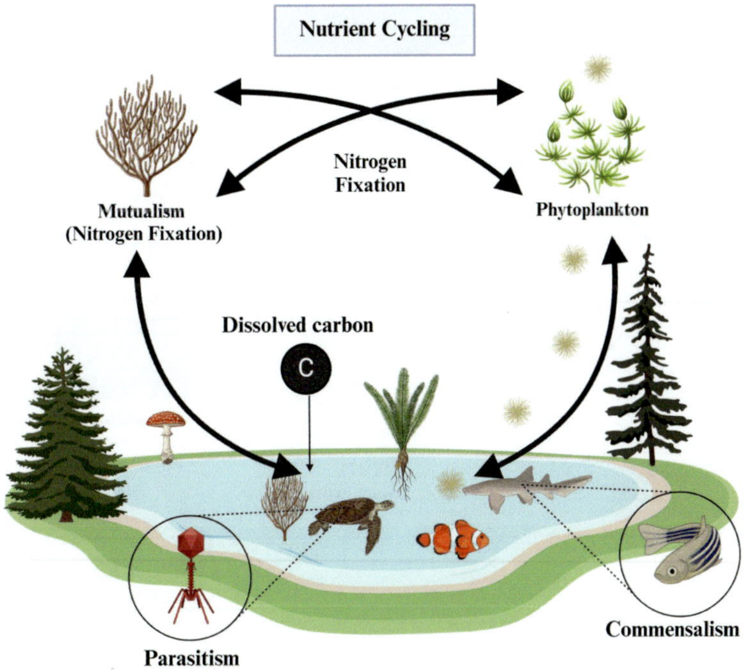

FIGURE 6.3 A model of nutrient cycling in marine environments with arrows showing how mutualism, commensalism, and parasitism influence the movement of nutrients. (Self-developed using Biorender.com.)

6.4 METHODOLOGIES FOR STUDYING SYMBIOTIC RELATIONSHIPS

6.4.1 Molecular and Genomic Approaches

Molecular and genomic approaches have revolutionized the study of symbiotic relationships in marine microbiomes. Techniques such as metagenomics, metatranscriptomics, and proteomics enable the comprehensive analysis of microbial communities and their interactions (Aguiar-Pulido et al., 2016). These methods provide insights into the genetic and functional capabilities of symbiotic microorganisms, revealing their roles in nutrient cycling, metabolic pathways, and ecological interactions. Also, advanced imaging techniques, including fluorescence *in situ* hybridization (FISH) and confocal microscopy, allow for the visualization of microbial interactions and the localization of specific microbial partners within host tissues (Figure 6.4) (Shakoori, 2017).

With the ability to examine the genetic makeup of individual microbial cells, single-cell genomics has become a valuable tool that helps scientists better comprehend the diversity of microbes and their potential for function within complex ecosystems. With the use of this technique, the genomes of uncultivated microorganisms can be recovered, exposing distinctive genetic characteristics and metabolic capacities that support symbiotic relationships and ecological processes in marine habitats. Single-cell genomics improves our understanding of how individual bacteria contribute to the overall dynamics of marine microbiomes by connecting genetic data to particular ecological roles (Woyke et al., 2017).

6.4.2 Experimental and Field Studies

Experimental and field studies are essential for investigating the functional implications of symbiotic relationships. Laboratory experiments, such as controlled co-cultivation studies and manipulation of environmental conditions, can provide insights into the mechanisms underlying microbial interactions and their effects on host organisms (Bosch et al., 2019). Field studies, including environmental sampling and monitoring of natural microbial communities, offer a broader perspective on the prevalence and ecological significance of symbiotic relationships in marine environments. Combining experimental and field approaches provides a holistic understanding of symbiosis roles and impacts in marine microbiomes (Leray et al., 2021). Longitudinal research on microbial communities provides valuable insights into the durability and stability of symbiotic relationships. The dynamics of symbiotic connections and its consequences for ecosystem health can be better understood by researchers by examining how these interactions alter in response to environmental changes or disruptions. This method helps identify critical boundaries or tipping points affecting the sustainability of these interactions and overall marine ecosystem health (Table 6.2) (Segaran et al., 2023).

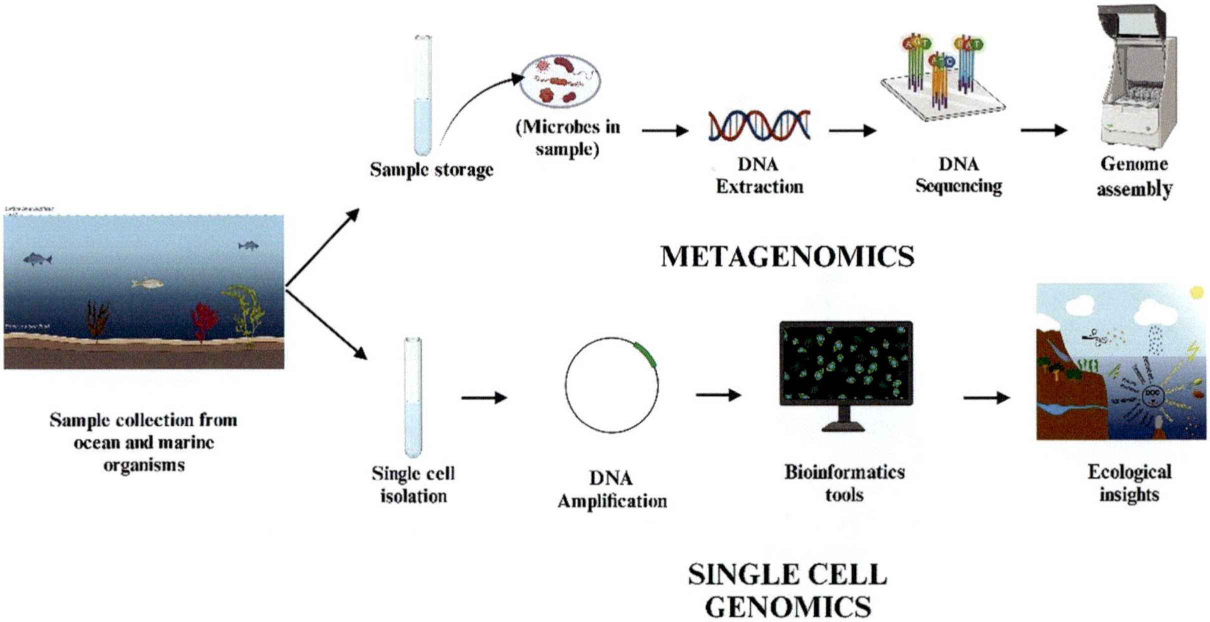

FIGURE 6.4 A diagram illustrating the various molecular and genomic approaches, such as metagenomics and single-cell genomics, alongside field study tools. (Self-developed using Biorender.com.)

TABLE 6.2
Methodologies for Studying Marine Symbiosis

Approach	Techniques	Insights	Reference
Molecular and genomic	Metagenomics, metatranscriptomics, single-cell genomics	Reveals genetic diversity, metabolic pathways, and symbiotic functions.	Mason et al. (2012)
Imaging	FISH, confocal microscopy	Visualizes microbial interactions and symbiont localization within host tissues.	Alesci et al. (2022)
Experimental and field studies	Controlled co-cultivation, environmental monitoring	Explores functional mechanisms and the prevalence of symbiotic relationships.	Selegato and Castro-Gamboa (2023)

6.5 SYMBIOSIS AND ENVIRONMENTAL CHANGE

6.5.1 Impact of Climate Change on Symbiotic Interactions

Climate change, including factors such as ocean warming, acidification, and nutrient enrichment, can alter symbiotic interactions in marine microbiomes. For instance, changes in seawater temperature and pH can affect the stability and function of mutualistic relationships, such as those between corals and their symbiotic dinoflagellates (Scanes et al., 2021). Ocean acidification may disrupt the ability of symbiotic microorganisms to perform essential metabolic functions, leading to shifts in microbial community structures and ecosystem dynamics. Understanding the impacts of environmental change on symbiosis is crucial for predicting and mitigating potential effects on marine ecosystems (Das & Mangwani, 2015). On marine food webs, the disruption of these symbiotic partnerships may have a domino effect. For instance, the entire reef ecosystem, which sustains a wide variety of marine life, may suffer if coral reefs deteriorate as a result of the loss of their symbiotic dinoflagellates. Fish populations and the means of subsistence for communities that depend on fishing and tourism linked to healthy coral reefs may be impacted by this loss in coral-dependent species (Figure 6.5) (Eddy et al., 2021).

6.5.2 Resilience and Adaptation of Symbiotic Relationships

The resilience and adaptation of symbiotic relationships to environmental change are key factors in maintaining ecosystem stability and function. Some marine microbes possess adaptive strategies that allow them to tolerate and thrive under changing conditions, such as the ability to modify metabolic pathways or engage in alternative symbiotic partnerships

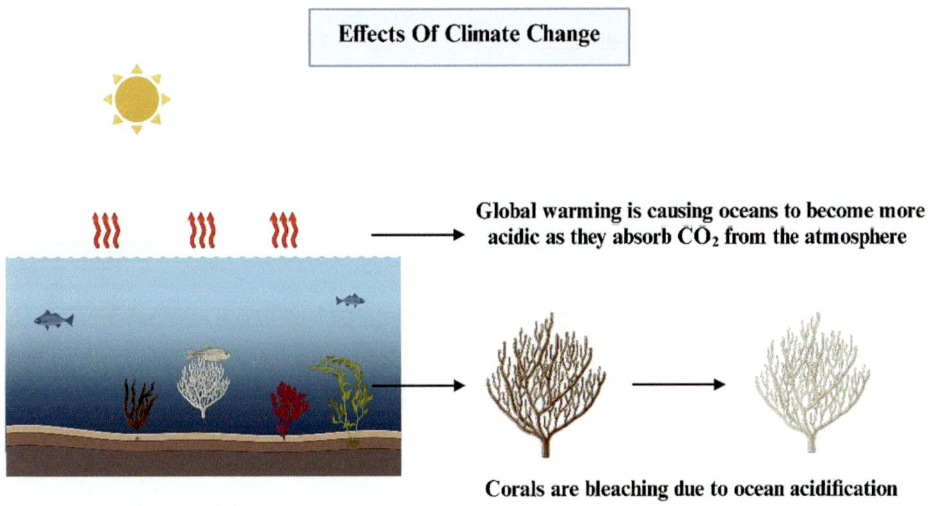

FIGURE 6.5 A visual showing the effects of climate change (e.g., ocean warming, acidification) on coral-dinoflagellate symbiosis. (Self-developed using Biorender.com.)

(Vanwonterghem & Webster, 2020). Investigating these adaptive responses provides insights into the potential for microbial resilience and the capacity of symbiotic relationships to withstand environmental stressors. Promoting the resilience of symbiotic interactions is essential for preserving the health and sustainability of marine ecosystems in the face of climate change (Philippot et al., 2021).

It is possible to forecast how marine microorganisms will adapt to changing environmental conditions by having a thorough understanding of their symbiotic interactions and evolutionary dynamics. Researchers can ascertain the durations over which adaptation takes place and find trade-offs between various evolutionary strategies by merging models of microbial adaptation with high-resolution ocean circulation data. This information is essential for predicting how the distribution of microbial traits and biogeochemical cycles may vary in response to climate change (Walworth et al., 2020).

6.6 FUTURE DIRECTIONS IN SYMBIOSIS RESEARCH

6.6.1 Emerging Trends and Technologies

Future research in symbiotic relationships within marine microbiomes will benefit from emerging trends and technologies. Advances in single-cell genomics, spatial transcriptomics, and high-resolution imaging techniques offer new opportunities for investigating microbial interactions at unprecedented levels of detail (Figure 6.6) (Wu et al., 2024). Integrating these technologies with traditional ecological and physiological approaches will enhance our understanding of symbiotic relationships and their roles in marine ecosystems. Also, the development of novel experimental models and techniques will facilitate the exploration of previously uncharacterized interactions and their functional implications (Stévenne et al., 2021).

Maximizing the potential of these new technologies will require interdisciplinary collaborations between ecologists, microbiologists, and bioinformaticians. This will allow for more thorough studies that can address intricate ecological issues and advance our knowledge of marine microbiomes (Parti & Szigeti, 2021).

6.6.2 Integration into Ecosystem Management

Integrating knowledge of symbiotic relationships into ecosystem management is crucial for addressing challenges related to marine conservation and sustainability. Understanding the ecological roles of symbiotic interactions can inform management strategies aimed at preserving marine biodiversity, enhancing ecosystem resilience, and mitigating the impacts of environmental change (Xu & Zhou, 2024). Collaborative efforts between researchers, policymakers, and stakeholders will be essential for translating scientific insights into effective management practices and promoting the long-term health of marine ecosystems (Lukambagire et al., 2024).

Furthermore, symbiotic interactions can be used to prioritize conservation efforts and identify essential areas in maritime spatial planning and zoning. To conserve these crucial ecological relationships, managers might identify and designate protected areas or carry out specific management measures in areas with high densities of symbiotic interactions or keystone symbiotic species (Podda & Porporato, 2023).

6.7 CONCLUSION

Symbiotic relationships within marine microbial communities are pivotal to the structure and function of oceanic ecosystems. These interactions—whether mutualistic, commensal, or parasitic—affect nutrient cycling, microbial diversity, and overall ecosystem dynamics. Recent advancements in

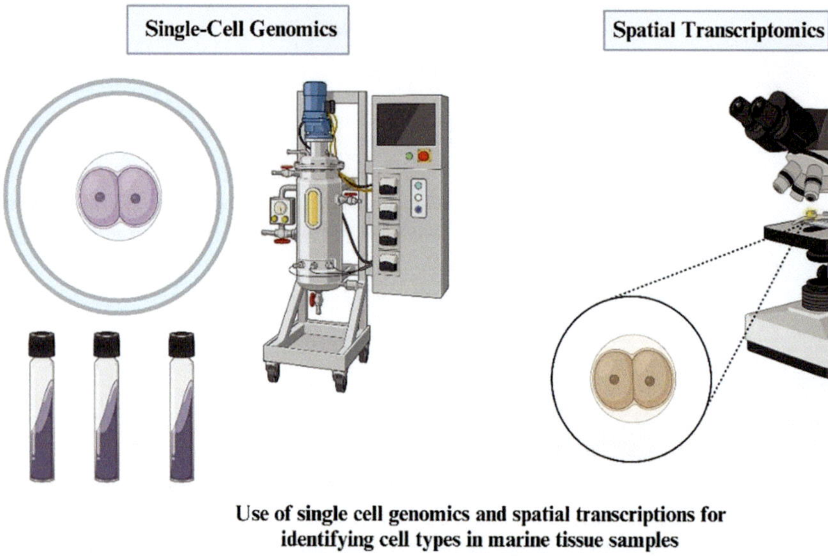

FIGURE 6.6 A graphic showcasing emerging trends and technologies like single-cell genomics and spatial transcriptomics. (Self-developed using Biorender.com.)

molecular, genomic, and high-resolution imaging techniques have significantly enhanced our understanding of these intricate relationships, revealing their ecological importance and functional roles. Techniques such as metagenomics, single-cell genomics, and advanced computational modeling have provided deeper insights into the resilience and adaptability of these symbiotic interactions under environmental stressors. This study is particularly important as it not only elucidates the mechanisms driving symbiotic resilience but also provides critical data for predicting the impacts of climate change and other anthropogenic pressures on marine ecosystems. By integrating these findings into ecosystem management strategies, we can better safeguard the health and sustainability of marine microbiomes. Future research, leveraging innovative technological approaches, will be essential for advancing our knowledge and ensuring the preservation of marine ecosystem integrity amidst ongoing environmental challenges.

ABBREVIATIONS

CO_2	Carbon dioxide
DNA	Deoxyribonucleic Acid
FISH	Fluorescence *in situ* Hybridization
N_2	Nitrogen

ACKNOWLEDGEMENT

The authors gratefully acknowledge the financial support provided by the Department of Biotechnology (DBT), Ministry of Science and Technology, Government of India, under Project No. DBT/2022–23/July/N/2240, as part of the DBT-RA Program, as well as the support from the School of Bioengineering and Biosciences, Lovely Professional University, Punjab, India.

REFERENCES

Aguiar-Pulido, V., Huang, W., Suarez-Ulloa, V., Cickovski, T., Mathee, K., & Narasimhan, G. (2016). Metagenomics, metatranscriptomics, and metabolomics approaches for microbiome analysis. *Evolutionary Bioinformatics*, *12*. https://doi.org/10.4137/EBO.S36436

esson, A., Curtsdotter, A., Eklöf, A., Ebenman, B., Norberg, J., & Barabás, G. (2021). The importance of species interactions in eco-evolutionary community dynamics under climate change. *Nature Communications*, *12*(1), 4759. https://doi.org/10.1038/s41467-021-24977-x

Alesci, A., Pergolizzi, S., Savoca, S., Fumia, A., Mangano, A., Albano, M., Messina, E., Aragona, M., Lo Cascio, P., Capillo, G., & Lauriano, E. R. (2022). Detecting intestinal goblet cells of the broadgilled hagfish *Eptatretus cirrhatus* (Forster, 1801): a confocal microscopy evaluation. *Biology*, *11*(9), 1366. https://doi.org/10.3390/biology11091366

Alkhalil, S. S. (2023). The role of bacteriophages in shaping bacterial composition and diversity in the human gut. *Frontiers in Microbiology*, *14*. https://doi.org/10.3389/fmicb.2023.1232413

Bang, C., Dagan, T., Deines, P., Dubilier, N., Duschl, W. J., Fraune, S., Hentschel, U., Hirt, H., Hülter, N., Lachnit, T., Picazo, D., Pita, L., Pogoreutz, C., Rädecker, N., Saad, M. M., Schmitz, R. A., Schulenburg, H., Voolstra, C. R., Weiland-Bräuer, N., … Bosch, T. C. G. (2018). Metaorganisms in extreme environments: do microbes play a role in organismal adaptation? *Zoology*, *127*, 1–19. https://doi.org/10.1016/j.zool.2018.02.004

Bosch, T. C. G., Guillemin, K., & McFall-Ngai, M. (2019). Evolutionary "experiments" in symbiosis: the study of model animals provides insights into the mechanisms underlying the diversity of host–microbe interactions. *BioEssays*, *41*(10). https://doi.org/10.1002/bies.201800256

Das, S., & Mangwani, N. (2015). Ocean acidification and marine microorganisms: responses and consequences. *Oceanologia*, *57*(4), 349–361. https://doi.org/10.1016/j.oceano.2015.07.003

Di Costanzo, F., Di Dato, V., & Romano, G. (2023). Diatom–bacteria interactions in the marine environment: complexity, heterogeneity, and potential for biotechnological applications. *Microorganisms*, *11*(12), 2967. https://doi.org/10.3390/microorganisms11122967

Drew, G. C., Stevens, E. J., & King, K. C. (2021). Microbial evolution and transitions along the parasite–mutualist continuum. *Nature Reviews Microbiology*, *19*(10), 623–638. https://doi.org/10.1038/s41579-021-00550-7

Ebert, D., & Fields, P. D. (2020). Host–parasite co-evolution and its genomic signature. *Nature Reviews Genetics*, *21*(12), 754–768. https://doi.org/10.1038/s41576-020-0269-1

Eddy, T. D., Lam, V. W. Y., Reygondeau, G., Cisneros-Montemayor, A. M., Greer, K., Palomares, M. L. D., Bruno, J. F., Ota, Y., & Cheung, W. W. L. (2021). Global decline in capacity of coral reefs to provide ecosystem services. *One Earth*, *4*(9), 1278–1285. https://doi.org/10.1016/j.oneear.2021.08.016

Fiore, C. L., Jarett, J. K., Olson, N. D., & Lesser, M. P. (2010). Nitrogen fixation and nitrogen transformations in marine symbioses. *Trends in Microbiology*, *18*(10), 455–463. https://doi.org/10.1016/j.tim.2010.07.001

Iwasa-Arai, T., Andrade, S. C. S., Miguel, C., Siqueira, S. G. L. B., Werneck, M. R., Leite, F. P. P., Moraes, L., Santos, M. R. D., Medeiros, L. S., Lopes, U. H., & Serejo, C. (2023). Coevolutionary implications of obligate commensalism in sea turtles: the case of the genus Hyachelia Barnard, 1967 (*Crustacea*, *Amphipoda*). *Zoological Studies*, *62*, e54. https://doi.org/10.6620/ZS.2023.62-54

Jiang, J., Wang, A., Deng, X., Zhou, W., Gan, Q., & Lu, Y. (2021). How *Symbiodiniaceae* meets the challenges of life during coral bleaching. *Coral Reefs*, *40*(4), 1339–1353. https://doi.org/10.1007/s00338-021-02115-9

Kern, L., Abdeen, S. K., Kolodziejczyk, A. A., & Elinav, E. (2021). Commensal inter-bacterial interactions shaping the microbiota. *Current Opinion in Microbiology*, *63*, 158–171. https://doi.org/10.1016/j.mib.2021.07.011

Krediet, C. J., Ritchie, K. B., Paul, V. J., & Teplitski, M. (2013). Coral-associated micro-organisms and their roles in promoting coral health and thwarting diseases. *Proceedings of the Royal Society B: Biological Sciences*, *280*(1755), 20122328. https://doi.org/10.1098/rspb.2012.2328

Leray, M., Wilkins, L. G. E., Apprill, A., Bik, H. M., Clever, F., Connolly, S. R., De León, M. E., Duffy, J. E., Ezzat, L., Gignoux-Wolfsohn, S., Herre, E. A., Kaye, J. Z., Kline, D. I., Kueneman, J. G., McCormick, M. K., McMillan, W. O., O'Dea, A., Pereira, T. J., Petersen, J. M., … Eisen, J. A. (2021). Natural experiments and long-term monitoring are critical to understand and predict marine host–microbe ecology and evolution. *PLOS Biology*, *19*(8), e3001322. https://doi.org/10.1371/journal.pbio.3001322

Li, Z. (2024). Editorial: Marine microbial symbioses: host-microbe interaction, holobiont's adaptation to niches and global climate change. *Frontiers in Microbiology*, *15*. https://doi.org/10.3389/fmicb.2024.1416897

Lukambagire, I., Matovu, B., Manianga, A., Bhavani, R. R., & S, A. (2024). Towards a collaborative stakeholder engagement pathway to increase ocean sustainability related to marine spatial planning in developing coastal states. *Environmental Challenges*, *15*, 100954. https://doi.org/10.1016/j.envc.2024.100954

Mason, O. U., Hazen, T. C., Borglin, S., Chain, P. S. G., Dubinsky, E. A., Fortney, J. L., Han, J., Holman, H.-Y. N., Hultman, J., Lamendella, R., Mackelprang, R., Malfatti, S., Tom, L. M., Tringe, S. G., Woyke, T., Zhou, J., Rubin, E. M., & Jansson, J. K. (2012). Metagenome, metatranscriptome and single-cell sequencing reveal microbial response to Deepwater Horizon oil spill. *The ISME Journal*, *6*(9), 1715–1727. https://doi.org/10.1038/ismej.2012.59

Maureaud, A., Gascuel, D., Colléter, M., Palomares, M. L. D., Du Pontavice, H., Pauly, D., & Cheung, W. W. L. (2017). Global change in the trophic functioning of marine food webs. *PLoS One*, *12*(8), e0182826. https://doi.org/10.1371/journal.pone.0182826

McLaren, M. R., & Callahan, B. J. (2020). Pathogen resistance may be the principal evolutionary advantage provided by the microbiome. *Philosophical Transactions of the Royal Society B: Biological Sciences*, *375*(1808), 20190592. https://doi.org/10.1098/rstb.2019.0592

Mohamed, A. R., Ochsenkühn, M. A., Kazlak, A. M., Moustafa, A., & Amin, S. A. (2023). The coral microbiome: towards an understanding of the molecular mechanisms of coral–microbiota interactions. *FEMS Microbiology Reviews*, *47*(2). https://doi.org/10.1093/femsre/fuad005

Mutalipassi, M., Riccio, G., Mazzella, V., Galasso, C., Somma, E., Chiarore, A., de Pascale, D., & Zupo, V. (2021). Symbioses of cyanobacteria in marine environments: ecological insights and biotechnological perspectives. *Marine Drugs*, *19*(4), 227. https://doi.org/10.3390/md19040227

Naureen, Z., Dautaj, A., Anpilogov, K., Camilleri, G., Dhuli, K., Tanzi, B., Maltese, P. E., Cristofoli, F., De Antoni, L., Beccari, T., Dundar, M., & Bertelli, M. (2020). Bacteriophages presence in nature and their role in the natural selection of bacterial populations. Acta Bio-Medica *: Atenei Parmensis*, *91*(13-S), e2020024. https://doi.org/10.23750/abm.v91i13-S.10819

Nyholm, S. V., & McFall-Ngai, M. J. (2021). A lasting symbiosis: how the Hawaiian bobtail squid finds and keeps its bioluminescent bacterial partner. *Nature Reviews Microbiology*, *19*(10), 666–679. https://doi.org/10.1038/s41579-021-00567-y

Parti, K., & Szigeti, A. (2021). The future of interdisciplinary research in the digital era: obstacles and perspectives of collaboration in social and data sciences—an empirical study. *Cogent Social Sciences*, *7*(1). https://doi.org/10.1080/23311886.2021.1970880

Perreau, J., & Moran, N. A. (2022). Genetic innovations in animal–microbe symbioses. *Nature Reviews Genetics*, *23*(1), 23–39. https://doi.org/10.1038/s41576-021-00395-z

Philippot, L., Griffiths, B. S., & Langenheder, S. (2021). Microbial community resilience across ecosystems and multiple disturbances. *Microbiology and Molecular Biology Reviews*, *85*(2). https://doi.org/10.1128/MMBR.00026-20

Podda, C., & Porporato, E. M. D. (2023). Marine spatial planning for connectivity and conservation through ecological corridors between marine protected areas and other effective area-based conservation measures. *Frontiers in Marine Science*, *10*. https://doi.org/10.3389/fmars.2023.1271397

Raina, J.-B., Eme, L., Pollock, F. J., Spang, A., Archibald, J. M., & Williams, T. A. (2018). Symbiosis in the microbial world: from ecology to genome evolution. *Biology Open*, *7*(2). https://doi.org/10.1242/bio.032524

Scanes, E., Parker, L. M., Seymour, J. R., Siboni, N., King, W. L., Danckert, N. P., Wegner, K. M., Dove, M. C., O'Connor, W. A., & Ross, P. M. (2021). Climate change alters the haemolymph microbiome of oysters. *Marine Pollution Bulletin*, *164*, 111991. https://doi.org/10.1016/j.marpolbul.2021.111991

Secor, P. R., & Dandekar, A. A. (2020). More than simple parasites: the sociobiology of bacteriophages and their bacterial hosts. *MBio*, *11*(2). https://doi.org/10.1128/mBio.00041-20

Segaran, T. C., Azra, M. N., Lananan, F., & Wang, Y. (2023). Microbe, climate change and marine environment: linking trends and research hotspots. *Marine Environmental Research*, *189*, 106015. https://doi.org/10.1016/j.marenvres.2023.106015

Selegato, D. M., & Castro-Gamboa, I. (2023). Enhancing chemical and biological diversity by co-cultivation. *Frontiers in Microbiology*, *14*. https://doi.org/10.3389/fmicb.2023.1117559

Shakoori, A. R. (2017). Fluorescence in situ hybridization (FISH) and its applications. In Bhat, T. A., & Wani, A. A., Eds., *Chromosome Structure and Aberrations* (pp. 343–367). Springer India. https://doi.org/10.1007/978-81-322-3673-3_16

Sheppard, C., Al-Husiani, M., Al-Jamali, F., Al-Yamani, F., Baldwin, R., Bishop, J., Benzoni, F., Dutrieux, E., Dulvy, N. K., Durvasula, S. R. V., Jones, D. A., Loughland, R., Medio, D., Nithyanandan, M., Pilling, G. M., Polikarpov, I., Price, A. R. G., Purkis, S. J., Riegl, B. M., … Zainal, K. (2012). Environmental concerns for the future of gulf coral reefs. In Riegl, B. M., & Purkis, S. J., Eds., *Coral Reefs of the World* (pp. 349–373). Springer. https://doi.org/10.1007/978-94-007-3008-3_16

Srivastava, P., Sahgal, M., & Dasila, H. (2023). Microbial symbiosis in marine ecosystem. In Morales-Oyervides, L., Fouillaud, M., Suyal, D. C., & Soni, R., Eds., *Current Status of Marine Water Microbiology* (pp. 33–44). Springer Nature Singapore. https://doi.org/10.1007/978-981-99-5022-5_3

Stévenne, C., Micha, M., Plumier, J.-C., & Roberty, S. (2021). Corals and sponges under the light of the holobiont concept: how microbiomes underpin our understanding of marine ecosystems. *Frontiers in Marine Science*, *8*. https://doi.org/10.3389/fmars.2021.698853

Vanwonterghem, I., & Webster, N. S. (2020). Coral reef microorganisms in a changing climate. *IScience*, *23*(4), 100972. https://doi.org/10.1016/j.isci.2020.100972

Walworth, N. G., Zakem, E. J., Dunne, J. P., Collins, S., & Levine, N. M. (2020). Microbial evolutionary strategies in a dynamic ocean. *Proceedings of the National Academy of Sciences of the United States of America*, *117*(11), 5943–5948. https://doi.org/10.1073/pnas.1919332117

Wiesmann, C. L., Wang, N. R., Zhang, Y., Liu, Z., & Haney, C. H. (2023). Origins of symbiosis: shared mechanisms underlying microbial pathogenesis, commensalism and mutualism of plants and animals. *FEMS Microbiology Reviews*, *47*(6). https://doi.org/10.1093/femsre/fuac048

Wilkins, L. G. E., Leray, M., O'Dea, A., Yuen, B., Peixoto, R. S., Pereira, T. J., Bik, H. M., Coil, D. A., Duffy, J. E., Herre, E. A., Lessios, H. A., Lucey, N. M., Mejia, L. C., Rasher, D. B., Sharp, K. H., Sogin, E. M., Thacker, R. W., Vega Thurber, R., Wcislo, W. T., … Eisen, J. A. (2019). Host-associated microbiomes drive structure and function of marine ecosystems. *PLOS Biology*, *17*(11), e3000533. https://doi.org/10.1371/journal.pbio.3000533

Woyke, T., Doud, D. F. R., & Schulz, F. (2017). The trajectory of microbial single-cell sequencing. *Nature Methods*, *14*(11), 1045–1054. https://doi.org/10.1038/nmeth.4469

Wu, Y., Zhuang, J., Song, Y., Gao, X., Chu, J., & Han, S. (2024). Advances in single-cell sequencing technology in microbiome research. *Genes & Diseases*, *11*(4), 101129. https://doi.org/10.1016/j.gendis.2023.101129

Xu, Y., & Zhou, M. (2024). Network-perspective marine ecosystem conservation and management, from concepts to applications. *HydroResearch*, *7*, 191–200. https://doi.org/10.1016/j.hydres.2024.03.002

7 Discovery, Characterization, and Utilization of Marine Enzymes

Anindita Biswas[1], Tanuja Mohanty[1], Ritu Kumari[2], and Gurmeen Rakhra[3]
[1]Department of Microbiology, School of Bioengineering and Biosciences, Lovely Professional University, Phagwara, India
[2]Department of Botany, School of Bioengineering and Biosciences, Lovely Professional University, Phagwara, India
[3]Department of Biochemistry, School of Bioengineering and Biosciences, Lovely Professional University, Phagwara, India

7.1 INTRODUCTION

Marine enzymes, derived from organisms inhabiting the diverse and extreme conditions of the ocean, exhibit a remarkable array of properties and functions that make them invaluable across various scientific and industrial domains (Honda et al., 1987). These enzymes, sourced from marine bacteria, fungi, algae, and invertebrates, often possess unique characteristics such as high stability under extreme temperatures, pressure, and salinity (Galperin et al., 2015). This adaptability enables their application in industries ranging from bioremediation and waste management to pharmaceuticals and food processing. Marine enzymes are particularly notable for their role in enhancing biotechnological processes, thanks to their robustness and efficiency in harsh conditions. Advances in techniques like metagenomics and structural biology continue to reveal novel enzymes and their potential, driving forward both fundamental research and practical applications (El-Gebali et al., 2019). The study of microorganisms' genomes that are taken straight out of the ocean is known as marine metagenomics. DNA extraction and cloning from a variety of organisms is a rapidly expanding topic of study. In addition to assisting in the development of new bioresources, marine metagenomics can teach us more about the biology and biochemistry of the ocean. The study of genetic material extracted directly from oceanic settings is known as marine metagenomics (Brazelton et al., 2012). By analysing the collective genome of all microorganisms in a sample, metagenomics allows researchers to get insights about the variety, function, and interactions of microbial communities in the ocean without the need to isolate and cultivate individual species (Harborne et al., 1992).

This field has revealed the enormous diversity of marine life, including many species that were previously unknown, and has completely changed our knowledge of marine ecosystems. Studying processes like carbon sequestration, nutrient cycling, and how marine life responds to environmental changes like pollution and climate change requires an understanding of marine metagenomics (Bianchi et al., 2018). Scientists can follow the evolution of microbial populations, uncover genes linked to certain roles, and find novel enzymes or bioactive substances with potential medical or biotechnological uses by examining DNA sequences from saltwater, sediments, or marine species (Tully et al., 2017). As exploration of marine environments progresses, the future holds promise for discovering even more diverse and useful enzymes, highlighting their growing importance in science and technology.

7.1.1 Objectives of This Chapter

- **Biodiversity Exploration:** Discover new enzymes from diverse marine organisms. Marine environments harbours a vast array of unique organisms, many of which produce enzymes with novel properties not found in terrestrial organisms (Sharma et al. 2016). Exploring this biodiversity can lead to the identification of enzymes with unique functionalities (Nguyen & Nguyen, 2017).
- **Biotechnological Applications:** Develop and optimize marine enzymes for industrial applications (Almaliti & Gerwick, 2023). Marine enzymes can be employed in various industries, including pharmaceuticals, agriculture, and environmental management. For instance, enzymes with extreme pH or temperature stability may be used in industrial processes, while others might be used in bioremediation or waste treatment (Hyde et al., 1998; Honda et al., 1987; Galperin et al., 2015).
- **Functional Characterization:** Understand the structure, function, and mechanism of marine enzymes (Logares et al., 2014). Detailed characterization helps in tailoring enzymes for specific applications, improving their efficiency and stability, and overcoming challenges related to enzyme performance in industrial settings (Guérard, 2007).
- **Environmental and Ecological Impact:** Assess the role of marine enzymes in ecological processes. Marine enzymes play crucial roles in nutrient cycling, organic matter degradation, and other

DOI: 10.1201/9781003509530-7

ecological functions. Understanding these roles can provide insights into marine ecosystems and their health (Rustad et al., 2011).

- **Sustainability and Resource Management:** Utilize marine enzymes to promote sustainable practices (Rohrmann et al., 1992). Marine enzymes can be used in processes that reduce environmental impact, such as green chemistry methods, sustainable agriculture practices, or bio-based products, contributing to resource conservation and environmental protection.
- **Innovation and Research Development:** Advance scientific knowledge and innovation through marine enzyme research. Research into marine enzymes can drive technological innovation, contribute to scientific discoveries, and foster interdisciplinary collaborations across fields such as marine biology, chemistry, and biotechnology (Qeshmi et al., 2020).

7.1.2 Importance of Marine Enzymes

Marine enzymes are becoming more and more known for their remarkable qualities and diverse range of uses. Their exceptional capacity to operate in harsh environments, including elevated pressure, salinity, and temperature, confers significant value in the biotechnological sectors (Nguyen & Nguyen, 2017). For example, these enzymes function as biocatalysts in industrial processes, accelerating chemical reactions in a way that is more ecologically benign than that of conventional chemical catalysts. Because of their distinct modes of action, marine enzymes present pharmaceutical companies with attractive avenues for the development of novel medications, such as antibiotics, antivirals, and anticancer medicines (Logares et al., 2015).

Marine enzymes are particularly advantageous to the food sector since they may be employed to create novel flavours, increase the nutritional content of products, and make bioactive peptides (Moisander et al., 2017). Furthermore, they play a big part in environmental protection, especially when it comes to bioremediation projects where they break down plastics and oil spills (Hewson et al., 2007). This skill is crucial for reducing the effects of industrial activity and cleaning up maritime ecosystems. Additionally, marine enzymes are employed to detoxify industrial waste, changing dangerous materials into less dangerous ones (Tiirola et al., 2002).

These enzymes increase the effectiveness of feed-based nutrient absorption, which promotes healthier aquatic organism growth and health in aquaculture and agriculture (Gildberg, 1993). Additionally, research is being done on their insecticidal qualities as potential natural substitutes for artificial pesticides. As far as science is concerned, marine enzymes are essential for furthering synthetic biology and genomic research (Suleria et al., 2016). They provide information on how organisms adapt to harsh conditions and can be engineered for particular uses, such as the production of biofuel.

Another significant benefit is the sustainable application of marine enzymes, which encourages greener industrial practices by lowering the requirement for harsh chemicals in a variety of operations (Salamone et al., 2012). Additionally, the study of these enzymes emphasizes how crucial it is to protect marine ecosystems, which are essential repositories of biological diversity and resources. Maritime enzymes are therefore essential for innovation and sustainability in many other fields, and they also serve as a stark reminder of the urgent need for conservation initiatives to save the maritime habitats that support these priceless biological resources (Jagtap & Manohar, 2021).

7.1.3 Importance of Marine Metagenomics

Directly recoverable genetic material from environmental samples in marine environments is studied by a field known as marine metagenomics. With the use of this method, scientists can examine the whole genomes of large oceanic microbial communities rather than isolating and growing individual species (Tiirola et al., 2002). Understanding the great diversity of marine microorganisms—many of which are uncultivable in lab settings—is possible through marine metagenomics. New genes, enzymes, and metabolic pathways that may be used in biotechnology, medicine, or environmental management are also discovered with its assistance (Addison et al., 2007). Understanding the functions of marine microorganisms in the health of ecosystems, the discovery of new biological resources, and global biogeochemical cycles depends on this field (Mohapatra et al., 2003).

The utilization of marine metagenomics has significantly transformed our comprehension of the diversity of microorganisms in the ocean and their influence on worldwide processes. The ability of marine metagenomics to uncover the enormous and until unknown genetic variety of marine microorganisms—many of which are incapable of being cultivated or studied conventionally—makes the technology significant (Mehrshad et al., 2018). Scientists may identify and characterize entire microbial communities and get insights into their involvement in marine ecosystems by directly examining DNA from environmental samples.

Finding novel genes and metabolic pathways is one of the main achievements of marine metagenomics (Buesseler et al., 2007). Since novel enzymes, antibiotics, and other bioactive substances can be found and possibly exploited in industrial, medicinal, and environmental applications, these results have a substantial impact on biotechnology. For example, the study of marine metagenomics has produced enzymes that can survive in harsh environments, which is useful for industrial processes that need to be highly stable and efficient (Barzkar & Sohail, 2020).

Our comprehension of the ecological roles played by marine microorganisms is further improved by marine metagenomics. These microorganisms are vital to the global biogeochemical cycles that sustain life on Earth and control its temperature, such as the nitrogen and carbon cycles (Snelgrove et al., 1997). Through examining the genetic composition of microbial communities, scientists can learn more about these processes and how environmental changes, such as pollution or climate change, may impact them. It makes it possible to identify alterations in microbial communities that could point to changes in the health of the ecosystem or the presence of contaminants. Managing and safeguarding maritime environments requires this knowledge, especially in light of growing environmental constraints and human activity (Douvere, 2008).

Additionally, marine metagenomics advances the understanding of marine viruses, which are prevalent in the ocean and have significant functions in controlling microbial populations and the cycling of nutrients (Borja et al., 2012). The stability of marine ecosystems and marine food webs can be understood by knowing the dynamics of viruses and how they interact with their hosts. All things considered, marine metagenomics is critical to expanding our understanding of marine biodiversity, finding new biological resources, and comprehending the intricate relationships that support marine life. It is an essential instrument for tackling global issues like environmental preservation, climate change mitigation, and the sustainable use of marine resources.

7.2 DISCOVERY OF MARINE ENZYMES

The discovery of marine enzymes represents a fascinating intersection of marine biology and biochemistry, underscoring the immense potential of the ocean's untapped resources (Zhang & Kim, 2010). Covering more than 70% of the Earth's surface, the marine environment is a treasure trove of biodiversity, with countless species adapted to a wide range of extreme and unique conditions. This immense diversity includes microorganisms, algae, and marine invertebrates that produce a variety of enzymes with distinctive properties (Zhao et al., 2016). The journey to discovering these enzymes began with the realization that the ocean harbours a vast array of organisms, many of which were previously unexplored. Early efforts in marine enzyme discovery often involved isolating enzymes from marine organisms and testing their properties in the laboratory (Coutinho et al., 2018). As techniques evolved, the discipline grew to encompass sophisticated molecular biology tools like metagenomics and proteomics, enabling the investigation of community composition and the discovery of undiscovered enzyme genes directly from samples obtained from the environment.

This section in the chapter explores the methods and milestones in the discovery of marine enzymes, detailing the innovative approaches used to uncover these biological treasures (Alma'abadi et al., 2015). It highlights the challenges and successes encountered in this ongoing quest and outlines the significant impact that marine enzymes have had on various scientific and industrial fields. By delving into the history and techniques of marine enzyme research, we gain insight into how these remarkable molecules continue to shape our understanding and utilization of marine resources (Kodzius & Gojobori, 2015).

The discovery of marine enzymes has been a significant and fascinating journey, blending scientific curiosity with practical applications. Here's a timeline highlighting key milestones in the field.

7.2.1 Early Discoveries and Foundations

Early 20th Century: Initial studies on marine microorganisms began, but the focus was not specifically on enzymes. Scientists were interested in the biodiversity of marine life and their potential applications (Birolli et al., 2019).

1950s and 1960s: As microbiology and marine biology advanced, researchers began isolating and characterizing enzymes from marine organisms (Bäumgen et al., 2021). This period saw the beginning of systematic exploration of marine environments for biochemical compounds.

1970s and 1980s: The discovery of extremophiles (organisms that thrive in extreme conditions) in marine environments, such as deep-sea hydrothermal vents, spurred interest in the unique enzymes they produce (Mineta et al., 2016). This era marked significant advancements in understanding marine microbial ecology and the biochemical capabilities of marine organisms.

1990s: With the advent of molecular biology techniques, including DNA sequencing and cloning, the study of marine enzymes accelerated. Scientists could now identify and characterize enzymes at the genetic level, leading to the discovery of many novel enzymes (Mizuno et al., 2013).

2000s Onwards: The advent of metagenomics along with additional methods with high throughput transformed marine enzyme discovery. These technologies enabled the direct examination of genetic code from samples obtained from the environment, eliminating the need to culture organisms. This era has seen a surge in the search for newly discovered marine enzymes with numerous and distinctive characteristics (Tseng & Tang, 2014).

7.2.2 Key Milestones in Marine Enzyme Research

7.2.2.1 Isolation and Characterization (1980s–1990s)

During this period, significant strides were made in isolating and understanding marine enzymes.

- **1980s:** Research focused on isolating enzymes from marine sources, including deep-sea organisms and extremophiles. These enzymes were found to exhibit remarkable stability and activity under extreme conditions, such as high pressure and salinity (Coutinho et al., 2018).
- **1988:** The publication of studies detailing the biochemical properties of marine enzymes

catalysed further research. One notable study was the isolation of marine proteases and their applications in industrial processes (Hurwitz et al., 2014).

- **1990s:** Advances in molecular biology techniques, such as gene cloning and sequencing, allowed scientists to delve deeper into the genetic and structural aspects of marine enzymes. The characterization of marine enzymes from diverse organisms, including bacteria, algae, and invertebrates, provided valuable insights into their unique properties (Popovic et al., 2015). The ability to examine the whole genomes of entire microbial communities from environmental samples became apparent to scientists as sequencing got more affordable and available. The field of marine metagenomics arose from the larger discipline of metagenomics, which is the direct investigation of genetic material extracted from environmental samples (Khare et al., 2015). This method, when applied to marine habitats, enables scientists to investigate the great diversity of microorganisms present there without the need to culture them in a lab, a procedure that is sometimes unfeasible for many marine species. In marine habitats, metagenomic research had its start in the 1990s and to sequence the genomes of the microbial communities, researchers first collected samples of ocean water and soil, separated the DNA from it, and then used it (Rastogi et al., 2017).
- **Environmental DNA (eDNA Studies):** The field of environmental DNA (eDNA) studies began to gain popularity in the 1980s and 1990s when it became possible to analyse DNA directly from environmental samples. Researchers started amplifying and studying genes (such as 16S rRNA genes) directly from environmental samples using molecular techniques like the polymerase chain reaction (PCR). As a result, microbiological species that had never been cultivated may be identified (Saeman et al., 1945).
- In 2003–2004, Craig Venter and colleagues carried out one of the seminal experiments in marine metagenomics. They embarked on the Sorcerer II Global Ocean Sampling Expedition, a project that required gathering water samples from all around the world. The findings showed an incredible diversity of genes and microbes that were previously unidentified. A noteworthy investigation conducted in the Sargasso Sea by the J. Craig Venter Institute revealed millions of new genes and thousands of new microbial species, highlighting the enormous genetic variety seen in marine environments (Molitoris & Schaumann, 1986).

Table 7.1 highlights the key developments in the field of marine enzyme discovery.

TABLE 7.1
The Key Developments and Methods in the Field of Marine Enzyme Discovery

Time Period	Key Milestone in Marine Enzyme Discovery	Details	References
Early 20th century	Initial studies on marine microorganisms	Exploration of marine biodiversity, and its potential applications	Birolli et al. (2019)
1950s–1960s	Isolation and characterization of marine enzymes	To isolate and characterize enzymes from marine organisms	Bäumgen et al. (2019)
1970s–1980s	Discovery of extremophiles and their enzymes	Deep-sea hydrothermal vents, leading to the study of their unique enzymes	Mineta et al. (2016).
1990s	Molecular biology techniques revolutionize study	Introduction of DNA sequencing and cloning technologies at the genetic level, uncovering many novel enzymes	Saeman et al. (1945)
2000s onwards	Metagenomics and high-throughput technologies	Enabling direct analysis of genetic material from environmental samples	Molitoris & Schaumann (1986) and Sharma et al. (2016)
2008s	Drug discovery	Marine enzymes in drug discovery and new therapies	Hewson et al. (2007)
2010s	Advances in structural biology	X-ray crystallography and cryo-electron microscopy and 3D structures of marine enzymes	Addison et al. (2007)
2015s	Marine enzyme innovations	Unique catalytic properties for agriculture, food processing, and other fields	Teira et al. (2006)

7.2.2.2 Genomic and Proteomic Breakthroughs (1990s–2000s)

The integration of genomics and proteomics revolutionized marine enzyme research.

- **1990s:** The advent of genomic and proteomic technologies enabled scientists to analyse marine organisms at a molecular level. Researchers began to identify and characterize a wide array of marine enzymes, uncovering novel enzymes with potential applications (Bianchi et al., 2018).
- **2000s:** Structural biology techniques like X-ray crystallography and cryo-electron microscopy (cryo-EM) have provided deep insights into the 3D structures of marine enzymes. This time it brought the finding of enzymes with extraordinary catalytic powers and stability, which broadened their applications (Sharma et al., 2016).

7.2.3 Recent Advances

2000s: Expansion into Industrial and Medical Applications: Marine enzymes began to be explored for a variety of industrial applications, including pharmaceuticals, biofuels, and environmental cleanup. Their stability and activity under extreme conditions made them particularly valuable (Moisander et al., 2017).

2008: Marine Enzymes in Drug Discovery: Researchers started investigating marine enzymes for their potential in drug discovery and development. Some marine enzymes showed promise in treating diseases and developing new therapies (Hewson et al., 2007).

2010s: Advances in Structural Biology: Structural biology techniques, such as X-ray crystallography and cryo-EM, enabled scientists to determine the 3D structures of marine enzymes. This understanding helped in designing more efficient industrial processes and drugs (Addison et al., 2007).

2015: Marine Enzyme Innovations: The discovery of novel marine enzymes with unique catalytic properties opened new avenues for applications in agriculture, food processing, and other fields.

The study of marine enzymes has evolved significantly from early observations to a vibrant field with diverse applications. Advances in technology and molecular biology continue to drive innovation and uncover new potential for these fascinating biological molecules (Teira et al., 2006).

7.3 CHARACTERIZATION OF MARINE ENZYMES

In view of their special adaptations to marine conditions, characterization of marine enzymes entails an analysis of their characteristics, uses, and functions.

7.3.1 Identification of the Source

Marine creatures are a common source of marine enzymes. These may consist of:

Marine Bacteria: *Pseudomonas* and *Vibrio* are two types of bacteria that are common in marine sediments and waters.

Marine Fungi: Species that grow well in marine conditions include *Aspergillus* and *Penicillium*.

Marine Algae: Because of their harsh environments, microalgae like *Dunaliella* and *Chlorella* create enzymes with special qualities (Ferraro et al., 2010).

Marine Invertebrates: These organisms, which have evolved to a variety of marine settings, include sponges, molluscs, and crustaceans.

Isolation: These organisms' enzymes are removed by:

Cell Lysis: Techniques to split open cells and release the enzymes, such as homogenization or chemical techniques like detergents.

Centrifugation and Filtering: To concentrate the enzyme and get rid of cell detritus (Suleria et al., 2016).

7.3.2 Enzyme Activity Assays

Specific Activity: This gauges an enzyme's ability to catalyse a reaction as effectively as possible; it is commonly measured as:

Activity Units: The quantity of substrate transformed into product in each amount of time.

Substance Particularity: Selecting a substrate should take into account the intended use or natural function of the enzyme. These could comprise proteins, lipids, other marine-derived substances, polysaccharides like agarose or chitosan, or marine enzymes.

Specificity Tests: Determine whether an enzyme has a wide range of molecular specificity or a small range by using various substrates (Gildberg et al., 1993).

Quantitative Assays: Determine the activity of the enzyme by observing how well it converts substrate into product. Assays that are colorimetric, fluorometric, or spectrophotometric may be used for this.

Qualitative Assays: For hydrolases and proteases in particular, agar plates or gels can be used to visualize enzyme activity.

Activity per milligram of protein: Indicates the effectiveness of the enzyme (Mehrshad et al., 2018).

Ideal Circumstances Assessed by Temperature: Identifying the highest temperature at which an enzyme is most active.

pH: Determining the range of pH in which the enzyme functions best.

Salinity: Conducting experiments with different salt concentrations to simulate maritime conditions (Shaaban et al., 2012).

7.3.3 Enzyme Kinetics

Kinetics of Michaelis-Menten: The enzyme functions at half its maximum velocity at a substrate concentration of K_m (Michaelis constant). Higher affinity for the substrate is indicated by a lower K_m. Maximum Reaction Rate, or V_{max}, is the highest rate of reaction that an enzyme can reach when the substrate reaches saturation.

Studies on Inhibitors: Examining how various inhibitors impact enzyme activity might shed light on how it is regulated and suggest possible therapeutic uses. Inhibitors that are competitive, non-competitive, and uncompetitive are common varieties (Loperena et al., 2012).

Types of Inhibitors in Enzyme Inhibition—Competitive Inhibition: Inhibitors vie for the active site with the substrate. This can be investigated by introducing different inhibitor doses and seeing changes in V_{max} and K_m.

Non-competitive Inhibition: Regardless of substrate concentration, inhibitors alter the activity of the enzyme by binding to a location other than the active site.

Uncompetitive Inhibition: Inhibitors only attach to the complex of the enzyme and substrate, influencing both V_{max} and K_m (Yelton et al., 2016).

7.3.4 Thermostability and Halotolerance

Thermostability—Heat Stability Tests: This entails assessing the enzyme's activity after it has been incubated at different temperatures. Since marine enzymes are adapted to thermophilic conditions, they frequently exhibit great thermostability.

Intolerant to Halotolerance:

Salt Tolerance Tests: Assessing the activity of enzymes in the presence of different salt concentrations. For marine enzymes to function in saline environments, high salinity is often the condition under which they show activity and stability (Salamone et al., 2012).

7.3.5 Structural Characterization

X-Ray Structure of Protein Crystallography: By examining the diffraction patterns of crystallized enzyme samples, this method yields an intricate three-dimensional structure of the enzyme at atomic resolution.

NMR Spectroscopy: This method is used to investigate proteins in solution and yields details on the kinetics and structure of the enzyme (Kamble et al., 2018).

Proteins can be seen at almost atomic resolution using cryo-EM, which involves freezing materials and using an electron microscope to examine. Mass spectrometry measures the mass-to-charge ratio of peptide fragments to determine the amino acid sequence and alterations. (Ulber et al., 2003).

Comparative Analysis: Determines structural and functional characteristics by aligning sequences with known enzymes.

7.3.6 Genetic Characterization

Expression and Gene Cloning: Cloning is the process of introducing a vector, such as a plasmid, containing the gene encoding the enzyme into a host organism, such as *Escherichia coli*, to generate the enzyme (Qeshmi et al., 2020).

Expression Systems: These comprise mammalian, yeast, or bacterial systems for the synthesis and purification of enzymes.

Sequence Databases in Bioinformatics: In order to predict function and structure, programs such as BLAST (Basic Local Alignment Search Tool) compare sequences with pre-existing databases.

Predictive modelling: Using sequence data, programs such as Phyre2 or SWISS-MODEL forecast three-dimensional structures (La Cono et al., 2018).

7.3.7 Functional Characterization

Substrate Specificity—Substrate Screening: Assessing the enzyme's efficiency and preference by subjecting it to a range of substrates.

Kinetic Parameters: Determining the specificity of an enzyme by measuring its activity with various substrates.

The mechanism of catalysis, known as mechanistic studies, entails examining the interactions between the enzyme and its substrates, including the identification of transition states and intermediates.

Research on Inhibitors and Activators: Comprehending the impact of compounds on the mechanism and activity of enzymes (Bonugli-Santos et al., 2015).

Characterizing marine enzymes involves a multifaceted approach, combining biochemical, genetic, structural, and functional analyses to fully understand their properties and applications.

7.3.8 Marine Metagenomics

Fortunately, until recently, extracting marine microorganisms and their derivatives of genes required a lot of effort and time using traditional cultivation techniques and biochemistry. Our understanding of the marine microbial world has advanced dramatically thanks to the great degree of complexity attained by molecular technology today (Salazar & Sunagawa, 2017). This technological advance has been greatly aided by the formation of global consortiums and ocean sampling programmes to investigate marine microbiological diversity. Not only are we learning more about the range of marine microbial diversity, but we are also beginning to understand the fundamentals of how microbes adapt molecularly to a variety of environments (Khare et al., 2015). This is related to our growing understanding of the variety of microbial gene products and their role in controlling the metabolism.

One of the biggest uncharted territories for humanity is the ocean (Zhao & Bajic, 2015). From extraordinarily high pressures in the ocean's deepest regions to surface air pressure that is typical, from low-saline conditions to salt-saturated saltwater brines, and from conditions above 300°C in hydrothermal deposits to below-freezing temperatures in sea ice, the surroundings incorporate numerous species of the microbes that have become accustomed to such challenging settings (Rastogi & Shrivastava, 2017). These microorganisms' metabolic diversity offers a supply of enzymes with special capabilities for use in industrial settings with challenging chemical and physical environments. Thousands of samples have been gathered and examined by these expeditions, yielding a wealth of data regarding the 16S rRNA DNA sequences and expanding our understanding of the diversity of microbes. There is, however, a dearth of related data regarding the innate functional biochemistry. Recent developments in sequencing technology, which enable large-scale, quick, low-cost sequencing of uncultivated biodiversity, are primarily responsible for the discrepancy between taxonomic and biochemical data (Hooker & Gerber, 2004). The most time-consuming stage in this process is bioinformatic analysis. But to fully utilize these sequences' biochemical or biotechnological potential, intensive wet lab effort is needed. This includes cloning large amounts of the desired genes, then expressing and characterizing the resulting enzymes (Honda et al., 1987).

Given that not each gene in a metagenome composition is fully expressed or can be adequately cloned and generated, this is not an easy operation (Mineta & Gojobori, 2016). The selection of choice of how to discover enzymes between naïve enzyme screening techniques and genetic data predictions is still up for debate. Naïve screens can identify genes with previously unknown functions, regardless of the enzyme's origin (Kodzius & Gojobori, 2015). However, sequencing combined with in silico predictions enables us to speed up the rate of discovery. It should be emphasized, of course, that novelty by itself does not imply improved enzymatic function or increased commercialization prospects. During its first stage, INMARE has helped to establish one of the biggest collections by creating sophisticated techniques for biotechnologically blueprinting marine microorganisms of metagenomic and genomic enzymes, with an emphasis on the most sought-after ones in the commercial sector (Mizuno et al., 2013). Now, this collection contains over 1,000 enzymes, of which 32% have been thoroughly characterized and approximately 94% are accessible in expression systems that are ready to use. The protein and amino acid sequences of data, geographical location, and phylogenetic sources of such enzymes are exceedingly diverse and non-redundant (Barone et al., 2014). They provide a collection of activities from not fewer than 193 geographically distinct places, at least 283 identified bacterial and archaeal taxa, and around 607 microorganisms with undetermined taxonomic affiliations. Identity can be as low as 5% when compared to database sequences (Alma'abadi et al., 2015).

Developing lead concepts and prototypes with an opportunity for speedy industrialization and intellectual property protection was made achievable by condensing and streamlining the tedious optimization process (Coutinho et al., 2018). This can be accomplished by putting enzyme candidates through a battery of tests to determine their commercial relevance in the early stages of discovery, as well as stabilization through the design, immobilization, and testing of environments that closely resemble those needed by industries (Woyke et al., 2009).

7.3.8.1 Culture Independent Method

Metagenomics is a very potent molecular biology technique. It speaks of cloning DNA recovered from mixed combinations and the direct extraction method of genetic investigation of microorganisms and collection of natural organisms (Bernan et al., 1997).

Metagenomics may be centred on the identification of biocatalysts for the synthesis and manufacture of secondary metabolites such as bioactive chemicals, as well as on the enzymatic coding of cluster (Martyniuk et al., 2020). Environmental conditions are commonly classified into three categories: extremely diversified settings (such as saline seawater); naturally or intentionally enriched environments for the target gene/biocatalyst; or extreme environments.

7.3.8.2 Screening, Sequencing, and Data Analysis of Metagenomic Libraries

There are two types of metagenomic library screening: function-based screening and sequence-based screening (El-Gebali et al., 2019). In these sequence-based screening, targets are identified using PCR amplifications genes from conserved sections of known genes, whereas robots are used in function-based screening to hunt for a well-defined phenotype.

A successful metagenomic technique may be hindered by the screening process because of the sensitivity of low signals and agar plate-based screening (López-Pérez et al., 2016). Additionally, the selection of a good recombinant host and a sufficient expression vector may have resulted in the expression of the target genes being the primary bottleneck. But in recent years, a number of these problems have been resolved because of ongoing advancements in robotic technology and sequencing technologies (Brazelton et al., 2012).

The creation of news-sequencing systems is an important link between the data analysis and screening processes. The idea behind this new platform is to sequence a dense ray of amplified DNA fragments using enzymatic modification cycles and imaging-based data gathering techniques. More contemporary technologies, referred to as "third

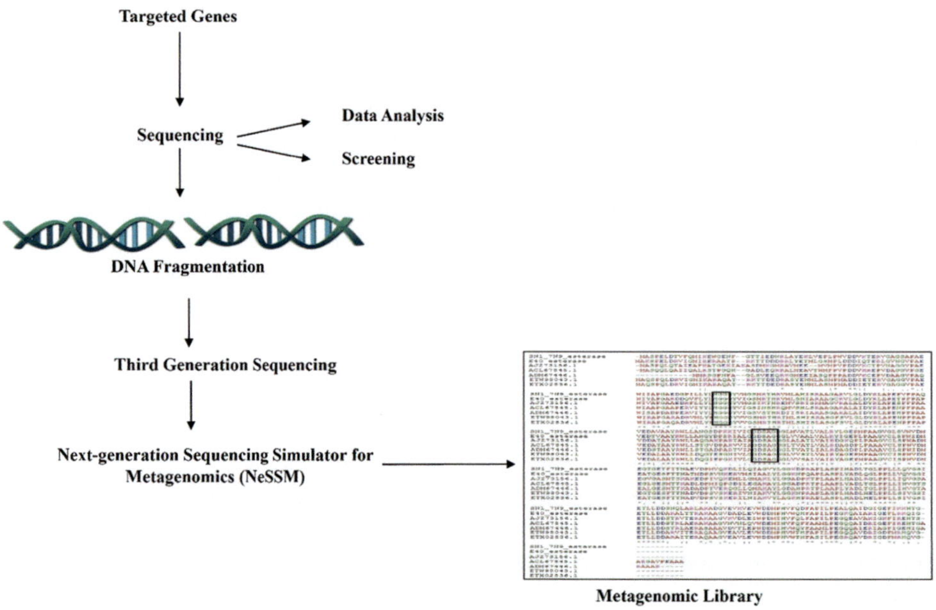

FIGURE 7.1 Data analysis of metagenomic libraries.

generation sequencing technologies," entail sequencing individual molecules and single cells (Harborne et al., 2022). Fluorescence detection-based third-generation sequencing technologies have already been introduced. The Next-generation Sequencing Simulator for Metagenomics (NeSSM) is one example of a quick simulation system for high-throughput metagenome sequencing (Bianchi et al., 2018).

It is necessary to process the vast amount of metagenomic information obtained by the sequencing platforms and to have the appropriate data-analysis tools (Figure 7.1).

The discovery of a novel cold active esterase known as Est97 was made possible by a metagenomic library screening of an Arctic intertidal zone (Tully et al., 2017). The mixture of enzymes demonstrated to maintain nearly 60% of relative activity at 20°C and extremely low thermostability point to its use in cold biotransformation. In reality, phospholipase A with lipase activity (henceforth referred to as MPlaG) is the first enzyme to be isolated from a metagenomic library from tidal flat sand on the Korean west coast. It displayed unique catalytic characteristics against phosphatidylcholine and olive oil in addition to demonstrating maximum activity at 25°C, indicating that MPlaG is a lipid-preferred phospholipase (Sharma et al., 2016).

7.4 UTILIZATION OF MARINE ENZYMES

The ocean is a huge storehouse of human existence, and the 21st century is the century of the ocean. These days, most nations deal with issues including pollution, excessive resource usage, and rising population. In the meantime, people now have new possibilities and a fresh source thanks to the advancements in marine biology. The importance of marine microbial enzymes, particularly marine extreme microbial enzyme in applications, has grown (Zhang & Kim, 2010). Due to the unique benefits of enzymes, a wide range of sectors are very interested in tailoring enzymatic techniques to the specific needs of their operations. Enzyme applications in medicine have also been growing. For instance, surgeons have reported encouraging clinical outcomes from the injection of specific enzymes, such as streptokinase, crystalline trypsin, and chymotrypsin, for the debridement of wounds. It is reasonable to anticipate rapid advancements in this sector given the rising and currently unanticipated medicinal usage of enzymes (Rao et al., 2017). The past few years have seen a significant increase in the uses of enzymes. There is a great chance that the number of applications for already accessible enzymes will continue to rise, as will the number of novel uses for existing enzymes (Trincone, 2011). Furthermore, certain emerging disciplines, such as metagenomics, provide a potent lens through which to examine the microbial world, which could fundamentally alter our comprehension of the enzymes that marine microbes produce. Russia, Finland, Spain, Canada, and other nations have also increased their study on marine enzymes (Muffler et al., 2015).

Global research into marine biology is still in its early stages, but it has enormous potential for advancement and applications that will help the industry due to the diversity of marine life and the uniqueness of biological metabolism (Debashish et al., 2005).

7.4.1 Some Industrially Important Marine Enzymes and Their Utilization in Various Aspects

7.4.1.1 Protease

Today's society makes extensive use of proteases (Debashish et al., 2005). Proteases are used in the pharmaceutical, leather, and detergent industries to make things like digestive and anti-inflammatory drugs. For the first time, in 1960, Dane discovered alkaline protease from *Bacillus licheniformis* (Loperena et al., 2012). Microorganisms continue to be the best available resource for the synthesis of proteases. Since the isolation of a novel kind of alkaline protease from marine Phytobacteria in 1972, several proteases have been consistently extracted from marine microbes (Sarkar et al., 2010). UV mutagenesis was used to extract the N1–35 strain of 30 different species of marine bacteria from seawater, mud, fish, and other sources. This strain generated protease with significant benefits above terrestrial ones. The Gland of Deshayes of a marine shipworm contained a mutually beneficial bacteria from which this alkaline protease has previously been identified (Qeshmi et al., 2020). In 2007, yeast strain (*Aureobasidium pullulans*) with an elevated yield of alkaline protease from the sea saltern of the China Yellow Sea was obtained. The enzyme's maximal production was 623.1 U/mg protein (7.2 U/mL). Haddar identified and extracted two detergent-stable alkaline serine proteases (BM1 and BM2) from seawater in 2009 using *Bacillus mojavensis* A21 generating alkaline proteases (Shi et al., 2013).

- **Bioremediation:** Proteases are used in bioremediation processes to break down proteins in oil spills and other marine pollutants. By degrading proteinaceous contaminants, proteases aid in the cleanup and restoration of polluted marine environments (Kennedy et al., 2010).
- **Food Processing:** In seafood processing, proteases are used to improve texture and flavour. For instance, they can help in tenderizing fish or enhancing the quality of fish sauces and other marine-based food products (Ferrer et al., 2019).
- **Pharmaceuticals:** Marine-derived proteases are studied for their potential use in pharmaceuticals due to their unique properties, such as stability and activity in extreme conditions. These enzymes can be used in drug development, particularly in the production of peptide-based drugs (Kennedy et al., 2011).

7.4.1.2 Lipase

The breakdown of fats and oils by lipases, which are widely distributed enzymes, releases free fatty acids, diacylglycerols, monoglycerols, and glycerol. In addition, lipases are effective in several other processes, including aminolysis, transesterification, and esterification. The amount of information regarding lipases that is currently available in the literature indicates that lipases have recently attracted more interest (Rasmussen & Morrissey, 2007). Additionally, many microbial lipases are sold commercially; these are mostly utilized in the manufacturing of detergents, paper, cosmetics, food flavouring, organic synthesis, and other industrial uses. In Europe, the market share of enzyme detergents has risen to 90%, whereas in Japan, it is approximately 80%. Lipases act as useful biocatalysts since they have broad substrate specificity, stable nature in organic solvents, and function under mild circumstances (Kim, 2019).

- **Pharmaceuticals and Biotechnology:** Marine lipases are studied for their potential applications in pharmaceuticals and biotechnology (Navvabi et al., 2018). Their ability to function in extreme marine conditions makes them valuable for developing specialized products, such as enzyme-based drug delivery systems or biocatalysts for synthetic reactions (Alderkamp et al., 2007).
- **Cosmetics:** Lipases from marine sources are also explored for use in the cosmetics industry. They can be incorporated into skincare products to help in the breakdown of fats and oils on the skin, contributing to cleansing and exfoliation (Martyniuk et al., 2020).
- **Aquaculture:** In aquaculture, lipases are used to improve the digestibility of lipid-rich fish feeds. By breaking down fats into simpler fatty acids and glycerol, lipases enhance nutrient absorption and growth rates in farmed fish and shellfish. This can lead to more efficient feed conversion and healthier aquatic organisms (Fernandes, 2014).

7.4.1.3 Extremozymes

The maritime environment is incredibly complicated, with strong acid and strong alkali, extremes in temperature, significant hydrostatic pressure, and extremely poor nutritional conditions (Martyniuk et al., 2020). A microbe that is extremophilic is one that can survive in extreme physical or geochemical conditions that are unfavourable to most other forms of life on Earth, and in some cases, even thrive in them (Kennedy et al., 2010).

In ecology, low-temperature microorganisms such as psychrotrophs and algoriphagus have clear benefits. It is difficult for the low-temperature microbes to become contaminated (Ferrer et al., 2019). These bacteria have easy growth conditions, and their enzymes have the benefits of high activity and high catalytic efficiency. Because of this, low-temperature microbial enzymes can help reduce processing times and reduce the need for costly heating and cooling systems, which can result in significant energy savings.

- **Oil Spill Cleanup:** Marine extremozymes, like thermostable and halotolerant lipases, are used to degrade hydrocarbon pollutants in oil spills. Their ability to function in high-salinity and varying temperature conditions makes them particularly effective in cleaning up marine oil spills (Sarkar et al., 2010).
- **Biosensors:** Extremozymes are used in the development of biosensors for detecting pollutants or biochemical markers in marine environments (Beygmoradi et al., 2018). Their ability to function under harsh conditions improves the robustness and reliability of these sensors.

- **Detergents:** Extremozymes, particularly those with protease or lipase activity, are used in the formulation of industrial detergents and cleaners. Their stability in high temperatures and alkaline conditions makes them effective for cleaning purposes in marine and industrial settings (Shahidi & Kamil, 2001).
- **Biocatalysis:** Extremozymes are used in various biotechnological processes, including the synthesis of pharmaceuticals, agrochemicals, and specialty chemicals (Chi et al., 2009). Their stability under extreme conditions can be harnessed for reactions that require high temperatures or salinities.

7.4.2 Enzymes Produced from Marine Microorganisms and Their Utilizations

Different enzymes are secreted by marine-dwelling bacteria and fungi according to their biological roles and habitats. The focus of the study has shifted to marine microbial enzymes, with a number of these enzymes attracting the interest of microbial prospectors. A few of these enzymes have been isolated from marine sediments and saltwater, purified, and their properties and uses described (Lima & Porto, 2016).

1. **Starch Hydrolysing Enzymes:** The following enzymes are involved in the breakdown of starch: α-amylase, β-amylase, glucoamylase, debranching enzymes (pullulanase), and α-glucosidase. These enzymes convert starch to low-molecular-weight molecules.

   Pullulanase, α-glucosidase, and α-amylase from archaea can all function at high temperatures and pH levels, making them suitable for application in a one-step process for the industrial bioconversion of starch (Hoppe et al., 2002). Using novel, thermoactivated enzymes to improve the starch-conversion process would drastically reduce the cost of producing sugar syrup.
2. **Agarase:** Because of its quick uses in gene technology, including the production of algal protoplasts and the elution and separation of DNA fragments from agarose gels after electrophoresis, agarase has been the focus of research for a considerable amount of time.

   Examples include the characterization of seaweed polysaccharides; the liquefaction of agar and agarose gels; the conversion of agarose to oligosaccharides; the synthesis of simple sugars such as neoagarobiose, neoagarotetraose, and neoagarohexaose; and the defouling of fermentors and bioreactors (Dhevagi & Poorani, 2006). Additionally, by contaminating complex polysaccharide layers, the purified enzyme might successfully manage contaminations caused by red algal blooms, stop biofouling of pipelines or submerged marine surfaces, or clean up biofouled surfaces after contamination.

   The activity of marine bacterial agarase is strong for the depolymerization of agarose and agar, among other complex polysaccharides (Barzkar & Sohail, 2020). A small number of bacteria produce agarase. Neoagarotetraose and neoagarobiose were produced at the endo-type β-agarase agarase-0107 which hydrolyses the β-1,4-linkage of agarose.

   A pH of about 8 has been identified from *Vibrio* sp. JT0107, a species that needs the salts of seawater to develop. Characterized agarase is produced by *Pseudomonas stutzeri*, *Aeromonas* sp., and *Vibrio* sp., which are isolated from seawater (Suleria et al., 2016).
3. **α-Galactosidase:** In the sugar business, α-galactosidase is used to hydrolyse α-galactosidase raffinose, increasing the yield of sucrose. Sponge, bacteria, and the algae *Polysiphonia* sp. were found to contain α-galactosidase. It was extracted from marine species of *Alteromonas* spp. (Cheng et al., 2020) connected to the scallop (*Patinopecten yessoensis*) and mussel (*Crenomytilus grayanus*).

   A new β-galactosidase-α-2,6-sialyltransferase is generated by ATCC 33539, ATCC 35083, or *Photobacterium damselae* JT016. The enzyme should be helpful in the industrial synthesis of sialosides and in the modification of glycoproteins (Rustad et al., 2011).
4. The creation of silage, flavours, alcohol, maize gluten, laundry detergents, and wastewater treatment all use cellulases and related enzyme products. *Aspergillus terreus* was isolated from the marine shipworm and a symbiotic bacteria discovered in the Deshayes gland. Saltwater generates cellulase (Debashish et al., 2005). Exo-1,4-β-cellobiohydrolase is produced by thermostable *Thermotoga* sp., a thermophilic bacterium (105°C).

7.4.3 Saccharification Process for Marine Sugars Derived from Marine Microorganisms

The utilization of algal biomass for fermentation processes and the generation of high-value fine chemicals and biofuels are made possible by the clarification of these marine polysaccharide utilization systems. Due to the fast growth rate of algae, algal biomass is now regarded as waste and builds up significantly, providing a low-cost and convenient supply of raw materials (Hoppe et al., 2002). Biorefinery concepts were primarily published for the effective saccharification and fermentation of carbohydrates from brown algae for use in biotechnological applications. One of these was the reported metabolic engineering of Mannitol and alginate. Later breakdown products are fermented to ethanol *by Saccharomyces cerevisiae.*

By means of metabolic engineering, an *E. coli* strain with the capacity to decompose was created, and absorb and metabolize alginate to produce bioethanol. The use of agar and carrageenan, two red algal polysaccharides, for the same purpose was made possible by the elucidation of the 3,6-anhydro-L-galactose metabolic route (Sarkar et al., 2010).

For *Defluviitalea phaphyphila*, it has been reported that brown algae can be directly bioconverted into ethanol. Engineering the yeast *S. cerevisiae* to hydrolyse laminarin from brown macroalgae enzymatically in order to produce bioethanol is another significant advancement in the generation of biofuel from algae. Biofuel feedstocks, such as macroalgal and associated genetic engineering techniques, are compiled in recent reviews (Shi et al., 2013).

Methyl-2,3-butanediol synthesis from glucose was achieved by *Bacillus licheniformis* metabolic engineering. The widespread glucanases found in a variety of animals can break down the brown algal laminarins and green algal glucans to create glucose. This permits the fermentation of 2,3-butanediol with marine polysaccharides as feedstock. Hydrogen is another possible energy source in addition to ethanol (Rasmussen & Morrissey, 2007). The evolution of fermentative hydrogen production was studied, illuminating the biochemical processes by which different bacteria produce hydrogen. Since the study of fermentative pathways begins with monosaccharides, which may be provided by either land plant or algae, the described hydrogen generation methods involving bacteria living on first- or second-generation plant sources can be simply converted to the utilization of algal biomass. Thus, *Thermotoga neapolitana*, a hyperthermophilic microbe, was used in the development of hydrogen on biomass of green alga *Chlamydomonas reinhardtii* and was studied (Barone et al., 2014).

The ability to completely break down each carbohydrate to the single-stranded level and the capacity to metabolize the associated monosaccharides that are generated by the breakdown of polysaccharides are, in theory, prerequisites for using carbohydrates in biotechnological procedures for fermentation (Bruno et al., 2019).

Therefore, more CAZymes are needed to break down a polysaccharide the more complex it is. Rare sugars found in polymers are frequently exclusively metabolized by a few bacteria. Yeasts with a high ethanol tolerance, such as *S. cerevisiae*, are commonly utilized for ethanol production. Even though yeasts are used to produce bioethanol from brown algae, they frequently lack the genes necessary for the metabolic pathways that code for the proteins needed to convert pentose sugars like xylose and arabinose (Gulder & Moore, 2009). The utilization of algal biomass in yeast fermentation procedures would become more difficult if metabolic engineering were needed to enable the utilization of these sugars. Rare sugars can be valuable products in and of themselves in addition to acting as a source of carbon for fermentation. Red algae can be used to isolate 3,6-anhydro-L-galactose, which has been shown to have anti-inflammatory and skin-whitening effects (Kamble and Vavilala, 2018).

7.4.4 Applications of Saccharification Process for Marine Industry

In the marine industry, saccharification can be applied in several innovative and impactful ways (Ferrer et al., 2019). Here's an overview of how this process is utilized in various marine-related applications (Figure 7.2):

7.4.4.1 Algae-Based Biofuels

Biomass Conversion: Polysaccharides are abundant in marine algae, including macro- and microalgae. Saccharification of algal biomass breaks down these polysaccharides into fermentable sugars (Bäumgen et al., 2021).

Bioethanol Production: Bioethanol, a sustainable fuel, may be created by microorganisms fermenting the resultant sugars (Kennedy et al., 2010).

Biodiesel Production: In some cases, algae are processed to extract lipids, which are then converted into biodiesel (Kodzius & Gojobori, 2015). Saccharification can be used in a pre-treatment step to improve the efficiency of lipid extraction.

7.4.4.2 Marine Bioproducts

Prebiotics and Dietary Fibres: Saccharification of marine algae can produce functional carbohydrates like fucoidans and alginates, which have health benefits and are used as dietary supplements or food additives (De Santi et al., 2016).

Biopolymers: Polysaccharides from marine sources can be saccharified and further processed to create biopolymers used in food, pharmaceuticals, and biodegradable materials (Klemetsen et al., 2018).

7.4.4.3 Marine Waste Utilization

Utilizing Waste Biomass: Marine waste (e.g., from fisheries or algae blooms) can be saccharified to convert it into valuable products. This approach helps reduce waste and adds economic value to otherwise discarded materials.

Biogas Production: Saccharified marine waste can be fermented to produce biogas, which is a renewable energy source.

7.4.4.4 Aquaculture

Fish Feed Ingredients: Saccharified marine biomass can be used to formulate fish feeds. The sugars or other byproducts from saccharification can serve as energy sources or attractants in aquaculture feeds.

Enhancing Feed Efficiency: Marine-derived enzymes used in saccharification can improve the digestibility of feed components, leading to better growth and health of aquaculture species.

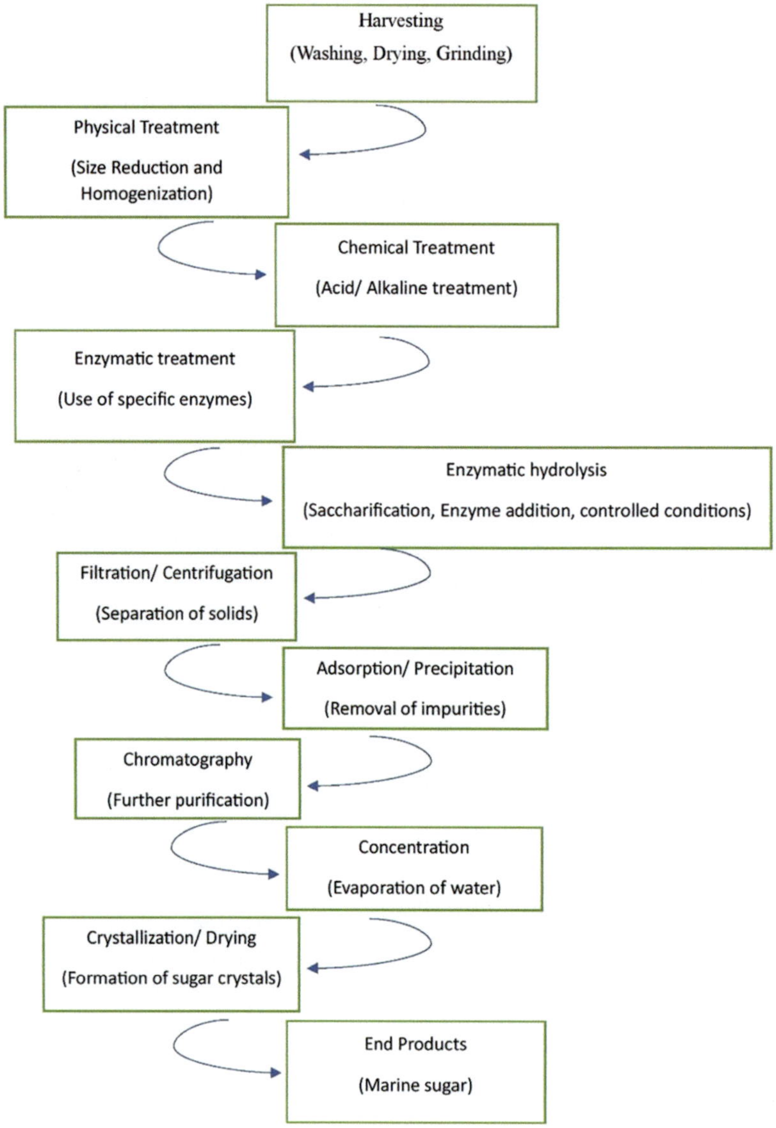

FIGURE 7.2 Saccharification process for marine sugars derived from marine microorganisms.

Key Considerations:

- **Enzyme Selection:** The choice of enzymes and their conditions for saccharification are critical for efficiency and effectiveness.
- **Cost and Sustainability:** The economic feasibility and environmental impact of saccharification processes should be carefully evaluated.
- **Regulatory Compliance:** Ensure that processes and products comply with relevant regulations and standards, particularly in food, pharmaceuticals, and environmental applications.

By leveraging saccharification in these ways, the marine industry can enhance sustainability, create valuable products, and contribute to innovative solutions across various sectors.

7.4.5 Industrial Application of Marine Enzymes

7.4.5.1 Application in the Biotechnology and Industrial Process

1. **Biofuels Production:** Biofuels are produced from biomass by using marine enzymes, especially those found in extremophiles, or organisms that live in harsh environments. Plant materials containing complex carbohydrates can be effectively broken down by these enzymes into simple sugars, which can subsequently be fermented to produce ethanol or other biofuels. Their exceptional resilience to adverse environments (such as high salinity and high temperatures) renders them very advantageous for the generation of industrial biofuels (Chasanah et al., 2009).

2. **Environmental Cleanup:** Pollutants such as oil spills are cleaned up using marine enzymes. Hydrocarbons and other persistent organic pollutants can be broken down by enzymes found in marine microorganisms, such as lipases and oxidases, into less dangerous compounds. This application is essential for reducing the harm from industrial operations and oil spills due to the environment (Barzkar & Sohail, 2020).
3. **Biocatalysis:** A variety of chemical processes use marine enzymes as biocatalysts. They are perfect for the synthesis of fine chemicals, medications, and specialized molecules due to their high selectivity and mild catalytic conditions. It is difficult to synthesize complicated molecules and optically pure chemicals using conventional chemical procedures; enzymes such as marine lipases and proteases are employed in this regard (Shaaban et al., 2012).
4. **Water Treatment and Effluent Processing:** Wastewater and industrial effluents are treated using marine-derived enzymes. They contribute to the breakdown of organic materials, lower pollution levels, and increase the efficacy of waste treatment facilities. Reducing the ecological footprint of industrial processes and upholding environmental requirements require this application (Shi et al., 2013).
5. **Paper and Pulp Industry:** In the pulp and paper business, marine enzymes—specifically, cellulases and hemicelluloses—are utilized to bleach pulp and take ink out of recycled paper. Because these enzymatic procedures use less energy and harsher chemicals than traditional chemical treatments, they are more environmentally friendly (Navvabi et al., 2018).
6. **Textile Sector and Fabric Processing:** The textile industry uses marine organism-derived enzymes for procedures like bio-polishing, which improves fabric quality by getting rid of fuzz and pills. Additionally, materials are desiccated and bleached using marine enzymes, which provide a more effective and environmentally friendly option than traditional chemical treatments (Birolli et al., 2019).

7.4.5.2 Application of Marine Enzymes in Pharmaceutical and Medicine Industry

1. **Drug Creation and Pharmaceutical Synthesis:** Complex medicinal compounds that are challenging to make with traditional chemical processes are synthesized using marine enzymes. Their efficiency and specificity contribute to the development of medications with minimal adverse effects and high purity. Enzymes obtained from marine sources, for example, can help produce antivirals, antibiotics, and anticancer drugs (Table 7.2) (Shahidi et al., 2001).

TABLE 7.2
The Applications of Marine Enzymes in Various Industries

Enzyme	Source	Application
Proteases	Fish, crustaceans	Protein hydrolysis, biocatalysis
		Peptide synthesis, wound debridement
		Meat tenderization, flavour enhancement
Lipases	Marine bacteria	Synthesis of biodiesel, bioremediation
		Drug formulation, biocatalysis
		Flavour enhancement, dairy product modification
Amylases	Marine algae	Starch hydrolysis, bioethanol production
		Digestive aids, drug delivery systems
		Brewing, baking, sweetener production
Cellulases	Marine fungi	Cellulose degradation, biofuel production
		Drug release systems, therapeutic agents
		Juice clarification, coffee processing
Chitinases	Crustacean shells	Chitin degradation, biopesticides
		Antifungal agents, drug development
		Food preservation, texturizers
Agarases	Red seaweed	Agarose production, DNA analysis
		Antitumor agents, drug delivery
		Gelatin substitutes, food gelling agents
Carrageenases	Marine algae	Carrageenan breakdown, polysaccharide analysis
		Drug delivery systems, antiviral agents
		Food thickening, stabilizers
Laccases	Marine fungi	Lignin degradation, dye decolorization
		Antimicrobial treatments, wound healing
		Wine stabilization, beverage clarification

(*Continued*)

TABLE 7.2 (*Continued*)
The Applications of Marine Enzymes in Various Industries

Enzyme	Source	Application
Alkaline phosphatases	Marine bacteria	Phosphate removal, biosensors
		Diagnostic reagents, bone health supplements
		Nutrient supplements, food additives
Polyphenol oxidases	Marine algae	Biosensors, antioxidant synthesis
		Anti-ageing formulations, drug synthesis
		Food preservation, browning control

2. **Discovery of New Drugs and Novel Compounds:** There are a lot of unusual bioactive compounds found in marine habitats. Marine microbes' enzymes can be used to discover and create novel medications. For instance, marine enzymes have enabled the discovery of new families of antibiotics and other compounds with promise for medicinal application (Kennedy et al., 2008).
3. **Enzyme-Based Treatment and the Enzyme Replacement Therapy:** In hereditary illnesses where individuals are deficient of enzymes, the use of marine enzymes is being investigated as a potential treatment option. Conditions like cystic fibrosis, certain types of muscular dystrophy, and lysosomal storage diseases can benefit from marine-derived enzymes that can function in the unique environments of human cells (Lee et al., 2010).
4. **Anti-inflammatory and Antioxidant Agents: Therapeutic Agents:** Some marine enzymes exhibit strong anti-inflammatory and antioxidant properties. These enzymes can be used to develop treatments for inflammatory diseases and conditions caused by oxidative stress, such as arthritis, cardiovascular diseases, and neurodegenerative disorders (Salamone et al., 2012).
5. **Wound Healing and Tissue Regeneration Regarding Biomaterials and Bioactive Dressings:** Marine enzymes are used in developing advanced wound care products. For example, enzymes that break down biofilms or promote the growth of new tissue are incorporated into dressings and gels to enhance wound healing and tissue regeneration (Nguyen & Nguyen, 2017).
6. **Diagnostic Tools, Biosensors and Diagnostics:** Marine enzymes are used in biosensors and diagnostic kits for detecting various diseases and conditions (Nguyen & Nguyen, 2017). Their stability and specificity make them valuable in developing highly sensitive and accurate diagnostic tools for clinical or medical uses (Martyniuk et al., 2020).

7.4.5.3 Application of Marine Enzymes in Food and Beverage Industry

1. **Food Processing and Preservation Regarding Texture and Flavour Enhancement:** Marine enzymes like proteases, lipases, and carbohydrases are used to improve the texture, flavour, and shelf life of food products. For instance, proteases from marine sources are used in the production of fish sauce, enhancing the fermentation process and resulting in a richer flavour profile (Dhevagi & Poorani, 2006).
2. **Clarification:** Enzymes such as amylases and pectinases are used to clarify fruit juices and wines by breaking down complex polysaccharides, leading to clearer, more visually appealing products.
3. **Fermentation of Soy Sauce and Fermented Products:** Marine enzymes aid in the fermentation process of various products, such as soy sauce and miso. They help break down proteins and carbohydrates, contributing to the development of distinct flavours and improving the nutritional value (Barzkar & Sohail, 2020).
4. **Nutraceuticals and Functional Foods:** Enzymes derived from marine sources are incorporated into functional foods and dietary supplements due to their health benefits. For example, marine-derived enzymes with anti-inflammatory and antioxidant properties are added to health drinks and food products aimed at promoting wellness (Navvabi et al., 2018).
5. **Dairy Industry and Cheese Production:** Marine enzymes, such as certain proteases, are used in the cheese-making process to enhance coagulation and flavour development (Beygmoradi et al., 2018). They can also help in producing low-lactose or lactose-free dairy products by breaking down lactose (Sarkar et al., 2010).
6. **Meat Tenderization and Improving Texture:** Marine-derived proteases are used to tenderize meat, making it more palatable and easier to cook.

This process breaks down muscle proteins, resulting in a softer texture and improved flavour (Parte et al., 2017).

7. **Baking Industry and Dough Conditioning:** Enzymes like amylases and proteases from marine sources are used in baking to improve dough making properties, enhance bread volume and the freshness of baked goods. These enzymes help break down starches and proteins in the flour, leading to better dough consistency and improved final product quality (Sarkar et al., 2010).
8. **Seafood Processing:** Marine enzymes are used in the production of surimi, a processed seafood product made from fish. Enzymes help in the gelation process, brushing up the texture and standardization of the final product (Shi et al., 2013).
9. **Byproduct Utilization:** Marine enzymes are employed to extract valuable proteins and oils from seafood processing byproducts, contributing to waste reduction and the creation of high-value ingredients (Suthindhiran et al., 2014).

7.5 CONCLUSIONS

There is a market need for active and customized enzymes. About 70% of the Earth's surface is made up of oceans, which are home to more species than rainforests. A vast variety of enzymes with potential uses in industry and medicine may be found in the marine environment. Recent years have seen the discovery of several proteases, lipases, antioxidants, and a vast diversity of enzymes from marine sources (Kamble & Vavilala, 2018). The exploration of marine enzymes has unveiled a wealth of novel biocatalysts with unique properties adapted to the diverse and often extreme conditions of marine environments. This journey from discovery to utilization encompasses several key stages:

Discovery includes marine environments, including coral reefs. Arctic locations and deep-sea vents provide a wealth of microbial variety. Through highly advanced techniques such as bioinformatics, metagenomics, and in situ sampling, scientists have identified a plethora of novel enzymes with potential biotechnological applications (Debashish et al., 2005).

Characterization of the detailed biochemical and structural analysis of these enzymes has revealed distinctive characteristics, such as stability under extreme temperatures, salinity, and pressure. These properties make marine enzymes particularly valuable for industrial processes that require robust and efficient catalysts (Ferrer et al., 2015).

Utilization includes the unique attributes of marine enzymes which have facilitated their application across various industries. In biotechnology, they are used in bioremediation, pharmaceuticals, and biofuel production. In industrial processes, their stability and efficiency enhance the production of food, beverages, and textiles. Furthermore, marine enzymes are integral to developing sustainable and eco-friendly technologies (Bäumgen et al., 2021).

The successful transition from discovery to application highlights the importance of marine biotechnology in addressing global challenges. Continued research and investment in this field promise to unlock further potential, driving innovation and sustainable development. By utilizing the power of marine enzymes, one can pave the way for advancements that benefit both industry and the environment.

REFERENCES

Addison, S. L., Foote, S. M., Reid, N. M., & Lloyd-Jones, G. (2007). *Novosphingobium nitrogenifigens* sp. Nov., a polyhydroxyalkanoate-accumulating diazotroph isolated from a New Zealand pulp and paper wastewater. *International Journal of Systematic and Evolutionary Microbiology*, *57*, 2467–2471.

Alderkamp, A. C., Van Rijssel, M., & Bolhuis, H. (2007). Characterization of marine bacteria and the activity of their enzyme systems involved in degradation of the algal storage glucan laminarin. *FEMS Microbiology Ecology*, *59*(1), 108–117.

Alma'abadi, A. D., Gojobori, T., & Mineta, K. (2015). Marine metagenome as a resource for novel enzymes. *Genomics, Proteomics and Bioinformatics*, *13*(5), 290–295.

Almaliti, J., & Gerwick, W. H. (2023). Methods in marine natural product drug discovery: what's new?. *Expert Opinion on Drug Discovery*, *18*(7), 687–691.

Barone, R., De Santi, C., Palma Esposito, F., Tedesco, P., Galati, F., Visone, M., ... & De Pascale, D. (2014). Marine metagenomics, a valuable tool for enzymes and bioactive compounds discovery. *Frontiers in Marine Science*, *1*, 38.

Barzkar, N., & Sohail, M. (2020). An overview on marine cellulolytic enzymes and their potential applications. *Applied Microbiology and Biotechnology*, *104*, 6873–6892.

Bäumgen, M., Dutschei, T., & Bornscheuer, U. T. (2021). Marine polysaccharides: occurrence, enzymatic degradation and utilization. *ChemBioChem*, *22*(13), 2247–2256.

Bernan, V. S., Greenstein, M., & Maiese, W. M. (1997). Marine microorganisms as a source of new natural products. *Advances in Applied Microbiology*, *43*, 57–90.

Beygmoradi, A., Homaei, A., Hemmati, R., Santos-Moriano, P., Hormigo, D., & Fernández-Lucas, J. (2018). Marine chitinolytic enzymes, a biotechnological treasure hidden in the ocean? *Applied Microbiology and Biotechnology*, *102*, 9937–9948.

Bianchi, D., Weber, T. S., Kiko, R. & Deutsch, C. (2018). Global niche of marine anaerobic metabolisms expanded by particle microenvironments. *Nature Geoscience*, *11*, 263–268.

Birolli, W. G., Lima, R. N., & Porto, A. L. (2019). Applications of marine-derived microorganisms and their enzymes in biocatalysis and biotransformation, the underexplored potentials. *Frontiers in Microbiology*, *10*, 1453.

Bonugli-Santos, R. C., dos Santos Vasconcelos, M. R., Passarini, M. R., Vieira, G. A., Lopes, V. C., Mainardi, P. H., ... & Sette, L. D. (2015). Marine-derived fungi: diversity of enzymes and biotechnological applications. *Frontiers in Microbiology*, *6*, 269.

Borja, Á., Dauer, D. M., & Grémare, A. (2012). The importance of setting targets and reference conditions in assessing marine ecosystem quality. *Ecological Indicators*, *12*(1), 1–7.

Brazelton, W. J., Nelson, B., & Schrenk, M. O. (2012). Metagenomic evidence for H_2 oxidation and H_2 production by serpentinite-hosted subsurface microbial communities. *Frontiers in Microbiology*, *2*, 268.

Bruno, S., Coppola, D., di Prisco, G., Giordano, D., & Verde, C. (2019). Enzymes from marine polar regions and their biotechnological applications. *Marine Drugs*, *17*(10), 544.

Buesseler, K. O., Antia, A. N., Chen, M., Fowler, S. W., Gardner, W. D., Gustafsson, O., … & Trull, T. (2007). An assessment of the use of sediment traps for estimating upper ocean particle fluxes. *Journal of Marine Research*, *65*, 345–416.

Cao, S., Zhang, W., Ding, W., Wang, M., Fan, S., Yang, B., … & Zhang, Y. Z. (2020). Structure and function of the Arctic and Antarctic marine microbiota as revealed by metagenomics. *Microbiome*, *8*, 1–12.

Chasanah, E., Zilda, D. S., & Uria, A. R. (2009). Screening and characterization of bacterial chitosanase from marine environment. *Journal of Coastal Development*, *12*(2), 64–72.

Cheng, T. H., Ismail, N., Kamaruding, N., Saidin, J., & Danish-Daniel, M. J. B. R. (2020). Industrial enzymes-producing marine bacteria from marine resources. *Biotechnology Reports*, *27*, e00482.

Chi, Z., Chi, Z., Zhang, T., Liu, G., Li, J., & Wang, X. (2009). Production, characterization and gene cloning of the extracellular enzymes from the marine-derived yeasts and their potential applications. *Biotechnology Advances*, *27*(3), 236–255.

Coutinho, F. H., Gregoracci, G. B., Walter, J. M., Thompson, C. C., & Thompson, F. L. (2018). Metagenomics sheds light on the ecology of marine microbes and their viruses. *Trends in Microbiology*, *26*(11), 955–965.

Curtin, R., & Prellezo, R. (2010). Understanding marine ecosystem-based management: a literature review. *Marine Policy*, *34*(5), 821–830.

De Santi, C., Leiros, H. K. S., Di Scala, A., de Pascale, D., Altermark, B., & Willassen, N. P. (2016). Biochemical characterization and structural analysis of a new cold-active and salt-tolerant esterase from the marine bacterium *Thalassospira* sp. *Extremophiles*, *20*, 323–336.

Debashish, G., Malay, S., Barindra, S., & Joydeep, M. (2005). Marine enzymes. *Marine Biotechnology I*, *96*, 189–218.

Demkina, A., Slonova, D., Mamontov, V., Konovalova, O., Yurikova, D., Rogozhin, V., … & Isaev, A. (2023). Benchmarking DNA isolation methods for marine metagenomics. *Scientific Reports*, *13*(1), 22138.

Dhevagi, P., & Poorani, E. (2006). Isolation and characterization of L-asparaginase from marine actinomycetes. *Indian Journal of Biotechnology*, *5*, 514–520.

Douvere, F. (2008). The importance of marine spatial planning in advancing ecosystem-based sea use management. *Marine Policy*, *32*(5), 762–771.

Duarte, A. W. F., Dos Santos, J. A., Vianna, M. V., Vieira, J. M. F., Mallagutti, V. H., Inforsato, F. J., … & Durães Sette, L. (2018). Cold-adapted enzymes produced by fungi from terrestrial and marine Antarctic environments. *Critical Reviews in Biotechnology*, *38*(4), 600–619.

El-Gebali, S., Mistry, J., Bateman, A., Eddy, S. R., Luciani, A., Potter, S. C., … & Finn, R. D. (2019). The Pfam protein families database in 2019. *Nucleic Acids Research*, *47*, D427–D432.

Farnelid, H., Andersson, A. F., Bertilsson, S., Al-Soud, W. A., Hansen, L. H., Sørensen, S., … & Riemann, L. (2011). Nitrogenase gene amplicons from global marine surface waters are dominated by genes of non-cyanobacteria. *PLoS One*, *6*, e19223.

Fernandes, P. (2014). Marine enzymes and food industry: insight on existing and potential interactions. *Frontiers in Marine Science*, *1*, 46.

Ferraro, V., Cruz, I. B., Jorge, R. F., Malcata, F. X., Pintado, M. E., & Castro, P. M. (2010). Valorisation of natural extracts from marine source focused on marine by-products: a review. *Food Research International*, *43*(9), 2221–2233.

Ferrer, M., Méndez-García, C., Bargiela, R., Chow, J., Alonso, S., García-Moyano, A., … & Inmare Consortium. (2019). Decoding the ocean's microbiological secrets for marine enzyme biodiscovery. *FEMS Microbiology Letters*, *366*(1), fny285.

Galperin, M. Y., Makarova, K. S., Wolf, Y. I., & Koonin, E. V. (2015). Expanded microbial genome coverage and improved protein family annotation in the COG database. *Nucleic Acids Research*, *43*, D261–D269.

Gilbert, J. A., & Dupont, C. L. (2011). Microbial metagenomics: beyond the genome. *Annual Review of Marine Science*, *3*(1), 347–371.

Gildberg, A. (1993). Enzymic processing of marine raw materials. *Process Biochemistry*, *28*(1), 1–15.

Griffiths, J. R., Kadin, M., Nascimento, F. J., Tamelander, T., Törnroos, A., Bonaglia, S., … & Winder, M. (2017). The importance of benthic–pelagic coupling for marine ecosystem functioning in a changing world. *Global Change Biology*, *23*(6), 2179–2196.

Guérard, F. (2007). Enzymatic methods for marine by-products recovery. In Fereidoon, S., Ed., *Maximising the Value of Marine By-Products* (pp. 107–143). Woodhead Publishing.

Gulder, T. A., & Moore, B. S. (2009). Chasing the treasures of the sea—bacterial marine natural products. *Current Opinion in Microbiology*, *12*(3), 252–260.

Harborne, N. R., Griffiths, L., Busby, S. J. W., & Cole, J. A. (1992). Transcriptional control, translation and function of the products of the five open reading frames of the *Escherichia coli* nir operon. *Molecular Microbiology*, *6*, 2805–2813.

Hewson, I., Moisander, P. H., Achilles, K. M., Carlson, C. A., Jenkins, B. D., Mondragon, E. A., … & Zehr, J. P. (2007). Characteristics of diazotrophs in surface to abyssopelagic waters of the Sargasso Sea. *Aquatic Microbial Ecology*, *46*, 15–30.

Honda, K., Ymamot, Y., & Tatsukawa, R. (1987). Antarctic marine ecosystem. In Proceedings *of the NIPR Symposium on Polar Biology* (Vol. 1, pp. 184–197).

Hooker, S. K., & Gerber, L. R. (2004). Marine reserves as a tool for ecosystem-based management: the potential importance of megafauna. *Bioscience*, *54*(1), 27–39

Hoppe, H. G., Arnosti, C., & Herndl, G. J. (2002). Ecological significance of bacterial enzymes in the marine environment. In Burns, R. G., & Dick, R. P., Eds., *Enzymes in the Environment: Activity, Ecology, and Applications* (pp. 73–107). CRC Press.

Hurwitz, B. L., Westveld, A. H., Brum, J. R., & Sullivan, M. B. (2014). Modeling ecological drivers in marine viral communities using comparative metagenomics and network analyses. *Proceedings of the National Academy of Sciences of the United States of America*, *111*(29), 10714–10719.

Hyde, K. D., Jones, E. G., Leaño, E., Pointing, S. B., Poonyth, A. D., & Vrijmoed, L. L. (1998). Role of fungi in marine ecosystems. *Biodiversity & Conservation*, *7*, 1147–1161.

Jagtap, A. S., & Manohar, C. S. (2021). Overview on microbial enzymatic production of algal oligosaccharides for nutraceutical applications. *Marine Biotechnology*, *23*(2), 159–176.

Kamble, P., & Vavilala, S. L. (2018). Discovering novel enzymes from marine ecosystems: a metagenomic approach. *Botanica Marina*, *61*(2), 161–175.

Kennedy, J., Flemer, B., Jackson, S. A., Lejon, D. P., Morrissey, J. P., O'gara, F., & Dobson, A. D. (2010). Marine metagenomics: new tools for the study and exploitation of marine microbial metabolism. *Marine Drugs*, *8*(3), 608–628.

Kennedy, J., O'leary, N. D., Kiran, G. S., Morrissey, J. P., O'Gara, F., Selvin, J., & Dobson, A. D. W. (2011). Functional metagenomic strategies for the discovery of novel enzymes and biosurfactants with biotechnological applications from marine ecosystems. *Journal of Applied Microbiology*, *111*(4), 787–799.

Khare, S. K., Pandey, A., & Larroche, C. (2015). Current perspectives in enzymatic saccharification of lignocellulosic biomass. *Biochemical Engineering Journal*, *102*, 38–44.

Kim, S. K. (2019). Marine microorganism resources and biotechnology. In Rotter, A., Barbier, M., Bertoni, F., Bones, A., Cancela, L., Carlsson, J., … & Vasquez, M., Eds., *Essentials of Marine Biotechnology* (pp. 381–415). Springer.

Klemetsen, T., Raknes, I. A., Fu, J., Agafonov, A., Balasundaram, S. V., Tartari, G., … & Willassen, N. P. (2018). The MAR databases: development and implementation of databases specific for marine metagenomics. *Nucleic Acids Research*, *46*(D1), D692–D699.

Kodzius, R., & Gojobori, T. (2015). Marine metagenomics as a source for bioprospecting. *Marine Genomics*, *24*, 21 30.

La Cono, V., Ruggeri, G., Azzaro, M., Crisafi, F., Decembrini, F., Denaro, R., … & Yakimov, M. M. (2018). Contribution of bicarbonate assimilation to carbon pool dynamics in the deep Mediterranean Sea and cultivation of actively nitrifying and CO_2-fixing bathypelagic prokaryotic consortia. *Frontiers in Microbiology*, *9*, 3.

Lee, H. S., Kwon, K. K., Kang, S. G., Cha, S. S., Kim, S. J., & Lee, J. H. (2010). Approaches for novel enzyme discovery from marine environments. *Current Opinion in Biotechnology*, *21*(3), 353–357.

Leu, S. Y., & Zhu, J. Y. (2013). Substrate related factors affecting enzymatic saccharification of lignocelluloses: our recent understanding. *Bioenergy Research*, *6*, 405–415.

Lidstrom, M. E. (1988). Isolation and characterization of marine methanotrophs. *Antonie Van Leeuwenhoek*, *54*, 189–199.

Lima, R. N., & Porto, A. L. M. (2016). Recent advances in marine enzymes for biotechnological processes. *Advances in Food and Nutrition Research*, *78*, 153–192.

Logares, R., Sunagawa, S., Salazar, G., Cornejo-Castillo, F. M., Ferrera, I., Sarmento, H., … & Acinas, S. G. (2014). Metagenomic 16S rDNA Illumina tags are a powerful alternative to amplicon sequencing to explore diversity and structure of microbial communities. *Environmental Microbiology*, *16*, 2659–2671.

Loperena, L., Soria, V., Varela, H., Lupo, S., Bergalli, A., Guigou, M., … & Batista, S. (2012). Extracellular enzymes produced by microorganisms isolated from maritime Antarctica. *World Journal of Microbiology and Biotechnology*, *28*, 2249–2256.

López-Pérez, M., Kimes, N. E., Haro-Moreno, J. M., & Rodríguez-Valera, F. (2016). Not all particles are equal: the selective enrichment of particle-associated bacteria from the Mediterranean Sea. *Frontiers in Microbiology*, *7*, 996.

Martyniuk, A. O., Skrotska, O. I., Shevchuk, T. A., & Pirog, T. P. (2020). Practically valuable metabolites of marine microorganisms. *Biotechnologia Acta*, *13*(3), 5–29.

Megrey, B. A., Link, J. S., Hunt Jr, G. L., & Moksness, E. (2009). Comparative marine ecosystem analysis: applications, opportunities, and lessons learned. *Progress in Oceanography*, *81*(1–4), 2–9.

Mehrshad, M., Rodríguez-Valera, F., Amoozegar, M. A., López-García, P., & Ghai, R. (2018). The enigmatic SAR202 cluster up close: shedding light on a globally distributed dark ocean lineage involved in sulfur cycling. *ISME Journal*, 12, 655–668.

Michel, G., & Czjzek, M. (2013). Polysaccharide-degrading enzymes from marine bacteria. In Trincone, A., Ed., *Marine Enzymes for Biocatalysis* (pp. 429–464). Woodhead Publishing.

Mineta, K., & Gojobori, T. (2016). Databases of the marine metagenomics. *Gene*, *576*(2), 724–728.

Mizuno, C. M., Rodriguez-Valera, F., Kimes, N. E., & Ghai, R. (2013). Expanding the marine virosphere using metagenomics. *PLoS Genetics*, *9*(12), e1003987.

Mohapatra, B. R., Bapuji, M., & Sree, A. J. A. B. (2003). Production of industrial enzymes (amylase, carboxymethylcellulase and protease) by bacteria isolated from marine sedentary organisms. *Acta Biotechnologica*, *23*(1), 75–84.

Moisander, P. H., Benavides, M., Bonnet, S., Berman-Frank, I., White, A. E., & Riemann, L. (2017). Chasing after non-cyanobacterial nitrogen fixation in marine pelagic environments. *Frontiers in Microbiology*, *8*, 1736.

Molitoris, H. P., & Schaumann, K. (1986). Physiology of marine fungi: a screening programme for growth and enzyme production. In Moss, S. T. Ed., *The Biology of Marine Fungi* (pp. 35–47). Springer.

Muffler, K., Sana, B., Mukherjee, J., & Ulber, R. (2015). Marine enzymes–production & applications. In Kim, S.-K., Ed., *Springer Handbook of Marine Biotechnology* (pp. 413–429). Springer.

Navvabi, A., Razzaghi, M., Fernandes, P., Karami, L., & Homaei, A. (2018). Novel lipases discovery specifically from marine organisms for industrial production and practical applications. *Process Biochemistry*, *70*, 61–70.

Nguyen, T. H., & Nguyen, V. D. (2017). Characterization and applications of marine microbial enzymes in biotechnology and probiotics for animal health. In Kim, S.-K., Ed., *Advances in Food and Nutrition Research* (Vol. 80, pp. 37–74). Academic Press.

Parte, S., Sirisha, V. L., & D'Souza, J. S. (2017). Biotechnological applications of marine enzymes from algae, bacteria, fungi, and sponges. In Kim, S.-K., Ed., *Advances in Food and Nutrition Research* (Vol. 80, pp. 75–106). Academic Press.

Popovic, A., Tchigvintsev, A., Tran, H., Chernikova, T. N., Golyshina, O. V., Yakimov, M. M., … & Yakunin, A. F. (2015). Metagenomics as a tool for enzyme discovery: hydrolytic enzymes from marine-related metagenomes. In Krogan, N. J., & Babu, M., Eds., *Prokaryotic Systems Biology* (pp. 1–20). Springer.

Qeshmi, F. I., Homaei, A., Fernandes, P., Hemmati, R., Dijkstra, B. W., & Khajeh, K. (2020). Xylanases from marine microorganisms: a brief overview on scope, sources, features and potential applications. *Biochimica et Biophysica Acta (BBA) – Proteins and Proteomics*, *1868*(2), 140312.

Rao, T. E., Imchen, M., & Kumavath, R. (2017). Marine enzymes: production and applications for human health. In Kim, S.-K., Ed., *Advances in Food and Nutrition Research* (Vol. 80, pp. 149–163). Academic Press.

Rasmussen, R. S., & Morrissey, M. T. (2007). Marine biotechnology for production of food ingredients. *Advances in Food and Nutrition Research*, *52*, 237–292.

Rastogi, M., & Shrivastava, S. (2017). Recent advances in second generation bioethanol production: an insight to pretreatment, saccharification and fermentation processes. *Renewable and Sustainable Energy Reviews*, *80*, 330–340.

Rohrmann, S., Lorenz, R., & Molitoris, H. P. (1992). Use of natural and artificial seawater for investigation of growth, fruit body production, and enzyme activities in marine fungi. *Canadian Journal of Botany*, *70*(10), 2106–2110.

Rustad, T., Storrø, I., & Slizyte, R. (2011). Possibilities for the utilisation of marine by-products. *International Journal of Food Science & Technology*, *46*(10), 2001–2014.

Saeman, J. F., Bubl, J. L., & Harris, E. E. (1945). Quantitative saccharification of wood and cellulose. *Industrial & Engineering Chemistry Analytical Edition*, *17*(1), 35–37.

Salamone, M., Cuttitta, A., Seidita, G., Mazzola, S., Bertuzzi, F., Ricordi, C., & Ghersi, G. (2012). Characterization of collagenolytic/proteolytic marine enzymes. *Chem Eng*, *27*, 1–6.

Salazar, G., & Sunagawa, S. (2017). Marine microbial diversity. *Currt Bio*, *27*(11), R489–R494.

Sarkar, S., Pramanik, A., Mitra, A., & Mukherjee, J. (2010). Bioprocessing data for the production of marine enzymes. *Marine Drugs*, *8*(4), 1323–1372.

Shaaban, M., Abd-Alla, H. I., Hassan, A. Z., Aly, H. F., & Ghani, M. A. (2012). Chemical characterization, antioxidant and inhibitory effects of some marine sponges against carbohydrate metabolizing enzymes. *Organic and Medicinal Chemistry Letters*, *2*, 1–12.

Shahidi, F., & Kamil, Y. J. (2001). Enzymes from fish and aquatic invertebrates and their application in the food industry. *Trends in Food Science & Technology*, *12*(12), 435–464.

Sharma, G., Khatri, I., & Subramanian, S. (2016). Complete genome of the starch-degrading myxobacteria *Sandaracinus amylolyticus* DSM 53668T. *Genome Biology and Evolution*, *8*, 2520–2529.

Shi, X., Song, E., & Zhang, C. (2013). Application of enzymes from marine microorganisms. In Kim, S.-K., Ed., *Marine Biomaterials: Characterization, Isolation and Applications* (pp. 787–796). CRC Press.

Snelgrove, P. V., Blackburn, T. H., Hutchings, P., Alongi, D. M., Grassle, J. F., Hummel, H., … & Solis-Weiss, V. (1997). The importance of marine sediment biodiversity in ecosystem processes. *Ambio*, *26*, 578–583.

Suleria, H. A. R., Masci, P., Gobe, G., & Osborne, S. (2016). Current and potential uses of bioactive molecules from marine processing waste. *Journal of the Science of Food and Agriculture*, *96*(4), 1064–1067.

Suthindhiran, K., Jayasri, M. A., Dipali, D., & Prasar, A. (2014). Screening and characterization of protease producing actinomycetes from marine saltern. *Journal of Basic Microbiology*, *54*(10), 1098–1109.

Teira, E., Lebaron, P., Van Aken, H. & Herndl, G. J. (2006). Distribution and activity of bacteria and archaea in the deep water masses of the North Atlantic. *Limnology and Oceanography*, *51*, 2131–2144.

Tiirola, M. A., Männistö, M. K., Puhakka, J. A. & Kulomaa, M. S. (2002). Isolation and characterization of *Novosphingobium* sp. Strain MT1, a dominant polychlorophenol-degrading strain in a groundwater bioremediation system. *Applied and Environmental Microbiology*, 68, 173–180.

Trincone, A. (2011). Marine biocatalysts: enzymatic features and applications. *Marine Drugs*, *9*(4), 478–499.

Tseng, C. H., & Tang, S. L. (2014). Marine microbial metagenomics: from individual to the environment. *International Journal of Molecular Sciences*, *15*(5), 8878–8892.

Tully, B. J., Sachdeva, R., Graham, E. D., & Heidelberg, J. F. (2017). 290 metagenome-assembled genomes from the Mediterranean Sea: a resource for marine microbiology. *PeerJ*, *5*, e3558.

Ulber, R., Arndt, M., Muffler, K., & Laroche, M. (2003). Enzymes from marine microorganisms. In *Actes de colloques-IFREMER* (pp. 9–16).

Wang, Y., Song, Q., & Zhang, X. H. (2016). Marine microbiological enzymes: studies with multiple strategies and prospects. *Marine Drugs*, *14*(10), 171.

Woyke, T., Xie, G., Copeland, A., Gonzalez, J. M., Han, C., Kiss, H., … & Stepanauskas, R. (2009). Assembling the marine metagenome, one cell at a time. *PLoS One*, *4*(4), e5299.

Wu, Y. L., Wang, S., Yang, D. F., Yang, L. Y., Wang, Q. Y., Yu, J., … & Pan, L. X. (2022). The discovery, enzymatic characterization and functional analysis of a newly isolated chitinase from marine-derived fungus *Aspergillus fumigatus* df347. *Marine Drugs*, *20*(8), 520.

Yelton, A. P., Acinas, S. G., Sunagawa, S., Bork, P., Pedrós-Alió, C., & Chisholm, S. W. (2016). Global genetic capacity for mixotrophy in marine picocyanobacteria. *ISME Journal*, *10*, 2946–2957.

Zhang, C., & Kim, S. K. (2010). Research and application of marine microbial enzymes: status and prospects. *Marine Drugs*, *8*(6), 1920–1934.

Zhao, F., & Bajic, V. B. (2015). The value and significance of metagenomics of marine environments. *Genomics, Proteomics and Bioinformatics*, *13*(5), 271–274.

Zhao, X. Q., Xu, X. N., & Chen, L. Y. (2016). Production of enzymes from marine actinobacteria. *Advances in Food and Nutrition Research*, *78*, 137–151.

8 Marine Microbiomes and Their Endless Potential for the Bioeconomy

Ashwini Kumar[1], *Puneet Raina*[2], *Rajan Bhandari*[3], *and Baljinder Singh*[4]
[1]Department of Chemistry, Government Post-Graduate College Una, Una, India
[2]Department of Zoology, Post-Graduate DAV College, Chandigarh, India
[3]Department of Geography, School of Environment and Earth Sciences, Central University of Punjab, Bathinda, India
[4]Department of Biochemistry, School of Basic Sciences, Central University of Punjab, Bathinda, India

8.1 INTRODUCTION

The production of biobased resources and their transformation into food, feed, bioenergy, and biobased products is known as the bioeconomy, sometimes termed as the "biobased economy." A biobased value chain consists of the initial generation of biobased resources, their processing, and market commercialization into higher-value items. This brings together several scientific disciplines and stakeholders from a wide range of sectors [1].

The Earth's oceans are vast, covering more than 70% of the planet's surface and hosting a rich diversity of life. Among the most fundamental, yet often overlooked, components of marine ecosystems are microorganisms. These microscopic organisms, collectively known as the marine microbiome, play pivotal roles in nutrient cycling, carbon sequestration, and maintaining ecosystem balance [2].

In recent years, advancements in molecular biology and sequencing technologies have shed light on the complexity and functional significance of marine microbiomes. Scientists increasingly focus on the broad and mostly unexplored field of marine microbiomes now, in their search for sustainable solutions. Replete with various microscopic creatures, these complex ecosystems can potentially revolutionize several bioeconomy industries. Marine microbiomes have numerous significant applications, ranging from biotechnology to environmental remediation.

Although less than 1% of all living microbes can be cultivated in a lab, even this tiny fraction has led to remarkable discoveries like antibiotics, which have saved millions of lives [3]. When the remainder of this vast universe of living genomes is awakened, we can only speculate what more amazing discoveries will be made. Since everything is stored in genes, current metagenomic research, which involves the large-scale sequencing of marine microorganisms' genetic material and uncovers similarities between the microbiomes of the ocean and the human gut, is not only offering compelling new insights into the reasons behind life on the planet Earth but also suggests potential causes and treatment options for metabolic illnesses like cancer [4].

Billions of genes within the marine microbiome can express and be a part of the metabolic pathways to produce an incredible array of small molecules and proteins, lipids, and other classes of products serving ecological purposes. These products can be "humanized" to discover novel biomaterials, enhanced enzymes, nutraceuticals, food and feedstocks, personal care and cosmetic products, and an ever-expanding list of marine natural products. Marine genomic resources is the new phrase for aquatic biodiversity [5].

Although "blue biotechnology" is still in its infancy, a report for the European DG Maritime Affairs and Fisheries (2014) [6] estimated that if a market growth of 7%–8% per year was maintained, blue biotechnology could generate €1 billion in revenue in Europe within five years and create 10,000 new jobs. According to a 2015 Smithers Rapra research analysis [4], this global market has the potential to grow from $4.8 billion in 2020 to $6.4 billion in 2025. Nevertheless, these estimates largely undervalue the significance of the marine microbiome because they are based on "classical" methods like aquaculture, bioenergy, and algal cosmetics.

With the emergence of transversal technologies aimed at utilizing other biotechnological advancements, like the development of bioinformatics, synthetic biology, molecular diagnostics and devices, biocatalysis, and the numerous OMICS technologies, market opportunities, and job creation are expected to rise dramatically in the future [7]. A crucial and inseparable component of the bioeconomy is the marine microbiome. Though big businesses are reluctant to invest in long-term innovations, there currently needs to be more coordination among policymakers, governments, academia, civil society organizations, and large industries, after the industry has duly recognized the enormous potential of the marine microbiome. This environment offers a great chance to establish highly specialized small and medium-sized businesses. A fresh wave of these "marine biotech enterprises," founded on contemporary bioprospecting, would employ highly skilled personnel and offer the opportunity to explore novel, "high-tech" concepts at the forefront of modern science [8].

DOI: 10.1201/9781003509530-8

Although there is no denying the promise of using marine microbiomes in the economy, there are several obstacles that need to be overcome to exploit these opportunities fully. The intricate dynamics of marine ecosystems and our limited knowledge of microbial interactions present formidable challenges to successfully using these species. Furthermore, there are significant hindrances in commercializing goods based on marine microbiomes, including scalability, standardization, and regulatory frameworks. Addressing these issues will take interdisciplinary cooperation, creative research techniques, and calculated policy measures.

This chapter explores the untapped potential of marine microbiomes in driving innovations in the bioeconomy, encompassing biotechnological applications, sustainable resource management, and environmental remediation.

8.2 MARINE MICROBIOME

The marine microbiome is a complex network of microorganisms inhabiting various marine environments, from surface waters to deep-sea sediments and hydrothermal vents. These microorganisms include bacteria, archaea, viruses, and protists, each contributing to the intricate web of interactions within marine ecosystems [9].

It also includes single-celled creatures, simple multicellular organisms, and symbiotic microbes. The "microbiome" of the ocean is made up of them all. This microbiome has evolved over billions of years into various adaptations and lifestyles that allow it to participate in the fluxes of almost all chemical components. There is no questioning the significance of the marine microbiome for human society and the health of our planet. The idea of microbial communities in the marine environment has grown in significance over the past few decades as a result of the diversity, abundance, and vital role that the marine microbiome plays in regulating the carbon and nitrogen cycles, maintaining the equilibrium of the world's marine biomass, and determining the structure and function of oceans [10]. A few noteworthy findings from the Census of Marine Life (2010) (http://www.coml.org/) show that there are more than a million living bacteria in every milliliter of seawater and that the number of bacteria in the open ocean surpasses 1,029, accounting for over 90% of all marine biomass. An estimated 20,000 species can be found in a single liter of saltwater, including an enormous number of eukaryotic cells, marine viruses, phages, and archaeal cells that equal those of bacteria. Recent studies utilizing high-throughput sequencing techniques have unveiled the staggering diversity of marine microbial communities, challenging our understanding of microbial ecology and evolution [11]. Understanding the composition, dynamics, and functional roles of marine microbiomes is essential for harnessing their biotechnological potential and promoting ecosystem health.

8.3 BIOCHEMICAL DIVERSITY

Biochemical diversity within marine microbiomes is a cornerstone of their potential for the bioeconomy. These diverse microbial communities encompass bacteria, archaea, viruses, and protists, inhabiting a variety of marine environments. They produce an extensive array of bioactive compounds with significant applications across various sectors. Marine-derived natural products, including antibiotics, anticancer agents, enzymes, and biopolymers, are of particular interest due to their unique chemical structures and pharmacological properties [12].

Recent studies have emphasized the importance of marine microbiomes as prolific sources of novel bioactive compounds. For instance, research by Blunt et al. [13] highlights the wealth of marine natural products, showcasing their potential for drug discovery and development. The discovery of bioactive compounds from marine microbes has led to the development of new pharmaceuticals, such as antivirals and immunosuppressants, with promising clinical applications. Furthermore, marine microbes produce enzymes with diverse catalytic activities, offering potential applications in biotechnology and industrial processes. Bull et al. [14] recognized the significance of marine actinobacteria as sources of industrially relevant enzymes, including proteases, lipases, and cellulases. These enzymes exhibit high specificity and stability under a range of conditions, making them valuable tools for various biotechnological applications, such as biocatalysis, food processing, and wastewater treatment. In addition to bioactive compounds and enzymes, marine microbiomes produce biopolymers with unique properties suitable for bioplastics, biodegradable packaging, and tissue engineering [15]. Focused efforts on understanding the biosynthetic pathways and metabolic capabilities of marine microbes have led to the identification of novel biopolymer-producing organisms and their corresponding genes [16]. By harnessing the biosynthetic potential of marine microbiomes, researchers aim to develop sustainable alternatives to conventional plastics derived from fossil fuels. The exploration of marine microbiomes for biochemical diversity holds significant promise for the bioeconomy. However, challenges such as accessing and culturing diverse microbial strains, as well as scaling up production processes, remain obstacles to fully realizing their potential. Collaborative efforts between academia, industry, and government agencies are crucial for overcoming these challenges and translating research findings into practical applications that contribute to economic growth and environmental sustainability.

In conclusion, marine microbiomes represent a vast reservoir of biochemical diversity, producing a myriad of bioactive compounds with potential applications in medicine, agriculture, and industry. Marine-derived natural products have yielded numerous pharmaceuticals, including antibiotics, antivirals, and anticancer agents. Moreover, marine microbes produce enzymes, surfactants, and biopolymers with industrial applications, ranging from biodegradable plastics to biofuels. Metagenomic approaches have revolutionized the discovery of novel biosynthetic gene clusters encoding bioactive compounds, offering promising avenues for drug discovery and biotechnological innovation.

8.4 BLUE BUSINESS

The concept of the blue economy emphasizes the sustainable utilization of ocean resources to foster economic growth while preserving marine ecosystems [17]. Marine microbiomes play a crucial role in the blue economy by offering biotechnological solutions to various industries. Bioprospecting efforts have led to the discovery of novel enzymes for industrial processes, biodegradable polymers for packaging, and bioactive compounds for pharmaceuticals. Furthermore, aquaculture operations can benefit from the application of probiotics and microbial additives to enhance productivity and mitigate environmental impacts.

The approval of the marine-derived pharmaceutical Yondelis® (trabectedin) for the treatment of specific types of soft tissue sarcomas and ovarian malignancies is an example of a successful "blue business" that provides "proof-of-concept" for harnessing marine chemical diversity in human health [18]. The biopharmaceutical business M/s. PharmaMar (Spain), which specializes in oncology and is dedicated to research and development, created this substance by drawing inspiration from the sea to find compounds with antitumor activity. Trabectedin was extracted from the tunicate *Ecteinascidia turbinata* and is a member of the Ecteinascidin family of anticancer chemicals. M/s. PharmaMar is also developing other intriguing marine-derived chemicals to treat different types of cancer, which shows that the sea might be a very rich source of novel medicines with unique modes of action.

One of the Earth's greatest concentrations of biological niches, biodiversity, and natural resources can be found in the Atlantic Ocean. The enormous resource's inherent value and economic significance are widely acknowledged, and appeals have been made for its sustainable exploitation and the scientific validation of management choices. International accords have been negotiated to promote cooperative science in the Atlantic Ocean, acknowledging the region's interconnectedness. Specifically, the European Union, Canada, and the United States of America signed the Galway Statement on Atlantic Ocean Cooperation in 2013 and the European Union, South Africa, and Brazil signed the Belém Statement on Atlantic Ocean Research and Innovation Cooperation in 2017 [19]. Working groups (WG) under the Atlantic Ocean Research Alliance (AORA) for the Galway Statement and joint pilot actions (JPA) under the All-Atlantic Cooperation for Ocean Research and Innovation (AANChOR; https://allatlanticocean.org/) for the Belém Statement were formed as governance structures to carry out these agreements. To promote development and innovation in fields such as seabed mapping, exploration, human capacity, ocean literacy, technology, infrastructure, aquaculture, biotechnology, and marine microbiomes, the WGs and JPAs established international networks to support research collaborations [20].

The Marine Microbiome Working Group (MMWG) was established as a component of the AORA Coordination and Support Action (CSA) with the goals of promoting the use of microbiomes in marine research, building a robust network of researchers studying marine microbiomes, utilizing microbiomes to aid in management, and informing the public about the significance of microbiomes in the operation of a healthy ocean. The Marine Biotechnology Initiative for the Atlantic (BIOTECMAR) was established under the AANChOR CSA to promote technology development and transfer for industrial, medicinal, and environmental uses incorporating marine biological resources [20]. In addition to the MMWG's objectives, BIOTECMAR set out to enhance sustainable aquaculture and food production, boost biodiscovery, and develop environmental monitoring technology.

The All-Atlantic Ocean Research and Innovation Alliance Declaration (All-Atlantic Declaration) was signed in 2022 to help ensure the long-term viability of these networks (https://allatlantic2022.com/). By adding additional partner countries, this pact broadens involvement and improves upon earlier accords. Under this declaration, the Republic of Cabo Verde, Morocco, and Argentina agreed to form the All-Atlantic Ocean Research and Innovation Alliance (AAORIA), joining the signatories of the Galway and Belém Statements (the European Union, Canada, the United States, Brazil, and South Africa). AAORIA seeks to promote the Atlantic Ocean's sustainable development by exchanging infrastructure, expertise, and human resources.

A wide awareness of the necessity of international collaborations to advance marine science research was demonstrated by the presence of signatories as well as representatives from Colombia, the Dominican Republic, Ghana, Norway, Portugal, and the United Kingdom at the signing of the All-Atlantic Declaration. In the context of the All-Atlantic pledge campaign of 2021, the AORA MMWG and AANChOR BIOTECMAR started collaborating to develop an integrated All-Atlantic network.

8.5 AQUACULTURE MICROBIOME

Currently, the fishing and aquaculture industries provide a partial or complete livelihood to more than 10% of the world's population [21]. The majority of those working in this field depend on wild fish. However, 90% of fish stocks are totally or excessively exploited, contributing to the depletion of fish stocks [21]. To fulfill the UN's Sustainable Development Goals (SDGs) number 2, 12, and 14, and feed a growing global population—which is predicted to reach ten billion people in about 30 years—with high-quality protein, a more productive and sustainable aquaculture sector is required [22]. The production of seafood (fish and plants combined) has been surpassed by the aquaculture sector in recent years [23]. From 2010 to 2014, employment in the fisheries sector fell by about one million, while employment in the aquaculture sector increased by 0.1 million people. There was a noticeable change from 1990 to 2014, with 83% of workers in fisheries and 17% in aquaculture, respectively [21]. In general, 67% of workers were employed in fisheries and 33% in aquaculture. According to projections made, the sector would produce 102 million tons by 2025 and up to 121 million tons by 2030, up from 74 million tons in 2014 [21]. Moreover, it has been proposed that the worldwide biological production potential for marine aquaculture is almost 100 times greater than for global seafood consumption. This suggests that adequate habitats do not hinder the sector's expansion [24]. Aquaculture, the farming of aquatic organisms, represents a vital sector of the global food industry. However, intensive aquaculture practices often lead to environmental degradation and disease outbreaks. The manipulation of the aquaculture microbiome offers a promising approach to address these challenges. Probiotics, prebiotics, and microbial-based bioremediation strategies can promote the health and resilience of aquaculture systems, reducing reliance on antibiotics and chemical treatments. Understanding the dynamics of the aquaculture microbiome is essential for optimizing production efficiency and sustainability.

8.6 BIOREMEDIATION AND SUSTAINABLE RESOURCE MANAGEMENT

Bioremediation and sustainable resource management are key aspects of harnessing the potential of marine microbiomes for the bioeconomy. Marine ecosystems face numerous threats, including pollution, habitat destruction, and climate change. However, marine microbiomes offer innovative solutions for mitigating these challenges through bioremediation and sustainable resource management strategies.

Microbes possess the metabolic versatility to degrade a wide range of pollutants, including hydrocarbons, plastics, heavy metals, and agricultural runoff. For example, studies have demonstrated the ability of marine bacteria to degrade oil spills through the production of enzymes such as alkane hydroxylases and esterases [25]. These microbe-driven degradation processes help to break down pollutants into harmless byproducts, thereby restoring the health and integrity of marine ecosystems. Furthermore, microbe-mediated nutrient cycling plays a vital role in sustainable resource management within marine ecosystems. For instance, nitrogen-fixing bacteria such as cyanobacteria and diazotrophic bacteria contribute to nitrogen fixation, facilitating the assimilation of nitrogen into organic matter and supporting primary productivity [23]. Additionally, microbial communities associated with marine sediments play essential roles in carbon sequestration and nutrient recycling, thereby influencing ecosystem productivity and resilience. Recent advances in bioremediation research have focused on the identification and characterization of microbial communities with specialized metabolic capabilities. Metagenomic and metatranscriptomic approaches have provided insights into the genetic diversity and functional potential of marine microbiomes for pollutant degradation [26]. By understanding the metabolic pathways and regulatory mechanisms involved in bioremediation processes, researchers can design targeted strategies to enhance microbial remediation efficiency and adaptability. In addition to bioremediation, sustainable resource management practices involve the conservation and responsible use of marine resources. Microbe-mediated processes, such as biological nitrogen fixation and denitrification, play essential roles in regulating nutrient cycles and ecosystem functioning [27]. Moreover, microbe-based technologies, such as microbial fuel cells and bioreactors, offer promising solutions for wastewater treatment, nutrient recovery, and energy production in marine environments [28].

Despite the potential of bioremediation and sustainable resource management, challenges such as microbial community resilience, ecosystem complexity, and regulatory constraints need to be addressed. Collaborative efforts between scientists, policymakers, and industry stakeholders are essential for implementing effective bioremediation strategies and promoting sustainable resource management practices in marine environments. By harnessing the biotechnological potential of marine microbiomes, we can address environmental challenges while fostering economic growth and resilience in coastal communities. In conclusion, bioremediation and sustainable resource management are essential components of harnessing the potential of marine microbiomes for the bioeconomy. With focused efforts at leveraging the metabolic diversity and ecological functions of marine microbes, scientists can help us develop innovative solutions to mitigate pollution, conserve marine resources, and promote environmental sustainability.

8.7 CHALLENGES AND FUTURE DIRECTIONS

Harnessing the true potential of marine microbiomes for the bioeconomy encompasses a range of interdisciplinary, technological, and regulatory considerations. Despite the promising applications of marine microbes in biotechnology, environmental remediation, and sustainable resource management, several hurdles must be overcome to realize their potential and ensure responsible and equitable utilization fully.

One significant challenge is the complexity and diversity of marine microbial communities. Marine environments harbor an immense variety of microorganisms with distinct metabolic capabilities, making it challenging to isolate and characterize specific strains or functional genes [29]. Additionally, the vast majority of marine microbes remain unculturable under laboratory conditions, limiting our ability to study their biology and exploit their biotechnological potential [30]. Technological limitations also pose obstacles to unlocking the full potential of marine microbiomes. Traditional culturing methods are often inadequate for capturing the diversity of marine microbes, necessitating the development of high-throughput sequencing techniques and bioinformatic tools for metagenomic and metatranscriptomic analysis [31]. Moreover, accessing remote marine environments for sampling and conducting field studies presents logistical challenges that require innovative solutions and collaborative efforts. Regulatory frameworks and intellectual property rights further complicate the translation of marine microbiome research into practical applications. Bioprospecting activities in marine environments are subject to international agreements and national regulations aimed at conserving biodiversity and ensuring fair and equitable benefit-sharing [32]. Balancing conservation objectives with commercial interests and intellectual property rights requires transparent governance mechanisms and stakeholder engagement to promote ethical and sustainable bioprospecting practices.

Despite these challenges, several promising avenues for future research and innovation in marine microbiomes exist. Advances in single-cell genomics, microfluidics, and synthetic biology hold promise for overcoming the limitations of traditional culturing methods and exploring the metabolic potential of individual microbial cells [33]. Additionally, interdisciplinary collaborations between microbiologists, ecologists, engineers, and social scientists are essential for addressing complex challenges and maximizing the societal benefits of marine microbiome research. Moreover, integrating omics-based approaches with ecological modeling and machine learning algorithms can enhance our understanding of marine microbial communities' dynamics and ecosystem functions [34]. By elucidating the ecological drivers and biogeochemical processes mediated by marine microbes, we can develop predictive models and management strategies to mitigate environmental threats and promote ecosystem resilience.

In a nutshell, it can be affirmed that while challenges abound in harnessing marine microbiomes for the bioeconomy, innovative research, technological advancements, and collaborative governance mechanisms offer opportunities to overcome these obstacles and unlock the full potential of marine microbial resources. Addressing these challenges and embracing interdisciplinary approaches, the scientific community can pave the way for sustainable and responsible utilization of marine microbiomes to find solutions to global challenges in healthcare, food security, and environmental sustainability.

REFERENCES

1. Lewandowski, I. (2018). *Bioeconomy: Shaping the Transition to a Sustainable, Biobased Economy* (p. 356). Springer Nature.
2. Dang, H., & Chen, C. T. A. (2017). Ecological energetic perspectives on responses of nitrogen-transforming chemolithoautotrophic microbiota to changes in the marine environment. *Frontiers in Microbiology*, *8*, 247341.
3. Bérdy, J. (2012). Thoughts and facts about antibiotics: where we are now and where we are heading. *The Journal of Antibiotics*, *65*(8), 385–395.
4. de la Calle, F. (2017). Marine microbiome as source of natural products. *Microbial Biotechnology*, *10*(6), 1293.
5. Theissinger, K., Fernandes, C., Formenti, G., Bista, I., Berg, P. R., Bleidorn, C., … & Zammit, G. (2023). How genomics can help biodiversity conservation. *Trends in Genetics*, *39*(7), 545–559.
6. Hadjimichael, M. (2018). A call for a blue degrowth: unravelling the European Union's fisheries and maritime policies. *Marine Policy*, *94*, 158–164.
7. Edrada-Ebel, R., Ævarsson, A., Polymenakou, P., Hentschel, U., Carettoni, D., Day, J., … & McNeil, B. (2018). SeaBioTech: from seabed to test-bed: harvesting the potential of marine biodiversity for industrial biotechnology. In Rampelotto, P. H., & Trincone, A., Eds., *Grand Challenges in Marine Biotechnology* (pp. 451–504). Springer.
8. Rotter, A., Barbier, M., Bertoni, F., Bones, A. M., Cancela, M. L., Carlsson, J., … & Vasquez, M. I. (2021). The essentials of marine biotechnology. *Frontiers in Marine Science*, *8*, 629629.
9. Munn, C. B. (2019). *Marine Microbiology: Ecology and Applications*. CRC Press.
10. Stal, L. J., & Cretoiu, M. S. (Eds.). (2022). *The Marine Microbiome*. Springer.
11. Ramesha, T., Kumaranb, R., Thangaraja, M., & Annaduraia, D. (2021). Marine microbial community studies using recent tools and techniques. In Shah, M., & Rodriguez-Couto, S., Eds., *Microbial Ecology of Wastewater Treatment Plants* (p. 359). Elsevier.
12. Saeed, A. F., Su, J., & Ouyang, S. (2021). Marine-derived drugs: recent advances in cancer therapy and immune signaling. *Biomedicine & Pharmacotherapy*, *134*, 111091.
13. Blunt, J. W., Copp, B. R., Keyzers, R. A., Munro, M. H., & Prinsep, M. R. (2015). Marine natural products. *Natural Product Reports*, *32*(2), 116–211.
14. Bull, A. T., Stach, J. E. M., Ward, A. C., & Goodfellow, M. (2005). Marine actinobacteria: perspectives, challenges, future directions. *Antonie Van Leeuwenhoek*, *87*, 65.
15. Rotter, A., Barbier, M., Bertoni, F., Bones, A. M., Cancela, M. L., Carlsson, J., … & Vasquez, M. I. (2021). The essentials of marine biotechnology. *Frontiers in Marine Science*, *8*, 629629.
16. Shera, S. S., & Banik, R. M. (2021). Metabolic engineering strategies to enhance microbial production of biopolymers. In Navanietha Krishnaraj, R., & Sani, R. K., Eds., *Biomolecular Engineering Solutions for Renewable Specialty Chemicals: Microorganisms, Products, and Processes* (pp. 247–285). John Wiley & Sons, Inc.
17. Sampaolo, G. (2022). The blue economy: a better stewardship of our seas and oceans as a long-term strategy to support sustainable growth. UniMC – Pubblicazioni Aperte Digitali.
18. Lu, W. Y., Li, H. J., Li, Q. Y., & Wu, Y. C. (2021). Application of marine natural products in drug research. *Bioorganic & Medicinal Chemistry*, *35*, 116058.
19. Kyriazi, Z., de Almeida, L. R., Marhadour, A., Kelly, C., Flannery, W., Murillas-Maza, A., … & Hynes, S. (2023). Conceptualising marine biodiversity mainstreaming as an enabler of regional sustainable blue growth: the case of the European Atlantic area. *Sustainability*, *15*(24), 16762.
20. Ortmann, A. C., Goodwin, K. D., Bolhuis, H., Groben, R., Makhalanyane, T., Bowman, J., … & Thompson, F. (2023). Catalyzing progress in the blue economy through joint marine microbiome research across the Atlantic. *Frontiers in Marine Science*, *10*, 1327719.
21. Action, S. I. (2020). *World Fisheries and Aquaculture* (pp. 1–244). Food and Agriculture Organization.
22. Nations, U. (2015). *Population Division*. Department of Economic and Social Affairs.
23. Naylor, R. L., Hardy, R. W., Buschmann, A. H., Bush, S. R., Cao, L., Klinger, D. H., … & Troell, M. (2021). A 20-year retrospective review of global aquaculture. *Nature*, *591*(7851), 551–563.

24. Galparsoro, I., Murillas, A., Pinarbasi, K., Sequeira, A. M., Stelzenmüller, V., Borja, Á., ... & Tett, P. (2020). Global stakeholder vision for ecosystem-based marine aquaculture expansion from coastal to offshore areas. *Reviews in Aquaculture*, *12*(4), 2061–2079.
25. Yusoff, D. F., Raja Abd Rahman, R. N. Z., Masomian, M., Ali, M. S. M., & Leow, T. C. (2020). Newly isolated alkane hydroxylase and lipase producing Geobacillus and Anoxybacillus species involved in crude oil degradation. *Catalysts*, *10*(8), 851.
26. Song, B., Li, Z., Li, S., Zhang, Z., Fu, Q., Wang, S., ... & Qi, S. (2021). Functional metagenomic and enrichment metatranscriptomic analysis of marine microbial activities within a marine oil spill area. *Environmental Pollution*, *274*, 116555.
27. Zhang, X., Ward, B. B., & Sigman, D. M. (2020). Global nitrogen cycle: critical enzymes, organisms, and processes for nitrogen budgets and dynamics. *Chemical Reviews*, *120*(12), 5308–5351.
28. Malaviya, P., Sharma, R., Sharma, S., & Pant, D. (2022). Role of microorganisms in environmental remediation and resource recovery through microbe-based technologies having major potentials. In de Bruijn, F. J., Smidt, H., Cocolin, L. S., Sauer, M., Dowling, D., & Thomashow, L., Eds., *Good Microbes in Medicine, Food Production, Biotechnology, Bioremediation, and Agriculture* (pp. 247–264). John Wiley & Sons Ltd.
29. Shu, W. S., & Huang, L. N. (2022). Microbial diversity in extreme environments. *Nature Reviews Microbiology*, *20*(4), 219–235.
30. Hosseini, H., Al-Jabri, H. M., Moheimani, N. R., Siddiqui, S. A., & Saadaoui, I. (2022). Marine microbial bioprospecting: exploitation of marine biodiversity towards biotechnological applications—a review. *Journal of Basic Microbiology*, *62*(9), 1030–1043.
31. Thangavelu, R., Ramachandraran, K., Thangaraj, M., & Annadurai, D. (2021). Marine microbial community studies using recent tools and techniques. In Shah, M., & Rodriguez-Couto, S., Eds., *Microbial Ecology of Wastewater Treatment Plants* (pp. 359–376). Elsevier.
32. Bhatia, P., & Chugh, A. (2015). Role of marine bioprospecting contracts in developing access and benefit sharing mechanism for marine traditional knowledge holders in the pharmaceutical industry. *Global Ecology and Conservation*, *3*, 176–187.
33. Leal-Alves, C., Deng, Z., Kermeci, N., & Shih, S. C. (2024). Integrating microfluidics and synthetic biology: advancements and diverse applications across organisms. *Lab on a Chip*, *24*, 2834–2860.
34. Louca, S., Jacques, S. M. S., Pires, A. P. F., Leal, J. S., Srivastava, D. S., Parfrey, L. W., Farjalla, V. F., & Doebeli, M. (2016). High taxonomic variability despite stable functional structure across microbial communities. *Nature Ecology & Evolution*, *1*(1), 0015.

9 Microorganism-Mediated Metal Solubilization or Bioleaching

Mamta Sharma[1] *and Davinder Singh*[2]
[1]Department of Biotechnology, School of Bioengineering and Biosciences, Lovely Professional University, Phagwara, India
[2]Department of Biotechnology, Thapar Institute of Engineering and Technology, Patiala, India

9.1 INTRODUCTION

Microorganism-mediated metal solubilization, commonly referred to as bioleaching or biomining, is a biotechnological process that utilizes specific microbes to extract metals from ores or mineral concentrates (Rendón-Castrillón et al., 2023). Bioleaching is a method that combines mining with biohydrometallurgy, which employs natural processes of interaction between microorganism and minerals. Microorganisms aid in the extraction of valuable metals from low-grade metal ore, industrial wastewater, or e-waste (Rohwerder et al., 2003). This technique utilizes the ability of certain microorganisms to oxidize metal ions, transforming them into soluble forms that facilitate easier recovery (Mishra et al., 2005). By harnessing these biological mechanisms, bioleaching offers an efficient and environmentally sustainable approach for metal extraction, waste treatment, and environmental remediation (Jaiswal and Srivastava, 2024; Zheng et al., 2024).

Bioleaching offers a cost-effective alternative to conventional pyrometallurgical and hydrometallurgical methods for recovering valuable metals (Rendón-Castrillón et al., 2023). Traditional pyrometallurgy involves high-temperature smelting to recover metals using reduction process following smelting. It's a simple process, but is energy-intensive and produces toxic gases and CO_2 emissions. On the other hand, hydrometallurgy employs chemical leaching and purification processes, which are more energy-efficient and cost-effective but rely on harmful inorganic acids like HCl, HNO_3, and H_2SO_4. These acids pose corrosion risks and create health hazards from toxic by-products and wastewater generation, in addition to increasing operational costs due to the use of reducing agents like H_2O_2. In contrast, bioleaching leverages biological processes to extract metals, potentially mitigating many of the environmental and safety challenges posed by traditional approaches (Rohwerder et al., 2003).

The demand for biomining has escalated due to the depletion of organic resources, the widespread occurrence of low-grade metallic ores, and the substantial volume of metallic waste generated during mining and beneficiation processes (Jaiswal and Srivastava, 2024). Numerous studies highlight the critical need to prioritize and optimize the use of microorganisms in microbial-mediated bioleaching to enhance the success of biomining operations. Throughout their growth phase, microorganisms secrete several organic acids along with enzymes. These acidic metabolites play a crucial role in the bioleaching process (Dusengemungu et al., 2021; Rendón-Castrillón et al., 2023).

This chapter provides a comprehensive analysis of bioleaching as an innovative and sustainable method for metal recovery and environmental remediation. It explores the microorganisms involved in the process, detailing their mechanisms of action and the various factors that influence the efficiency of metal solubilization. The chapter also examines the broad range of bioleaching applications across diverse sectors, while critically assessing its limitations, challenges, and future potential. By highlighting both its current impact and future prospects, this review emphasizes bioleaching's promise as a key technology for advancing sustainable resource management and environmental sustainability.

9.2 BIOLEACHING OF METALS

For many years, bioleaching has been extensively explored as an effective method for metals recovery from low-grade ores that were not economically viable for processing using conventional techniques. Copper, zinc, nickel, cobalt, uranium, silver, and gold are among the valuable metals that have been successfully extracted using bioleaching. Biooxidation is a process used to solubilize interfering metal sulfides from precious metal ores. In this process, sulfides containing iron and arsenic are dissolved, allowing the gold, which remains insoluble, to be recovered through cyanidation (Rohwerder et al., 2003). Chemolithotrophic bacteria, particularly *Thiobacillus ferrooxidans* and *Thiobacillus thiooxidans*, are essential for the recovery of metals from sulfide minerals because they help transform insoluble metal sulfides into soluble metal sulfates. On the other hand, heterotrophic bacteria and fungi can be used to process non-sulfide ores and minerals (Olson et al., 2003; Bosecker, 1997).

Bioleaching has also been investigated for recovering nickel, cobalt, lithium, and manganese from spent lithium-ion batteries (Rohwerder et al., 2003). Anaerobic reductive and

DOI: 10.1201/9781003509530-9

aerobic bioleaching using microorganisms that produce biogenic acid are the two methods for bioleaching oxidized minerals (Mishra et al., 2005). The most extensively researched method for extracting heavy metals is aerobic bioleaching, either through complexolysis or acidolysis. By means of aerobic fermentation and carbon sources, *Aspergillus niger*, for instance, metabolizes a variety of organic acids that be utilized as leaching agents (Zheng et al., 2024).

9.3 MICROORGANISMS INVOLVED IN BIOLEACHING

9.3.1 Bacteria

Microorganisms, particularly acidophilic bacteria like *Acidithiobacillus* and *Pseudomonas*, play a crucial role in bioleaching, effectively extracting these metals while minimizing environmental impact. Through processes like bioconcentration and biotransformation, these bacteria enhance metal recovery from tailings, making it possible to convert hazardous waste into valuable resources. In bioleaching, the most active bacteria come from the genus *Thiobacillus*, which are rod-shaped, spore-free, gram-negative bacteria that thrive in aerobic environments (Rendón-Castrillón et al., 2023). As chemolithoautotrophic organisms, most thiobacilli use atmospheric carbon dioxide to create new cell material. By oxidizing reduced or partially reduced sulfur compounds, such as sulfides, elemental sulfur, and thiosulfate, they produce sulfate as their final product and get energy (Olson et al., 2003; Bosecker, 1997; Rohwerder et al., 2003).

In this process, *Leptospirillum ferrooxidans* is another crucial bacterium, an acidophilic, obligately chemolithotrophic organism that oxidizes ferrous iron. In contrast to *T. ferrooxidans*, *L. ferrooxidans* can withstand larger quantities of uranium, molybdenum, and silver as well as lower pH levels. It cannot oxidize sulfur or its derivatives, though, and is more sensitive to copper. Therefore, *L. ferrooxidans* needs to work with other microbes to attack mineral sulfides; it cannot do this on its own (Bosecker, 1997; Rohwerder et al., 2003). Furthermore, at temperatures of about 50°C, *Thiobacillus*-like bacteria, also known as Th-bacteria, are moderately thermophilic and flourish on pyrite, pentlandite, and chalcopyrite. These bacteria need yeast extract to develop, but they use ferrous iron as an energy source. One such organism that can oxidize ferrous iron, elemental sulfur, and metal sulfides is *Acidianus brierleyi*, a facultatively aerobic, chemolithoautotrophic archaeon (Olson et al., 2003). Anaerobic conditions cause elemental sulfur to be reduced to H_2S. While *Sulfobacillus thermosulfidooxidans* depends on these molecules but needs yeast extract for growth, members of the genus *Sulfolobus* also oxidize comparable compounds for energy (Bosecker, 1997).

Chen et al. (2022) investigated how low-grade copper sulfide ore bioleaching was affected by agricultural waste fermentation (AWF) residues. Adding 5 g/L AWF enhanced copper recovery to 78.35% and increased bacterial concentration to 9.56×10^7 cells/mL. Microbial community analysis revealed significant changes, particularly in the proportions of *Acidithiobacillus ferrooxidans*, which peaked at 28.63% with AWF. The findings suggest that utilizing agricultural wastes can improve copper recovery and influence microbial communities in bioleaching processes. In order to improve the removal of cadmium and zinc, Ge et al. (2022) concentrated on a microbial consortium of sulfur-oxidizing bacteria in their study of heavy metal contamination of agricultural soil. The bioleaching group achieved the highest removal efficiencies—79.3% for cadmium and 45.2% for zinc—compared to acid leaching and sulfur treatments. Following remediation, total phosphorus and available potassium decreased by 53.7% and 41.2%, respectively, with exogenous microbes becoming the dominant population by the end of the study.

Bulaev et al. (2017) investigated the microbial population involved in biooxidation of gold-bearing sulfide ore using cultural and molecular methods. The key strains isolated were *Acidithiobacillus thiooxidans* and *Ferroplasma* species, with optimal growth at 38°C–40°C. Metagenomic analysis revealed significant representation of *Acidiferrobacter* and *Ferroplasma*, indicating their important roles in industrial bioleaching processes. Pangayao et al. (2015) examined the bioleaching of zinc, manganese, chromium, iron, and copper from coal ash using mixed cultures of *Acidithiobacillus albertensis* and *A. thiooxidans*. Optimized parameters included bacterial inoculum, pulp density, and sulfur weight. Maximum leaching efficiencies reached 85.01% for chromium and over 70% for other metals under optimized conditions. Watling et al. (2014) explored the bioleaching of a polymetallic ore containing organic and inorganic carbon using mixed cultures of microorganisms from an auto-heating coal mine. The porous structure of the ore facilitated metal extraction, with higher recoveries of cobalt, nickel, copper, and zinc in inoculated tests. Pyrite oxidation generated acids that reduced acid consumption during leaching. The fact that between 40% and 60% of the metals were acid-soluble emphasizes how crucial leachate-free acidity is to extraction. The microbial community included five bacterial and three archaeal species, with *Ferrimicrobium acidiphilum* being predominant at higher temperatures. The findings suggest that microbial activity significantly enhances metal recovery in bioleaching processes.

Williamson et al. (2021) explored the potential of siderophores for sustainable metal recovery from iron oxide-rich residues and low-grade nickel ore. *Pseudomonas putida* produced the highest yield of a mixed-type siderophore, characterized as hydroxamate and catecholate. Experiments revealed its ability to selectively leach hard metal ions like Mn and Zn, exceeding its natural affinity for Fe^{3+}. Zhang et al. (2019) investigated how ferric iron enhances arsenopyrite bioleaching by the strain *S. thermosulfidooxidans* YN-22, using various spectroscopic techniques. Adding 6 g/L of Fe^{3+} significantly boosted bioleaching by promoting cell growth and degrading sulfur and orpiment intermediates. Additionally, it resulted in the creation of Fe(III) products like jarosite, scorodite, and schwertmannite. The function of pyrite in augmenting bornite bioleaching by *L. ferriphilum* was examined by Wang et al. (2018). Wang et al. (2018) investigated the role of pyrite in enhancing bornite bioleaching by *L. ferriphilum*. The redox potential rose to 360 mV and the copper extraction efficiency climbed to 95.9% with the addition of pyrite, as opposed to only 19% without it. Pyrite also reduced acid consumption significantly. Electrochemical tests revealed a galvanic effect that facilitated bornite oxidation. The presence of pyrite prevented passivation by elemental sulfur and jarosite, leading to a proposed mechanism for its accelerated dissolution of bornite. Nicolova et al. (2017) also studied the bioleaching of uranium and heavy metals from polluted cinnamon forest soil using acidophilic chemolithotrophic bacteria. Over 170 days, optimal conditions reduced pollutants below permissible levels. The treated soil exhibited significantly increased biomass production in various plants compared to untreated soil, demonstrating effective bioremediation. González et al. (2015) explored the potential of *Acidiphilium cryptum* for bioleaching iron and manganese ores. Using different media, the presence of *A. cryptum* was crucial for dissolving hematite and goethite. Adding oxalate significantly enhanced microbial dissolution rates, while thionine also improved hematite dissolution in column assays, highlighting its effectiveness.

9.3.2 Fungi

Due to their ability to release organic acids that aid in the solubilization of metal ions from solution, filamentous fungi are becoming more widely acknowledged for their function in bioleaching. The bioleaching potential of different fungal species for metal extraction from various sources is presented in Table 9.1. Many organic acids, such as citric acid, lactic acid, gluconic acid, fumaric acid, oxalic acid, tartaric acid, malic acid, ketoglutaric acid, itaconic acid, and aconitic acid, can be produced by fungi. Fungal bioleaching

TABLE 9.1
Bioleaching Potential of Various Fungal Species for Metal Extraction from Different Sources

Species	Metal Targeted	Extracted From	Leaching Capacity (%)	Potential Mechanism	Reference
Penicillium simplicissimum	V and Ni	Residual ash of power plant	40% Ni, 100% V	Indirect leaching mechanism	Rasoulnia et al. (2016)
Cladosporium sp. A	Mg and Ag	Mine tailings	67% Ag, 74% Mg	Lixiviation	Huerta-Rosas et al. (2020)
Aspergillus niger	Ur	Uranium mine	69.4% Ur	Production of organic acids	Sun et al. (2020)
Penicillium simplicissimum	Cu and Ni	Printed circuit boards	89% Ni, 90% Cu	Variation in carbon source	Arshadi et al. (2019)
Aspergillus niger	Ni, Fe, Sn Zn, Al, Co, Cu, and Pb	Fine-grained electronic scrap	95% Cu and Co, 15%–35% Al, Zn, Pb, and Sn <10% for Fe and Ni	Citric and oxalic acid formation	Ahn et al. (2005)
Aspergillus niger (MXPE6)	Au, Cu, Ni	Computer motherboards and cellular phone printed circuit boards	42% Au, 5% Cu, 0.8% Ni	Production of organic acids	Madrigal-Arias et al. (2015)
Cladosporium sp. B	Mg and Ag	Mine tailings	58% Mg, 40% Ag	Lixiviation	Huerta-Rosas et al. (2020)
Penicillium simplicissimum and *Aspergillus niger*	Al	Bauxite (ore)	≥98%	Enhanced metal tolerance	Shah et al. (2020)
Aspergillus niger	Al, Cu, Co, Ni, Li, and Mn	Mobile phone batteries (lithium ion)	65% Al, 100% Cu, 45% Co, 38% Ni, 95% Li, 70% Mn	Production of citric acid	Horeh et al. (2016)
Aspergillus nomius JAMK1	Ni, Fe, Cu, and Pb	Spent batteries	99.8% Ni, 50% Fe 75.14% Cu, 73.2% Pb	Bioaccumulation and biosorption	Chatterjee et al. (2020)
Aspergillus niger	Co, Cu, Li, Mn, Ni, and Al	Mobile phone (lithium ion) batteries	38% Co, 94% Cu, 100% Li, 72% Mn, 45% Ni, 62% Al	Organic acids production	Bahaloo-Horeh et al. (2018)
Aspergillus niger (M1DGR)	Cd and Cr	Contaminated soil	98% Cd, 43% Cr	Production of organic acids	Khan et al. (2019)
Aspergillus niger SY1	Cu, Cd, Zn, and Pb	Contaminated sediments	62.3% Cu, 93.5% Cd, 68.2% Zn, 11.5% Pb	One step bioleaching	Zeng et al. (2015)

(*Continued*)

TABLE 9.1 (*Continued*)
Bioleaching Potential of Various Fungal Species for Metal Extraction from Different Sources

Species	Metal Targeted	Extracted From	Leaching Capacity (%)	Potential Mechanism	Reference
Aspergillus niger, A. foetidus, and *A. carbonarius*	Al	Spent catalyst	88.43%	Production of organic acids	Das et al. (2019)
Penicillium chrysogenum strains	Mg and Ag	Mine tailings	69% Mg, 53% Ag	Lixiviation	Huerta-Rosas et al. (2020)
Aspergillus niger (MXPE6+MX7)	Au, Cu, and Ni	Computer motherboards and cellular phone printed circuit boards	87% Au, 0.2% Cu, 0.6% Ni	Production of organic acids	Madrigal-Arias et al. (2015)
Aspergillus nomius JAMK1	Fe, Ni, Cu, and Pb	Spent batteries	50% Fe, 99.8% Ni, 75.14% Cu, 73.2% Pb	Bioaccumulation and biosorption	Chatterjee et al. (2020)
Aspergillus niger strain SY1	Cd, Cu, Zn, and Pb	Contaminated sediments	99.5% Cd, 71.9% Cu, 76.4% Zn, 65.4% Pb	Two step bioleaching	Zeng et al. (2015)

primarily occurs through the action of these organic acids generated during fermentation, which facilitate the mobilization of metals via ligand-induced solubilization. Fungi can be rapidly cultivated using simple fermentation techniques and cost-effective growth media, making fungal bioleaching an accessible and efficient method for metal recovery (Rendón-Castrillón et al., 2023).

Fungal bioleaching is a complex process that integrates multiple mechanisms like acidolysis, complexolysis, redoxolysis, alkaloysis, bioaccumulation, and biosorption which work synergistically to enhance the bioleaching process (Dusengemungu et al., 2021). *A. niger* uses acidolysis, complexolysis, redoxolysis, and bio-accumulation as its primary bioleaching processes (Zheng et al., 2024). Studies have demonstrated the efficacy of fungi in bioleaching metals from a variety of sources, including low-grade ores, spent catalysts, mine tailings, and electronic waste. The most important and efficient contributors to biological leaching are fungi from *Aspergillus* and *Penicillium* genera (Dusengemungu et al., 2021). Filamentous fungi exhibit remarkable adaptability and tolerance to environmental stresses, making them valuable assets in enhancing bioleaching of low-grade minerals. Regardless of their ability to produce organic acids, many filamentous fungi exhibit intrinsic resistance to metals due to the distinctive characteristics of their cell wall membranes. In bioleaching, various species of fungi such as *Aspergillus* sp. (Ghosh et al., 2020; Madrigal-Arias et al., 2015); *Cladosporium* sp. (Huerta-Rosas et al., 2020); and *Penicillium* sp. (Arshadi et al., 2019; Huerta-Rosas et al., 2020; Shah et al., 2020) have been used to mobilize and leach metals from mine tailings, e-waste, contaminated sediments, and low-quality bauxite ores.

Rasoulnia et al. (2016) used one-, two-, and spent-medium bioleaching techniques at a pulp density of 1% (w/v) to examine the bioleaching capacity of *Penicillium simplicissimum* for the recovery of metals from power plant residual ash (PPR ash). The removal of volatile and carbonaceous components from the ash, which impacted fungal growth and metal leachability, was another way that the study looked at how thermal pre-treatment at different temperatures affected the leaching of vanadium, nickel, and iron. The spent-medium bioleaching of the original PPR ash produced the highest extraction yields, with 40% nickel and nearly 100% vanadium. For pre-treated PPR ash at 400°C, the maximum extraction yield of iron was 48.3% using spent-medium bioleaching.

Huerta-Rosas et al. (2020) found that when incubated in a 9k medium, three distinct fungal strains—*Cladosporium* sp. A, *Cladosporium* sp. B, and *Penicillium chrysogenum*—effectively leached manganese and silver from solid mine wastes. Under these conditions, the leaching efficiencies for manganese and silver ranged from 58%–74% to 40%–67%, respectively, due to a mixed biotic-abiotic process. In contrast, when grown in LMM medium, the fungal strains exhibited lower manganese leaching efficiencies of 0.17%–0.24%, while bacterial strains in PDB medium achieved 1.42%–1.73%. Silver leaching was significantly reduced to <0.1% under these conditions. Sun et al. (2020) investigated the effects of operational parameters on uranium bioleaching by *A. niger*, achieving over 80% recovery. Morphological changes revealed more reactive surfaces, with bilayer structures enhancing bioleaching efficiency. Filamentous fungi effectively recover metals from mineral ores, but high pulp densities can reduce yields due to metal toxicity (Shah et al., 2020). This study adapted *A. niger* and *P. simplicissimum* to enhance aluminum recovery from low-grade bauxite. Using a liquid medium with 5% pulp density and molasses, the adapted strains achieved over 98% aluminum bioleaching.

Arshadi and his co-workers (Arshadi et al., 2019) aimed to identify a suitable alternative carbon substrate for bioleaching copper and nickel from printed circuit boards (PCBs) using *P. simplicissimum*. The findings revealed that sugar was identified as the most economical and effective medium for copper recovery, yielding a 90% recovery at pH 7 with 3.3×10^7 spores. Conversely, the highest nickel recovery (89%) occurred in molasses at pH 2 with 10^6

spores. Ahn et al. (2005) reported *A. niger*, a promising fungal candidate for the effective leaching (95%) of copper and cobalt from fine grated electronic scrap (50 g/L). In contrast, aluminum, zinc, lead, and tin were leached at rates of approximately 15%–35%, while nickel and iron were found in the leachate at levels below 10%. In order to investigate alternative techniques for recovering metals from waste electrical and electronic equipment (WEEE), Madrigal-Arias et al. (2015) used two strains of *A. niger* to assess the bioleaching efficiency of gold, copper, and nickel on PCBs for cell phones and gold-plated finger integrated circuits (GFICMs). The study highlighted the commercial value and industrial applications of these metals. When using *A. niger* strain MXPE6, the bioleaching of gold ranged from 42% to 1%; however, when MXPE6 and MX7 were combined, the results for PCBs and GFICMs were 87% and 28%, respectively. The combined strains demonstrated 0.2% and 29% leaching for PCBs and GFICMs, while MXPE6 demonstrated 24% and 5% leaching for copper. There was only nickel leaching in PCBs.

Horeh et al. (2016) also developed a bio-hydrometallurgical method using the fungal activity of *A. niger* to detoxify and recover metals from spent lithium-ion mobile phone batteries. One-step, two-step, and spent medium bioleaching were among the conditions that were investigated in this study. At a pulp density of 1% in spent medium bioleaching, the highest recovery rates were 100% for copper, 95% for lithium, 70% for manganese, 65% for aluminum, 45% for cobalt, and 38% for nickel. It concluded that bioleaching demonstrated higher heavy metal removal efficiency than chemical leaching. Similarly, Chatterjee et al. (2020) also investigated the use of *Aspergillus nomius* (JAMK1) to leach heavy metals from e-waste. The organism's metal uptake involved bioaccumulation and biosorption, analyzed via atomic absorption spectroscopy. Results indicated its potential for effective heavy metal recovery (75.14% copper, 73.2% lead, 99.8% nickel, and 50% iron) from aqueous solutions and electronic waste. In contrast to unadapted fungi, adapted *A. niger* showed improved metal leaching and organic acid generation, according to a different study by Bahaloo-Horeh et al. (2018). At 1% pulp density, 100% lithium, 94% copper, and a notable recovery of other metals were obtained in the presence of spent lithium-ion battery powder.

Three native fungal strains—*Penicillium rubens*, *A. niger*, and *Aspergillus fumigatus*—were isolated and identified by Khan et al. (2019) from soil contaminated with heavy metals in Pakistan's Hattar Industrial Estate. *A. niger* demonstrated the highest bioaccumulation efficiency, achieving 98% for cadmium and 43% for chromium removal, while *A. fumigatus* showed lower efficiencies of 79% and 69%, respectively. Using *A. niger* strain SY1, Zeng et al. (2015) attempted to explore the bioleaching of heavy metals from contaminated sediments. One-step bioleaching resulted in only 11.5% lead extraction, while two-step bioleaching reached 65.4%. Other metals showed high leaching efficiencies, with 99.5% cadmium, 56% copper, and 76.4% zinc extracted in the two-step bioleaching process. Das et al. (2019) used three *Aspergillus* strains (*A. niger*, *A. foetidus*, and *A. carbonarius*) to investigate the bioleaching of aluminum from spent catalysts. The study was conducted in batch culture with varying catalyst loadings. *A. foetidus* achieved the highest leaching efficiency of 88.43% at 0.8% catalyst loading. Additionally, using molasses as a low-cost carbon source, the optimal concentration of 40 g/L resulted in 60% bioleaching efficiency. Overall, *A. foetidus* demonstrated significant potential for aluminum recovery from spent catalysts.

9.4 BIOLEACHING MECHANISMS

Bioleaching is primarily classified into two types: direct leaching and indirect leaching (Mishra et al., 2005). Direct leaching entails direct contact between microorganisms and the material being treated, with oxidation occurring through enzymatically catalyzed reactions. This process can be conducted in two distinct phases: the first phase involves adding waste during the inoculation stage, while the second phase introduces waste after the formation of the initial metabolic acid (Bosecker, 1997). In contrast, indirect leaching does not require direct interaction between minerals and microorganisms. Instead, this method relies on bacteria to produce strong oxidizing agents, which facilitate the leaching process. Indirect leaching also follows a two-step approach: first, metabolic acids or used media are collected, and then these substances are employed to dissolve metallic components without the involvement of organisms (Jaiswal and Srivastava, 2024).

The primary mechanisms involved in bioleaching include acidolysis, complexolysis, redoxolysis, bioaccumulation, and biosorption (Zheng et al., 2024) (Mishra et al., 2005). Each of these mechanisms plays a crucial role in metal solubilization and recovery. The protonation of oxygen atoms in organic acids produced by microorganisms facilitates the indirect leaching process known as acidolysis. By weakening the bonds and interacting with the metal-ore interface, protons help remove metal ions from the ore. The protonated oxygen's reaction with water, which causes the metal oxides to separate from the solid surface, completes this process. Complexolysis is a process achieved through chelation, which works in tandem with acidolysis to balance metal ions in solution. This mechanism stabilizes the reaction by facilitating complexation between metal ions and organic acids. Fungi release amino acids that contribute to complexolysis, enhancing metal solubility despite their limited excretion capacity (Srichandan et al., 2019). Organic acids act as natural chelating agents, effectively forming metallic complexes with metal ions. Redoxolysis, also known as an oxide reduction reaction, is a process by which fungi use microbial oxidation and reduction reactions to obtain energy from solid-phase minerals. By deaminating amino acids and hydrolyzing urea enzymatically, bacteria use alkaloysis to recover metals while utilizing amino acids as energy sources. This mechanism can efficiently mobilize metals from silicate or aluminosilicate complexes,

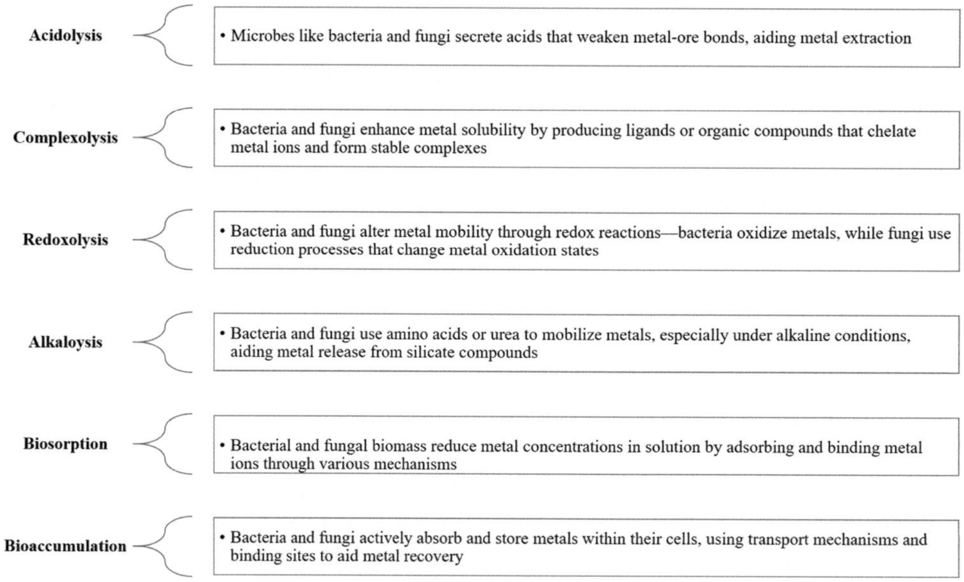

FIGURE 9.1 Comparison of bacterial and fungal bioleaching mechanisms.

facilitating bioleaching at high pH levels (Zheng et al., 2024).

The active uptake of metals by living cells is known as bioaccumulation, where soluble metal ions are transported across the cell membrane and accumulate in vacuoles. The functional groups (sulfate, hydroxyl, carboxyl, phosphate, and amine) found in fungal mycelium aid in this process by binding metal ions and functioning as cation exchangers (Srichandan et al., 2019). Unlike biosorption, bioaccumulation does not require metabolites and occurs more slowly, involving both active metabolic reactions and passive adsorption. Biosorption is the process through which fungal biomass adsorbs metal ions from leaching solutions, thereby reducing metal concentrations. In bioleaching, organic acids dissolve metals from the substrate while the biomass captures these ions via adsorption, complexation, ion exchange, and precipitation. Studies have demonstrated that during fungal bioleaching, biosorption takes place; for instance, Pb, Zn, Al, and Fe were efficiently leached by *A. niger* from fly ash from incineration of municipal solid waste that had been treated with gluconic acid. The biosorption kinetics of lead followed a pseudo-second-order model, while iron, lead, and zinc adhered to the Langmuir isotherm model. A comparison of bacterial and fungal bioleaching mechanisms is shown in Figure 9.1.

9.5 FACTORS AFFECTING BIOLEACHING

The effectiveness of bioleaching is influenced by several bio-physico-chemical factors (Figure 9.2), including pH levels, temperature, nutrient availability, media, oxygen supply, pre-culture period and type of inoculum used, the size of the substrate, carbon source, duration of the bioleaching process, and development of effective microbial strains (Zheng et al., 2024; Bosecker, 1997). Both the chemical and mineralogical composition of the ore, as well as the efficiency of the microorganisms, play a crucial role in influencing the effectiveness of bioleaching. Optimal metal extraction yields can only be attained when leaching conditions align with the ideal growth parameters for the bacteria.

The supply of O_2 and CO_2 plays a crucial part in facilitating microbial growth and the organic acids secretion, which are vital for bioleaching (Bosecker, 1997). pH significantly influences microbial growth and community structure, while temperature impacts bioleaching by enhancing microbial activity and altering the bacterial species. It's important to note that the organic acid production is not directly linked to total biomass, underscoring the need for careful selection of inoculum size to achieve optimal bioleaching outcomes. Additionally, redox potential plays a crucial role in bioleaching since it significantly influences the bacterial populations involved in the process. Together, these factors interact to enhance the efficiency of bioleaching processes (Zheng et al., 2024).

9.6 ADVANTAGES AND APPLICATIONS

As natural resources are becoming scarce, researchers are increasingly paying attention to the metals recovery from secondary wastes, which are not only abundant but also hazardous, posing significant risks to ecosystems. These wastes can exhibit dangerous properties like reactivity, toxicity, ignitability, and corrosivity (Saxena, 2015; Devi et al., 2017). There is a pressing need for sustainable mining technologies worldwide to effectively address the challenges associated with metal resource extraction and minimize its environmental impact. Furthermore, the depletion of significant essential metals and ore reserves underscores the urgent need for alternative technologies to exploit other

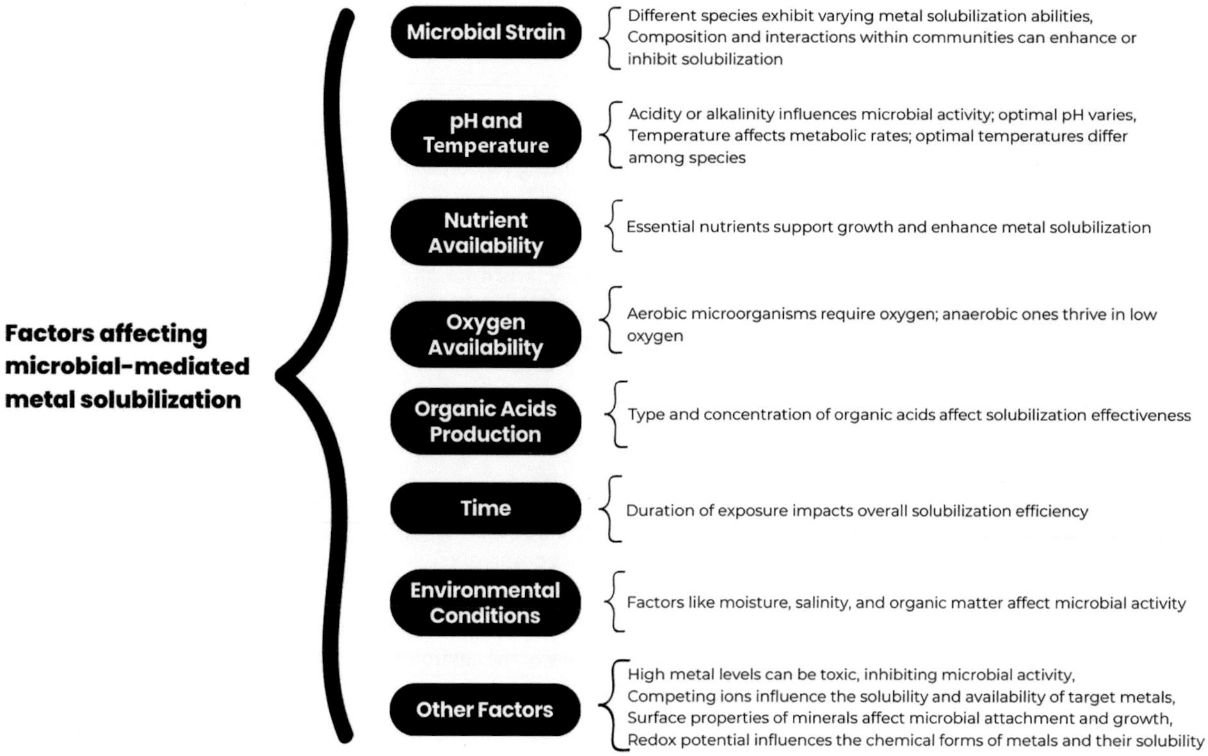

FIGURE 9.2 Factors affecting microbial-mediated metal solubilization.

mineral sources, waste materials, and low-grade ores (Dusengemungu et al., 2021). Toxic metals like molybdenum, nickel, copper, vanadium, cobalt, and lead are frequently found in secondary solid wastes, such as fly ash, slag, e-waste, and metal-containing sludge, which can be hazardous to the environment. Since these metals can harm biotic communities, including humans, it is essential to treat these wastes. Direct disposal without treatment risks leaching metals into land and water bodies.

Bioleaching presents several significant advantages, positioning it as a sustainable and cost-efficient method for metal extraction (Figure 9.3). By utilizing microorganisms to solubilize metals from ores and waste materials, bioleaching eliminates the need for hazardous chemicals, thereby reducing environmental contamination (Zheng et al., 2024). It operates under ambient temperatures and pressures, resulting in lower energy consumption compared to traditional methods. Furthermore, bioleaching is particularly effective for extracting metals from complex materials and low-grade ores. Its versatility extends to a wide range of metals, including those in electronic waste and mining tailings, contributing to recycling efforts and minimizing dependence on primary mining. In comparison to conventional extraction techniques, bioleaching has several advantages, such as environmentally sustainable metal recovery, lower energy consumption, and less environmental effect (Srichandan et al., 2019).

Bioleaching encompasses a wide array of applications across various industries, particularly in the extraction of metals from ores, waste materials, electronic waste, and environmental remediation (Figure 9.4). In the mining sector, it is widely used to recover valuable metals such as copper, gold, nickel, and zinc from low-grade ores and complex mineral deposits that are challenging for traditional methods to process (Arya and Kumar, 2020; Yun et al., 2024; Rendón-Castrillón et al., 2023). Bioleaching is also instrumental in extracting metals from mining tailings, reducing environmental hazards associated with mine waste. Additionally, bioleaching is increasingly employed in recycling e-waste, where it enables the recovery of precious metals such as gold, silver, and palladium from discarded electronic devices (Chatterjee et al., 2020; Jaiswal and Srivastava, 2024). By recovering valuable resources and reducing the environmental impact of disposing of e-waste, bioleaching supports the circular economy. Moreover, bioleaching is being explored for the recovery of rare earth elements and other critical materials, further expanding its potential in resource recovery and waste management.

Bioleaching utilizes various microorganisms to effectively extract metals from ores and concentrates. Key species involved in this process include *A. thiooxidans*, *L. ferrooxidans*, *Thiobacillus thioparus*, *A. ferrooxidans*, *S. thermosulfidooxidans*, *Sulfolobus acidocaldarius*, *Sulfolobus brierley*, and *Acidiphilium acidophilum* (Rohwerder et al., 2003; Olson et al., 2003). These microorganisms play a critical role in enhancing metal recovery rates, thereby contributing to more sustainable and environmentally friendly mining. *Aspergillus*, *Penicillium*, and *Acidithiobacillus* are some of the most common taxa used in fungal bioleaching (Rendón-Castrillón et al., 2023).

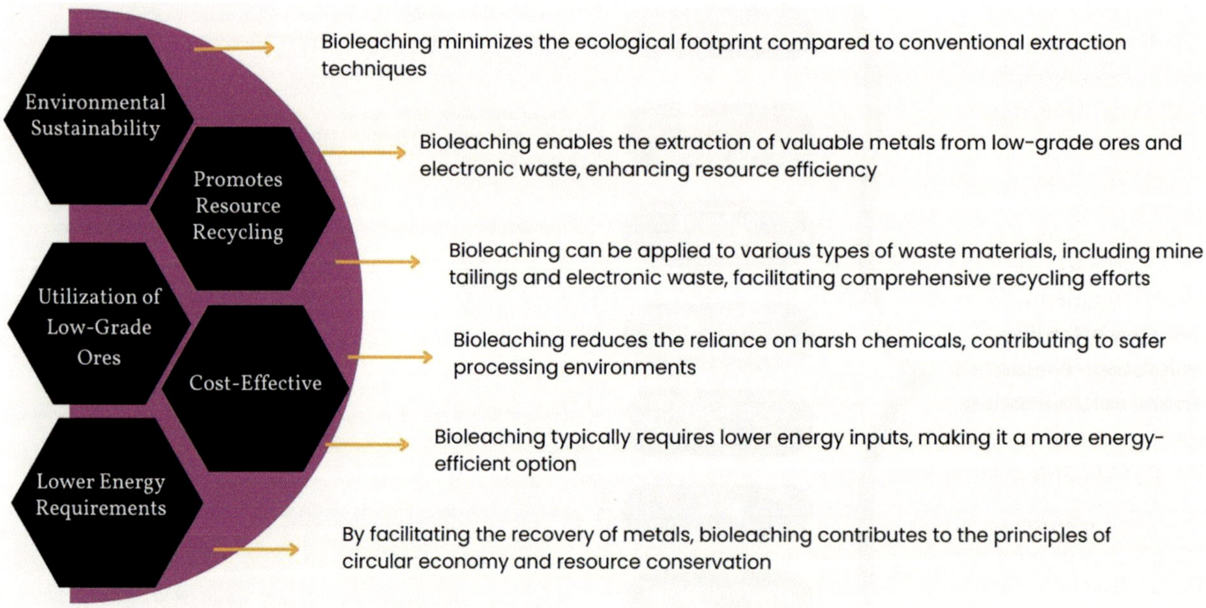

FIGURE 9.3 Benefits of microbial-mediated bioleaching—an overview of the environmental, economic, and operational advantages of utilizing microbial processes for metal extraction.

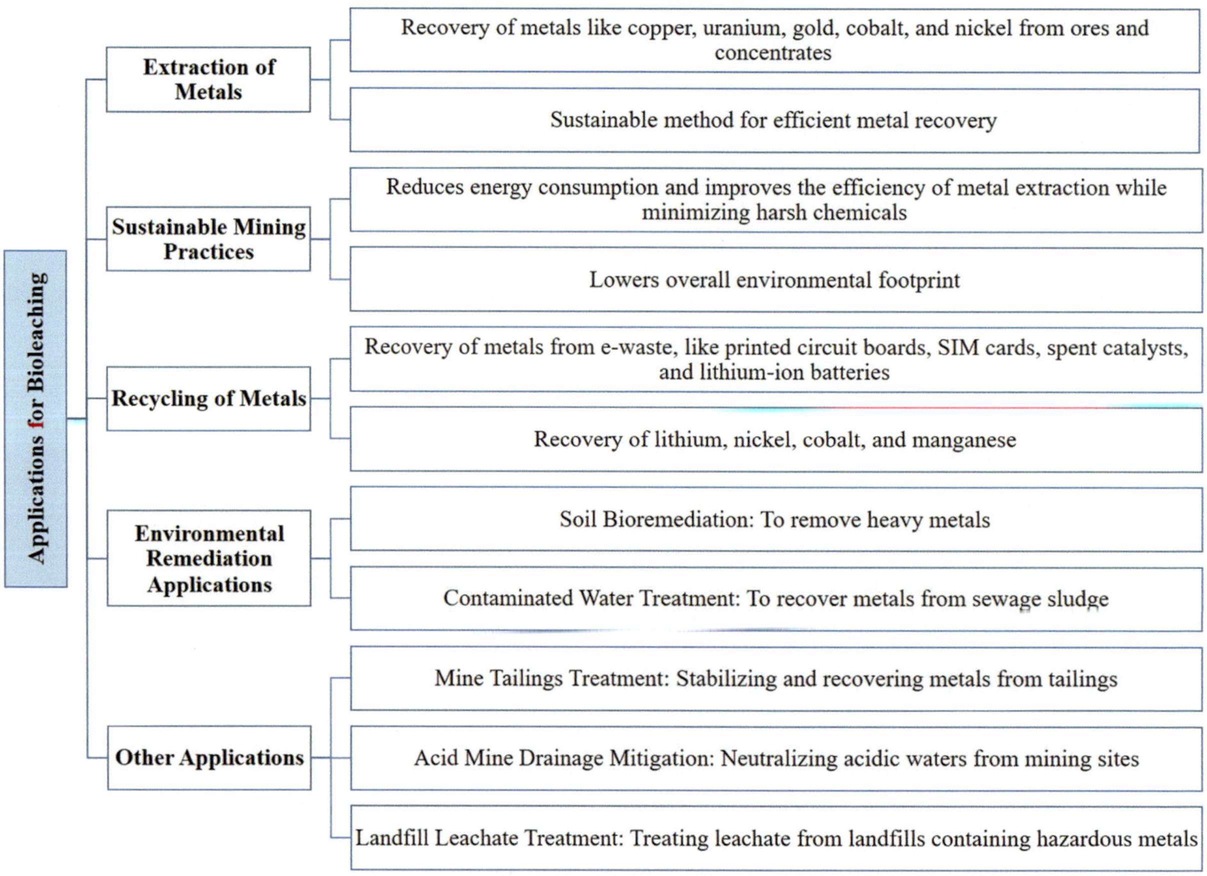

FIGURE 9.4 Applications of bioleaching, highlighting its role in metal extraction from low-grade ores, sustainable mining practices, and environmental remediation.

The increasing prevalence of heavy metal pollution has emerged as a critical global environmental issue, given its significant implications for public health, ecological integrity, and economic stability (Roy et al., 2024). Microbial activity has proven effective in remediating heavy metal polluted soil. In bioleaching, microbes not only generate H_2SO_4 or organic acids to leach heavy metals but also alter metallic compounds. This metabolic activity transforms these compounds into soluble and extractable forms, facilitating remediation. The microbes immobilize heavy metals through processes such as biosorption, bioaccumulation, biomineralization, and bioleaching (Zheng et al., 2024).

9.7 CHALLENGES AND FUTURE PERSPECTIVE

Bioleaching represents a transformative approach for sustainable metal recovery, leveraging the natural processes of microorganisms. These organisms facilitate the solubilization of metals from low-grade ores and secondary wastes, including electronic waste and fly ash, through various biochemical pathways. As industrialization continues to generate significant amounts of solid waste laden with heavy metals like lead, copper, and zinc, bioleaching offers an eco-friendly alternative to traditional recovery methods, improving extraction efficiency while reducing environmental pollution.

Despite its potential, the widespread application of bioleaching faces several challenges. The complex mechanisms of metal extraction, coupled with difficulties in metal recovery after biosorption, hinder its effectiveness. Additionally, the need for optimization in microbial-mineral interactions and conditions such as pH and temperature is critical for enhancing performance (Figure 9.5). Recent advancements in omics technologies hold promise for overcoming these hurdles by facilitating the development of novel microbial strains tailored for specific remediation tasks.

As the need for sustainable waste management solutions grows, particularly in the context of climate change, interest in scaling up bioleaching has intensified. Successful full-scale implementation will require integrated approaches that combine various leaching technologies and strategies. By addressing the current limitations and investing in research, bioleaching can not only mitigate the environmental risks posed by toxic metal(loid) accumulation but also contribute to a circular economy in metal recovery. Ultimately, with ongoing innovation and optimization, bioleaching stands poised to play a pivotal role in sustainable resource management and environmental remediation.

9.8 CONCLUSIONS

Bioleaching is advantageous for processing low-grade ores, where it proves more economical than conventional chemical methods. Additionally, bioleaching reduces reliance on harmful chemicals and minimizes environmental damage, while allowing for the selective extraction of specific metals. Microbes oxidize metal sulfides to release metals in a soluble form, metabolizing sulfur or iron compounds to create acidic conditions that enhance metal solubility. As a by-product, *A. ferrooxidans* generates sulfuric acid (H_2SO_4), further aiding in the dissolution of metal ions. The solubilized metal ions are then recovered using ion exchange, solvent extraction, and precipitation.

Bioleaching finds significant applications across various fields, especially in the extraction of metals like copper, gold, and nickel from ores that are not economically viable for traditional processing methods. Moreover, bioleaching

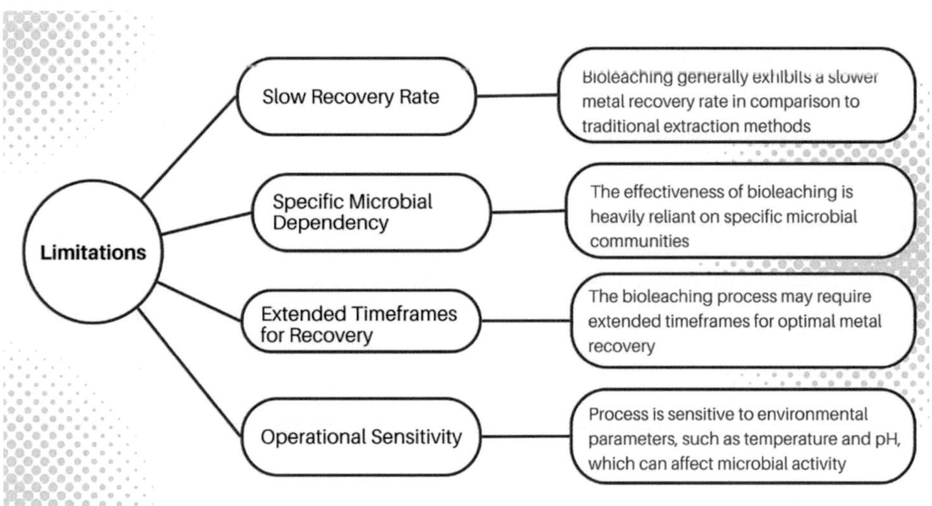

FIGURE 9.5 Limitations of microbial-mediated metal recovery: an illustration of the challenges and constraints associated with the bioleaching process.

techniques can be applied to treat waste and contaminated soils, promoting the recovery of precious metals while addressing environmental pollution. Additionally, bioleaching plays a vital role in the remediation of metal-contaminated sites by effectively mobilizing and removing toxic metals from the environment. However, bioleaching does have limitations, including a slower processing time compared to traditional methods, its effectiveness being confined to specific metals, and the necessity for precise environmental conditions—such as temperature and pH—to optimize microbial activity. Overall, despite these challenges, bioleaching holds significant potential for enhancing sustainable practices in resource management and environmental remediation, contributing to a more circular economy.

REFERENCES

Ahn, J. W., Jeong, J. K., Lee, J. C., Kim, D. G., 2005. Bioleaching of valuable metals from electronic scrap using fungi (*Aspergillus niger*) as a microorganism. *Resourc. Recycl.* 14(5), 24–31.

Arshadi, M., Nili, S., Yaghmaei, S., 2019. Ni and Cu recovery by bioleaching from the printed circuit boards of mobile phones in non-conventional medium. *J. Environ. Manage.* 250, 109502. https://doi.org/10.1016/j.jenvman.2019.109502

Arya, S., Kumar, S., 2020. Bioleaching: Urban mining option to curb the menace of E-waste challenge. *Bioengineered* 11, 640–660. https://doi.org/10.1080/21655979.2020.1775988

Bahaloo-Horeh, N., Mousavi, S. M., Baniasadi, M., 2018. Use of adapted metal tolerant Aspergillus niger to enhance bioleaching efficiency of valuable metals from spent lithium-ion mobile phone batteries. *J. Clean. Prod.* 197, 1546–1557. https://doi. org/10.1016/j.jclepro.2018.06.299

Bosecker, K., 1997. Bioleaching: Metal solubilization by microorganisms. *FEMS Microbiol. Rev.* 20, 591–604. https://doi.org/10.1016/S0168-6445(97)00036-3

Bulaev, A., Belyi, A., Panyushkina, A., Solopova. N., Pivovarova. T., 2017. Microbial population of industrial biooxidation reactors. *Solid State Phenom.* 262, 48–52. https://doi.org/10.4028/www.scientific.net/SSP.262.48

Chatterjee, A., Das, R., Abraham, J., 2020. Bioleaching of heavy metals from spent batteries using Aspergillus nomius JAMK1. Int. J. Environ. Sci. Technol. 17, 49–66. https://doi. org/10.1007/s13762-019-02255-0

Chen, W., Yin, S., Song, Q., Wang, L., Chen, X., 2022. Enhanced copper recovery from low grade copper sulfide ores through bioleaching using residues produced by fermentation of agricultural wastes. *Int. J. Miner. Metall. Mater*. 29, 2136–2143. https://doi.org/10.1007/s12613-021-2392-1

Das, S., Naik Deshavath, N., Goud, V. V., Dasu, V. V., 2019. Bioleaching of Al from spent fluid catalytic cracking catalyst using Aspergillus species. *Biotechnol. Reports.* 23, e00349. https://doi.org/10.1016/j.btre.2019.e00349

Devi, K. S., Sujana, O., Singh, T. C., 2017. Hazardous waste management in India. *Int. J. Adv. Eng. Res. Dev.* 4. https://doi.org/10.21090/ijaerd.13164

Dusengemungu, L., Kasali, G., Gwanama, C., Mubemba, B., 2021. Overview of fungal bioleaching of metals. *Environ. Adv.* 5, 100083. https://doi.org/10.1016/j.envadv.2021.100083

Ge, Y., Cheng, H., Zhou, W., Zhou, H., Wang, Y., 2022. Enhancing Zn and cd removal from heavy metal-contaminated paddy soil using an artificial microbial consortium. *J. Soils Sediments* 22, 218–228. https://doi.org/10.1007/s11368-021-03066-y

Ghosh, S. K., Sen, R., Chanakya, H. N., Agamuthu, P. (Eds.), 2020. *Bioresource Utilization and Bioprocess, Bioresource Utilization and Bioprocess*. Springer. https://doi.org/10.1007/978-981-15-1607-8

González, E., González, F., Muñoz, J., Blázquez, M. L., Ballester, A., 2015. Reductive dissolution of iron oxides and manganese bioleaching by Acidiphilium cryptum JF-5. *Adv. Mater. Res.* 1130, 347–350. https://doi.org/10.4028/www.scientific.net/AMR.1130.347

Horeh, N. B., Mousavi, S. M., Shojaosadati, S. A., 2016. Bioleaching of valuable metals from spent lithium-ion mobile phone batteries using Aspergillus Niger. *J. Power Sources* 320, 257–266. https://doi.org/10.1016/j.jpowsour.2016.04.104

Huerta-Rosas, B., Cano-Rodríguez, I., Gamiño-Arroyo, Z., Gómez-Castro, F. I., Carrillo-Pedroza, F. R., Romo-Rodríguez, P., Gutiérrez-Corona, J. F., 2020. Aerobic processes for bioleaching manganese and silver using microorganisms indigenous to mine tailings. *World J. Microbiol. Biotechnol.* 36, 1–16. https://doi.org/10.1007/s11274-020-02902-6

Jaiswal, M., Srivastava, S., 2024. A review on sustainable approach of bioleaching of precious metals from electronic wastes. *J. Hazard. Mater. Adv.* 14, 100435. https://doi.org/10.1016/j.hazadv.2024.100435

Khan, I., Aftab, M., Shakir, S. U., Ali, M., Qayyum, S., Rehman, M. U., Haleem, K. S., Touseef, I., 2019. Mycoremediation of heavy metal (Cd and Cr)–polluted soil through indigenous metallotolerant fungal isolates. *Environ. Monit. Assess.* 191. https://doi.org/10.1007/s10661-019-7769-5

Madrigal-Arias, J. E., Argumedo-Delira, R., Alarcón, A., Mendoza-López, M. R., García-Barradas, O., Cruz-Sánchez, J. S., Ferrera-Cerrato, R., Jiménez-Fernández, M., 2015. Bioleaching of gold, copper and nickel from waste cellular phone PCBs and computer goldfinger motherboards by two *Aspergillus niger* strains. *Braz. J. Microbiol.* 46, 707–713. https://doi.org/10.1590/S1517-838246320140256

Mishra, D., Kim, D. J., Ahn, J. G., Rhee, Y. H., 2005. Bioleaching: A microbial process of metal recovery: A review. *Met. Mater. Int.* 11, 249–256. https://doi.org/10.1007/BF03027450

Nicolova, M., Spasova, I., Georgiev, P., Groudev, S., 2017. Microbial removal of toxic metals from a heavily polluted soil. *J. Geochem. Explor.* 182, 242–246. https://doi.org/10.1016/j.gexplo.2016.11.003

Olson, G. J., Brierley, J. A., Brierley, C. L., 2003. Bioleaching review part B: Progress in bioleaching: Applications of microbial processes by the minerals industries. *Appl. Microbiol. Biotechnol.* 63, 249–257. https://doi.org/10.1007/s00253-003-1404-6

Pangayao, D. C., Van Hullebusch, E. D., Gallardo, S. M., Bacani, F. T., 2015. Bioleaching of trace metals from coal ash using mixed culture of Acidithiobacillus albertensis and Acidithiobacillus thiooxidans. *J. Eng. Sci. Technol.* 10, 36–45.

Rasoulnia, P., Mousavi, S. M., Rastegar, S. O., Azargoshasb, H., 2016. Fungal leaching of valuable metals from a power plant residual ash using Penicillium simplicissimum: Evaluation of thermal pretreatment and different bioleaching methods. *Waste Management* 52, 309–317.

Rendón-Castrillón, L., Ramírez-Carmona, M., Ocampo-López, C., Gómez-Arroyave, L., 2023. Bioleaching techniques for sustainable recovery of metals from solid matrices. *Sustainability* 15, 10222. https://doi.org/10.3390/su151310222

Rohwerder, T., Gehrke, T., Kinzler, K., Sand, W., 2003. Bioleaching review part A: Progress in bioleaching: Fundamentals and mechanisms of bacterial metal sulfide oxidation. *Appl. Microbiol. Biotechnol.* 63, 239–248. https://doi.org/10.1007/s00253-003-1448-7

Roy, R., Samanta, S., Pandit, S., Naaz, T., Banerjee, S., Rawat, J.M., Chaubey, K.K., Saha, R.P., 2024. An overview of bacteria-mediated heavy metal bioremediation strategies. *Appl. Biochem. Biotechnol.* 196, 1712–1751. https://doi.org/10.1007/s12010-023-04614-7

Saxena, N., 2015. Hazardous waste: An emerging concern a review. *Int. J. Lat. Tren. Eng. Technol.* 6, 255–259.

Shah, S. S., Palmieri, M. C., Sponchiado, S. R. P., Bevilaqua, D., 2020. Enhanced bio-recovery of aluminum from low-grade bauxite using adapted fungal strains. *Braz. J. Microbiol.* 51, 1909–1918. https://doi.org/10.1007/s42770-020-00342-w

Srichandan, H., Mohapatra, R. K., Parhi, P. K., Mishra, S., 2019. Bioleaching approach for extraction of metal values from secondary solid wastes: A critical review. *Hydrometallurgy* 189, 105122. https://doi.org/10.1016/j.hydromet.2019.105122

Sun, J., Li, G., Li, Q., Wang, Y., Ma, J., Pang, C., Ma, J., 2020. Impacts of operational parameters on the morphological structure and uranium bioleaching performance of bio-ore pellets in one-step bioleaching by Aspergillus niger. *Hydrometallurgy*, 195, 105378. https://doi.org/10.1016/j.hydromet.2020.105378

Wang, X., Liao, R., Zhao, H., Hong, M., Huang, X., Peng, H., Wen, W., Qin, W., Qiu, G., Huang, C., Wang, J., 2018. Synergetic effect of pyrite on strengthening bornite bioleaching by Leptospirillum ferriphilum. *Hydrometallurgy* 176, 9–16. https://doi.org/10.1016/j.hydromet.2017.12.003

Watling, H., Collinson, D., Fjastad, S., Kaksonen, A., Li, J., Morris, C., Perrot, F. A., Rea, S. M., Shiers, D. W. 2014. Column bioleaching of a polymetallic ore: effects of pH and temperature on metal extraction and microbial community structure. *Miner. Eng.* 58, 90–99. https://doi.org/10.1016/j.mineng.2014.01.022

Williamson, A. J., Folens, K., Matthijs, S., Cortes, Y. P., Varia, J., Du Laing, G., Boon, N., Hennebel, T., 2021. Selective metal extraction by biologically produced siderophores during bioleaching from low-grade primary and secondary mineral resources. *Miner. Eng.* 163,106774. https://doi.org/10.1016/j.mineng.2021.106774

Yun, S., Jung, H., Lee, H. J., Yang, Y., Lee, J. S., Hur, M., Lee, B., Ahn, J., Hwang, G., 2024. Bioleaching of valuable metals from three cathode active materials comprising lithium nickel cobalt manganese (NCM) oxide using indigenous microorganisms. *J. Ind. Eng. Chem.* 135, 552–560.

Zeng, X., Wei, S., Sun, L., Jacques, D. A., Tang, J., Lian, M., Ji, Z., Wang, J., Zhu, J., Xu, Z., 2015. Bioleaching of heavy metals from contaminated sediments by the Aspergillus niger strain SY1. *J. Soils Sediments* 15, 1029–1038. https://doi.org/10.1007/ s11368-015-1076-8.

Zhang, D. R., Xia, J. L., Nie, Z. Y., Chen, H. R., Liu, H. C., Deng, Y., Zhao, Y. D., Zhang, L. L., Wen, W., Yang, H. Y., 2019. Mechanism by which ferric iron promotes the bioleaching of arsenopyrite by the moderate thermophile Sulfobacillus thermosulfidooxidans. *Process Biochem.* 81, 11–21. https://doi.org/10.1016/j.procbio.2019.03.004

Zheng, X., Lin, H., Du, D., Li, G., Alam, O., Cheng, Z., Liu, X., Jiang, S., Li, J., 2024. Remediation of heavy metals polluted soil environment: A critical review on biological approaches. *Ecotoxicol. Environ. Saf.* 284, 116883. https://doi.org/10.1016/j.ecoenv.2024.116883

10 Microbes as Biocatalysts of Marine Micropollutants

Nisha Shankhwar[1]*, Awadhesh Kumar Verma*[2]*, Sofia Shaik*[2]*, Kalhan Pandita*[2]*, Karishma Vivek Kathpalia*[2]*, Tanya Singh*[2]*, and Neeta Raj Sharma*[2]

[1]Special Centre for Nanoscience, Jawaharlal Nehru University, New Delhi, India

[2]Department of Biotechnology and Bioinformatics, School of Bio-engineering and Biosciences, Lovely Professional University, Phagwara, India

10.1 INTRODUCTION TO MARINE MICROPOLLUTANTS

10.1.1 Definition and Classification

Marine micropollutants are trace concentrations of contaminants that persist in the marine environment despite their low concentrations (<1 μg/L). These pollutants being mainly anthropogenic include pharmaceutical waste, waste accumulated by personal care products, use of pesticides in agriculture, industrial chemical effluents, micro-plastics, nano-plastics, and heavy metals waste. Their persistent toxicity can pose significant risks to marine ecosystems and to human health as well (Abbasi et al., 2022). Understanding the role of microbes as biocatalysts in the degradation and transformation of these micropollutants is essential for analyzing their environmental impact and to develop strategies for management (Figure 10.1) (Bilal et al., 2022).

10.1.2 Types of Marine Micropollutants

Marine micropollutants encompass a broad range of chemical substances found in minute concentrations in aquatic environments. These pollutants are basically classified on the basis of their origin and chemical properties (Table 10.1):

1. **Pharmaceuticals, Personal Care Products (PPCPs):** These include antibiotics, painkillers, and endocrine-disrupting compounds (EDCs) like hormones. PPCPs are commonly found in coastal waters due to their widespread use and incomplete removal in wastewater treatment plants (WWTPs). Studies have shown that pharmaceuticals, even at trace levels, can affect the behavior, reproduction, and growth of marine species such as fish, mollusks, and crustaceans. For instance, anti-inflammatory drugs like diclofenac have been linked to the decline in marine invertebrate populations (Gaw et al., 2014; Khalid & Abdollahi, 2021).
2. **Pesticides:** Pesticides such as herbicides, insecticides, and fungicides enter the marine environment via surface runoff, leaching, and atmospheric deposition. These substances disrupt marine ecosystems by affecting photosynthetic organisms like algae, which are foundational to marine food webs. Herbicides like atrazine have been found to impair photosynthesis in phytoplankton, reducing primary production and impacting species that rely on these microorganisms for food (Tudi et al., 2021).
3. **Industrial Chemicals:** Polychlorinated biphenyls (PCBs), polycyclic aromatic hydrocarbons (PAHs), and per- and polyfluoroalkyl substances (PFAS) are introduced into the marine environment through industrial discharges, accidental spills, and atmospheric deposition. These compounds are highly persistent and can bioaccumulate in marine organisms, leading to chronic health effects such as immune suppression, reproductive failures, and cancers in marine species (Hermabessiere et al., 2017).
4. **Microplastics and Nanoplastics:** One of the most pervasive forms of marine micropollutants is plastic. Microplastics (<5 mm) and nanoplastics (<100 nm) result from the breakdown of larger plastic debris or are directly released from products like cosmetics, textiles, and paints. These particles not only pose a physical hazard to marine organisms that ingest them but also act as vectors for other toxic chemicals, including persistent organic pollutants (POPs). Microplastics have been found in deep-sea sediments and marine organisms, suggesting their widespread distribution and impact on ecosystems (Chaudhari et al., 2024).
5. **Heavy Metals:** Mercury, lead, and cadmium are often by-products of industrial processes, mining, and the burning of fossil fuels. These toxic metals are transported to marine environments through atmospheric deposition and runoff. Once in the marine system, they tend to accumulate in sediments and aquatic organisms. Mercury, for example, is converted by microbes into methylmercury, a highly toxic form that bioaccumulates in marine food chain, and poses serious health risks to humans who consume seafood (Mitra et al., 2022).

DOI: 10.1201/9781003509530-10

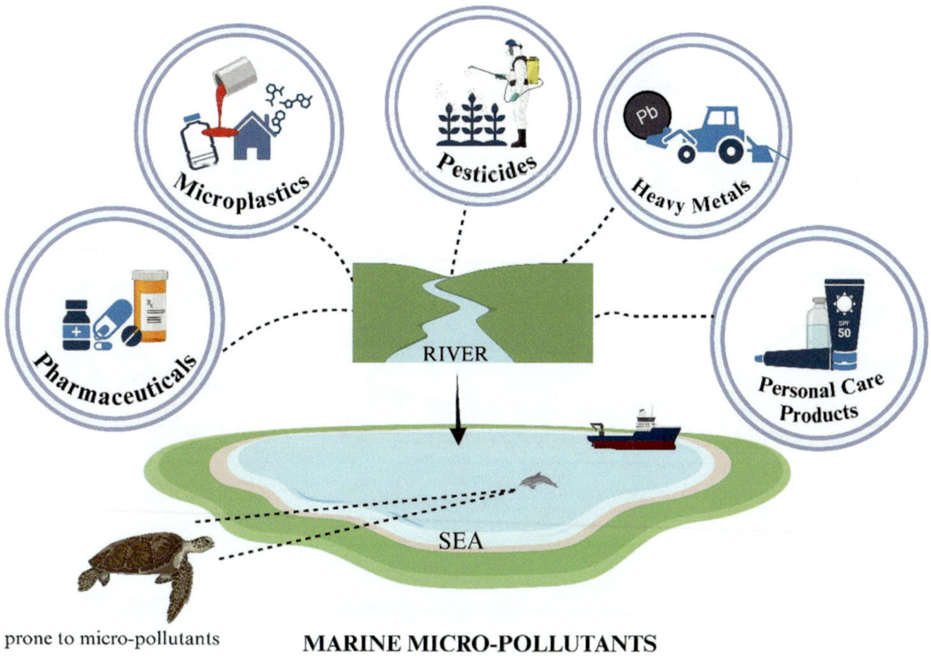

FIGURE 10.1 Diagram describing sources of marine micropollutants endangering marine diversity. (Self-developed using Biorender.com.)

TABLE 10.1
Classification and Impact of Marine Micropollutants

Micropollutant Classifications	Sources	Environmental Impact	Examples	Reference
Pharmaceuticals	Excretion, improper disposal	Bioaccumulation, toxicity to marine organisms, hormonal disruption,	Antibiotics, hormones, painkillers	Gaw et al. (2014)
Personal care products	Cosmetic and hygiene products	Disruption of marine life, bioaccumulation	Parabens, fragrances, UV filters	Khalid and Abdollahi (2021)
Pesticides	Agricultural runoff	Disruption of food chains, toxicity to aquatic life	Herbicides, insecticides, fungicides	(Tudi et al., 2021)
Industrial chemicals	Manufacturing processes, urban runoff	Persistent pollution, bioaccumulation, toxicity to marine organisms	Flame retardants, plasticizers, solvents	Hermabessiere et al. (2017)
Microplastics, nanoplastics	Degradation of plastic debris	Ingestion by marine organisms, physical/chemical harm, bioaccumulation	Plastic beads, fibers, fragments	Chaudhari et al. (2024)
Heavy metals	Mining, industrial waste	Toxicity, bioaccumulation, neurological damage in marine organisms and humans	Mercury, lead, cadmium, arsenic	Mitra et al. (2022)

10.1.3 Sources and Environmental Impact

Marine micropollutants enter the marine environment through various pathways, including runoff from land, atmospheric deposition, and direct discharge from industrial or municipal sources and have reached depths of ocean up to 3,000–10,000 m when they are being studied and investigated (Spindola Vilela et al., 2022).

1. **Runoff from Land:** Agricultural runoff containing pesticides and fertilizers, as well as urban runoff from roads and industrial sites, introduces pollutants into rivers and streams that eventually reach the ocean (Müller et al., 2020).
2. **Atmospheric Deposition:** Pollutants can be transported through the atmosphere and deposited into the ocean via precipitation or dry fallout. This process is particularly important for volatile organic compounds and particulate matter (Thiagarajan et al., 2024).
3. **Direct Discharge:** Industrial and municipal WWTPs, often inadequate at removing micropollutants, directly discharge treated or untreated

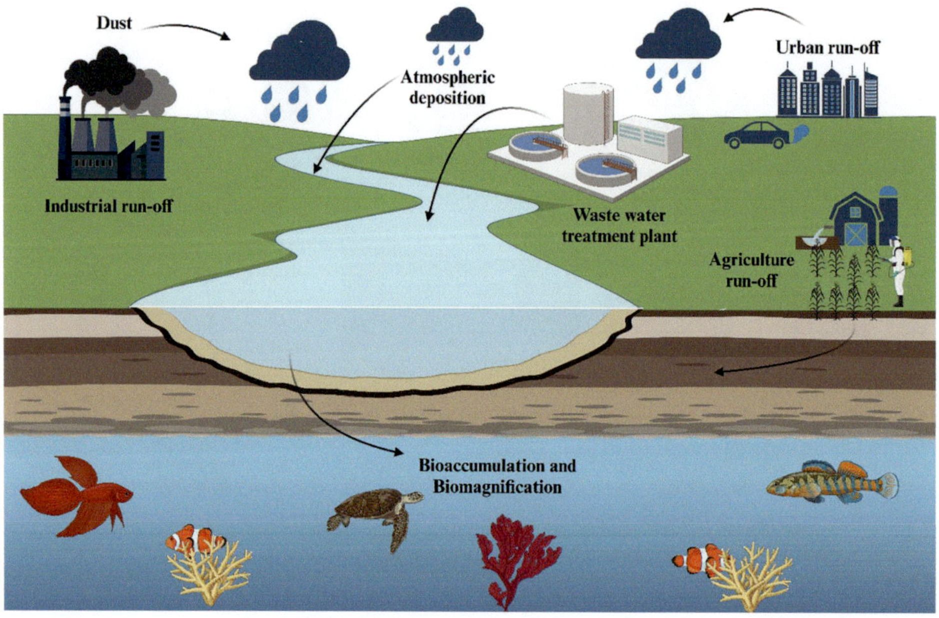

FIGURE 10.2 Diagram showing sources and environmental pathways of marine micropollutants. (Self-developed using Biorender.com.)

effluents containing various contaminants into water bodies (El Hammoudani et al., 2024).

Once introduced, due to their low degradability and ability to resist breakdown by natural processes, they can persist in the environment. Over time, micropollutants accumulate in marine sediments, organisms, and the water column, causing multiple adverse effects:

1. **Toxicity to Marine Organisms:** Some micropollutants, such as pharmaceuticals and pesticides, are toxic to marine life even at low concentrations. They can disrupt hormonal, reproductive, and immune systems, affecting population dynamics and species diversity (Kataoka & Kashiwada, 2021).
2. **Bioaccumulation and Bio-magnification:** Perpetual pollutants could accumulate in tissues of marine organisms of higher trophic levels. These pollutants become more concentrated as they move up the food chain through bio-magnification, ultimately affecting top predators and humans who consume seafood (Baek et al., 2022).
3. **Ecosystem Disruption:** Micropollutants can alter the natural functions of ecosystems, such as nutrient cycling, energy flow, and habitat structure. For example, antibiotics in the marine environment can disrupt microbial communities essential for nutrient cycling and biodegradation. Once in the marine environment, these pollutants can accumulate in sediments, biota, and the water column, potentially causing adverse effects on marine life and altering ecosystem functions. The environmental impact of micropollutants is influenced by their chemical properties, persistence, and interactions with marine organisms (Warner et al., 2019).

Micropollutants and their metabolites have broad physicochemical properties which help them interact with biotic and abiotic environments. This can be used to gain information about the flow paths and boundary conditions, like coexisting redox conditions or temperature fields and gradients (Figure 10.2) (Liu et al., 2021).

10.2 MICROBIAL BIODEGRADATION OF MARINE MICROPOLLUTANTS

10.2.1 Mechanisms of Microbial Degradation

Microbes have diverse enzymatic systems that help them to degrade numerous organic pollutants. Key mechanisms of microbial biodegradation include oxidative degradation, hydrolysis, and reductive dechlorination. Oxidative enzymes, such as cytochrome P450 monooxygenases and peroxidases, facilitate the breakdown of complex pollutants into simpler compounds (Ahmad, 2022). Hydrolytic enzymes, such as esterases and amidases, cleave chemical bonds in micropollutants, while reductive dechlorination involves the removal of chlorine atoms from organohalides. These enzymatic activities contribute to the detoxification and mineralization of micropollutants.

Microbes degrade marine micropollutants through a variety of enzymatic processes, each involving different classes of enzymes tailored to the pollutant type (Figure 10.3). The key mechanisms include (Table 10.2):

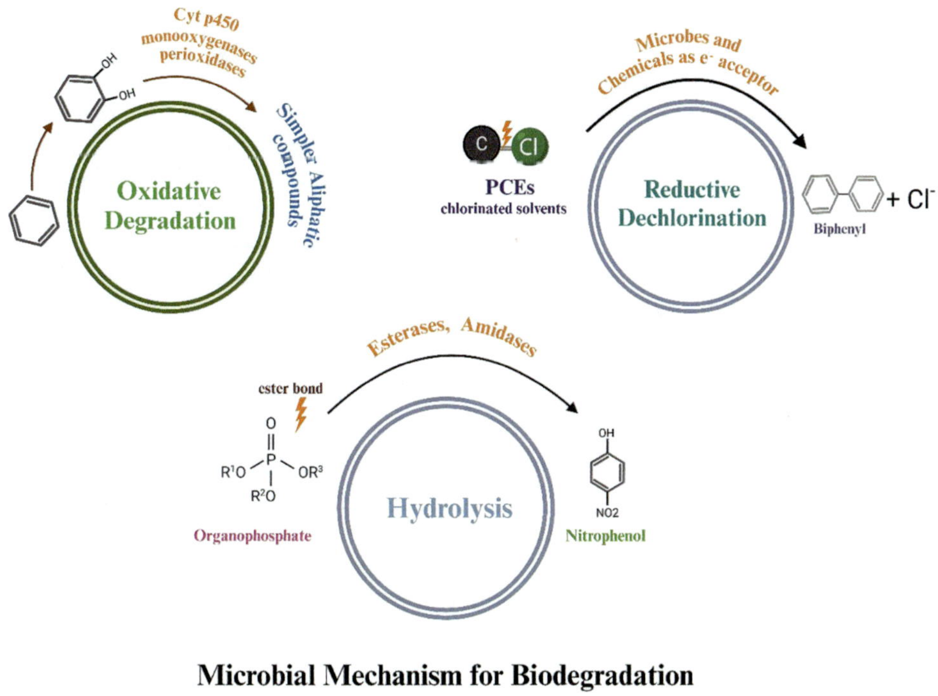

FIGURE 10.3 Figure showing microbial mechanisms for the biodegradation of marine micropollutants. (Self-developed using Biorender.com.)

TABLE 10.2
Microbial Biodegradation Mechanisms and Applications

Degradation Mechanism	Enzymes Involved	Pollutants Targeted	Application in Bioremediation	Reference
Oxidative degradation	Cytochrome P450, peroxidases	Hydrocarbons, pharmaceuticals	Used for degrading aromatic hydrocarbons and detoxifying organic pollutants	Li et al. (2001)
Hydrolysis	Esterases, amidases	Pesticides, pharmaceuticals	Effective in breaking ester/amide bonds in organic micropollutants	Lai et al. (2023)
Reductive dechlorination	Reductive dehalogenases	PCBs, chlorinated solvents	Important for detoxifying chlorinated organic pollutants	Peng et al. (2019)
Primary metabolism	Oxidative and hydrolytic enzymes	Hydrocarbons, industrial chemicals	Complete mineralization to non-toxic products	Saravanan et al. (2021)
Secondary metabolism	Microbial enzymes	Pharmaceuticals, pesticides	Transformation to intermediate metabolites that may require further degradation	Huang et al. (2018)

1. **Oxidative Degradation:** Enzymes like cytochrome P450, monooxygenases and peroxidases help in oxidative breakdown of complex organic pollutants. They introduce oxygen atoms into the chemical structure of pollutants, making them more soluble and easier to degrade. In the case of hydrocarbons and pharmaceuticals, oxidation can lead to the breakdown of aromatic rings and other stable structures (Ahmad, 2022).
2. **Hydrolysis:** Hydrolytic enzymes such as esterases and amidases catalyze the cleavage of chemical bonds through the addition of water molecules. This mechanism is particularly effective for breaking down ester and amide bonds in micropollutants like pesticides and pharmaceuticals (Esteves et al., 2021).
3. **Reductive Dechlorination:** This process is essential for degrading chlorinated compounds, such as PCBs and chlorinated solvents. Reductive dechlorination involves removing chlorine atoms from the molecule, which can lead to the detoxification of these highly persistent pollutants (Esteves et al., 2021).
4. **Synergistic Degradation:** In many marine environments, microbial communities consisting of different species work together to degrade

pollutants more effectively than any single strain could. For instance, certain bacteria may initiate the breakdown of a pollutant, while others complete the degradation process. This cooperative metabolism is especially important for recalcitrant pollutants like PAHs and PCBs, which require multiple steps for complete degradation (Esteves et al., 2021).

Also, microbial consortia between microalgae and heterotrophic bacteria coexist in aquatic ecosystems and cooperatively affect each another's metabolism and physiology, often enhancing biodegradation efficiency through synergistic interactions, where different species work together to degrade complex substrates more effectively than individual strains alone. This cooperative metabolism is mostly important in degrading recalcitrant pollutants such as PAHs and PCBs (Esteves et al., 2021).

10.2.2 Microbial Taxa Involved in Biodegradation

Microbial taxa of bacteria, fungi, and archaea are involved in breakdown of marine micropollutants. For example, the bacterial genera *Pseudomonas*, *Burkholderia*, and *Sphingomonas* have the ability to degrade hydrocarbons, pesticides, and pharmaceuticals (Lee et al., 2019). *Aspergillus* and *Penicillium* have also demonstrated capacity for degrading complex organic pollutants (Deshmukh et al., 2016). Additionally, certain marine archaea are involved in the depletion of chlorinated compounds and other persistent pollutants (Krzmarzick et al., 2018).

Numerous microbial taxa have been identified as key players in the biodegradation of marine micropollutants:

1. **Bacteria:** Genera such as *Pseudomonas*, *Burkholderia*, and *Sphingomonas* are known for their ability to break hydrocarbons, pesticides, and pharmaceuticals at microscopic level. These bacteria utilize various metabolic pathways to break down organic pollutants and can be found in many marine environments, from coastal sediments to deep sea habitats.
2. **Fungi:** Marine fungi, including species of *Aspergillus* and *Penicillium*, are also important for breaking down complex organic pollutants. Fungi produce a wide range of enzymes of extracellular nature which are capable of degrading lignin-like structures, such as laccases and peroxidases.
3. **Archaea:** Although less studied, certain marine archaea have demonstrated the ability of depletion of chlorinated compounds and other POPs. These extremophiles are particularly active in high-salinity or high-temperature environments, such as hydrothermal vents, where they contribute to the breakdown of pollutants in extreme conditions.

Marine archaea such as haloarchaea closely related to *Haloarcula vallismortis* have been isolated from a salt marsh in France with a strain EH4. A strain of *Halobacteriaceae* L1 was also isolated from the Dead Sea, *Natrialba* sp. of C21 strain has also been isolated from oil-contaminated saline water in Ain Salah, Algeria, which is able to degrade various aliphatic and aromatic compounds such as phenol, naphthalene, and pyrene, along with crude oil and xenobiotic pollutants (Krzmarzick et al., 2018). The identification and characterization of these microbial taxa provide insights into their roles in pollutant degradation and their potential applications in bioremediation (Figure 10.4).

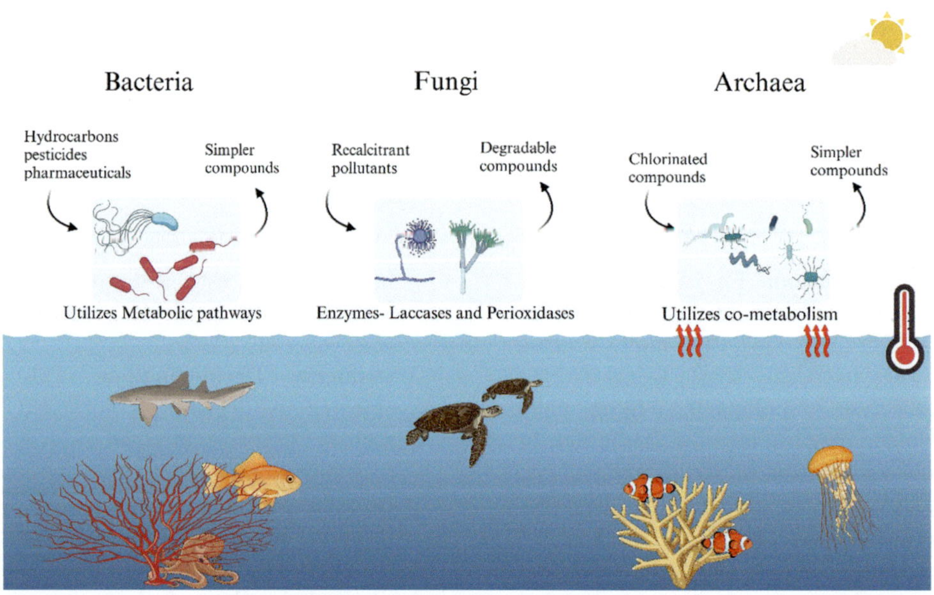

FIGURE 10.4 Diagram showing microbial taxa involved in the biodegradation of marine micropollutants. (Self-developed using Biorender.com.)

10.3 BIOTRANSFORMATION PATHWAYS AND PRODUCTS

10.3.1 Primary and Secondary Metabolism

Microbial biotransformation of micropollutants involves both primary and secondary metabolic pathways. Primary metabolism typically leads to the detoxification and mineralization of pollutants, resulting in the formation of non-toxic end products such as CO_2, inorganic salts, and water. Secondary metabolism may produce intermediate metabolites, which can be further transformed by other microorganisms or enzymatic systems. Understanding these metabolic pathways is crucial for assessing the efficacy of microbial biocatalysts in pollutant degradation and identifying potential by-products with environmental or toxicological significance (Desiante et al., 2021). Basic chemical reactions like isomerization, condensation, oxidation, reduction, functional group's introduction, hydrolysis, and new carbon bond formation are the basis of microbial transformations. The characteristics of microbes like high surface-to-volume ratio, expansion rate, and metabolism rate make them an ideal choice for toxic pollutant biotransformation (Tyoker Kukwa et al., 2022).

10.3.2 Identification of Metabolic By-products

The identification and characterization of metabolic by-products generated during microbial degradation are essential for understanding the complete degradation pathways and potential environmental risks. Analytical techniques, such as gas chromatography-mass spectrometry (GC-MS), high-performance liquid chromatography (HPLC), and nuclear magnetic resonance (NMR) spectroscopy, are employed to identify and quantify these by products (Tang, 2011). High-molecular-weight (HMW) PAHs and their metabolites are mostly analyzed by GC with flame ionization and MS detection and HPLC with ultraviolet, fluorescence, and MS detection. HMW PAH metabolites have different characteristics such as polarity, acidity, size, stability, volatility, solubility, sorption, and isomeric characteristics. MS analysis is helpful in allowing a full or partial identification of unknown compounds in complex environmental samples (Méndez García & García de Llasera, 2021). This information aids in evaluating the effectiveness of microbial biocatalysts and the potential persistence or toxicity of intermediate and final degradation products.

10.4 APPLICATIONS IN BIOREMEDIATION

10.4.1 Designing Bioremediation Strategies

Microbial biocatalysts are living organisms, primarily bacteria, fungi, or algae, that catalyze biochemical reactions to transform harmful pollutants into less or non-toxic substances. These biocatalysts are increasingly employed in bioremediation strategies aimed at reducing the environmental burden of marine micropollutants like hydrocarbons, heavy metals, and POPs. The versatility and efficiency of microbial systems make them particularly valuable for addressing the complex mixture of contaminants found in marine environments. Microbial bioremediation strategies can be broadly classified into two main approaches: *in-situ* bioremediation and *ex-situ* bioremediation. Each approach has its own advantages, limitations, and specific applications depending on the nature and extent of the pollution, as well as the environmental conditions. Optimizing conditions such as temperature, nutrient availability, and pH can enhance microbial degradation rates. Bio-augmentation and bio-stimulation are common techniques employed to improve bioremediation efficiency (Azubuike et al., 2016).

- ***In-situ* Bioremediation:** It involves the application of microbial cultures or the stimulation of indigenous microbial communities directly at the contaminated site. This method is particularly useful for addressing large-scale marine pollution events, such as oil spills or wastewater discharge zones, where it may be impractical or impossible to remove the contaminated material. The goal is to enhance the natural biodegradation processes by providing optimal conditions for microbial activity or by introducing microorganisms with specific degradative capabilities. Key techniques in in-situ bioremediation include:
  1. **Bioventing:** This technique entails the controlled introduction of oxygen into the subsurface environment to stimulate aerobic microorganism activity. Bioventing is especially effective for degrading organic volatile and semi-volatile compounds.
  2. **Bioslurping:** A method that combines bioventing with vacuum-enhanced extraction to remove and degrade contaminants from the soil or water. It is often used for the remediation of petroleum hydrocarbons.
  3. **Biosparging:** Like bioventing, this technique involves injecting air or oxygen directly into the groundwater, promoting the aerobic breakdown of pollutants.
  4. **Phytoremediation:** This method involves using plants to stimulate the biodegradation of pollutants. Plants can enhance microbial activity in the rhizosphere—the area of soil around their roots—by releasing root exudates that provide nutrients for microorganisms. Certain plants can take up and accumulate contaminants, aiding in their removal from the environment.

In-situ bioremediation is often favored due to its cost-effectiveness and minimal disruption to the environment. However, its success hinges on several factors, including the bioavailability of contaminants, the availability of suitable microbial communities, and environmental conditions such as temperature, pH, and oxygen levels. The effectiveness of in-situ bioremediation may be restricted in extreme environments or when contaminants are deeply buried or trapped within the sediment matrix.

- ***Ex-situ* Bioremediation:** Ex-situ bioremediation entails the extraction of contaminated sediments or water from the marine environment to a controlled facility, where microbial cultures are utilized to break down the pollutants. This method is typically employed when pollutants are present in high concentrations or when the environmental conditions at the site are unsuitable for effective in-situ treatments. Ex-situ bioremediation offers enhanced control over environmental factors such as temperature, oxygen levels, and nutrient availability, which can be optimized to improve microbial degradation rates (Azubuike et al., 2016). Key techniques in ex-situ bioremediation include:
  1. **Biopiles:** Contaminated soil or sediment is dug up and placed in above-ground piles, where it is aerated and treated with nutrients and microbial cultures. The process is monitored and controlled to optimize conditions for microbial degradation.
  2. **Windrow Composting:** Similar to biopiles, this method involves forming contaminated material into long rows (windrows) and periodically turning them to maintain aerobic conditions. Windrow composting is effective for the treatment of organic pollutants.
  3. **Bioreactors:** Contaminated water or sediment is placed in a reactor vessel where conditions are tightly controlled to maximize microbial activity. Bioreactors can be designed to operate in either aerobic or anaerobic conditions, depending on the type of pollutants being treated. This method is highly efficient but may be more costly due to the need for specialized equipment and infrastructure.
  4. **Land Farming:** This method involves spreading contaminated soil in a thin layer on the ground and regularly turning it to improve air circulation and encourage microbial activity. Nutrients and water are also added to support microbial growth. Land farming is a common technique for cleaning up soil contaminated with petroleum.

Ex-situ bioremediation is generally more efficient for treating highly concentrated pollutants or when precise control over the treatment process is necessary. However, it is typically more resource-intensive and can involve significant logistical challenges, particularly when large volumes of contaminated material need to be transported and treated. Additionally, *ex-situ* methods may disrupt local ecosystems and require careful management to prevent secondary contamination (Figure 10.5) (Azubuike et al., 2016).

10.4.2 Case Studies of Successful Bioremediation

Several case studies highlight the successful application of microbial biocatalysts in the bioremediation of marine micropollutants. For example, the use of genetically engineered bacteria to degrade oil spills and synthetic polymers has demonstrated significant reductions in pollutant concentrations and environmental impact (Rafeeq et al., 2023). Other research has focused on using microbial consortia to remove pharmaceuticals and pesticides from marine ecosystems (Qian et al., 2020). These case studies provide important insights into the practical use of microbial biocatalysts and the factors that affect successful bioremediation outcomes.

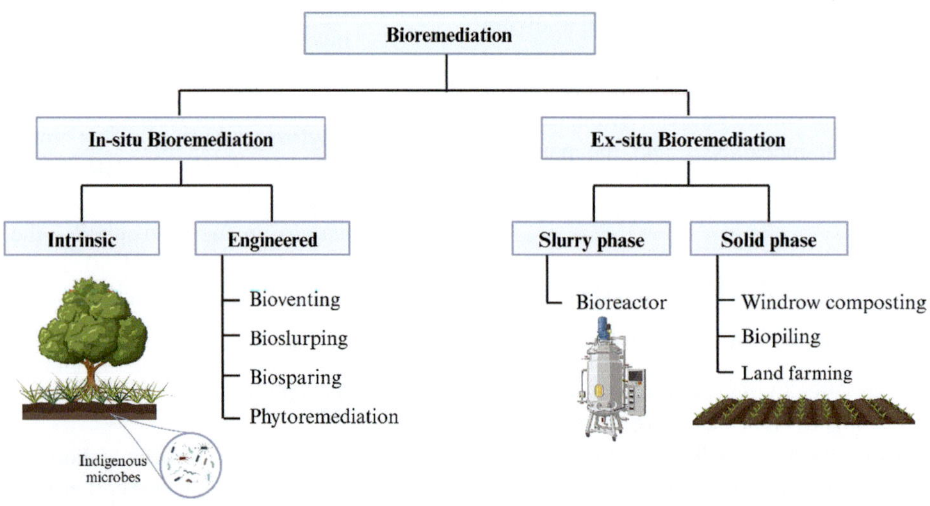

FIGURE 10.5 Bioremediation strategies exploring about *in-situ* and *ex-situ* bioremediation. (Self-developed using Biorender.com.)

Various groups of microorganisms have been associated with hydrocarbon degradation in marine environments, including filamentous fungi such as *Aspergillus* and *Penicillium*, as well as yeast like *Candida*. Nevertheless, bacteria are typically regarded as the main contributors to hydrocarbon biodegradation in aquatic environments. Specifically, Gammaproteobacteria, Alphaproteobacteria, and Actinobacteria sourced from seawater and sediment samples have demonstrated considerable potential for hydrocarbon bioremediation (Perdigão et al., 2021).

10.5 CHALLENGES AND LIMITATIONS

10.5.1 Factors Affecting Biodegradation Efficiency

The effectiveness of microbial biodegradation of marine micropollutants is influenced by several factors, including the chemical properties of the pollutants, the availability of suitable microbial strains, and environmental conditions such as temperature, salinity, and nutrient availability (Shaji et al., 2024). Pollutant bioavailability and the presence of co-contaminants can also affect microbial degradation rates. Tackling these challenges necessitates a thorough comprehension of the interactions between pollutants and microbial systems, alongside the creation of strategies to enhance degradation processes.

The bioavailability of pollutants and the presence of co-contaminants can influence these degradation processes. To improve biodegradation, strategies such as bioaugmentation, biostimulation, in-situ bioremediation, and ex-situ bioremediation can be utilized. Grasping the interplay among these factors is crucial for formulating effective bioremediation methods to combat marine pollution (Singh & Kumar Mishra, 2021).

10.5.2 Regulatory and Environmental Considerations

The application of microbial biocatalysts in bioremediation must consider regulatory and environmental factors. Regulatory guidelines for deploying genetically modified microorganisms must be comprehensive, addressing potential risks such as harmful by-product production and impacts on non-target organisms (Akpasi et al., 2023). It is crucial to consider the long-term consequences of bioremediation on ecosystem health and stability, preventing unintended environmental harm from remediation efforts. The use of microbial biocatalysts in marine bioremediation must be carefully regulated to prevent unintended environmental impacts. Issues include:

- **Release of Genetically Engineered Microorganisms (GEMs):** The use of GEMs in open marine environments raises concerns about horizontal gene transfer, the potential spread of antibiotic resistance, and unintended effects on non-target organisms (Rafeeq et al., 2023).
- **Formation of Harmful By-products:** Incomplete degradation of micropollutants can cause the accumulation of harmful by-products, emphasizing the importance of rigorous monitoring and control in bioremediation processes (Panigrahy et al., 2022).
- **Long-Term Ecological Effects:** The introduction of large numbers of microorganisms into marine ecosystems may disrupt natural microbial communities and ecological processes. Long-term monitoring is necessary to ensure that bioremediation efforts do not cause unintended harm to marine biodiversity (Jain et al., 2023).

Introducing bioremediating microbes directly to contaminated sites can face challenges due to environmental stresses such as extreme temperatures, pH levels, salinity, and pollutant toxicity. To tackle these issues, enhancing the selected microorganisms can improve the efficacy of bioaugmentation. Two primary strategies for enhancement include cell immobilization and genetic modification. For example, Antarctic bacteria *Pseudoalteromonas haloplanktis* TAC125 have been genetically modified to express the toluene-o-xylene monooxygenase (ToMO) gene derived from mesophilic bacteria *Pseudomonas spp.* OX1. This modification enabled the enzyme to effectively convert various aromatic compounds, demonstrating the potential of genetic engineering to improve microbial performance in adverse conditions (Agustina Tan et al., 2023).

10.6 FUTURE DIRECTIONS AND RESEARCH OPPORTUNITIES

10.6.1 Innovative Approaches in Biocatalysis

Future research in microbial biocatalysis will focus on developing innovative approaches to enhance the degradation of marine micropollutants. Advances in synthetic biology, genomics, and metabolomics offer opportunities to design and optimize microbial strains with improved degradation capabilities. The combination of computational modeling and high-throughput screening technologies can enhance the identification of novel biocatalysts and aid in the design of effective bioremediation strategies (González-Plaza et al., 2022).

Computer simulations can help predict the behavior of microorganisms and their interactions with pollutants, aiding in the design of more efficient bioremediation processes (Gayathiri et al., 2023). Automated screening methods can rapidly evaluate large numbers of microbial strains to identify those with the best degradation capabilities (Sarnaik et al., 2020).

10.6.2 Integration with Environmental Management Practices

Integrating microbial biocatalysis with environmental management practices is essential for addressing the challenges of marine micropollutants. Joint efforts among researchers, policymakers, and industry stakeholders will be crucial for developing and implementing effective bioremediation strategies (Kuppan et al., 2024). Applying microbial biocatalysis to specific marine environments and pollutants requires a deep understanding of local conditions and microbial communities. Microbial biocatalysis offers versatile solutions for a range of marine pollution issues. Bacteria and fungi can degrade oil spills, remove heavy metals, break down plastics, and reduce the impact of PPCPs and eutrophication (Deshmukh et al., 2016). Also, incorporating microbial biocatalysis into broader pollution management frameworks can enhance the sustainability and effectiveness of remediation efforts. To achieve large-scale reductions in marine pollution, microbial biocatalysis must be integrated into broader environmental management practices. This includes:

- **Collaboration Between Researchers, Policymakers, and Industry:** The successful implementation of microbial bioremediation technologies requires close collaboration between different stakeholders. Policymakers must create regulations that support the development and use of these technologies, while industry must adopt environmentally sustainable practices that minimize the release of micropollutants (Bala et al., 2022).
- **Site-Specific Bioremediation Solutions:** Marine environments are highly variable, with different pollutant profiles, ecological conditions, and microbial communities. Developing site-specific bioremediation strategies that account for these factors will be key to the success of microbial biocatalysis in real-world applications (Tedesco et al., 2024).
- **Monitoring and Assessment:** Long-term monitoring of bioremediation efforts is essential to assess their effectiveness and ensure that they do not cause unintended environmental harm. New tools for monitoring microbial activity, pollutant concentrations, and ecosystem health will be critical for the success of these efforts (Ayilara & Babalola, 2023).

10.7 CONCLUSION

Microbial biocatalysts offer a promising and environmentally sustainable solution to marine pollution. These microorganisms can degrade hazardous micropollutants, such as pharmaceuticals, pesticides, plastics, and industrial chemicals, transforming them into less harmful compounds. By harnessing their natural capabilities, we can mitigate the environmental and health risks associated with these pollutants, particularly in vulnerable marine ecosystems. However, successful application in real-world scenarios requires addressing challenges such as pollutant bioavailability, environmental conditions, and the presence of multiple contaminants. Introducing engineered microorganisms also carries risks, including ecological imbalances, the spread of antibiotic resistance, and the formation of toxic by-products. Advancements in microbial biotechnology, such as synthetic biology, metagenomics, and immobilization of microbial enzymes, offer opportunities to enhance bioremediation. These innovations can improve the effectiveness, practicality, and scalability of pollutant removal in contaminated marine areas. To fully realize the potential of microbial bioremediation, it must be integrated into broader environmental management strategies, combining scientific research, policy development, and industrial collaboration. Sustainable practices, supportive regulations, and long-term monitoring are essential to ensure these interventions are effective and environmentally safe. In conclusion, microbial biocatalysts are a powerful tool for mitigating marine pollution. By embracing microbial bioremediation with innovation and responsible environmental stewardship, we can protect marine ecosystems and ensure ocean health for future generations.

ABBREVIATIONS

CO_2	Carbon dioxide
EDCs	Endocrine-Disrupting Compounds
GC-MS	Gas Chromatography-Mass Spectrometry
GEMs	Genetically Engineered Microorganisms
HMW PAHs	High-Molecular-Weight Polycyclic Aromatic Hydrocarbons
HPLC	High-Performance Liquid Chromatography
NMR	Nuclear Magnetic Resonance
PAHs	Polycyclic Aromatic Hydrocarbons
PCBs	Polychlorinated Biphenyls
PFAS	Per- and Polyfluoroalkyl Substances
pH	Potential of Hydrogen
POPs	Persistent Organic Pollutants
PPCPs	Pharmaceuticals and Personal Care Products
ToMO	Toluene-o-Xylene Monooxygenase
UV	Ultraviolet
WWTPs	Wastewater Treatment Plants

ACKNOWLEDGMENT

One of the authors extends profound gratitude to the Department of Biotechnology (DBT), Ministry of Science and Technology, Government of India, for their substantial financial support under project No. DBT/2022–23/July/N2240, provided through the prestigious DBT-RA Program. This support has been instrumental in advancing this research.

REFERENCES

Abbasi, N. A., Shahid, S. U., Majid, M., & Tahir, A. (2022). Ecotoxicological risk assessment of environmental micropollutants. In Hashmi, M. Z., Wang, S., & Ahmed, Z., Eds., *Environmental Micropollutants* (pp. 331–337). Elsevier. https://doi.org/10.1016/B978-0-323-90555-8.00004-0

Agustina Tan, W., Celina, G., & Pranawijaya, S. (2023). Bioremediation in the marine environment: challenges and prospective methods for enhancement. *E3S Web of Conferences*, *374*, 00038. https://doi.org/10.1051/e3sconf/202337400038

Ahmad, I. (2022). Microalgae–bacteria consortia: a review on the degradation of polycyclic aromatic hydrocarbons (PAHs). *Arabian Journal for Science and Engineering*, *47*(1), 19–43. https://doi.org/10.1007/s13369-021-06236-9

Akpasi, S. O., Anekwe, I. M. S., Tetteh, E. K., Amune, U. O., Shoyiga, H. O., Mahlangu, T. P., & Kiambi, S. L. (2023). Mycoremediation as a potentially promising technology: current status and prospects—a review. *Applied Sciences*, *13*(8), 4978. https://doi.org/10.3390/app13084978

Ayilara, M. S., & Babalola, O. O. (2023). Bioremediation of environmental wastes: the role of microorganisms. *Frontiers in Agronomy*, *5*. https://doi.org/10.3389/fagro.2023.1183691

Azubuike, C. C., Chikere, C. B., & Okpokwasili, G. C. (2016). Bioremediation techniques–classification based on site of application: principles, advantages, limitations and prospects. *World Journal of Microbiology and Biotechnology*, *32*(11), 180. https://doi.org/10.1007/s11274-016-2137-x

Baek, S.-S., Yun, D., Pyo, J., Kang, D., Cho, K. H., & Jeon, J. (2022). Analysis of micropollutants in a marine outfall using network analysis and decision tree. *Science of the Total Environment*, *806*, 150938. https://doi.org/10.1016/j.scitotenv.2021.150938

Bala, S., Garg, D., Thirumalesh, B. V., Sharma, M., Sridhar, K., Inbaraj, B. S., & Tripathi, M. (2022). Recent strategies for bioremediation of emerging pollutants: a review for a green and sustainable environment. *Toxics*, *10*(8), 484. https://doi.org/10.3390/toxics10080484

Bilal, M., Lam, S. S., & Iqbal, H. M. N. (2022). Biocatalytic remediation of pharmaceutically active micropollutants for environmental sustainability. *Environmental Pollution*, *293*, 118582. https://doi.org/10.1016/j.envpol.2021.118582

Chaudhari, S., Parmar, H., & Samnani, P. (2024). Microplastics in personal care products and cosmetics. In Kumar, A., & Singh, V., Eds., *Microplastics Pollution and Its Remediation* (pp. 171–188). Springer Nature Singapore. https://doi.org/10.1007/978-981-97-4068-0_9

Deshmukh, R., Khardenavis, A. A., & Purohit, H. J. (2016). Diverse metabolic capacities of fungi for bioremediation. *Indian Journal of Microbiology*, *56*(3), 247–264. https://doi.org/10.1007/s12088-016-0584-6

Desiante, W. L., Minas, N. S., & Fenner, K. (2021). Micropollutant biotransformation and bioaccumulation in natural stream bio films. *Water Research*, *193*, 116846. https://doi.org/10.1016/j.watres.2021.116846

ElHammoudani, Y., Dimane, F., Haboubi, K., Benaissa, C., Benaabidate, L., Bourjila, A., Achoukhi, I., El Boudammoussi, M., Faiz, H., Touzani, A., Moudou, M., & Esskifati, M. (2024). Micropollutants in wastewater treatment plants: a bibliometric—bibliographic study. *Desalination and Water Treatment*, *317*, 100190. https://doi.org/10.1016/j.dwt.2024.100190

Esteves, F., Rueff, J., & Kranendonk, M. (2021). The central role of cytochrome P450 in xenobiotic metabolism—a brief review on a fascinating enzyme family. *Journal of Xenobiotics*, *11*(3), 94–114. https://doi.org/10.3390/jox11030007

Gaw, S., Thomas, K. V., & Hutchinson, T. H. (2014). Sources, impacts and trends of pharmaceuticals in the marine and coastal environment. *Philosophical Transactions of the Royal Society B: Biological Sciences*, *369*(1656), 20130572. https://doi.org/10.1098/rstb.2013.0572

Gayathiri, E., Prakash, P., Kumaravel, P., Jayaprakash, J., Ragunathan, M. G., Sankar, S., Pandiaraj, S., Thirumalaivasan, N., Thiruvengadam, M., & Govindasamy, R. (2023). Computational approaches for modeling and structural design of biological systems: a comprehensive review. *Progress in Biophysics and Molecular Biology*, *185*, 17–32. https://doi.org/10.1016/j.pbiomolbio.2023.08.002

González-Plaza, J. J., Furlan, C., Rijavec, T., Lapanje, A., Barros, R., Tamayo-Ramos, J. A., & Suarez-Diez, M. (2022). Advances in experimental and computational methodologics for the study of microbial-surface interactions at different omics levels. *Frontiers in Microbiology*, *13*. https://doi.org/10.3389/fmicb.2022.1006946

Hermabessiere, L., Dehaut, A., Paul-Pont, I., Lacroix, C., Jezequel, R., Soudant, P., & Duflos, G. (2017). Occurrence and effects of plastic additives on marine environments and organisms: a review. *Chemosphere*, *182*, 781–793. https://doi.org/10.1016/j.chemosphere.2017.05.096

Huang, Y., Xiao, L., Li, F., Xiao, M., Lin, D., Long, X., & Wu, Z. (2018). Microbial degradation of pesticide residues and an emphasis on the degradation of cypermethrin and 3-phenoxy benzoic acid: a review. *Molecules*, *23*(9), 2313. https://doi.org/10.3390/molecules23092313

Jain, R., Gaur, A., Suravajhala, R., Chauhan, U., Pant, M., Tripathi, V., & Pant, G. (2023). Microplastic pollution: understanding microbial degradation and strategies for pollutant reduction. *Science of the Total Environment*, *905*, 167098. https://doi.org/10.1016/j.scitotenv.2023.167098

Kataoka, C., & Kashiwada, S. (2021). Ecological risks due to immunotoxicological effects on aquatic organisms. *International Journal of Molecular Sciences*, *22*(15), 8305. https://doi.org/10.3390/ijms22158305

Khalid, M., & Abdollahi, M. (2021). Environmental distribution of personal care products and their effects on human health. *Iranian Journal of Pharmaceutical Research : IJPR*, *20*(1), 216–253. https://doi.org/10.22037/ijpr.2021.114891.15088

Krzmarzick, M. J., Taylor, D. K., Fu, X., & McCutchan, A. L. (2018). Diversity and niche of archaea in bioremediation. *Archaea*, *2018*, 3194108. https://doi.org/10.1155/2018/3194108

Kuppan, N., Padman, M., Mahadeva, M., Srinivasan, S., & Devarajan, R. (2024). A comprehensive review of sustainable bioremediation techniques: eco friendly solutions for waste and pollution management. *Waste Management Bulletin*, *2*(3), 154–171. https://doi.org/10.1016/j.wmb.2024.07.005

Lai, J., Huang, H., Lin, M., Xu, Y., Li, X., & Sun, B. (2023). Enzyme catalyzes ester bond synthesis and hydrolysis: the key step for sustainable usage of plastics. *Frontiers in Microbiology*, *13*. https://doi.org/10.3389/fmicb.2022.1113705

Lee, Y., Lee, Y., & Jeon, C. O. (2019). Biodegradation of naphthalene, BTEX, and aliphatic hydrocarbons by *Paraburkholderia aromaticivorans* BN5 isolated from petroleum-contaminated soil. *Scientific Reports*, *9*(1), 860. https://doi.org/10.1038/s41598-018-36165-x

Li, Q.-S., Ogawa, J., Schmid, R. D., & Shimizu, S. (2001). Engineering cytochrome P450 BM-3 for oxidation of polycyclic aromatic hydrocarbons. *Applied and Environmental Microbiology*, *67*(12), 5735–5739. https://doi.org/10.1128/AEM.67.12.5735-5739.2001

Liu, Z., Liang, Z., Li, K., Huang, T., Ma, J., & Wen, G. (2021). Degradation of micropollutants and formation of oxidation by-products during the ozone/peroxymonosulfate system: a critical review. *Water*, *13*(21), 3126. https://doi.org/10.3390/w13213126

Méndez García, M., & García de Llasera, M. P. (2021). A review on the enzymes and metabolites identified by mass spectrometry from bacteria and microalgae involved in the degradation of high molecular weight PAHs. *Science of the Total Environment*, *797*, 149035. https://doi.org/10.1016/j.scitotenv.2021.149035

Mitra, S., Chakraborty, A. J., Tareq, A. M., Emran, T. Bin, Nainu, F., Khusro, A., Idris, A. M., Khandaker, M. U., Osman, H., Alhumaydhi, F. A., & Simal-Gandara, J. (2022). Impact of heavy metals on the environment and human health: novel therapeutic insights to counter the toxicity. *Journal of King Saud University—Science*, *34*(3), 101865. https://doi.org/10.1016/j.jksus.2022.101865

Müller, A., Österlund, H., Marsalek, J., & Viklander, M. (2020). The pollution conveyed by urban runoff: a review of sources. *Science of the Total Environment*, *709*, 136125. https://doi.org/10.1016/j.scitotenv.2019.136125

Panigrahy, N., Priyadarshini, A., Sahoo, M. M., Verma, A. K., Daverey, A., & Sahoo, N. K. (2022). A comprehensive review on eco-toxicity and biodegradation of phenolics: recent progress and future outlook. *Environmental Technology & Innovation*, *27*, 102423. https://doi.org/10.1016/j.eti.2022.102423

Peng, P., Schneidewind, U., Haest, P. J., Bosma, T. N. P., Danko, A. S., Smidt, H., & Atashgahi, S. (2019). Reductive dechlorination of 1,2-dichloroethane in the presence of chloroethenes and 1,2-dichloropropane as co-contaminants. *Applied Microbiology and Biotechnology*, *103*(16), 6837–6849. https://doi.org/10.1007/s00253-019-09985-8

Perdigão, R., Almeida, C. M. R., Magalhães, C., Ramos, S., Carolas, A. L., Ferreira, B. S., Carvalho, M. F., & Mucha, A. P. (2021). Bioremediation of petroleum hydrocarbons in seawater: prospects of using lyophilized native hydrocarbon-degrading bacteria. *Microorganisms*, *9*(11), 2285. https://doi.org/10.3390/microorganisms9112285

Qian, X., Chen, L., Sui, Y., Chen, C., Zhang, W., Zhou, J., Dong, W., Jiang, M., Xin, F., & Ochsenreither, K. (2020). Biotechnological potential and applications of microbial consortia. *Biotechnology Advances*, *40*, 107500. https://doi.org/10.1016/j.biotechadv.2019.107500

Rafeeq, H., Afsheen, N., Rafique, S., Arshad, A., Intisar, M., Hussain, A., Bilal, M., & Iqbal, H. M. N. (2023). Genetically engineered microorganisms for environmental remediation. *Chemosphere*, *310*, 136751. https://doi.org/10.1016/j.chemosphere.2022.136751

Saravanan, A., Kumar, P. S., Vo, D.-V. N., Jeevanantham, S., Karishma, S., & Yaashikaa, P. R. (2021). A review on catalytic-enzyme degradation of toxic environmental pollutants: microbial enzymes. *Journal of Hazardous Materials*, *419*, 126451. https://doi.org/10.1016/j.jhazmat.2021.126451

Sarnaik, A., Liu, A., Nielsen, D., & Varman, A. M. (2020). High-throughput screening for efficient microbial biotechnology. *Current Opinion in Biotechnology*, *64*, 141–150. https://doi.org/10.1016/j.copbio.2020.02.019

Shaji, A., Kamalesh, R., Dinakarkumar, Y., Saravanan, A., Arokiyaraj, S., Mani, H. P., Veera, H. M., Muthu, D. B., Ramakrishnan, G., & Ivo Romauld, S. (2024). Microbial degradation of marine plastic debris: a comprehensive review on the environmental effects, disposal, and biodegradation. *Biochemical Engineering Journal*, *201*, 109133. https://doi.org/10.1016/j.bej.2023.109133

Singh, A., & Kumar Mishra, V. (2021). Biodegradation of organic pollutants for its effective remediation from the environment and the role of various factors affecting the biodegradation process. In Kumar Mishra, V., & Kumar, A., Eds., *Sustainable Environmental Clean-up* (pp. 1–27). Elsevier. https://doi.org/10.1016/B978-0-12-823828-8.00001-3

Spindola Vilela, C. L., Damasceno, T. L., Thomas, T., & Peixoto, R. S. (2022). Global qualitative and quantitative distribution of micropollutants in the deep sea. *Environmental Pollution*, *307*, 119414. https://doi.org/10.1016/j.envpol.2022.119414

Tang, J. (2011). Microbial metabolomics. *Current Genomics*, *12*(6), 391–403. https://doi.org/10.2174/138920211797248619

Tedesco, P., Balzano, S., Coppola, D., Esposito, F. P., de Pascale, D., & Denaro, R. (2024). Bioremediation for the recovery of oil polluted marine environment, opportunities and challenges approaching the blue growth. *Marine Pollution Bulletin*, *200*, 116157. https://doi.org/10.1016/j.marpolbul.2024.116157

Thiagarajan, V., Nah, T., & Xin, X. (2024). Impacts of atmospheric particulate matter deposition on phytoplankton: a review. *Science of the Total Environment*, *950*, 175280. https://doi.org/10.1016/j.scitotenv.2024.175280

Tudi, M., Daniel Ruan, H., Wang, L., Lyu, J., Sadler, R., Connell, D., Chu, C., & Phung, D. T. (2021). Agriculture development, pesticide application and its impact on the environment. *International Journal of Environmental Research and Public Health*, *18*(3), 1112. https://doi.org/10.3390/ijerph18031112

Tyoker Kukwa, D., Omolara Afolabi, F., Kweinor Tetteh, E., Michael Smarte Anekwe, I., & Chetty, M. (2022). Bioremediation of hazardous wastes. In Jeyakumar, R. B., Sankarapandian, K., & Ravi, Y. K., Eds., *Hazardous Waste Management* (pp. 25–47). IntechOpen. https://doi.org/10.5772/intechopen.102458

Warner, W., Licha, T., & Nödler, K. (2019). Qualitative and quantitative use of micropollutants as source and process indicators. A review. *Science of the Total Environment*, *686*, 75–89. https://doi.org/10.1016/j.scitotenv.2019.05.385

11 Mighty Microbes
Nature's Cleanup Crew for Petroleum Hydrocarbons in the Deep Blue Sea

Awadhesh Kumar Verma[1], *Madhusmita Kalita*[1], *Lipakshi Gautam*[1], *Tanya Singh*[1], *Karishma Vivek Kathpalia*[1], *Nisha Shankhwar*[2], *and Neeta Raj Sharma*[1]

[1]Department of Biotechnology & Bioinformatics, School of Bioengineering & Biosciences, Lovely Professional University, India

[2]Special Centre for Nanoscience, Jawaharlal Nehru University, New Delhi, India

11.1 INTRODUCTION TO PETROLEUM HYDROCARBONS IN MARINE ENVIRONMENTS

11.1.1 Nature and Sources of Petroleum Hydrocarbons

Hydrocarbons of petroleum (PHs) are known for being complex organic mixture compounds which are collected from crude oil and its refined products. These hydrocarbons include alkanes, cycloalkanes, aromatic hydrocarbons, and various polycyclic aromatic hydrocarbons (PAHs). PHs are usually extracted from biomasses of plants and animals which form geologically through millions of years and more, making a significant contribution to the development of the economy. But, due to human activities which exploit, the extraction and processing of petroleum cause dangerous petrochemical pollution in the environment. Saturated hydrocarbons (SHs) and PAHs are known as major constituents of PHs, where PAHs are considered as continual pollutants causing strong toxicity to our ecosystem and humans (Liu et al., 2022) (Table 11.1).

In marine environments, PHs enter the ocean through natural seeps, accidental oil spills, runoff from land-based sources, and operational discharges from ships and offshore drilling activities. Their persistence and toxicity make them significant environmental contaminants with the potential to disrupt marine ecosystems and pose risks to human health (Sharma et al., 2024). Important sources involve ship discharges, tanker oil spills and non-tank vessels, and operational discharges from aircrafts (Carpenter & Reichelt-Brushett, 2023).

11.1.2 Environmental Impact of Petroleum Hydrocarbons

The release of PHs into marine environments could have detrimental effects on marine life and ecosystems. These impacts include toxicity to marine organisms, bioaccumulation and biomagnification of hydrocarbons in food

TABLE 11.1
Composition of Petroleum Hydrocarbons

Type of Hydrocarbon	General Characteristics	Chemical Structure	Examples	References
Alkanes	Saturated hydrocarbons; less reactive; major component of crude oil	Straight or branched chains of single-bonded carbon atoms (C_nH_{2n+2})	Methane (CH_4), ethane (C_2H_6), octane (C_8H_{18})	Joshi and Mishra (2022)
Cycloalkanes	Saturated rings; less reactive; contribute to the complexity of crude oil	Ring structure single-bonded carbon atoms (C_nH_{2n})	Cyclohexane (C_6H_{12}), methylcyclohexane	Kraus et al. (2020)
Aromatic Hydrocarbons	Contains one or more benzene rings; often more toxic; found in gasoline	Benzene rings with alternating double bonds (C_6H_6)	Benzene (C_6H_6), toluene (C_7H_8), xylene (C_8H_{10})	Laskar and Hashisho (2020)
PAHs (Polycyclic Aromatic Hydrocarbons)	Major organic pollutants with high toxicity; can bioaccumulate in the environment	Multiple fused benzene rings (anthracene, pyrene)	Anthracene ($C_{14}H_{10}$), pyrene ($C_{16}H_{10}$)	El-Bouhy et al. (2024)

DOI: 10.1201/9781003509530-11

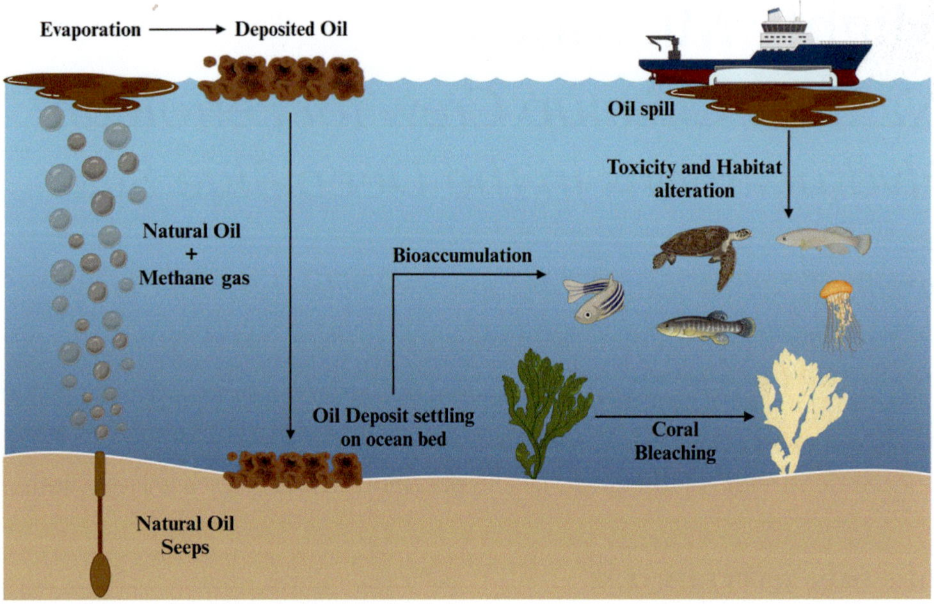

FIGURE 11.1 The combined diagram shows both the sources of petroleum hydrocarbons (natural seeps, oil spills) and their environmental impact on marine life (toxicity, bioaccumulation, habitat alteration). (Self-developed using Biorender.com.)

webs, and alterations to habitat structures and ecosystem functions. This is dependent on concentration and types of hydrocarbons, duration of exposure, and characteristics of the affected environment. The toxicity caused by PHs to marine organisms is influenced by their persistence, bioavailability, and the ability of organisms to metabolize them. Weathering processes can change the composition and toxicity of the oil, with PAHs becoming more significant contributors as monoaromatic compounds decrease. Photodegradation and photoactivation can also affect toxicity (Honda & Suzuki, 2020). These factors collectively determine the ecological impact of oil releases, which in turn destroys the insulation property of furry mammals like sea otters, and the property of water repellence in bird's feathers, which can result in death due to hypothermia. Marine wildlife are especially susceptible to oil spills when the contamination affects their habitats or preys. Chronic exposures to hydrocarbons may also affect their reproduction, resulting in dwindling numbers of their species (Honda & Suzuki, 2020) (Figure 11.1).

11.2 MICROBIAL DEGRADATION OF PETROLEUM HYDROCARBONS

11.2.1 Mechanisms of Microbial Degradation

Microbes in marine environments and their function in processing PH bioremediation by degrading the complex compounds are interrelated and can be the solution to solve most of the environmental problems. One such property is the digestive enzymes that help to digest anything, including organic substances naturally found which cannot be effectively eradicated by microorganisms (Mahmoud, 2022). Degradation mechanisms involve several processes, including:

1. **Alkane Degradation:** Alkanes are degraded through a process of oxidation catalyzed by enzymes such as alkane hydroxylases and monooxygenases, which introduce oxygen into the hydrocarbon molecule to form alcohols and subsequent oxidation products (Guo et al., 2023).

   The alkane hydroxylases go through the first reaction by a rate-limiting step, which involves introduction of oxygen atoms. Examples of these include AlkB, AlkM, cytochrome P450, flavoprotein alkane, and dioxygenase. Among these, cytochrome P450 and integrated membrane alkane hydroxylases have been studied and are structurally more mature with respect to others (Guo et al., 2023).
2. **Integrated Membrane Alkane Hydroxylases:**

   From the species *Pseudomonas putida* of strain GPO1 and the strain ADP-1 from *Acinetobacter calcoaceticus*, they relate to Alk-B and Alk-M as two major system enzymes, respectively. The difference in their function is that Alk-M helped in oxidating long-chain n-alkanes (C_{16}–C_{30}) and Alk-B helped in oxidation of short-chain alkanes (C_5–C_{12}) (Whyte et al., 2002).
3. **Cytochrome P450:**

   The family of cytochrome P450 consisting of alkane hydroxylase are considered as heme-thiolate proteins. It is further categorized into Class I and Class II for its constituents. Class I has CYP153 enzymes from *A. calcoaceticus* strain

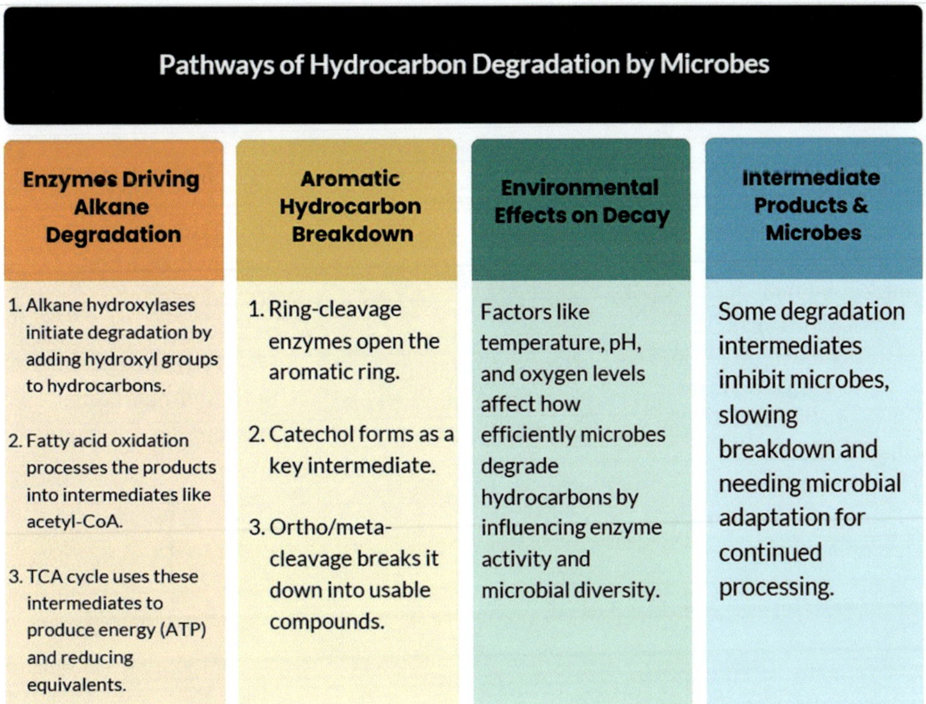

FIGURE 11.2 The flowchart illustrates the microbial degradation pathways for both alkanes and aromatic hydrocarbons. It highlights key enzymes such as alkane hydroxylases for alkanes and ring-cleavage enzymes for aromatic hydrocarbons. (Self-developed using Biorender.com.)

EB104, which oxidizes n-alkanes (C_6–C_{11}) from *Novosphingobium aromaticivorans*. This could be helpful in analyzing metabolic engineering strategies in the direction of synthetic biology (Wang et al., 2023).

4. **Aromatic Hydrocarbon Degradation:**
   Aromatic hydrocarbons, including PAHs, are degraded through processes such as ring cleavage, facilitated by dioxygenases and other specialized enzymes. These pathways lead to the formation of non-toxic compounds along with CO_2 and water (Zylstra & Gibson, 1991). Aromatic hydrocarbons are universally distributed and majority of these compounds form through organic matter decomposition. At high temperature this process leads to formation of non-substituted PAHs. At low temperatures where crude petroleum is formed, this process leads to formation of alkyl-substituted aromatic hydrocarbons (Zylstra & Gibson, 1991).
5. **Reductive Dichlorination:** In the case of chlorinated hydrocarbons, reductive dechlorination involves removing chlorine atoms through dehalogenases, which convert these compounds into non-chlorinated forms. Generally, the remediation technologies can be divided into two categories according to the site difference, *in-situ* and *ex-situ*. The localization of sub-surface pollution usually requires the application of *in-situ* approaches which applies contaminant removal in a natural setting. Bioremediation appears most suitable for *in-situ* treatment that uses the metabolisms of microorganisms for contaminant detoxification as an environmentally effective and feasible approach, as compared to physicochemical techniques (Xiao et al., 2020) (Figure 11.2).

11.2.2 Microbial Taxa Involved in Hydrocarbon Degradation

A diverse array of microbial taxa contributes to the PHs in marine biology. Here are the key players (Figure 11.3):

- **Bacteria:** Genera such as *Pseudomonas*, *Alcanivorax*, and *Bacillus* are known for their ability to diminish several hydrocarbons including alkanes and aromatics. These bacteria often possess specialized enzymatic systems that enable them to metabolize complex hydrocarbons. PH-degrading bacteria are naturally existing everywhere which convert PHs to efficient energy resources. Their feature can be utilized for bioremediation of ecosystems polluted by petroleum. A wide range of bacterial species have been used as bio-degraders to degrade PHs. Degrading aliphatics and polyaromatics demonstrate oxidative reactions, but they differ greatly because different bacterial species have specific oxygenases (Xu et al., 2018).

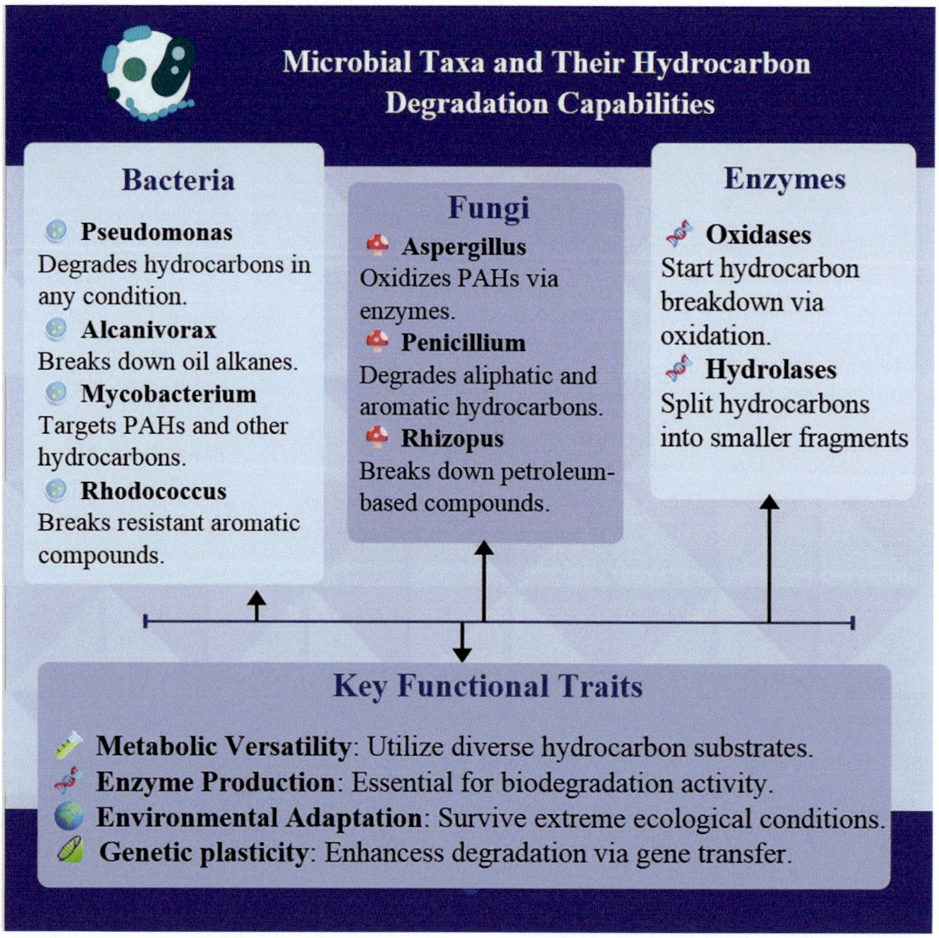

FIGURE 11.3 A schematic showing different microbial taxa (*e.g., Pseudomonas, Aspergillus*) and their hydrocarbon degradation capabilities. This figure integrates information on specific genera, their associated enzymes, and key characteristics. (Self-developed using Biorender.com.)

- **Fungi:** Marine fungi, such as species from the genera *Aspergillus* and *Penicillium*, have demonstrated capabilities for degrading aromatic hydrocarbons and other organic pollutants through oxidative and hydrolytic processes (Mayans et al., 2024).

  Fungi can also help in bioremediation of hydrocarbons for interaction with environmental conditions, including the indigenous population of microbes. It is necessary that treatability assays should be incorporated into initial site characterization process. Through cytochrome P450 monooxygenase complex, fungi can break down hydrocarbons under aerobic conditions, which also catalyzes the aliphatic or aromatic compounds' hydroxylation along with casual oxidation by extracellular oxidases. Edible fungi such as *Agaricus bisporus, Pleurotus ostreatus,* or *Pleurotus eryngii* also produce extracellular oxidases like Lac or MnP, which allow them to oxidize many organic substances along with lignin. The spent mushroom substrate is useful as a waste product rich in organic matter and nutrients after harvesting, which makes it a valuable bioremediation material (Mayans et al., 2024).
- **Archaea:** Marine archaea, particularly those from phylum Euryarchaeota, are known as important contributors in degrading hydrocarbons present in deep-sea environments, where the process of anaerobic degradation happens. Through enrichment cultures, microorganisms that deplete aliphatic-aromatic hydrocarbons in the environments of high-salt concentrations are isolated and identified as aerobic halophilic archaea of the class Halobacteria. Such isolates were obtained from salt marshes and salt ponds in Southern France, coastal sabkhas in Kuwait, Çamalti Saltern in Turkey, Lake Urmia in Iran, an oil field in Tatarstan, and other hypersaline environments (Oren, 2019). Through phylogenetic relationship these isolates occur in *Halobacterium*, *Haloarcula*, *Halococcus*, *Haloferax*, *Halorubrum*, and *Natrialba* genera, which grow at salt concentrations from >10% up to saturation. For them, compounds used as growth substrates are crude oil, C_{10}-C_{34} *n*-alkanes, pristane; and aromatic hydrocarbons such as benzene,

TABLE 11.2
Microbial Taxa and Their Hydrocarbon Degradation Capabilities

Microbial Taxon	Specific Genera	Degradation Pathways/Enzymes	Key Characteristics	References
Bacteria	*Pseudomonas*, *Alcanivorax*, *Bacillus*	Alkane hydroxylases, aromatic dioxygenases, various oxidases	Universal, and help in degrading vast types of hydrocarbons	Brzeszcz and Kaszycki (2018)
Fungi	*Aspergillus*, *Penicillium*	Cytochrome P450 monooxygenases, extracellular oxidases (laccases, MnP)	Effective in degrading aromatic hydrocarbons; thrive in aerobic conditions	Essabri et al. (2019)
Archaea	*Halobacterium*, *Haloarcula*, *Halococcus*	Hydrocarbon utilization pathways; enzymes for high-salt environments	Adapted to extreme conditions; degrade hydrocarbons in anaerobic/hypersaline environments	Oren (2019)

toluene, naphthalene, acenaphthene, phenanthrene, anthracene, and pyrene. But none of the strains are biochemically documented, and none of the isolates are currently available from open access cultures (Oren, 2019) (Table 11.2).

11.3 FACTORS INFLUENCING MICROBIAL DEGRADATION

11.3.1 Environmental Conditions

The efficiency of microbial degradation of PHs is influenced by various environmental conditions, including:

1. **Temperature:** Temperature has a significant role in metabolic activity of marine microbes and the solubility of hydrocarbons. Optimal temperature ranges enhance microbial degradation rates, while extreme temperatures can inhibit microbial activity and have a marked effect on hydrocarbon degradation. The effects of temperature are interactive with other factors, such as the quality of the hydrocarbon mixture and the composition of the microbial community. Hydrocarbon biodegradation can also occur at low temperatures (Atlas, 1981).
2. **Oxygen Availability:** Oxygen is critical for aerobic degradation of hydrocarbons. Anoxic environments, such as deep-sea sediments, may require alternative electron acceptors for microbial degradation, such as nitrate or sulfate. The importance of oxygen for hydrocarbon degradation can be seen in most degradative pathways in saturated aromatic hydrocarbons that involve oxygenases and molecular oxygen (Atlas, 1981).
3. **Nutrient Availability**: Nitrogen and phosphorus influence the depletion of hydrocarbons. Nutrient limitations can slow down microbial growth and activity, whereas nutrient enrichment can enhance microbial degradation processes. In some cases, nutrient amendments are used to stimulate microbial activity and accelerate bioremediation efforts. The available concentrations of these gases after major oil spills in sea severely limit the extent of degradation of hydrocarbons. Insufficient nutrient replenishing rate cannot support frequent biodegradation of oil spills in large amount. Fertilizers having N and P could be used to stimulate hydrocarbon degradation at microbial level (Atlas, 1981).
4. **Salinity:** The salt concentration in marine environments can impact microbial community composition and the availability of hydrocarbons. Some microorganisms are adapted to high-salinity conditions and can effectively degrade hydrocarbons in these environments. Adsorption method for treatment of salinity has been widely used, along with membrane separation, ion exchange, or electrodialysis. But due to high cost and secondary pollution these technologies are only applied in limited conditions. Other biological treatments like activated sludge process can be favored because of their better economic performance and sustainability (He et al., 2017) (Figure 11.4).

11.4 BIOREMEDIATION STRATEGIES FOR PETROLEUM HYDROCARBONS

11.4.1 *In-Situ* Bioremediation

This process involves bioremediation techniques application directly in the contaminated environment. By definition, it involves biologically treating groundwater and contaminated soil without soil excavation or treatment of groundwater above soil. It relies on the use of plants and to immobilize the contaminants *in-situ*, bringing together a combination of microbiological, chemical, geological, and engineering sciences (Jørgensen, 2007).

These strategies include:

1. **Nutrient Addition:** This helps in stimulating the growth and activity of hydrocarbon-degrading microorganisms. The rationale behind nutrient addition is to provide the necessary nutrients that support the growth and metabolic activity of

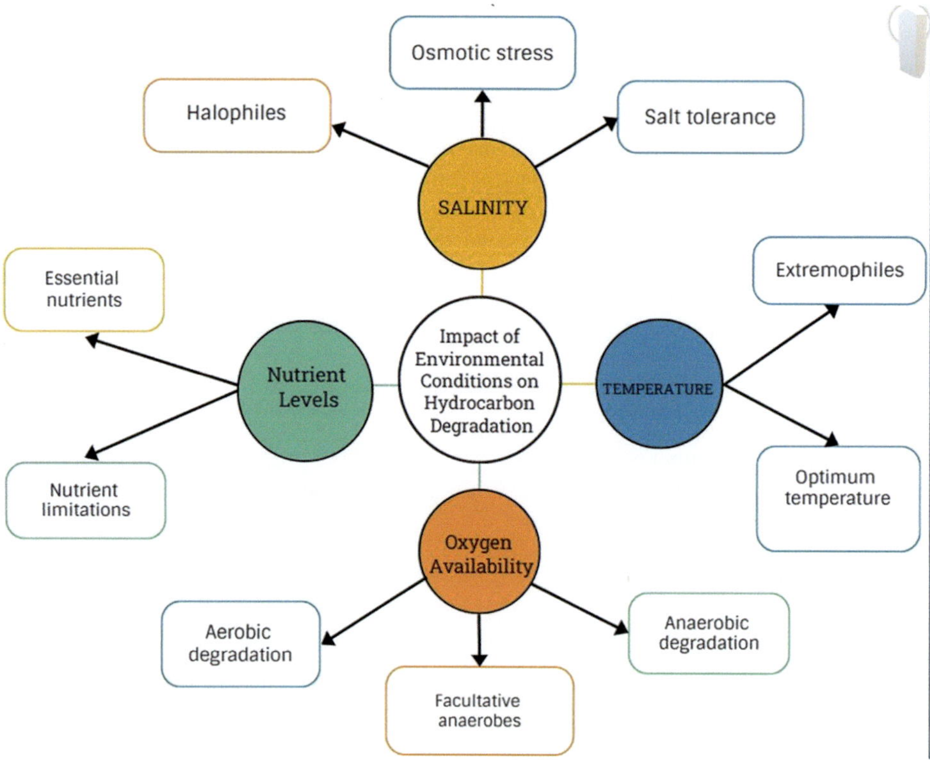

FIGURE 11.4 A diagram illustrating how different environmental factors (temperature, salinity, oxygen availability, nutrient levels) affect microbial hydrocarbon degradation. This figure could be a flowchart or a set of interconnected graphs. (Self-developed using Biorender.com.)

native or introduced microorganisms to accelerate the breakdown of contaminants (Wu et al., 2019).

2. **Bioaugmentation:** Bio-stimulation involves the modification of the environment to stimulate existing microbes capable of bioremediation. This is done by addition of various forms of limiting nutrients and electron acceptors, such as phosphorus, nitrogen, oxygen, or carbon (in the form of molasses), which are otherwise available in quantities low enough to constrain microbial activity (Omokhagbor Adams et al., 2020).
3. **Biostimulation:** It is the process of modifying environmental conditions for promoting growth of indigenous hydrocarbon-degrading microbes, such as adjusting pH or oxygen levels. The rationale for this approach is that indigenous microbial populations may not be capable of degrading the wide range of potential substrates present in complex mixtures such as petroleum or that they may be in a stressed state as a result of the recent exposure to the spill. For this approach to be successful in the field, the seed microorganisms must be able to degrade most petroleum components, maintain genetic stability and viability during storage, survive in foreign and hostile environments, effectively compete with indigenous microorganisms, and move through the pores of the sediment to the contaminants (Omokhagbor Adams et al., 2020).

11.4.2 *Ex-Situ* Bioremediation

This process involves treating attenuated materials in a controlled setting, such as bioreactors or composting facilities. It involves extracting pollutants from resources which are contaminated and transporting them to another site for treatment. This process is usually based on the cost of treatment, type of pollutant, degree of pollution, geographical location, and geology of the polluted site (Paul et al., 2021).

This technique includes:

1. **Bioreactors:** They use controlled reactors to speed up hydrocarbon degradation through optimized environmental conditions and microbial inoculation. Bioreactors offer a controlled environment for bioremediation, allowing for precise manipulation of factors like temperature, pH, and aeration. This control enhances pollutant degradation efficiency and reduces treatment time. While bioreactors can be costly, they provide several advantages, including the ability to treat various pollutants, such as BTEX, and the potential to use genetically modified organisms safely. However, optimizing bioreactor performance requires a careful balance of factors and may be challenging due to the complexity of biological processes (Paul et al., 2021).
2. **Slurry Bioreactors:** They are highly controlled systems used for soil bioremediation. They offer

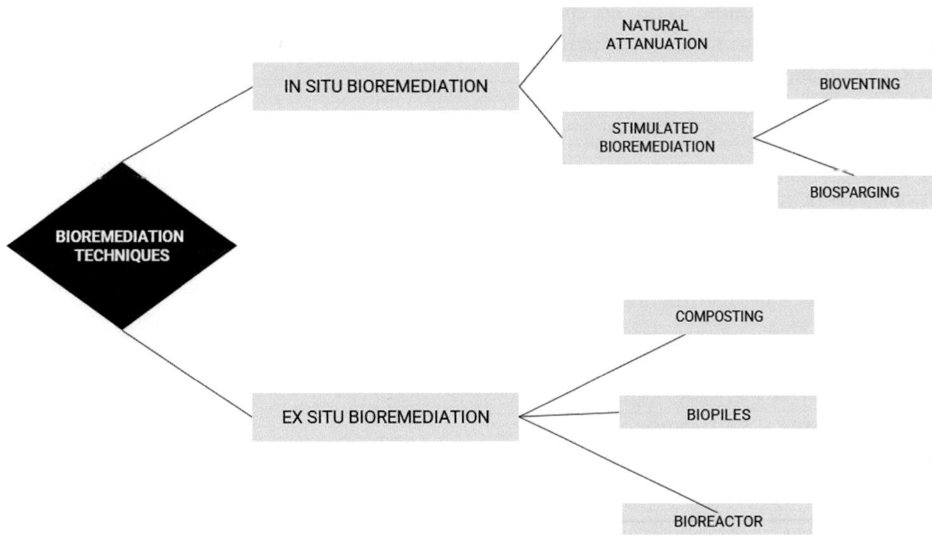

FIGURE 11.5 A flowchart that outlines the different techniques of bioremediation like *in-situ* and *ex-situ*, and their specific applications and processes. (Self-developed using Biorender.com.)

precise control over operating parameters but are expensive. These reactors mix water and contaminated soil to create a slurry, then it is treated aerobically, anoxically, or anaerobically. They are often used in batch or sequencing batch modes. Pre-treatment of soil is necessary to improve efficiency. Slurry bioreactors have been successfully used for removing various contaminants, including PAHs. Anaerobic processes are gaining popularity for PAH removal, especially in soils with low oxygen permeability. Rotating drum bioreactors (RDBs) offer improved mixing and mass transfer efficiency. Slurry bioreactors have also been applied to treat other contaminants like herbicides, pesticides, and explosives (Tomei & Daugulis, 2013). Slurry bioreactors have been used successfully to treat various contaminants, including herbicides, pesticides, and explosives. Studies on pendimethalin and 2,4-dichlorophenoxyacetic acid have shown their effectiveness in bioremediation. Lindane has also been successfully treated using slurry bioreactors, with different operating conditions and co-substrates affecting efficiency. Nitroaromatics like TNT have been effectively removed in slurry bioreactors compared to landfarming. Treating arsenic-contaminated soil using a dissimilatory arsenate-reducing bacterium is considered as an unconventional application (Romantschuk et al., 2023).

3. **Composting:** Composting utilizes aerobic decomposition processes to degrade hydrocarbons in contaminated soils or sediments. The principle involves mixing the contaminated soil with non-hazardous organic elements or solid wastes like manure, agricultural waste suitable for composting applications, to encouraging bacterial population development to degrade pollutants in the soil. The process happens under aerobic conditions and uses heat generated during oxidative exothermic reactions to cause temperature increase of about 50°C–60°C. The production of mature compost can be used in land restoration for industrial, municipal, or housing work as well (Tomei & Daugulis, 2013).

11.4.3 Bioremediation Case Studies

Several case studies illustrate the successful application of microbial bioremediation techniques for PHs (Gogoi et al., 2003). Examples include the treatment of oil spills using oil-degrading bacteria, bioremediation of contaminated sediments through bioreactors, and using bioaugmentation to enhance hydrocarbon degradation in polluted marine environments (Gouda et al., 2008) (Figure 11.5).

11.5 CHALLENGES AND FUTURE PERSPECTIVES

11.5.1 Challenges in Bioremediation

Several challenges impact effectiveness of microbial bioremediation of PHs. They are very stable and not easy to degrade, can remain for longer periods which could damage our ecosystem. Biodegradability of hydrocarbons is also a challenge because being hydrophobic and insoluble in water, they can be non-bioavailable to microbes (Radhakrishnan et al., 2023). Bioremediation could be a useful, cost-effective, and sustainable technique to degrade these contaminants, including:

1. **Complexity of Hydrocarbons:** The complexity and variability of PHs can make it difficult to identify and target specific degradation pathways (Radhakrishnan et al., 2023).
2. **Environmental Variability:** The diverse and often harsh conditions of marine environments can affect microbial activity and the success of bioremediation efforts.
3. **Formation of Toxic By-products:** The degradation of hydrocarbons can lead to intermediate by-product formation that may be toxic to the environment (Pandolfo et al., 2023).
4. **Limited Degradability of Certain Pollutants:**
   - **Persistent Organic Pollutants (POPs):** Metal contaminants, chlorinated compounds, and PAHs are resistant to degradation. These compounds may not be easily broken down by microorganisms or plants (Tufail et al., 2022).
   - **Complexity of Mixed Contaminants:** Sites often contain a mixture of pollutants, which can interact in ways that hinder the degradation process (Goutam Mukherjee et al., 2022).
5. **Bioavailability of Contaminants:**
   - **Limited Accessibility:** Contaminants may be trapped in soil particles or within the pores of rocks, making them inaccessible to microorganisms or plants. This reduces the efficiency of bioremediation efforts (Padhye et al., 2023).
   - **Hydrophobic Pollutants:** Certain pollutants, like PAHs, are hydrophobic and tend to adsorb strongly to soil particles, further limiting their bioavailability (Patel et al., 2020).

11.6 DISCUSSION

Microbes have an important role in bioremediation of PHs in marine environments, serving as nature's cleanup crew for these persistent pollutants.

- **Petroleum hydrocarbons** pose a significant threat to marine ecosystems. Their release into the environment, primarily through accidental spills, can have devastating consequences. The nature and sources of these hydrocarbons vary, but their detrimental effects on organisms and ecosystems are undeniable (M. Alzahrani & Rajendran, 2019).
- **Microbial degradation** offers promise to mitigate the impact of PH pollution. However, factors such as nutrient availability, oxygen levels, and temperature can influence the efficiency of this process. Bioremediation strategies, including bio-stimulation and bioaugmentation, have been developed to enhance microbial activity and accelerate hydrocarbon degradation (Chunyan et al., 2023). Through a report, it has been suggested that locally produced kenaf core (Ifeken-100) is helpful in the bioremediation of soils polluted by crude oil (Omenna et al., 2024).

11.7 FUTURE RESEARCH DIRECTIONS

Future research in microbial bioremediation of PHs will focus on:

1. **Enhanced Strain Development:** Engineering microbial strains with improved degradation capabilities and resistance to environmental stressors (Rafeeq et al., 2023).
2. **Integrated Approaches:** Combining bioremediation with other pollution control methods, such as physical and chemical treatments, to address complex contamination scenarios (Koshlaf & S Ball, 2017).
3. **Monitoring and Assessment:** Developing advanced monitoring and assessment tools to evaluate the effectiveness of bioremediation and the potential impacts on marine ecosystems (Das & Chandran, 2011).
4. **Consortia Development:** Future research will explore the use of microbial consortia, where different species work together to degrade PHs more effectively than individual strains. The focus will be on understanding the synergistic interactions between these microorganisms (Cao et al., 2022).
5. **Quorum Sensing and Communication:** Understanding and manipulating microbial communication (quorum sensing) to enhance cooperative behavior in microbial communities will be a significant research focus (Stephens & Bentley, 2020).

11.8 CONCLUSION

The complexity of PHs, the variability of environmental conditions, and the potential for long-term effects on ecosystems necessitate ongoing research and development. Future perspectives include exploring naive microbial strains with enhanced degradation capabilities, optimization of bioremediation techniques, and integration of advanced technologies for more effective pollution control. By addressing such challenges and harnessing the benefits of microbial degradation, a more sustainable and resilient marine environment can be developed. Use of bacteria which are genetically altered represents a research direction with broad implications. But their need may be questionable, considering that indigenous species perform effectively but full potential of wild species is not realized fully due to limited understanding of various phytoremediation approaches, including the systems of enzymes that help degrade pollutants. Through their diverse enzymatic systems and metabolic pathways, microbes contribute to the degradation and transformation of hydrocarbons, mitigating their environmental impact. Addressing the challenges of bioremediation and exploring future research directions will enhance our ability to manage PH contamination and protect marine ecosystems. The continued advancement of microbial bioremediation technologies holds promise for improving environmental health and sustainability in the deep blue sea.

ABBREVIATIONS

AlkB, AlkM	Alkane Hydroxylases
BTEX	Benzene, Toluene, Ethylbenzene, and Xylene
CYP153	Cytochrome P450 alkane hydroxylase
N and P	Nitrogen and Phosphorus
PAHs	Polycyclic Aromatic Hydrocarbons
PHs	Petroleum Hydrocarbons
RDBs	Rotating drum bioreactors
SHs	Saturated Hydrocarbons
TNT	2,4,6-trinitrotoluene

ACKNOWLEDGEMENT

The authors gratefully acknowledge the financial support provided by the Department of Biotechnology (DBT), Government of India, under Project No. DBT/2022-23/July/N/2240, as part of the DBT-RA Program. We also extend our appreciation to the School of Bioengineering and Biosciences, Lovely Professional University, Punjab, India, for their essential support and resources throughout this research.

REFERENCES

Atlas, R. M. (1981). Microbial degradation of petroleum hydrocarbons: an environmental perspective. *Microbiological Reviews*, *45*(1), 180–209. https://doi.org/10.1128/mr.45.1.180-209.1981

Alzahrani, A. M., & Rajendran, P. (2019). Petroleum hydrocarbon and living organisms. In Muharrem Ince, Olcay Kaplan Ince (eds.), *Hydrocarbon Pollution and its Effect on the Environment*. IntechOpen. https://doi.org/10.5772/intechopen.86948

Brzeszcz, J., & Kaszycki, P. (2018). Aerobic bacteria degrading both n-alkanes and aromatic hydrocarbons: an undervalued strategy for metabolic diversity and flexibility. *Biodegradation*, *29*(4), 359–407. https://doi.org/10.1007/s10532-018-9837-x

Cao, Z., Yan, W., Ding, M., & Yuan, Y. (2022). Construction of microbial consortia for microbial degradation of complex compounds. *Frontiers in Bioengineering and Biotechnology*, *10*, 1051233. https://doi.org/10.3389/fbioe.2022.1051233

Carpenter, A., & Reichelt-Brushett, A. (2023). Oil and gas. In Amanda Reichelt-Brushett (ed.), *Marine Pollution – Monitoring, Management and Mitigation* (pp. 129–153). Springer Nature. https://doi.org/10.1007/978-3-031-10127-4_6

Chunyan, X., Qaria, M. A., Qi, X., & Daochen, Z. (2023). The role of microorganisms in petroleum degradation: current development and prospects. *Science of the Total Environment*, *865*, 161112. https://doi.org/10.1016/j.scitotenv.2022.161112

Das, N., & Chandran, P. (2011). Microbial degradation of petroleum hydrocarbon contaminants: an overview. *Biotechnology Research International*, 2011, 1–13. https://doi.org/10.4061/2011/941810

El-Bouhy, Z., Reda, R., Mohamed, F., Elashhab, M., & Abdel Rahman, A. (2024). Polycyclic aromatic hydrocarbons (PAHs) pollution approaches in aquatic ecosystems: perils and remedies using green technologies. *Zagazig Veterinary Journal*, *52*(1), 49–87. https://doi.org/10.21608/zvjz.2024.239217.1221

Essabri, A. M. A., Aydinlik, N. P., & Williams, N. E. (2019). Bioaugmentation and biostimulation of total petroleum hydrocarbon degradation in a petroleum-contaminated soil with fungi isolated from olive oil effluent. *Water, Air, & Soil Pollution*, 230(3), 76. https://doi.org/10.1007/s11270-019-4127-8

Gogoi, B. K., Dutta, N. N., Goswami, P., & Krishna Mohan, T. R. (2003). A case study of bioremediation of petroleum-hydrocarbon contaminated soil at a crude oil spill site. *Advances in Environmental Research*, *7*(4), 767–782. https://doi.org/10.1016/S1093-0191(02)00029-1

Gouda, M. K., Omar, S. H., Nour Eldin, H. M., & Chekroud, Z. A. (2008). Bioremediation of kerosene II: a case study in contaminated clay (laboratory and field: scale microcosms). *World Journal of Microbiology and Biotechnology*, *24*(8), 1451–1460. https://doi.org/10.1007/s11274-007-9634-x

GoutamMukherjee, A., Ramesh Wanjari, U., Eladl, M. A., El-Sherbiny, M., Elsherbini, D. M. A., Sukumar, A., Kannampuzha, S., Ravichandran, M., Renu, K., Vellingiri, B., Kandasamy, S., & Valsala Gopalakrishnan, A. (2022). Mixed contaminants: occurrence, interactions, toxicity, detection, and remediation. *Molecules*, *27*(8), 2577. https://doi.org/10.3390/molecules27082577

Guo, X., Zhang, J., Han, L., Lee, J., Williams, S. C., Forsberg, A., Xu, Y., Austin, R. N., & Feng, L. (2023). Structure and mechanism of the alkane-oxidizing enzyme AlkB. *Nature Communications*, *14*(1), 2180. https://doi.org/10.1038/s41467-023-37869-z

He, H., Chen, Y., Li, X., Cheng, Y., Yang, C., & Zeng, G. (2017). Influence of salinity on microorganisms in activated sludge processes: a review. *International Biodeterioration & Biodegradation*, *119*, 520–527. https://doi.org/10.1016/j.ibiod.2016.10.007

Honda, M., & Suzuki, N. (2020). Toxicities of polycyclic aromatic hydrocarbons for aquatic animals. *International Journal of Environmental Research and Public Health*, *17*(4), 1363. https://doi.org/10.3390/ijerph17041363

Jørgensen, K. S. (2007). In situ bioremediation. In Allen I. Laskin, Sima SariaslanI and Geoffrey M. Gadd (eds.), *Advances in applied microbiology*, *61*, (pp. 285–305). https://doi.org/10.1016/S0065-2164(06)61008-3

Joshi, S., & Mishra, S. (2022). Recent advances in biofuel production through metabolic engineering. *Bioresource Technology*, *352*, 127037. https://doi.org/10.1016/j.biortech.2022.127037

Koshlaf, E., & S Ball, A. (2017). Soil bioremediation approaches for petroleum hydrocarbon polluted environments. *AIMS Microbiology*, *3*(1), 25–49. https://doi.org/10.3934/microbiol.2017.1.25

Kraus, F., Martins Oliveira, A. A., Bueno Bontorin, A. C., Rouge dos Santos, A., & Cancino, L. R. (2020). i-pentane and i-pentene as gasoline surrogates for the transport industry: a numerical analysis on ignition delay times using detailed chemical kinetics. *Proceedings of the 18th Brazilian Congress of Thermal Sciences and Engineering,* Brazil. https://doi.org/10.26678/ABCM.ENCIT2020.CIT20-0512

Laskar, I. I., & Hashisho, Z. (2020). Insights into modeling adsorption equilibria of single and multicomponent systems of organic and water vapors. *Separation and Purification Technology*, *241*, 116681. https://doi.org/10.1016/j.seppur.2020.116681

Liu, Q., Xia, C., Wang, L., & Tang, J. (2022). Fingerprint analysis reveals sources of petroleum hydrocarbons in soils of different geographical oilfields of China and its ecological assessment. *Scientific Reports*, *12*(1), 4808. https://doi.org/10.1038/s41598-022-08906-6

Mahmoud, D. A. R. (2022). Mechanism of microbial biodegradation: secrets of biodegradation. In Gomaa A. M. Ali, Abdel Salam H. Makhlouf (eds.), *Handbook of Biodegradable Materials* (pp. 1–15). Springer International Publishing. https://doi.org/10.1007/978-3-030-83783-9_6-1

Mayans, B., Antón-Herrero, R., García-Delgado, C., Delgado-Moreno, L., Guirado, M., Pérez-Esteban, J., Escolástico, C., & Eymar, E. (2024). Bioremediation of petroleum hydrocarbons polluted soil by spent mushroom substrates: microbiological structure and functionality. *Journal of Hazardous Materials*, *473*, 134650. https://doi.org/10.1016/j.jhazmat.2024.134650

Omenna, E. C., Omage, K., Ezaka, E., & Azeke, M. A. (2024). Bio-augmentation and bio-stimulation with kenaf core enhanced bacterial enzyme activities during bio-degradation of petroleum hydrocarbon in polluted soil. *Scientific Reports*, *14*(1), 8. https://doi.org/10.1038/s41598-023-50882-y

OmokhagborAdams, G., Tawari Fufeyin, P., Eruke Okoro, S., & Ehinomen, I. (2020). Bioremediation, biostimulation and bioaugmention: a review. *International Journal of Environmental Bioremediation & Biodegradation*, *3*(1), 28–39. https://doi.org/10.12691/ijebb-3-1-5

Oren, A. (2019). Aerobic hydrocarbon-degrading archaea. In Terry J. McGenity (ed.), *Taxonomy, Genomics and Ecophysiology of Hydrocarbon-Degrading Microbes* (pp. 41–51). Springer International Publishing. https://doi.org/10.1007/978-3-030-14796-9_5

Padhye, L. P., Srivastava, P., Jasemizad, T., Bolan, S., Hou, D., Shaheen, S. M., Rinklebe, J., O'Connor, D., Lamb, D., Wang, H., Siddique, K. H. M., & Bolan, N. (2023). Contaminant containment for sustainable remediation of persistent contaminants in soil and groundwater. *Journal of Hazardous Materials*, *455*, 131575. https://doi.org/10.1016/j.jhazmat.2023.131575

Pandolfo, E., Barra Caracciolo, A., & Rolando, L. (2023). Recent advances in bacterial degradation of hydrocarbons. *Water*, *15*(2), 375. https://doi.org/10.3390/w15020375

Patel, A. B., Shaikh, S., Jain, K. R., Desai, C., & Madamwar, D. (2020). Polycyclic aromatic hydrocarbons: sources, toxicity, and remediation approaches. *Frontiers in Microbiology*, 11, 562813. https://doi.org/10.3389/fmicb.2020.562813

Paul, O., Jasu, A., Lahiri, D., Nag, M., & Ray, R. R. (2021). In situ and ex situ bioremediation of heavy metals: the present scenario. *Journal of Environmental Engineering and Landscape Management*, *29*(4), 454–469. https://doi.org/10.3846/jeelm.2021.15447

Radhakrishnan, A., Balaganesh, P., Vasudevan, M., Natarajan, N., Chauhan, A., Arora, J., Ranjan, A., Rajput, V. D., Sushkova, S., Minkina, T., Basniwal, R. K., Kapardar, R., & Srivastav, R. (2023). Bioremediation of hydrocarbon pollutants: recent promising sustainable approaches, scope, and challenges. *Sustainability*, *15*(7), 5847. https://doi.org/10.3390/su15075847

Rafeeq, H., Afsheen, N., Rafique, S., Arshad, A., Intisar, M., Hussain, A., Bilal, M., & Iqbal, H. M. N. (2023). Genetically engineered microorganisms for environmental remediation. *Chemosphere*, *310*, 136751. https://doi.org/10.1016/j.chemosphere.2022.136751

Romantschuk, M., Lahti-Leikas, K., Kontro, M., Galitskaya, P., Talvenmäki, H., Simpanen, S., Allen, J. A., & Sinkkonen, A. (2023). Bioremediation of contaminated soil and groundwater by in situ biostimulation. *Frontiers in Microbiology*, *14*, 1258148. https://doi.org/10.3389/fmicb.2023.1258148

Sharma, K., Shah, G., Singh, H., Bhatt, U., Singhal, K., & Soni, V. (2024). Advancements in natural remediation management techniques for oil spills: challenges, innovations, and future directions. *Environmental Pollution and Management*, *1*, 128–146. https://doi.org/10.1016/j.epm.2024.08.003

Stephens, K., & Bentley, W. E. (2020). Synthetic biology for manipulating quorum sensing in microbial consortia. *Trends in Microbiology*, *28*(8), 633–643. https://doi.org/10.1016/j.tim.2020.03.009

Tomei, M. C., & Daugulis, A. J. (2013). Ex situ bioremediation of contaminated soils: an overview of conventional and innovative technologies. *Critical Reviews in Environmental Science and Technology*, *43*(20), 2107–2139. https://doi.org/10.1080/10643389.2012.672056

Tufail, M. A., Iltaf, J., Zaheer, T., Tariq, L., Amir, M. B., Fatima, R., Asbat, A., Kabeer, T., Fahad, M., Naeem, H., Shoukat, U., Noor, H., Awais, M., Umar, W., & Ayyub, M. (2022). Recent advances in bioremediation of heavy metals and persistent organic pollutants: a review. *Science of the Total Environment*, *850*, 157961. https://doi.org/10.1016/j.scitotenv.2022.157961

Wang, M., Ding, M., & Yuan, Y. (2023). Bioengineering for the microbial degradation of petroleum hydrocarbon contaminants. *Bioengineering*, *10*(3), 347. https://doi.org/10.3390/bioengineering10030347

Whyte, L. G., Smits, T. H. M., Labbé, D., Witholt, B., Greer, C. W., & van Beilen, J. B. (2002). Gene cloning and characterization of multiple alkane hydroxylase systems in Rhodococcus strains Q15 and NRRL B-16531. *Applied and Environmental Microbiology*, *68*(12), 5933–5942. https://doi.org/10.1128/AEM.68.12.5933-5942.2002

Wu, M., Wu, J., Zhang, X., & Ye, X. (2019). Effect of bioaugmentation and biostimulation on hydrocarbon degradation and microbial community composition in petroleum-contaminated loessal soil. *Chemosphere*, *237*, 124456. https://doi.org/10.1016/j.chemosphere.2019.124456

Xiao, Z., Jiang, W., Chen, D., & Xu, Y. (2020). Bioremediation of typical chlorinated hydrocarbons by microbial reductive dechlorination and its key players: a review. *Ecotoxicology and Environmental Safety*, *202*, 110925. https://doi.org/10.1016/j.ecoenv.2020.110925

Xu, X., Liu, W., Tian, S., Wang, W., Qi, Q., Jiang, P., Gao, X., Li, F., Li, H., & Yu, H. (2018). Petroleum hydrocarbon-degrading bacteria for the remediation of oil pollution under aerobic conditions: a perspective analysis. *Frontiers in Microbiology*, *9*, 2885. https://doi.org/10.3389/fmicb.2018.02885

Zylstra, G. J., & Gibson, D. T. (1991). Aromatic hydrocarbon degradation: a molecular approach. In Jane K. Setlow (ed.), *Genetic Engineering* (pp. 183–203). Springer US. https://doi.org/10.1007/978-1-4615-3760-1_8

12 Unleashing the Power of Omics
Revolutionizing Bioremediation for a Greener Tomorrow

Anupriya Eswaran[1], *Ponnulakshmi Rajagopal*[2], *Vishnu Priya Veeraraghavan*[1], *and Selvaraj Jayaraman*[1]

[1]Centre of Molecular Medicine and Diagnostics (COMManD), Department of Biochemistry, Saveetha Dental College & Hospital, Saveetha Institute of Medical & Technical Sciences, Saveetha University, Chennai, India

[2]Department of Central Research Laboratory, Meenakshi Ammal Dental College and Hospitals, Meenakshi Academy of Higher Education and Research, Deemed to be University, Chennai, India

12.1 INTRODUCTION

12.1.1 Overview of Bioremediation: Principles and Practices

Bioremediation is a green technique that breaks down, detoxifies, or changes hazardous environmental pollutants using biological organisms, especially microbes. This approach makes use of the natural metabolic processes of microbes, which have developed to use contaminants as sources of energy and nutrients, converting them into chemicals that are either less hazardous or even innocuous. As a sustainable and affordable substitute for conventional remediation approaches like chemical or physical ones, bioremediation has grown in importance over time as a vital instrument in environmental management (Vidali, 2001). A key component of bioremediation's success is bioavailability, which measures how easily a contaminant can be broken down by microorganisms.

For bioremediation to be successful, the pollutant must be in a form that microorganisms can absorb and metabolize (Boopathy, 2000). Generally, bioavailability is impacted due to various factors including the contaminant solubility, existence of competing substances, and the characteristics of the contaminated site. Precisely, biostimulation and bioaugmentation are two common strategies employed to enhance bioremediation (Table 12.1).

12.1.2 Types of Bioremediation

Bioremediation methods are broadly classified into *in-situ* and *ex-situ* techniques, depending on whether the treatment occurs at the site of contamination or if the contaminated material is removed for treatment elsewhere (Vidali, 2001). *In-situ* bioremediation is a method which involves treating the contaminated material in its original location without excavation. It involves various techniques including bioaugmentation, bioventing, biosparging, biostimulation, bioslurping, and phytoremediation. Biostimulation involves modifying the environment to stimulate the existing microbial population,

TABLE 12.1
Overview of Omics Technologies in Bioremediation

Omics Technology	Key Features	Applications in Bioremediation	Examples of Case Studies	References
Genomics	DNA sequencing, genome mapping	Identification of pollutant-degrading microorganisms	Oil spill cleanup using microbial genomics	Kumar et al. (2023)
Metagenomics	Analysis of microbial communities	Assessment of microbial diversity in contaminated sites	Metagenomic profiling of wastewater treatment	Herrera-Calderon et al. (2024)
Proteomics	Protein expression analysis	Identification of enzymes involved in pollutant degradation	Enzyme-based bioremediation of heavy metals	Singh et al. (2020)
Metabolomics	Metabolic profiling	Mapping of biochemical pathways in pollutant degradation	Metabolomic analysis of PCB degradation	Pomfret et al. (2019)
Transcriptomics	Gene expression studies	Understanding microbial response to environmental stress	Transcriptomic analysis of arsenic-resistant bacteria	Tripathi and Poluri (2021)

DOI: 10.1201/9781003509530-12

often by adding nutrients or altering environmental conditions like pH and temperature (Gentry et al., 2004). For example, in the case of oil spills, adding fertilizers can enhance the degradation of hydrocarbons by indigenous microorganisms, thereby enhancing indigenous microbes by adding nutrients to contaminated soil or groundwater, effectively treating hydrocarbons and metal contamination (Boopathy, 2000; Kao et al., 2008). In contrast, bioaugmentation entails the introduction of particular strains of microorganisms that have been shown to break down particle pollutants, particularly in situations when the local microbial community is deficient in the requisite metabolic processes. This technique is frequently used to remediate microbiological wastes in soil containing aromatic and chlorinated hydrocarbons as well as municipal wastewater (Gentry et al., 2004; Sharma, 2012). Additionally, another method called "bioventing" includes adding oxygen to the soil to boost aerobic microbial activity, and "biosparging" entails creating a saturated zone in order to encourage the aerobic biodegradation of pollutants in groundwater (Singh and Ward, 2004). Bioslurping uses vacuum-enhanced pumping, bioventing, and soil vapor extraction to boost oxygen and microbial activity, removing contaminants from soil and groundwater. Its efficiency may be diminished by low soil permeability. For organic pollutants that are volatile or semivolatile, this technique is utilized (Vidali, 2001). An additional *in-situ* technique called phytoremediation employs plants to absorb and purify pollutants from water and soil (Gentry et al., 2004). *Ex-situ* bioremediation methods involve the physical removal of contaminated material from its original location for treatment in a separate, controlled environment. This approach allows for more precise management of the remediation process and can be tailored to the specific characteristics of the contaminated material. Several *ex-situ* techniques are commonly used, including biopiling, bioreactors, biofilters, and land farming. Using biopiling excavated soil is stacked into heaps and given several treatments to speed up the bioremediation process. By using biostimulation techniques, biopiling aims to maximize the environment for the microbial breakdown of pollutants. This frequently entails enriching the soil pile with nutrients like phosphorus and nitrogen in order to stimulate microbial activity. The soil is routinely mixed to maintain adequate aeration and moisture, essential for supporting microbial activity and contaminant degradation. The efficiency of biopiling can be affected by various conditions such as soil content and composition, pollutant characteristics, and environmental regulations (Vidali, 2001). Bioreactors are engineered systems designed to treat contaminated material under controlled conditions. These systems can handle various types of waste, including soil, sludge, and liquid waste, and provide a controlled environment that maximizes the efficiency of pollutant degradation. In a bioreactor, factors such as temperature, pH, oxygen levels, and nutrient concentrations can be precisely regulated to enhance microbial activity and optimize the degradation of contaminants (Singh and Ward, 2004). This controlled setting allows for continuous monitoring and adjustment of the remediation process, ensuring that it operates at peak efficiency. An efficient *ex-situ* bioremediation method for gaseous pollutants such as volatile organic compounds (VOCs) is the use of biofilters. They entail passing contaminated air through a column that is filled with a porous material, such as compost or activated carbon. Microorganisms infiltrate the medium of the biofilter when air passes through it, breaking down the contaminants. Optimizing variables including air flow rate, filter media composition, and microbe selection can increase the effectiveness of biofilters (Boopathy, 2000; Vidali, 2001). Land farming is often utilized for treating hydrocarbon-contaminated soil and other organic pollutants. Each *ex-situ* bioremediation technique comes with unique advantages and limitations, with its effectiveness largely determined by the properties of the contaminated material and the type of pollutants present.

By removing the contaminated material from its original location and treating it under controlled conditions, *ex-situ* methods offer the advantage of more precise management of the remediation process and can often achieve higher levels of pollutant degradation compared to *in-situ* methods. However, they also involve additional costs and logistical considerations related to the excavation, transport, and treatment of contaminated material (Vidali, 2001) (Figure 12.1).

12.2 ENVIRONMENTAL CHALLENGES AND THE NEED FOR BIOREMEDIATION

Environmental contamination due to industrialization, urbanization, and agricultural practices contains challenges to ecosystem stability and human health. Environmental compartments such as soil, water, and air are loaded with contaminants such as pesticides, petroleum hydrocarbons, heavy metals, and industrial solvents. Pollutants like pesticides, petroleum hydrocarbons, heavy metals, and industrial solvents are ubiquitous in soil, water, and the air, among other environmental compartments (Sharma et al., 2022). Traditional remediation methods, while effective in some cases, are often costly, resource-intensive, and can cause secondary pollution (Boopathy, 2000). This highlights the need for sustainable and eco-friendly remediation approaches, where bioremediation plays a critical role. Traditional remediation methods like soil excavation, chemical oxidation, and thermal desorption, while effective, are associated with significant drawbacks. Soil excavation and off-site disposal, for instance, are effective but can be costly and environmentally disruptive, potentially leading to habitat destruction and secondary pollution during transport (Vidali, 2001). Chemical treatments can result in toxic byproducts that require further treatment, while physical methods often

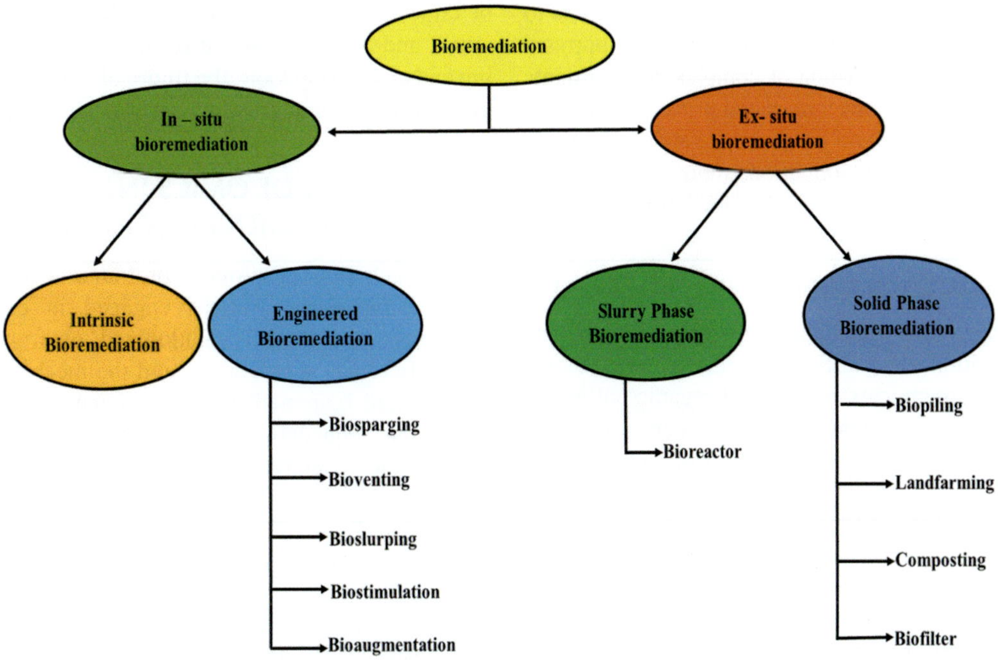

FIGURE 12.1 Categorizing the bioremediation into *in-situ* and *ex-situ* approaches. *In-situ* bioremediation includes intrinsic bioremediation, which relies on natural processes, and engineered bioremediation, which enhances microbial activity. *Ex-situ* bioremediation involves removing contaminated material for treatment, and is further divided into slurry phase and solid phase techniques. This classification highlights the various strategies employed to degrade or detoxify environmental pollutants using biological methods.

fail to restore the natural ecological balance of the contaminated site (Boopathy, 2000).

12.2.1 Advantages of Bioremediation over Traditional Methods

Bioremediation presents several distinct advantages compared to conventional remediation techniques. At its core, bioremediation a natural process leverages metabolic abilities of microorganisms to address environmental contamination. This intrinsic reliance on biological processes makes bioremediation a more sustainable option for managing pollutants. Unlike traditional methods, which often rely on synthetic chemicals and energy-intensive procedures, bioremediation exploits the natural abilities of microorganisms to break down pollutants, reducing the need for additional resources and minimizing the environmental footprint of the remediation process (Singh and Ward, 2004; Sharma et al., 2022).

12.2.2 Cost-Effectiveness and Environmental Impact

One of the notable benefits of bioremediation is its cost-effectiveness, particularly when compared to traditional remediation methods. *In-situ* bioremediation techniques, which treat contaminants at their original site, are generally less disruptive and less costly than *ex-situ* methods that involve the excavation and transport of contaminated materials. By relying on the natural metabolic processes of microorganisms, *in-situ* bioremediation reduces the need for expensive chemical treatments and energy-intensive processes. For instance, techniques such as bioventing and biosparging can enhance the activity of indigenous microorganisms without the need for additional chemical inputs, thereby lowering overall remediation costs (Gentry et al., 2004). Furthermore, bioremediation is less invasive than traditional methods, which often require significant physical disruption of the contaminated site. Techniques like land farming and biopiling can be implemented with minimal disturbance to the surrounding environment, preserving ecological balance and reducing the risk of secondary pollution (Vidali, 2001). This approach is particularly advantageous in sensitive or urban areas where extensive excavation or chemical treatments would have harmful effects on environment and local communities.

12.2.3 Versatility and Targeted Degradation

Bioremediation's versatility is another significant advantage. It can effectively target a broad spectrum of contaminants, including hydrocarbons, heavy metals, and persistent organic pollutants (POPs). Microorganisms have evolved to metabolize various types of pollutants, making bioremediation suitable for diverse contamination scenarios. For example, bioremediation techniques have been successfully applied to address petroleum spills, chlorinated solvents, and heavy metal contamination in soil and water (Sharma et al., 2022).

Bioremediation's ability to target specific contaminants is improved through techniques like bioaugmentation, which involves introducing specialized microorganisms to break down particulate pollutants. This focused approach enables the efficient degradation of complex contaminants that may be resistant to native microbial processes. By customizing the bioremediation process to match the unique characteristics of the contamination, more effective and faster remediation results can be achieved further.

12.2.4 Complete Mineralization and End Products

A key advantage of bioremediation is its potential for complete mineralization of organic contaminants. During bioremediation, microorganisms can convert organic pollutants into simpler, non-toxic end products such as carbon dioxide, water, and biomass. This process eliminates the need for further treatment or disposal of hazardous residues, as the contaminants are transformed into harmless substances. For example, hydrocarbon degradation by microorganisms results in the conversion of complex organic compounds into carbon dioxide and water, which are natural components of the environment. The complete mineralization of contaminants not only addresses the immediate pollution issue but also contributes to long-term environmental health. By converting harmful pollutants into benign substances, bioremediation helps restore the ecological balance of contaminated sites and reduces the risk of secondary pollution. This aspect of bioremediation makes it a valuable tool for achieving sustainable and effective environmental management (Vidali, 2001). On the whole, bioremediation provides numerous benefits over traditional remediation methods, including its sustainability, cost-effectiveness, versatility, and potential for complete mineralization of contaminants. Therefore, by exploiting the natural abilities of microorganisms, bioremediation provides an environmentally friendly and efficient solution to addressing a wide range of pollution challenges.

12.3 ENVIRONMENTAL APPLICATIONS AND FUTURE PROSPECTS

Bioremediation has been successfully applied in various environmental settings, such as oil spill cleanups, industrial wastewater treatment, and contaminated soil restoration. A notable example is the Exxon Valdez oil spill in 1989, where biostimulation was used to accelerate crude oil degradation by indigenous microorganisms (Singh and Ward, 2004). Bioremediation has also been effective in treating groundwater contaminated with chlorinated solvents, using anaerobic microorganisms to degrade these toxic compounds (Sharma et al., 2022). Looking forward, the integration of omics technologies, including genomics, transcriptomics, proteomics, and metabolomics, is expected to enhance the precision and effectiveness of bioremediation strategies (Sharma et al., 2022). These technologies offers beneficial insights to microbial communities and their metabolic capabilities, enabling the design of targeted bioremediation efforts. With advancements in omics and the development of more sophisticated modeling and monitoring tools, the future of bioremediation holds great promise for addressing complex challenges.

12.4 ACTION OF OMICS IN ENVIRONMENTAL SCIENCE

Omics technologies provide profound insights into the intricate biological systems engaged in environmental processes, and have revolutionized environmental science. Microbial communities and the mechanisms behind contaminant bioremediation have been clarified by next-generation sequencing, protein and metabolic profiling, and advanced analytics. Genomic analysis, proteomics, metabolomics, and other fields all fall under the umbrella word "omics" and offer unique perspectives on the molecular processes underlying environmental events (Sharma et al., 2022). Genomics focuses on the study of an organism's complete set of DNA, including all its genes, allowing researchers to identify and characterize the genetic basis of organismal responses to environmental changes. For instance, metagenomics, a branch of genomics, enables the examination of microbial communities in their natural habitats without the need for cultivation, thereby uncovering the diversity and functional potential of these communities in processes like bioremediation (Kumavath and Pratap, 2012; Kumar et al., 2023). By examining the entire collection of proteins that are expressed by a cell, tissue, or organism—referred to as the proteome—proteomics expands on this knowledge. Since proteins are the primary executors of biological functions, proteomics allows scientists to link genes to their functional outcomes, offering insights into the enzymatic pathways involved in the degradation of pollutants. This is crucial for bioremediation efforts, where identifying the key enzymes involved in the breakdown of contaminants can lead to more effective environmental cleanup strategies. Metabolomics takes this a step further by studying the complete set of metabolites. The small molecules are involved in metabolism within a biological system. Metabolomics provides a snapshot of the metabolic state of an organism or community at a given time, reflecting the real-time biochemical activities in response to environmental stresses or the presence of pollutants. This allows for the identification of biomarkers for environmental monitoring and the optimization of bioremediation processes by understanding the metabolic pathways that are active during the degradation of specific contaminants (Pomfret et al., 2019; Sharma et al., 2022). Beyond these, other omics disciplines such as transcriptomics, which studies RNA transcripts, and epigenomics, which examines modifications to the genome that do not involve changes to the DNA sequence, further enhance our understanding of how organisms interact with and adapt to their environments. The integration of data from these various omics approaches, known as multi-omics, provides a holistic view

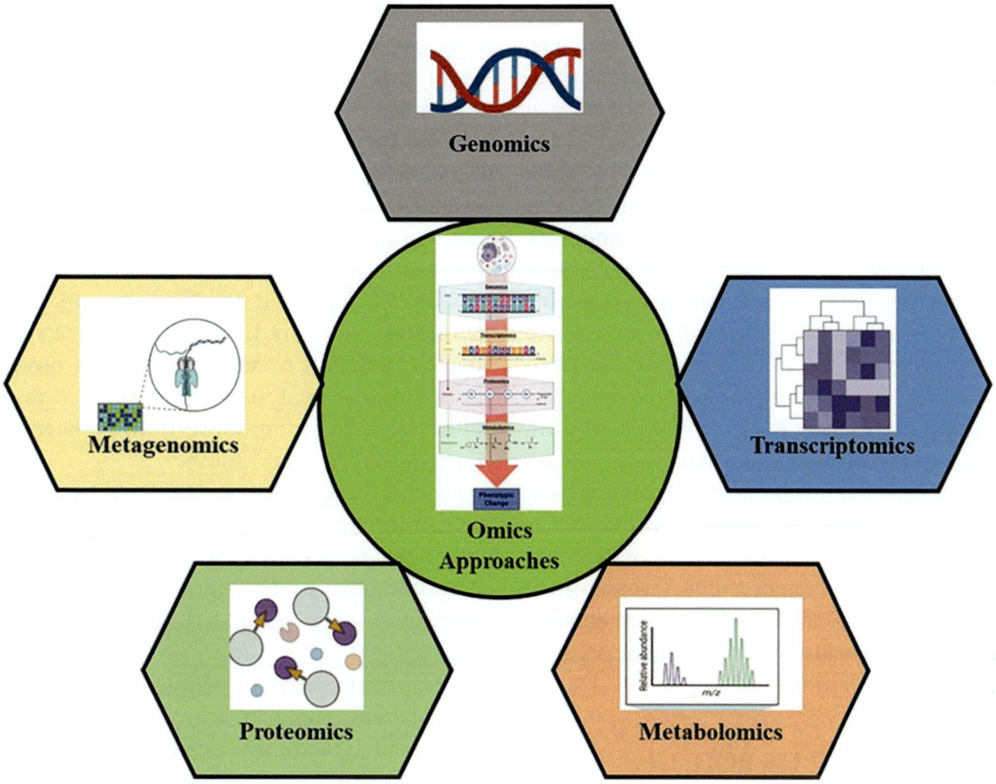

FIGURE 12.2 Multiple omics approaches. This figure illustrates the integration of various omics approaches, including genomics, metagenomics, transcriptomics, proteomics, and metabolomics, to provide a comprehensive understanding of biological systems. Each approach focuses on different molecular components: genomics on DNA, metagenomics on microbial communities, transcriptomics on RNA, proteomics on proteins, and metabolomics on metabolites. The central concept emphasizes the interconnectedness of these layers to decipher complex biological processes and phenotypic changes in environmental and bioremediation studies.

of biological systems, enabling a deeper understanding of environmental processes at multiple levels of organization. This integrative approach has transformed traditional environmental studies, which often relied on isolated observations of specific genes or proteins, by enabling the study of biological systems as interconnected networks (Du et al., 2021). Omics technologies have significantly advanced environmental science by offering comprehensive and integrative insights into the molecular underpinnings of environmental processes. This transformation allows for more accurate and effective interventions in environmental management and conservation, marking a shift from traditional studies to a more systems-oriented approach that can address the complexity of environmental challenges (Sharma et al., 2022) (Figure 12.2).

12.5 GENOMICS AND METAGENOMICS IN BIOREMEDIATION

Genomics and metagenomics have dramatically transformed the field of bioremediation, enabling a deeper knowledge on the complex of microbial communities responsible for the degradation of environmental pollutants. By leveraging genomics, researchers can delve into the genetic architecture of individual microorganisms, unraveling the specific genes, regulatory networks, and metabolic pathways that are activated during bioremediation processes. For instance, the identification of genes transcribing for enzymes such as oxygenases and dehydrogenases, which are crucial for the breakdown of hydrocarbons, has paved the way for the genetic engineering of microorganisms to enhance their bioremediation efficiency (Bhandari et al., 2021). Additionally, genomics has facilitated the study of horizontal gene transfer among microbial communities, a mechanism that significantly contributes to the spread of bioremediation capabilities across different species, thereby increasing the resilience and adaptability of microbial consortia in polluted environments (Aminov, 2011).

Metagenomics, which involves sequencing and analyzing DNA directly from environmental samples, has further expanded our ability to explore the vast diversity of microbial life, particularly those organisms that cannot be cultured in the lab. This technique has proven invaluable in identifying novel bioremediation candidates, including previously unknown species and microbial consortia that exhibit unique metabolic capabilities for degrading complex pollutants such as polycyclic aromatic hydrocarbons (PAHs), chlorinated compounds, and heavy metals (Bastida et al., 2017). For example, metagenomic studies conducted in oil-contaminated marine environments have not only identified key hydrocarbon-degrading bacteria but have also revealed the presence of functional genes related to sulfur and nitrogen cycling, which are essential for the complete mineralization of hydrocarbons and the restoration of ecological balance (Chandran et al., 2020).

The practical applications of genomics in bioremediation are highlighted by numerous case studies. In one notable example, the genomic analysis of microbial communities in uranium-contaminated groundwater identified bacteria capable of reducing soluble uranium(VI) to insoluble uranium(IV), thereby preventing the spread of radioactive contamination (Williams et al., 2013). Similarly, metagenomic approaches have been applied to study the microbial communities in sediments from the Exxon Valdez oil spill, leading to the discovery of indigenous microbial populations that played a vital role in the arrest of environment spoilation by breaking down the residual hydrocarbons (Bruckberger et al., 2019). These successful applications underscore the potential of genomics and metagenomics to not only enhance our understanding of microbial ecology in polluted environments but also to develop more effective and targeted bioremediation strategies.

12.6 PROTEOMICS AND ITS ROLE IN ENVIRONMENTAL SYSTEMS

Proteomics is the study of proteins, offering deep insights into cellular functions and biological systems. Since proteins are the products of gene expression and drive most cellular activities, proteomics enables the identification, quantification, and analysis of their modifications, interactions, and functions within living systems (Aebersold and Mann, 2016). In the context of bioremediation, understanding the proteome of organisms capable of degrading pollutants is crucial for optimizing and enhancing these natural processes. Bioremediating organisms, such as certain bacteria, fungi, and plants, have evolved complex enzymatic systems that enable them to metabolize and detoxify various environmental pollutants, including hydrocarbons, pesticides, and heavy metals (Lacerda and Reardon, 2009).

The variety of proteins and enzymes that directly contribute to the breakdown of pollutants is reflected in these organisms' proteomes. Enzymes like oxygenases, for instance, are essential in the initial breakdown of complex organic pollutants in microbial bioremediation because they incorporate oxygen into their molecular structure, making them more water-soluble and facilitate further degradation. Lactases and peroxidases, found in fungi, are pivotal in the degradation of lignin and similar aromatic compounds, showcasing the diversity of enzymatic activities involved in bioremediation (Lacerda and Reardon, 2009). Additionally, proteins involved in stress response and detoxification, such as heat shock proteins and metallothioneins, are often upregulated in bioremediating organisms exposed to toxic environments, highlighting the adaptive mechanisms that enhance survival and functionality in polluted habitats. To identify and characterize these key enzymes and proteins, environmental proteomics leverages advanced analytical techniques. Mass spectrometry (MS) is essential in proteomics, providing high-throughput identification and quantification of proteins with outstanding sensitivity and precision (Aebersold and Mann, 2016). Techniques such as tandem mass spectrometry (MS/MS) provide detailed information on protein sequences and post-translational modifications, which are essential for understanding how proteins are regulated and function under different environmental conditions. Liquid chromatography (LC) is often coupled with MS to separate complex protein mixtures, facilitating the analysis of proteomes in diverse biological samples, including those from polluted environments (Lacerda and Reardon, 2009).

Beyond these core techniques, recent advances in metaproteomics have expanded the scope of environmental proteomics by allowing the study of protein expression in entire microbial communities rather than isolated species. This approach is particularly valuable in understanding bioremediation processes in situ, as it captures the collective enzymatic activities of microbial consortia that often work synergistically to degrade pollutants (Wilmes and Bond, 2006). Bioinformatics tools are vital for analyzing data from these techniques, aiding in protein identification, function prediction, and modeling metabolic pathways for pollutant degradation. The integration of these techniques and tools in environmental proteomics not only enhances our understanding of the molecular mechanisms underlying bioremediation but also provides a platform for developing more effective strategies for environmental cleanup. By identifying key enzymes and proteins that can be harnessed or engineered for improved pollutant degradation, proteomics offers the potential to optimize bioremediation processes, making them faster, more efficient, and applicable to a wider range of environmental contaminants. This approach represents a significant step toward sustainable environmental management, where the natural capacity of organisms to detoxify pollutants can be fully realized and exploited (Lacerda and Reardon, 2009).

12.7 METABOLOMICS: AN IN-DEPTH LOOK INTO METABOLIC NETWORKS

Metabolomics is the comprehensive study of metabolites within a biological system, and offers a powerful approach to understanding biochemical pathways, especially in the context of environmental bioremediation. By studying the entire set of metabolites, researchers can uncover the metabolic networks and pathways organisms employ to degrade pollutants. Mapping metabolic pathways in pollutant degradation involves identifying key metabolites and enzymes involved in the conversion of harmful compounds into less toxic forms. This mapping is crucial for understanding how microorganisms, plants, or other biological entities transform environmental contaminants. For instance, metabolic profiling can reveal the pathways used by bacteria to degrade PAHs, helping to optimize conditions for their biodegradation (Liao et al., 2023).

Metabolomics also enacts a crucial role in monitoring and optimizing bioremediation processes. By tracking changes in the metabolome of an organism exposed to pollutants, scientists can assess the effectiveness of bioremediation strategies in real time. This allows for the adjustment of environmental conditions, such as nutrient availability or pH, to enhance the degradation process. Moreover, metabolomics can help identify biomarkers of pollutant exposure and degradation, offering a more precise measure of bioremediation progress compared to traditional methods (Patankar et al., 2022). Recent innovations in environmental metabolomics have further expanded its application in bioremediation. Advanced analytical techniques, such as MS and nuclear magnetic resonance (NMR) spectroscopy, coupled with bioinformatics tools, have improved the resolution and accuracy of metabolite detection and quantification. These innovations enable the detailed characterization of metabolic pathways involved in the degradation of complex pollutants, such as heavy metals and POPs. Additionally, the connection of metabolomics with other omics technologies, such as genomics and proteomics, provides a holistic view of the cellular response to pollutants, facilitating the development of more effective bioremediation strategies (Singh et al., 2020; Sharma et al., 2022).

12.8 TRANSCRIPTOMICS: ASSESSING ENVIRONMENTAL RESPONSES

Transcriptomics is the study of the complete set of RNA transcripts produced by the genome under specific circumstances, and is crucial for decoding the environmental response of organisms. This powerful tool allows researchers to understand how gene expression changes in response to various environmental stressors, providing insights into the molecular mechanisms underlying these responses. In the context of bioremediation, transcriptomics plays a vital role by revealing how microorganisms adjust their gene expression to degrade pollutants, thus offering a deeper understanding of the biological processes involved in contaminant breakdown. For instance, transcriptomic analysis has been employed to identify key genes and regulatory networks in bacteria that are upregulated during the degradation of heavy metals and organic pollutants, enabling the development of more effective bioremediation strategies (Tripathi and Poluri, 2021).

Techniques in environmental transcriptomics, such as RNA sequencing (RNA-seq), have significantly advanced our ability to analyze the transcriptome with high precision and depth. RNA-seq allows for the quantification of gene expression levels and the discovery of novel transcripts, providing a comprehensive view of the transcriptional landscape under various environmental conditions (Ge et al., 2013). This technique has been instrumental in identifying stress-responsive genes in microorganisms that are critical for survival in polluted environments, thereby aiding in the optimization of microbial consortia used in bioremediation. Integrating transcriptomics with other omics approaches, such as proteomics and metabolomics, offers a holistic approach to understanding environmental responses. This multi-omics integration allows for the correlation of transcriptomic data with protein and metabolite levels, leading to a more complete picture of the biological processes involved in pollutant degradation. By combining these datasets, researchers can identify the functional proteins and metabolic pathways that are most active during bioremediation, leading to more targeted and efficient environmental cleanup strategies (Shi et al., 2024).

12.9 INTEGRATIVE OMICS AND BIOREMEDIATION

Integrative omics is transforming the field of bioremediation by providing a comprehensive framework to understand the complex dynamics of microbial communities and their functional roles in degrading environmental pollutants. This holistic approach involves the convergence of various omics technologies, including genomics, transcriptomics, proteomics, and metabolomics, which together offer a detailed view of the genetic, transcriptional, protein, and metabolic landscapes within contaminated ecosystems. Systems biology is central to this approach, as it enables the integration and interpretation of vast omics datasets, linking them to the underlying biochemical and regulatory networks that govern microbial activity. By decoding these networks, researchers can uncover the molecular mechanisms and metabolic pathways that are critical for the degradation of pollutants, such as hydrocarbons, heavy metals, and organic contaminants (Beale et al., 2016; Dangi et al., 2019; Sharma et al., 2022). The integration of multi-omics data not only enhances our understanding of these processes but also facilitates the identification of key microbial players and their interactions, which are essential for optimizing bioremediation strategies. For instance, through multi-omics integration, it becomes possible to construct predictive models that simulate the behavior of microbial consortia under various environmental conditions, enabling the design of tailored interventions that maximize pollutant degradation. Such models are invaluable for developing efficient, site-specific bioremediation strategies that minimize environmental impact while maximizing remediation efficiency (Chakraborty et al., 2012). Several successful case studies underscore the potential of integrative omics in environmental cleanup. For example, in the remediation of petroleum-contaminated soils, integrative omics approaches have revealed intricate microbial networks and novel metabolic pathways involved in hydrocarbon degradation, leading to more effective bioremediation strategies. Similarly, in the context of heavy metal contamination, integrative omics has identified microbial species with unique

TABLE 12.2
Tools and Techniques in Omics-Driven Bioremediation

Tool/Technique	Omics Type	Description	Application in Bioremediation	References
CRISPR/Cas9	Genomics	Gene editing tool for precise DNA modifications	Engineering microorganisms for enhanced pollutant degradation	Zhang (2019)
Mass Spectrometry	Proteomics	Analyzes protein structure and function	Identification of key proteins in biodegradation pathways	Han et al. (2008)
NMR Spectroscopy	Metabolomics	Determines chemical structure of metabolites	Analyzing metabolic shifts during bioremediation processes	Emwas et al. (2019)
RNA-Seq	Transcriptomics	Sequencing technique for studying gene expression	Profiling microbial responses to toxic environments	Wang et al. (2009)
Bioinformatics Tools	Multi-omics	Software for analyzing and integrating omics data	Comprehensive analysis of multi-omics data for bioremediation strategies	Gioti et al. (2024)

metal-resistant properties and the associated gene clusters responsible for detoxification, providing new avenues for bioremediation applications. These case studies highlight the power of integrative omics in uncovering previously hidden biological processes, paving the way for innovative and more effective approaches to environmental remediation (Singh and Ward, 2004: Malla et al., 2018).

12.10 ADVANCED TOOLS AND TECHNOLOGIES IN OMICS FOR BIOREMEDIATION

Advanced omics tools have revolutionized bioremediation, providing enhanced precision and efficiency in environmental cleanup. Cutting-edge technologies such as CRISPR (Clustered Regularly Interspaced Short Palindromic Repeats), Artificial Intelligence (AI), and Big Data are at the forefront of this transformation. CRISPR, a revolutionary genome-editing tool, allows for precise modifications in the genetic makeup of microorganisms, enhancing their ability to degrade pollutants more efficiently. For instance, CRISPR has been employed to edit bacterial genomes to increase their capacity to metabolize toxic compounds in contaminated environments. AI plays a crucial role in predicting the behavior of microorganisms under various environmental conditions, optimizing the selection of bioremediation agents, and improving the overall process efficiency. Big Data analytics enables the integration and analysis of vast omics datasets, uncovering complex patterns and interactions within microbial communities that are critical for effective bioremediation strategies (Ye et al., 2020; Dey et al., 2023). Bioinformatics has become indispensable in managing and analyzing the enormous amounts of data generated from omics studies. Innovative bioinformatics tools make it easier to process, interpret, and visualize omics data, offering insights into the functional capacities of microbial communities and how they interact with contaminants. These resources are crucial for locating important genes and bioremediation pathways so that focused treatments can be made (Bansal, 2005). Looking ahead, omics future in environmental sciences appears with ongoing advancements likely to drive even greater innovations in bioremediation. Integrating multi-omics approaches like genomics, transcriptomics, proteomics, and metabolomics provides a deeper understanding of microbial ecosystems, leading to more effective and sustainable bioremediation solutions. Additionally, emerging technologies like synthetic biology could pave the way for the design of novel microorganisms tailored for specific bioremediation tasks, further enhancing the efficiency and scope of environmental cleanup efforts (Malla et al., 2018) (Table 12.2).

12.11 CHALLENGES AND OPPORTUNITIES IN OMICS-DRIVEN BIOREMEDIATION

The integration of omics technologies into bioremediation presents both significant challenges and promising opportunities. Technical and practical challenges are among the most pressing issues in this field. The intricacy of environmental samples, which frequently include a wide variety of microorganisms interacting in complex ways, is one of the main obstacles. This complexity makes it worse to accurately search and characterize the microbial communities involved in bioremediation (Bansal, 2005). Additionally, the sheer volume of data generated by omics studies, coupled with the need for high computational power and advanced bioinformatics tools, can be overwhelming. The translation of laboratory findings into field applications also poses practical challenges, as controlled environments in the lab may not accurately reflect the dynamic and variable conditions in natural settings (Malla et al., 2018). Beyond technical hurdles, ethical, legal, and social implications (ELSI) are critical considerations in omics-driven bioremediation. The use of genetically modified organisms (GMOs), especially those engineered through CRISPR and other genome-editing technologies, raises concerns about their potential impact on ecosystems and human health. Regulatory frameworks governing the release of such organisms into the environment are still evolving, and there is ongoing debate about the long-term consequences of their use (Dey et al., 2023). Additionally, there is a need for greater transparency and public engagement in bioremediation projects to ensure that societal concerns are adequately addressed.

Despite these challenges, the future of omics-driven bioremediation is filled with opportunities for innovation and collaboration. Advances in multi-omics approaches, such as integrating genomics, proteomics, and metabolomics, offer the potential for more comprehensive and effective bioremediation strategies. These approaches enhance our understanding of microbial communities and their interactions with pollutants, allowing for the development of targeted solutions (Malla et al., 2018). Furthermore, collaborative efforts between scientists, industry stakeholders, and regulatory bodies can foster the responsible development and deployment of omics technologies in bioremediation. By addressing the technical, ethical, and practical challenges, omics-driven bioremediation has the potential to revolutionize environmental cleanup efforts and contribute to a more sustainable future (Maganathan et al., 2020; Ye et al., 2020).

12.12 CASE STUDIES IN OMICS-DRIVEN BIOREMEDIATION

The application of omics technologies in bioremediation has led to several notable success stories around the globe, demonstrating their potential to address environmental contamination challenges effectively. In the case of the contaminated mining sites in Australia, metagenomics and proteomics were employed to analyze and enhance the capabilities of indigenous microbial communities. This approach identified key bacterial strains with high metal tolerance and provided insights into their metabolic pathways, leading to successful remediation of heavy metals and improved soil health (Gruszecka-Kosowska et al., 2022). Similarly, in the United States, a study focused on the cleanup of pesticide-polluted soils in agricultural regions utilized a combination of transcriptomics and metabolomics. This integrated approach allowed researchers to identify and optimize microbial strains with high pesticide degradation efficiency, resulting in significant reductions in pesticide residues and enhanced soil fertility (Lee-Ann Ataikiru and Ajuzieogu, 2023). These case studies underscore several key lessons and best practices in the field of omics-driven bioremediation. Firstly, site-specific analyses using omics technologies are crucial for selecting the most effective microorganisms tailored to the unique contamination profile of each site. Secondly, integrating various omics approaches, such as combining metagenomics with proteomics or transcriptomics, provides a holistic view of microbial functions and interactions, leading to more targeted and efficient remediation strategies. Lastly, ongoing monitoring and adaptive management, informed by continuous omics data, are essential for optimizing bioremediation processes and addressing dynamic environmental conditions. These insights not only highlight the success of omics in practical applications but also offer valuable guidance for future bioremediation projects (Malla et al., 2018; Sharma et al., 2022).

12.13 CONCLUSION

The future of bioremediation is poised to play a pivotal role in the green economy, offering sustainable solutions to environmental pollution while fostering economic growth. As global awareness of environmental issues intensifies, the integration of bioremediation with green economic strategies becomes increasingly critical. Bioremediation's potential to restore polluted environments aligns with the principles of a green economy, which emphasizes reducing environmental risks and ecological scarcities while promoting sustainable development. Innovations in omics technologies, particularly integrative omics approaches, are set to revolutionize bioremediation practices. By leveraging more on omics technologies and types and its deep fields, researchers can create more efficient and targeted bioremediation strategies. For example, integrating multi-omics data provides a comprehensive understanding of microbial communities involved in pollutant degradation, allowing for the design of customized bioremediation processes that enhance environmental restoration. However, successfully implementing bioremediation within a green economy requires addressing several challenges.

Technical hurdles such as the scalability of laboratory findings to field applications, and the need for robust monitoring systems to track bioremediation progress, remain significant obstacles. Moreover, ethical, legal, and social considerations, such as the potential release of GMOs into the environment, require careful regulation and public engagement. Looking ahead, the future of bioremediation in a green economy will likely involve increased collaboration between academia, industry, and policymakers. Such partnerships are essential for translating scientific advancements into practical applications that benefit both the environment and the economy. Furthermore, the development of policies that incentivize green technologies, including bioremediation, will be crucial for driving widespread adoption and ensuring long-term environmental sustainability. In conclusion, bioremediation is poised to become a cornerstone of the green economy, offering a viable path toward sustainable environmental management. By addressing current challenges and leveraging technological innovations, bioremediation can contribute significantly to the transition toward a greener, more resilient economy.

REFERENCES

Aebersold, R., & Mann, M. (2016). Mass-spectrometric exploration of proteome structure and function. *Nature*, *537*(7620), 347–355.

Aminov, R. I. (2011). Horizontal gene exchange in environmental microbiota. *Frontiers in Microbiology*, *2*, 158.

Bansal, A. K. (2005). Bioinformatics in microbial biotechnology – a mini review. *Microbial Cell Factories*, *4*(1), 19.

Bastida, F., Torres, I. F., Hernández, T., & García, C. (2017). The impacts of organic amendments: Do they confer stability against drought on the soil microbial community? *Soil Biology & Biochemistry*, *113*, 173–183.

Beale, D. J., Karpe, A. V., & Ahmed, W. (2016). Beyond metabolomics: A review of multi-omics-based approaches. In D. J. Beale, K. A. Kouremenos, & E. A. Palombo (Eds.), *Microbial Metabolomics* (pp. 289–312). Springer International Publishing.

Bhandari, S., Poudel, D. K., Marahatha, R., Dawadi, S., Khadayat, K., Phuyal, S., Shrestha, S., Gaire, S., Basnet, K., Khadka, U., & Parajuli, N. (2021). Microbial enzymes used in bioremediation. *Journal of Chemistry*, *2021*, 1–17.

Boopathy, R. (2000). Factors limiting bioremediation technologies. *Bioresource Technology*, *74*(1), 63–67.

Bruckberger, M. C., Morgan, M. J., Walsh, T., Bastow, T. P., Prommer, H., Mukhopadhyay, A., Kaksonen, A. H., Davis, G., & Puzon, G. J. (2019). Biodegradability of legacy crude oil contamination in Gulf War damaged groundwater wells in Northern Kuwait. *Biodegradation*, *30*(1), 71–85.

Chakraborty, R., Wu, C. H., & Hazen, T. C. (2012). Systems biology approach to bioremediation. *Current Opinion in Biotechnology*, *23*(3), 483–490.

Chandran, H., Meena, M., & Sharma, K. (2020). Microbial biodiversity and bioremediation assessment through omics approaches. *Frontiers in Environmental Chemistry*, *1*, 570326.

Dangi, A. K., Sharma, B., Hill, R. T., & Shukla, P. (2019). Bioremediation through microbes: systems biology and metabolic engineering approach. *Critical Reviews in Biotechnology*, *39*(1), 79–98.

Dey, S., Anand, U., Pandey, D. K., & Ghorai, M. (2023). Microbial community and their role in bioremediation of polluted e-waste sites. In V. Kumar, M. Bilal, L. F. R. Ferreira, & H. M. N. Iqbal (Eds.), *Genomics Approach to Bioremediation*, 1st ed. (pp. 363–377). Wiley.

Du, Y., Fan, K., Lu, X., & Wu, C. (2021). Integrating multi-omics data for gene-environment interactions. *Biotech*, *10*(1), 3.

Emwas, A.-H., Roy, R., McKay, R. T., Tenori, L., Saccenti, E., Gowda, G. A. N., Raftery, D., Alahmari, F., Jaremko, L., Jaremko, M., & Wishart, D. S. (2019). NMR spectroscopy for metabolomics research. *Metabolites*, *9*(7), 123.

Gentry, T., Rensing, C., & Pepper, I. (2004). New approaches for bioaugmentation as a remediation technology. *Critical Reviews in Environmental Science and Technology*, *34*(5), 447–494.

Ge, Y., Wang, D.-Z., Chiu, J.-F., Cristobal, S., Sheehan, D., Silvestre, F., Peng, X., Li, H., Gong, Z., Lam, S. H., Wentao, H., Iwahashi, H., Liu, J., Mei, N., Shi, L., Bruno, M., Foth, H., & Teichman, K. (2013). Environmental OMICS: current status and future directions. *Journal of Integrated OMICS*, *3*(2), 75–87. https://doi.org/10.5584/jiomics.v3i2.141

Gioti, A., Theodosopoulou, D., Bravakos, P., Magoulas, A., & Kotoulas, G. (2024). The bioinformatics landscape in environmental omics: Lessons from a national ELIXIR survey. *iScience*, *27*(6), 110062.

Gruszecka-Kosowska, A., Ampatzoglou, A., & Aguilera, M. (2022). Integration of omics approaches enhances the impact of scientific research in environmental applications. *International Journal of Environmental Research and Public Health*, *19*(14), 8758.

Han, X., Aslanian, A., & Yates, J. R., 3rd. (2008). Mass spectrometry for proteomics. *Current Opinion in Chemical Biology*, *12*(5), 483–490.

Herrera-Calderon, A. C., Leal, L., Suárez-Bautista, J. D., Manotas-Viloria, H. S., Muñoz-García, A., Franco, D., Arenas, N. E., & Vanegas, J. (2024). Metagenomic and genomic analysis of heavy metal-tolerant and -resistant bacteria in resource islands in a semi-arid zone of the Colombian Caribbean. *Environmental Science and Pollution Research International*, *31*(4), 5596–5609.

Kao, C. M., Chen, C. Y., Chen, S. C., Chien, H. Y., & Chen, Y. L. (2008). Application of in situ biosparging to remediate a petroleum-hydrocarbon spill site: field and microbial evaluation. *Chemosphere*, *70*(8), 1492–1499.

Kumar, V., Bilal, M., Ferreira, L. F. R., & Iqbal, H. M. N. (2023). *Genomics Approach to Bioremediation: Principles, Tools, and Emerging Technologies*. https://books.google.com/books?hl=en&lr=&id=pQGqEAAAQBAJ&oi=fnd&pg=PA23&dq=Chapter+25+-+Genomics+in+understanding+bioremediation+of+inorganic+pollutants&ots=GCkY39-Um2&sig=Hr3AmVYG0D6FcaFChh0mcnvO7CQ

Kumavath, R., & Pratap, D. (2012). Omics in bioremediation: a consolidated overview. *Indian Journal of Biotechnology*, *6*(8,9), 242–253.https://www.tsijournals.com/articles/omics-in-bioremediation-a-consolidated-overview.pdf

Lacerda, C. M. R., & Reardon, K. F. (2009). Environmental proteomics: applications of proteome profiling in environmental microbiology and biotechnology. *Briefings in Functional Genomics & Proteomics*, *8*(1), 75–87.

Lee-Ann Ataikiru, T., & Ajuzieogu, C. A. (2023). Enhanced bioremediation of pesticides contaminated soil using organic (compost) and inorganic (NPK) fertilizers. *Heliyon*, *9*(12), e23133.

Liao, J., Goodrich, J., Walker, D. I., Lin, Y., Lurmann, F., Qiu, C., Jones, D. P., Gilliland, F., Chazi, L., & Chen, Z. (2023). Metabolic pathways altered by air pollutant exposure in association with lipid profiles in young adults. *Environmental Pollution*, *327*(121522), 121522.

Maganathan, T., Senthilkumar, S., & Balakrishnan, V. (2020). Machine Learning and Data Analytics for Environmental Science: a review, prospects and challenges. *IOP Conference Series. Materials Science and Engineering*, *955*(1), 012107.

Malla, M. A., Dubey, A., Yadav, S., Kumar, A., Hashem, A., & Abd Allah, E. F. (2018). Understanding and designing the strategies for the microbe-mediated remediation of environmental contaminants using omics approaches. *Frontiers in Microbiology*, *9*, 1132. https://www.frontiersin.org/journals/microbiology/articles/10.3389/fmicb.2018.01132/full

Patankar, R. S., Reshi, N., & Kutty, R. (2022). Bioremediation: Gaining insights through metabolomics. In V. Kumar & I. S. Thakur (Eds.), *Omics Insights in Environmental Bioremediation* (pp. 37–46). Springer Nature Singapore.

Pomfret, S. M., Brua, R. B., Izral, N. M., & Yates, A. G. (2019). Metabolomics for biomonitoring: an evaluation of the metabolome as an indicator of aquatic ecosystem health. *Environmental Reviews*, *28*(1), 89–98.

Sharma, P., Singh, S. P., Iqbal, H. M. N., & Tong, Y. W. (2022). Omics approaches in bioremediation of environmental contaminants: an integrated approach for environmental safety and sustainability. *Environmental Research*, *211*(113102), 113102.

Sharma, S. (2012). *Bioremediation: Features, Strategies and Applications*.https://www.scirp.org/(S(ny23rubfvg45z345vbrepxrl))/reference/referencespapers?referenceid=2360534

Shi, C., Cheng, L., Yu, Y., Chen, S., Dai, Y., Yang, J., Zhang, H., Chen, J., & Geng, N. (2024). Multi-omics integration analysis: tools and applications in environmental toxicology. *Environmental Pollution*, *360*(124675), 124675.

Singh, A., Ward, O. P. (2004). Biotechnology and Bioremediation — An Overview. In Singh, A., Ward, O.P. (Eds.), *Biodegradation and Bioremediation. Soil Biology*, vol 2. Springer, Berlin, Heidelberg. https://doi.org/10.1007/978-3-662-06066-7_1.

Singh, D., Geat, N., Mehriya, M., Rajawat, M. V. S., Prasanna, R., Kumar, A., Kumari, G., & Jha, M. N. (2020). Omics (genomics, proteomics, metabolomics, etc.) tools to study the environmental microbiome and bioremediation. In B. K. Kashyap, M. K. Solanki, D. V. Kamboj, & A. K. Pandey (Eds.), *Waste to Energy: Prospects and Applications* (pp. 235–260). Springer Singapore.

Tripathi, S., & Poluri, K. M. (2021). Heavy metal detoxification mechanisms by microalgae: insights from transcriptomics analysis. *Environmental Pollution*, *285*(117443), 117443.

Vidali, M. (2001). Bioremediation – an overview. *Pure and Applied Chemistry*, *73*(7), 1163–1172.

Wang, Z., Gerstein, M., & Snyder, M. (2009). RNA-Seq: a revolutionary tool for transcriptomics. *Nature Reviews. Genetics*, *10*(1), 57–63.

Williams, K. H., Bargar, J. R., Lloyd, J. R., & Lovley, D. R. (2013). Bioremediation of uranium-contaminated groundwater: a systems approach to subsurface biogeochemistry. *Current Opinion in Biotechnology*, *24*(3), 489–497.

Wilmes, P., & Bond, P. L. (2006). Metaproteomics: studying functional gene expression in microbial ecosystems. *Trends in Microbiology*, *14*(2), 92–97.

Ye, Z., Yang, J., Zhong, N., Tu, X., Jia, J., & Wang, J. (2020). Tackling environmental challenges in pollution controls using artificial intelligence: a review. *The Science of the Total Environment*, *699*(134279), 134279.

Zhang, F. (2019). Development of CRISPR-Cas systems for genome editing and beyond. *Quarterly Reviews of Biophysics*, *52*, e6. https://doi.org/10.1017/S0033583519000052

13 Exploring the Potential of Microbial Bioprospecting for Functional Foods

Jayenth Jayachandran[1], *Diptimayee Das*[2], *Rithi Adhilakshmi Thangamani*[1], *Arunkumar Radhakrishnan*[1], *Surajit Pathak*[2], *and Antara Banerjee*[2]

[1]Department of Pharmacology, Chettinad Hospital and Research Institute (CHRI), Chettinad Academy of Research and Education (CARE), Chennai, India

[2]Medical Biotechnology Lab, Faculty of Allied Health Sciences, Chettinad Academy of Research and Education (CARE), Chettinad Hospital and Research Institute (CHRI), Chennai, India

13.1 INTRODUCTION TO MICROBIAL BIOPROSPECTING

Over centuries, the discovery and utilization of microorganisms in producing fermented foods like bread, beer, and cheese marked some of the earliest forms of bioprospecting, even though the underlying biological principles were not yet understood (Benkeblia 2022). The concept of microbial bioprospecting as we understand it today began to take shape with the advent of microbiology in the late 19th and early 20th centuries. Scientists like Louis Pasteur and Robert Koch's revolutionary research established the foundations for future research for identifying and characterizing microorganisms, unlocking the potential to explore their roles beyond traditional applications (Elsawey et al. 2020). Pasteur's work on fermentation and Koch's postulates for linking specific microbes to diseases revolutionized the field, opening doors to a more systematic approach to exploring microbial diversity.

Microbial bioprospecting traces its roots back to the early 1900s when Alexander Fleming made the ground-breaking discovery of penicillin in 1928, which marked the beginning of the modern antibiotic era. This discovery ignited a global search for other microorganisms capable of producing medically significant compounds, particularly antibiotics (Kapoor et al. 2020). Over the decades, scientists explored soil, marine, and other natural environments, isolating microbes that yielded a variety of bioactive substances, including streptomycin and tetracycline (Abdel-Razek et al. 2020). As biotechnology advanced, particularly in the late 20th century, microbial bioprospecting evolved with the integration of genetic and molecular biology techniques. The advent of genome sequencing, metagenomics, and synthetic biology allowed researchers to explore microbial diversity on a deeper level, including previously unculti-vable organisms (Alam et al. 2021). This shift expanded the scope of microbial bioprospecting beyond medicine to fields such as agriculture and functional foods, where microbes are now recognized for their role in producing probiotics, enzymes, and other health-promoting bioactive.

Today, microbial bioprospecting is recognized as a crucial avenue for discovering new functional ingredients that can enhance food quality and contribute to human health. The exploration of microorganisms from diverse and often extreme environments continues to yield novel bioactive compounds, enzymes, and probiotics, driving innovation in the functional food industry (Kumar et al. 2021). As the demand for health-promoting foods grows, so does the significance of microbial bioprospecting in meeting these needs, making it a key focus in the ongoing quest to develop functional foods that cater to modern dietary preferences and health requirements.

Microbial bioprospecting is the process of exploring and studying microorganisms, particularly from diverse or extreme environments, to discover new bioactive compounds that can have beneficial applications in various industries (Verma and Shah 2022). Microorganisms, encompassing bacteria, fungi, and archaea, possess diverse metabolic capacities, enabling them to generate a broad range of chemicals, including enzymes, antibiotics, and other bioactive compounds. The scope of microbial bioprospecting extends across fields like pharmaceuticals, agriculture, and functional foods, where there is a growing demand for natural, sustainable ingredients (Pushpangadan et al. 2018). This practice is particularly relevant today as it holds the potential to uncover new compounds that can be harnessed for improving health, nutrition, and food safety, while also addressing issues such as antibiotic resistance and the need for novel therapeutic agents, whereas through advanced biotechnological tools like genome mining, metagenomics, and synthetic biology, microbial bioprospecting enables the efficient identification and development of new, potent bioactive substances, making it an essential part of modern biotechnological innovation (Aware and Jadhav 2022).

Foods that provide extra health advantages beyond basic nutrition are known as functional foods, and microbial bioprospecting is a key component in their creation, whereas by exploring the vast diversity of microorganisms, researchers can identify new strains capable of producing bioactive compounds like prebiotics, probiotics,

DOI: 10.1201/9781003509530-13

antioxidants, enzymes, and peptides (Chandra et al. 2020). These substances are essential for increasing the nutritional content of functional food since they boost the immune system, lower inflammation, and promote intestinal health. Microbes used in fermentation, such as certain yeast and lactic acid bacteria (LAB), help transform raw ingredients into nutrient-rich products like yogurt, kefir, and kimchi (Chen 2021). Through microbial bioprospecting, novel microbes can be discovered and utilized to create new or improved functional foods, helping meet the growing consumer demand for healthier, natural products.

In summary, microbial bioprospecting is a dynamic and evolving field that plays a pivotal role in biotechnological innovation. Microbial bioprospecting has consistently broadened its horizons, revealing the potential of microbes to improve nutrition, health, and sustainability, from its early beginnings in the discovery of antibiotics to its current uses in the production of functional foods. Furthermore, advancements in technology have continued to deepen our understanding of microbial diversity, the role of microbial bioprospecting in discovering novel bioactive compounds will only grow in importance, offering promising solutions for addressing current and future challenges in food and medicines.

13.2 DIVERSITY OF MICROBIAL SOURCES

13.2.1 Bacterial and Archaea Species in Functional Foods

LAB have been crucial in various food production processes for centuries, dating back to the early consumption of fermented foods. However, these microorganisms have long been involved in traditional food production, in which the main metabolic activity is lactic acid fermentation (Damián et al. 2022). LAB is a broad class of Gram-positive bacteria used in many different industrial applications (Sharma et al. 2023). LAB, a diverse group of Gram-positive bacteria with a wide range of industrial applications, encompasses eight main species commonly used to impart desired flavor and texture to fermented foods. In conjunction with *Bifidobacteria* (including *Bifidobacterium animalis, Bifidobacterium longum*, and *Bifidobacterium lactis*) or *Lactobacilli* (including *Acidophilus, Lactobacillus, Lactobacillus rhamnosus, Lactobacillus johnsonii*, and *Lacticaseibacillus casei*), *Streptococcus thermophilus* is present (Ljungh et al. 2006; Harzallah and Belhadj 2023). These species are crucial in fermenting dairy products like yogurt, kefir, and cheese. Their probiotic properties contribute to improved digestion, enhanced immune function, and potentially reduced gastrointestinal issues (Ashaolu 2020). Probiotics, live microorganisms that confer health benefits when consumed in adequate amounts, encompass beneficial bacteria and yeasts. These microbes support the immune system, combat harmful bacteria, and help maintain a balanced gut microbiota (Anee et al. 2021).

Probiotics can eliminate harmful microbes through various molecular mechanisms and by modulating the host's immune response (Mazziotta et al. 2023). Also, certain *Bacillus* species including *Bacillus subtilis* and *Bacillus coagulans* are used in functional foods for their spore-forming ability, which allows them to live through challenging conditions such as the stomach's acidic environment (Payne et al. 2024). However, they are known to produce enzymes that aid in digestion and have potential anti-inflammatory and immune-modulating effects. *Propionibacterium freudenreichii,* used in Swiss cheese production, is another bacterium with probiotic potential. It generates metabolites that are advantageous and include vitamin B12, adding to the nutritional content of functional foods (Piwowarek et al. 2018). Also, certain archaeal species hold potential in the functional food industry like methanogens and halophilic archaea. These archaea, which include species like *Methanobrevibacter smithii,* are involved in the human gut's digestion process (Nkamga et al. 2017). Although not directly used in food production, their role in modulating the gut microbiota and improving energy harvest from the diet suggests potential for future applications in functional foods (Bauer-Estrada et al. 2023). Halophilic archaea, such as those from the *Halobacterium* genus, have been investigated for their unique bioactive compounds (Corral et al. 2019). Hence these organisms produce carotenoids, antioxidants, and other metabolites that could be harnessed to develop functional food ingredients, particularly in the creation of naturally derived colorants or preservatives.

13.2.2 Role of Fungi and Yeasts in Functional Foods

Specific environmental stressors can trigger the production of fungal bioactive compounds, including primary and secondary metabolites (Łysakowska et al. 2023). Certain fungi like *Penicillium* and *Aspergillus* species are included for their ability to produce antioxidants like flavonoids and phenolic acids. Studies discovered that *Aspergillus terreus* endophytic fungal extracts, which were separated from *Hibiscus sabdariffa* leaves, had significant antioxidant activity, as shown by a *2,2-diphenyl-1-picrylhydrazyl* (DPPH) radical scavenging assay result of 91.88 ± 0.17 µg/mL (Khalil et al. 2020). Fungi are also rich sources of enzymes such as proteases, lipases, and amylases, which are used in the food industry to improve the nutritional profile of foods. Liu et al. (2022) have reviewed that *Aspergillus oryzae* is widely used in the fermentation of soy products like miso and soy sauce, where it breaks down the complex proteins and carbohydrates, making these foods easier to digest and increasing their bioavailability of nutrients. However, fungi produce various antimicrobial compounds, including antibiotics like penicillin, which can be incorporated into functional foods to enhance their safe shelf life (Mishra et al. 2021).

Yeasts, such as *Saccharomyces cerevisiae*, are widely used in the fermentation of bread, beer, and wine. During fermentation, they produce bioactive compounds like B-complex vitamins, enhancing the nutritional value of these foods (Jach et al. 2022). Some yeast strains also exhibit probiotic properties, contributing to gut health and producing beta-glucans, which have been shown to lower cholesterol and boost immunity (Singh and Bhardwaj 2023). Furthermore, yeasts generate organic acids like lactic and citric acid, which act as natural preservatives and enhance flavor and nutrient bioavailability (Tofalo et al. 2020). Certain yeast strains can also produce bioactive peptides with antihypertensive, antioxidant, and antimicrobial properties, further increasing their value in functional food development.

13.2.3 Marine and Extremophilic Microorganisms

Marine and extremophilic microorganisms thrive in harsh environments such as deep-sea vents, polar regions, and acidic hot springs. These organisms produce unique bioactive compounds, including enzymes, pigments, polysaccharides, and secondary metabolites, that possess exceptional stability and bioactivity (Cyske et al. 2021; Rani et al. 2021). Their ability to function under extreme conditions makes them valuable for enhancing the nutritional profile, shelf life, and processing efficiency of functional food ingredients, offering novel solutions for health and wellness.

13.3 BIOACTIVE COMPOUNDS FROM MICROORGANISMS

Microorganisms that produce probiotics and prebiotics play a crucial role in promoting human health, as illustrated in Table 13.1. Probiotics are live beneficial bacteria found in fermented foods like sauerkraut and yogurt. They contribute to gut health by balancing the microbiome, improving digestion, and potentially alleviating conditions like inflammatory bowel disease and irritable bowel syndrome (Dahiya and Nigam 2022). Prebiotics, on the other hand, are indigestible fibers that serve as nourishment for these beneficial microorganisms, promoting their growth and activity. Foods rich in prebiotics, such as garlic, onions, and whole grains, can enhance the effectiveness of probiotics by fostering a healthy gut environment (Aziz et al. 2024). Together, probiotics and prebiotics support immune function, metabolic health, and may even influence mental well-being through the gut-brain axis.

Microbial enzymes such as amylases, proteases, and lipases, derived from various microorganisms, are utilized in food processing to enhance the nutritional value, flavor, and texture of food products (Jan et al. 2024). These enzymes are favored over plant and animal-derived enzymes due to their increased stability and efficiency, enabling cost-effective and consistent production in industrial applications (Niyonzima et al 2020). Additionally, secondary metabolites, including compounds like antibiotics, pigments, and enzyme inhibitors, are produced during the stationary phase of microbial growth (Ofoedum et al. 2024).

TABLE 13.1
Overview of Microorganisms and Their Bioactive Compounds in Functional Foods

Microorganisms	Source	Bioactive Compounds	Potential Application in Functional Foods	References
Lactobacillus rhamnosus	Dairy products, fermented foods	Prebiotics, lactic acid	It enhances gut health, boosts immune function, and supports digestive health.	Gamallat et al. (2019)
Saccharomyces cerevisiae	Yeast fermentation, alcoholic beverages	β-glucans, enzymes	Improves immune function, lowers cholesterol, and supports cardiovascular health	Najm and Muhsen (2023)
Bifidobacterium longum	Human gut, fermented dairy	Short-chain fatty acids (SCFAs)	Promotes gut health, modulates immune response, and anti-inflammatory effects.	Yao et al. (2021)
Aspergillus oryzae	Fermented soybeans, miso, soy sauce	Proteases, amylases, lipases	Aids in digestion; improve protein and carbohydrate breakdown and enhance food flavor	Murthy and Kusumoto (2015)
Streptomyces sp.	Soil, extreme environments	Antibiotics, enzymes, antioxidants	Adds antimicrobial properties to food, and natural preservative potential.	Kumar et al. (2021)
Rhodotorula sp.	Marine environments soil	Carotenoids, lipids	Enhances the antioxidant content in foods, and provides natural pigmentation.	Ochoa-Viñals et al. (2024)
Spirulina platensis	Algae, fresh water	Phycocyanin, proteins, vitamins	Rich in antioxidants, and protein fortification, improves overall nutrition.	Kumar et al. (2023)
Bacillus subtilis	Soil, fermented foods	Surfactin, antimicrobial peptides	A natural preservative in fermented foods enhances safety and shelf-life	Mnif et al. (2012)
Pediococcus pentosaceus	Fermented vegetables, dairy products.	Bacteriocins	Inhibits pathogenic bacteria in food, and contributes to food preservation.	Jiang et al. (2021)
Kluyveromyces lactis	Whey, dairy products	Lactase, β-galactosidase	Improves lactose digestion, and enhances dairy product nutritional value.	Akyüz and Mazı (2020)

These metabolites contribute to food preservation, flavor enhancement, and may even offer health benefits, making them valuable assets in the food industry. The production of these metabolites is influenced by nutrient availability and environmental conditions, highlighting their potential for biotechnological applications in food and nutrition.

Natural preservatives and antimicrobial peptides (AMPs) are essential for improving food safety and extending shelf life. AMPs, bioactive molecules produced by various organisms, exhibit potent antimicrobial activity by disrupting microbial cell membranes and inhibiting essential cellular functions (Datta and Roy 2021). The biocompatibility and biodegradability of AMPs make them attractive alternatives to synthetic preservatives, which can raise health concerns due to potential adverse effects (Motelica et al. 2020). Moreover, AMPs can maintain food quality by preventing spoilage and foodborne illnesses. They have applications in food packaging and can significantly inhibit microbial growth and contamination even at low concentrations. Additionally, AMPs can reduce lipid oxidation in certain meat products (Liu et al. 2021).

13.4 TECHNOLOGICAL APPROACHES IN MICROBIAL BIOPROSPECTING

The discovery and characterization of microbial strains with potential applications in functional foods have been significantly advanced by genomics and metagenomics. Genomics, involving the analysis of an organism's entire DNA sequence, allows scientists to investigate the genetic blueprint of individual microbes. This enables the identification of the genes responsible for producing bioactive compounds, conferring resistance to environmental stressors, and regulating metabolic pathways (Satam et al. 2023). Metagenomics expands this exploration to entire microbial communities, including those that are unculturable or exist in symbiotic relationships (Suman et al. 2022). By analyzing the collective genomes of microbial communities from environmental samples, metagenomic provides insights into their functional potential, diversity, and structure. Metagenomics has been instrumental in identifying microbial strains that contribute to health benefits, such as probiotics with specific metabolic activities that enhance gut health or immune function (Srinivas et al. 2022). Also, metagenomics data can be used to track the dynamics of microbial communities during food fermentation processes, enabling the optimization of these processes for improved product quality and safety.

Advancements in fermentation methods have revolutionized the generation of functional components from microbes, improving the productivity and scale of bioactive chemical production (Sun et al. 2022). Modern fermentation techniques such as submerged fermentation (SmF) and solid-state fermentation (SSF), allow for precise control over microbial growth conditions, optimizing the yield of desired compounds like enzymes, vitamins, probiotics, and bioactive peptides (Manfredini et al. 2021; Nemes et al. 2022). The diverse roles of food microbes extend beyond functional foods, encompassing probiotics for gut health, live biotherapeutics for disease treatment, soil and ocean bioremediation for environmental cleanup, biofertilizers for agriculture, and animal feed for enhancing livestock and pet health, as shown in Figure 13.1.

Furthermore, innovations in bioreactor design, continuous fermentation, and metabolic engineering have further enhanced the sustainability and scalability of these processes (Boodhoo et al. 2022). By manipulating microorganisms' metabolic pathways, scientists can enhance the synthesis of specific functional components. Additionally, the integration of microbial consortia and co-cultures in fermentation processes enables the synthesis of complex chemicals that are challenging to produce with single organisms (Geraldi et al. 2021). Advances in omics technologies, such as genomics and metabolomics, provides insights into microbial metabolic networks, aiding the identification of novel microorganisms from extreme environments. These microorganisms offer the potential for producing unique and robust bioactive compounds suitable for functional foods and nutraceuticals (Zhang et al. 2024).

Innovative bioreactor technologies have significantly advanced the optimization of microbial growth and productivity, particularly for the production of functional ingredients (Augustin et al. 2024). Modern bioreactor designs emphasize precise control of environmental factors, such as oxygen content, pH, temperature, and nutrition availability, to ensure ideal conditions for microbial metabolism and the production of bioactive compounds (Kowalczyk et al. 2022). One key innovation is the development of single-use bioreactors, which reduce contamination risks and improve operational flexibility (Takao et al. 2023). Modular bioreactors offer customization and adaptation to the specific needs of different microorganisms, facilitating improved yields. The adoption of continuous bioreactors, which operate continuously and provide steady-state conditions that boost microbial production, represents another significant advancement (Tripathi and Shrivastava 2019). These methods lower manufacturing costs, enhance resource utilization, and minimize downtime. Furthermore, sophisticated sensors and automation systems integrated into bioreactors enable real-time monitoring and control of fermentation parameters, improving product quality (Gargalo et al. 2022). Hybrid bioreactors, combining elements of both SSF and SmF, are gaining popularity, offering a versatile platform for cultivating diverse microbial species (Ezeorba et al. 2024). These systems optimize nutrient distribution and oxygen transfer, which is particularly crucial in the cultivation of both aerobic and anaerobic microorganisms.

13.5 APPLICATIONS IN FUNCTIONAL FOOD DEVELOPMENT

13.5.1 Fermented Foods

Microbial fermentation enhances the bioavailability of nutrients such as organic acids, vitamin B, and amino acids.

FIGURE 13.1 Food microbes: a multifaceted function.

Fermented foods like kimchi, kefir, and yogurt often contain prebiotics, which promote gut health, digestion, and immune function. The fermentation process breaks down complex compounds like lactose and fiber, making fermented foods more digestigble, particularly for those with sensitivities (Shah et al. 2023). Additionally, fermentation generates organic acids and other antibacterial compounds, enhancing the shelf life of foods like kimchi without the need for chemical preservatives. Moreover, microbial fermentation produces compounds with antibacterial, anti-inflammatory, and antioxidant properties, contributing to overall health and disease prevention (Shahbazi et al. 2021).

13.5.2 Nutraceuticals

Microbial derivatives of prebiotics and probiotics are formulated into supplements to promote gut health, enhance immunity, and support mental well-being (Zawistowska-Rojek et al. 2022). Microorganisms are also utilized to produce bioactive peptides with antihypertensive, antioxidant, and anti-inflammatory activities, which are incorporated into nutraceuticals (Olvera-Rosales et al. 2023). Furthermore, microbial processes can enhance the bioavailability and potency of plant-based extracts, improving their effectiveness in nutraceutical products. Microorganisms are also employed to produce enzymes like lactase and vitamins like B12 and K2, which are integrated into nutraceuticals designed to address specific deficiencies (Porzi et al. 2021).

13.5.3 Fortified Foods

During fermentation, bacteria and yeasts can produce vitamins such as B12, D, and K2 which are then used to fortify foods, particularly plant-based and vegan options. Fermentation also enhances the bioavailability of minerals like calcium, iron, and magnesium, supporting bone health, reducing anemia risk, and improving overall nutrient absorption (Manzoor et al. 2024). Certain microorganisms can produce omega-3 fatty acids, which can be added to fortified foods to promote heart and brain health. Additionally, microbial fermentation can generate functional fats and oils rich in healthy lipids, further fortifying foods and contributing to cardiovascular health and reduced inflammation (Karageorgou et al. 2023; Mazzocchi et al. 2021). Moreover, fermentation can improve the sensory qualities of fortified foods, enhancing taste and texture while preserving their health benefits, as shown in Figure 13.2.

13.6 CHALLENGES AND LIMITATIONS

Regulatory hurdles pose a significant challenge in the commercialization of microbial products. Organizations like the European Food Safety Authority (EFSA) and the Food and Drug Administration (FDA) mandate extensive safety evaluations and testing before these products can reach the market. This involves a rigorous assessment of potential allergenicity, toxicity, and long-term health effects. Even for microbial species generally recognized as safe (GRAS),

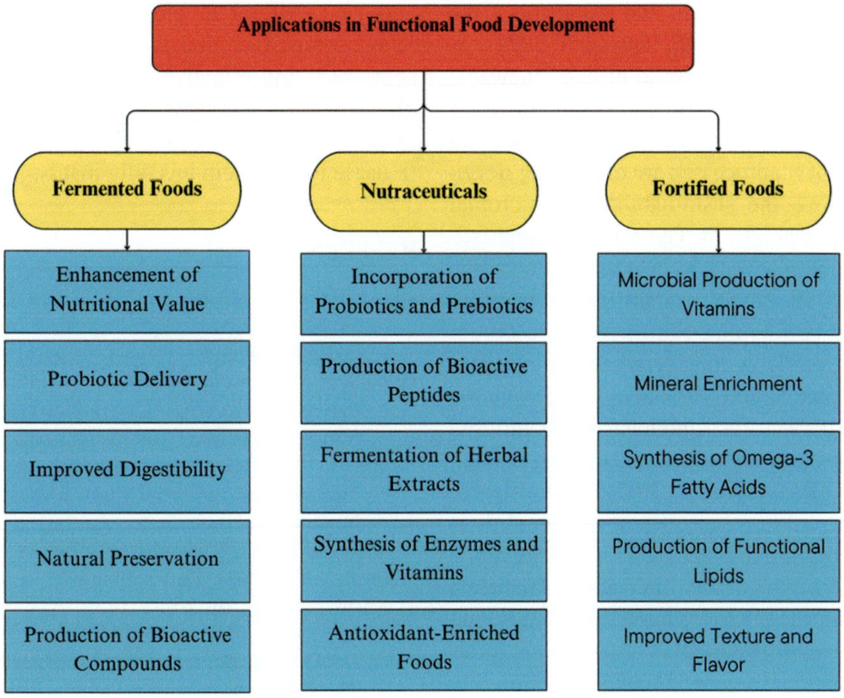

FIGURE 13.2 Applications in functional food development.

specific strains used in fermentation may require individual approval, leading to prolonged and costly commercialization processes (Grujović et al. 2022). Additionally, strict regulations govern health claims associated with microbial products in functional foods, limiting the ability to fully market their benefits. Moreover, the risk of contamination with pathogenic microbes during fermentation necessitates stringent quality control measures to ensure consumer safety (Mendonca et al. 2020).

Scaling up microbial fermentation from laboratory to industrial scale presents numerous challenges. Optimizing bioreactor conditions to maintain consistent microbial performance and prevent contamination at a large scale is a primary difficulty (Kumar et al. 2021). This demands precise control over environmental parameters and careful management of microbial populations. Furthermore, producing microbial-derived compounds can be expensive, particularly when specialized media, equipment, and containment measures are required for large-scale productions. Costs can further escalate due to the need for a stable and reliable supply of raw materials, including specific microbial strains and fermentation substrates (Ye et al. 2024). Supply chain fluctuations can add economic strain, impacting the overall feasibility of microbial product manufacturing. Industrial-scale fermentation also consumes substantial energy, water, and other resources, potentially hindering sustainability and profitability. A successful large-scale operation often depends on adopting efficient technologies that minimize resource consumption and enhance overall process efficiency (Kehrein et al. 2020).

Despite the health benefits of microbial products, some consumers may hesitate to embrace them due to misconceptions about safety, hygiene, or a perception of them being unnatural. Unfamiliarity with ingredients or acceptance of microbial products can vary significantly across regions and cultures, with some populations more receptive to fermented foods than others, often based on dietary traditions. Scientific terms, like "fermented by microbes" or "probiotic cultures", can also influence purchasing decisions. Acceptance of microbial products can vary widely across regions and cultures, with some populations more receptive to fermented foods than others, often based on dietary traditions (Yotova et al. 2022). However, a lack of public understanding regarding the health benefits and safety of microbial-derived functional foods can hinder market adoption, highlighting the need for better consumer education and effective marketing strategies.

13.7 FUTURE PERSPECTIVES

One of the most exciting future developments in microbial bioprospecting for functional foods is the development of next-generation microbial strains through advances in synthetic biology and genomics. These innovations are enabling the creation of microorganisms with enhanced capabilities to produce functional ingredients, such as novel probiotics, enzymes, and bioactive compounds (Neekhra et al. 2022). Additionally, the use of microbial consortia, where multiple species work together to produce more complex and diverse compounds, is gaining traction. This approach not only enhances the health benefits of functional foods but also improve the robustness and efficiency of fermentation processes. Precision fermentation is another emerging trend, allowing for the targeted production of specific compounds.

This increases yield, reduces waste, and streamlines the development of functional ingredients (Day et al. 2022).

Furthermore, efforts are underway to utilize renewable resources, such as agricultural by-products and waste, as fermentation substrates, thus minimizing environmental impact. Energy-efficient bioprocesses are also being developed to further improve the sustainability of microbial production (Singhania et al. 2022). These include sophisticated bioreactor designs that recycle heat and water while conserving energy. Additionally, scientists are working to engineer microbial strains that can thrive in resource-limited environments, reducing the overall ecological footprint of large-scale fermentation. Advances in microbiome research are paving the way for functional foods tailored to individual gut microbiota compositions, offering personalized health benefits (Kashyap et al. 2017). Customizable probiotics designed to address specific health conditions like digestive disorders, immune deficiencies, or metabolic diseases are likely to emerge as a key innovation (Abouelela and Helmy 2024). Integrating nutrigenomics with microbial bioprospecting could enable the creation of functional foods that align with an individual's genetic makeup, optimizing nutrition and health outcomes based on personal metabolic and microbiome data.

13.8 CONCLUSION

Exploration and application of microorganisms in functional foods represent a significant advancement in biotechnology. Since microbial bioprospecting has evolved from its early focus on medicinal compounds to a broader scope, encompassing nutrition, health, and food safety, there have been new avenues for enhancing food quality and human health. Microbial bioprospecting stands as a testament to the incredible potential harbored within the microscopic world. From its origins in the discovery of life-saving antibiotics to its current role in shaping the future of food, this field has consistently pushed the boundaries of scientific exploration and innovation. The progressive convergence of genomics, metagenomics, and advanced fermentation techniques signifies a promising frontier in the exploration of microbial diversity. This synergy not only facilitates a deeper understanding of microbial genomes and community dynamics but also empowers the targeted production of bioactive compounds and functional ingredients. This ongoing integration holds the potential to revolutionize the development of functional foods, expanding their scope beyond mere nutritional sustenance to actively contribute to human health and well-being. This progress underscores the importance of microbial bioprospecting in developing sustainable and innovative functional foods that cater to the growing demand for health-promoting products.

While challenges persist in terms of regulatory hurdles, scalability, and consumer acceptance, the potential rewards are immense. As we delve deeper into the intricate world of microbial interactions and metabolic pathways, we are poised to harness their power to create a new generation of functional foods that not only nourish but also protect and heal. Microbial bioprospecting has the potential to revolutionize the food industry, offering sustainable, personalized, and effective solutions to address the growing global health challenges. The future of food is undoubtedly microbial, and the journey toward a healthier and more sustainable food system has only just begun.

REFERENCES

Abdel -Razek AS, El-Naggar ME, Allam A, Morsy OM, Othman SI. Microbial natural products in drug discovery. *Processes*. 2020;8(4):470.

Abouelela ME, Helmy YA. Next-generation probiotics as novel therapeutics for improving human health: current trends and future perspectives. *Microorganisms*. 2024;12(3):430.

Akyüz G, Mazı BG. Physicochemical and sensory characterization of bread produced from different dough formulations by *Kluyveromyces lactis*. *Journal of Food Processing and Preservation*. 2020;44(6):e14498.

Alam K, Abbasi MN, Hao J, Zhang Y, Li A. Strategies for natural products discovery from uncultured microorganisms. *Molecules*. 2021;26(10):2977.

Anee IJ, Alam S, Begum RA, Shahjahan RM, Khandaker AM. The role of probiotics on animal health and nutrition. *The Journal of Basic and Applied Zoology*. 2021;82:1–6.

Ashaolu TJ. Immune boosting functional foods and their mechanisms: a critical evaluation of probiotics and prebiotics. *Biomedicine & Pharmacotherapy*. 2020;130:110625.

Augustin MA, Hartley CJ, Maloney G, Tyndall S. Innovation in precision fermentation for food ingredients. *Critical Reviews in Food Science and Nutrition*. 2024;64(18):6218–38.

Aware C, Jadhav J. Bioprospecting potential of microbes for the therapeutic application. In Verma P, Shah MP, editors *Bioprospecting of Microbial Diversity Challenges and Applications in Biochemical Industry, Agriculture and Environment Protection* (2022, pp. 223–55). Amsterdam (NL): Elsevier.

Aziz T, Hussain N, Hameed Z, Lin L. Elucidating the role of diet in maintaining gut health to reduce the risk of obesity, cardiovascular and other age-related inflammatory diseases: recent challenges and future recommendations. *Gut Microbes*. 2024;16(1):2297864.

Baeyens J, Zhang H, Nie J, Appels L, Dewil R, Ansart R, Deng Y. Reviewing the potential of bio-hydrogen production by fermentation. *Renewable and Sustainable Energy Reviews*. 2020 Oct 1;131:110023.

Bauer-Estrada K, Sandoval-Cuellar C, Rojas-Muñoz Y, Quintanilla-Carvajal MX. The modulatory effect of encapsulated bioactives and probiotics on gut microbiota: improving health status through functional food. *Food & Function*. 2023;14(1):32–55.

Benkeblia N. Fungi in food processing. In Benkeblia N, editor *Mycoagroecology* (2022, pp. 243–63). Boca Raton (FL): CRC Press.

Boodhoo KV, Flickinger MC, Woodley JM, Emanuelsson EA. Bioprocess intensification: a route to efficient and sustainable biocatalytic transformations for the future. *Chemical Engineering and Processing-Process Intensification*. 2022;172:108793.

Chandra P, Sharma RK, Arora DS. Antioxidant compounds from microbial sources: a review. *Food Research International*. 2020;129:108849.

Chen P. Lactic acid bacteria in fermented food. In Dhanasekaran D, Sankaranarayanan A, editors *Advances in Probiotics: Microorganisms in Food and Health* (2021, pp. 397–416). Academic Press, Elsevier.

Corral P, Amoozegar MA, Ventosa A. Halophiles and their biomolecules: recent advances and future applications in biomedicine. *Marine Drugs*. 2019;18(1):33.

Cyske Z, Jaroszewicz W, Żabińska M, Lorenc P, Sochocka M, Bielańska P, Grabowski Ł, Gaffke L, Pierzynowska K, Węgrzyn G. Unexplored potential: biologically active compounds produced by microorganisms from hard-to-reach environments and their applications. *Acta Biochimica Polonica*. 2021;68(4):565–74.

Dahiya D, Nigam PS. Probiotics, prebiotics, synbiotics, and fermented foods as potential biotics in nutrition improving health via microbiome-gut-brain axis. *Fermentation*. 2022;8(7):303.

Damián MR, Cortes-Perez NG, Quintana ET, Ortiz-Moreno A, Garfias Noguez C, Cruceño-Casarrubias CE, Sánchez Pardo ME, Bermúdez-Humarán LG. Functional foods, nutraceuticals, and probiotics: a focus on human health. *Microorganisms*. 2022;10(5):1065.

Datta S, Roy A. Antimicrobial peptides as potential therapeutic agents: a review. *International Journal of Peptide Research and Therapeutics*. 2021; 27:555–77.

Day L, Altermann E, Chanyi R, Hicks T, Knowles S, Mullaney J. *Fermentation for Future Food Systems–What Are the Opportunities for New Zealand?* (2022). Lincoln, New Zealand: AgResearch Ltd.

Elsawey H, Patz S, Nemr RA, Sarhan MS, Hamza MA, Youssef HH, Abdelfadeel MR, Daanaa HS, El-Tahan M, Abbas M, Fayez M. Plant broth-(not bovine-) based culture media provide the most compatible vegan nutrition for in vitro culturing and in situ probing of plant microbiota. *Diversity*. 2020;12(11):418.

Ezeorba TP, Okeke ES, Mayel MH, Nwuche CO, Ezike TC. Recent advances in biotechnological valorization of agro-food wastes (AFW): optimizing integrated approaches for sustainable biorefinery and circular bioeconomy. *Bioresource Technology Reports*. 2024:101823.

Gamallat Y, Ren X, Meyiah A, Li M, Ren X, Jamalat Y, Song S, Xie L, Ahmad B, Shopit A, Mousa H. The immune-modulation and gut microbiome structure modification associated with long-term dietary supplementation of Lactobacillus rhamnosus using 16S rRNA sequencing analysis. *Journal of Functional Foods*. 2019;53:227–36.

Gargalo CL, Lopez PC, Hasanzadeh A, Udugama IA, Gernaey KV. On-line monitoring of process parameters during fermentation. In Pandey A, Negi S, Soccol CR, editors. *Current Developments in Biotechnology and Bioengineering*. Amsterdam (NL): Elsevier; 2022. p. 117–64.

Geraldi A, Khairunnisa F, Farah N, Bui LM, Rahman Z. Synthetic scaffold systems for increasing the efficiency of metabolic pathways in microorganisms. *Biology*. 2021;10(3):216.

Grujović MŽ, Mladenović KG, Semedo-Lemsaddek T, Laranjo M, Stefanović OD, Kocić-Tanackov SD. Advantages and disadvantages of non-starter lactic acid bacteria from traditional fermented foods: potential use as starters or probiotics. *Comprehensive Reviews in Food Science and Food Safety*. 2022;21(2):1537–67.

Harzallah D, Belhadj H. Lactic acid bacteria as probiotics: characteristics, selection criteria and role in immunomodulation of human GI mucosal barrier. In Kongo M, ed., *Lactic Acid Bacteria— R & D for Food, Health, and Livestock Purposes* (2013, pp. 197–216). Rijeka (Croatia): IntechOpen.

Jach ME, Serefko A, Ziaja M, Kieliszek M. Yeast protein as an easily accessible food source. *Metabolites*. 2022;12(1):63.

Jan B, Abass S, Ahmad S. Application of microbial enzymes in food industry. In Ahmad F, Mohammad ZH, Ibrahim SA, Zaidi S, editors. *Microbial Biotechnology in the Food Industry: Advances, Challenges, and Potential Solutions* (2024, pp. 323–36). Cham: Springer International Publishing.

Jiang S, Cai L, Lv L, Li L. *Pediococcus pentosaceus*, a future additive or probiotic candidate. *Microbial Cell Factories*. 2021;20:1–4.

Kapoor D, Sharma P, Sharma MM, Kumari A, Kumar R. Microbes in pharmaceutical industry. *Microbial Diversity, Interventions and Scope*. 2020:259–99.

Karageorgou D, Rova U, Christakopoulos P, Katapodis P, Matsakas L, Patel A. Benefits of supplementation with microbial omega-3 fatty acids on human health and the current market scenario for fish-free omega-3 fatty acid. *Trends in Food Science & Technology*. 2023;136:169–80.

Kashyap PC, Chia N, Nelson H, Segal E, Elinav E. Microbiome at the frontier of personalized medicine. *Mayo Clinic Proceedings*. 2017;92(12):1855–64. Rochester (MN): Elsevier.

Kehrein P, Van Loosdrecht M, Osseweijer P, Garfí M, Dewulf J, Posada J. A critical review of resource recovery from municipal wastewater treatment plants–market supply potentials, technologies, and bottlenecks. *Environmental Science: Water Research & Technology*. 2020;6(4):877–910.

Khalil D, El-Zayat SA, El-Sayed MA. Phytochemical screening and antioxidant potential of endophytic fungi isolated from Hibiscus sabdariffa. *Journal of Applied Biotechnology Reports*. 2020;7(2):116–24.

Kowalczyk T, Merecz-Sadowska A, Picot L, Brčić Karačonji I, Wieczfinska J, Śliwiński T, Sitarek P. Genetic manipulation and bioreactor culture of plants as a tool for industry and its applications. *Molecules*. 2022;27(3):795.

Kumar A, Mukhia S, Kumar R. Industrial applications of cold-adapted enzymes: challenges, innovations and future perspective. *3 Biotech*. 2021;11(10):426.

Kumar M, Kumar P, Das P, Solanki R, Kapur MK. Protection of surplus food from fungal spoilage using Streptomyces spp.: a green approach. *Archives of Microbiology*. 2021;203:941–50.

Kumar R, Sharma V, Das S, Patial V, Srivatsan V. Arthrospira platensis (Spirulina) fortified functional foods ameliorate iron and protein malnutrition by improving growth and modulating oxidative stress and gut microbiota in rats. *Food & Function*. 2023;14(2):1160–78.

Kumar V, Ahluwalia V, Saran S, Kumar J, Patel AK, Singhania RR. Recent developments on solid-state fermentation for production of microbial secondary metabolites: challenges and solutions. *Bioresource Technology*. 2021;323:124566.

Liu L, Chen X, Hao L, Zhang G, Jin Z, Li C, Yang Y, Rao J, Chen B. Traditional fermented soybean products: processing, flavor formation, nutritional and biological activities. *Critical Reviews in Food Science and Nutrition*. 2022;62(7):1971–89.

Liu Y, Sameen DE, Ahmed S, Dai J, Qin W. Antimicrobial peptides and their application in food packaging. *Trends in Food Science & Technology*. 2021; 112:471–83.

Ljungh A, Wadström T. Lactic acid bacteria as probiotics. *Current Issues in Intestinal Microbiology*. 2006;7(2):73–90.

sakowska P, Sobota A, Wirkijowska A. Medicinal mushrooms: their bioactive components, nutritional value and application in functional food production—a review. *Molecules*. 2023;28(14):5393.

Manfredini PG, Cavanhi VA, Costa JA, Colla LM. Bioactive peptides and proteases: characteristics, applications and the simultaneous production in solid-state fermentation. *Biocatalysis and Biotransformation*. 2021;39(5):360–77.

Manzoor MF, Ali A, Ain HB, Kausar S, Khalil AA, Aadil RM, Zeng XA. Bioaccessibility mechanisms, fortification strategies, processing impact on bioavailability, and therapeutic potentials of minerals in cereals. *Future Foods*. 2024:100425.

Mazziotta C, Tognon M, Martini F, Torreggiani E, Rotondo JC. Probiotics mechanism of action on immune cells and beneficial effects on human health. *Cells*. 2023;12(1):184.

Mazzocchi A, De Cosmi V, Risé P, Milani GP, Turolo S, Syrén ML, Sala A, Agostoni C. Bioactive compounds in edible oils and their role in oxidative stress and inflammation. *Frontiers in Physiology*. 2021;12:659551.

Mendonca A, Thomas-Popo E, Gordon A. Microbiological considerations in food safety and quality systems implementation. In Gordon A, editor. *Food Safety and Quality Systems in Developing Countries* (2020, pp. 185–260). Amsterdam (Netherlands): Academic Press.

Mishra B, Mishra AK, Kumar S, Mandal SK, Nsv L, Kumar V, Baek KH, Mohanta YK. Antifungal metabolites as food bio-preservative: innovation, outlook, and challenges. *Metabolites*. 2021;12(1):12.

Mnif I, Besbes S, Ellouze R, Ellouze-Chaabouni S, Ghribi D. Improvement of bread quality and bread shelf-life by Bacillus subtilis biosurfactant addition. *Food Science and Biotechnology*. 2012;21:1105–12.

Motelica L, Ficai D, Ficai A, Oprea OC, Kaya DA, Andronescu E. Biodegradable antimicrobial food packaging: trends and perspectives. *Foods*. 2020;9(10):1438.

Murthy PS, Kusumoto KI. Acid protease production by Aspergillus oryzae on potato pulp powder with emphasis on glycine releasing activity: a benefit to the food industry. Food and Bioproducts Processing. 2015;96:180–8.

Najm M, Muhsen A. Extracting β-glucan from Saccharomyces cerevisiae, and its biological effect in reducing cholesterol levels in the laboratory and studying its effect. *Euphrates Journal of Agricultural Science*. 2023;15(2):424–33.

Neekhra S, Pandith JA, Mir NA, Manzoor A, Ahmad S, Ahmad R, Sheikh RA. Innovative approaches for microencapsulating bioactive compounds and probiotics: an updated review. *Journal of Food Processing and Preservation*. 2022;46(11):e16935.

Nemes SA, Călinoiu LF, Dulf FV, Fărcas AC, Vodnar DC. Integrated technology for cereal bran valorization: perspectives for a sustainable industrial approach. *Antioxidants*. 2022;11(11):2159.

Niyonzima FN, Veena SM, More SS. Industrial production and optimization of microbial enzymes. *Microbial Enzymes: Roles and Applications in Industries*. 2020:115–35.

Nkamga VD, Henrissat B, Drancourt M. Archaea: essential inhabitants of the human digestive microbiota. *Human Microbiome Journal*. 2017;3:1–8.

Ochoa-Viñals N, Alonso-Estrada D, Pacios-Michelena S, García-Cruz A, Ramos-González R, Faife-Pérez E, Michelena-Álvarez LG, Martínez-Hernández JL, Iliná A. Current advances in carotenoid production by *Rhodotorula* sp. *Fermentation*. 2024;10(4):190.

Ofoedum AF, Uyanwa NC, Chikelu EC, Iroagba LN, Ugwoezuonu JN, Anaeke EJ, Odeyemi TA, Okezie FP, Nwuka MU. Primary and secondary metabolites as products of microbial metabolism: uses and application in foods, pharmaceutical and allied industries. A review. *European Journal of Applied Science, Engineering and Technology*. 2024;2(2):4–16.

Olvera-Rosales LB, Cruz-Guerrero AE, García-Garibay JM, Gómez-Ruíz LC, Contreras-López E, Guzmán-Rodríguez F, González-Olivares LG. Bioactive peptides of whey: obtaining, activity, mechanism of action, and further applications. *Critical Reviews in Food Science and Nutrition*. 2023;63(30):10351–81.

Payne J, Bellmer D, Jadeja R, Muriana P. The potential of bacillus species as probiotics in the food industry: a review. *Foods*. 2024;13(15):2444.

Piwowarek K, Lipińska E, Hać-Szymańczuk E, Kieliszek M, Ścibisz I. Propionibacterium spp.—source of propionic acid, vitamin B12, and other metabolites important for the industry. *Applied Microbiology and Biotechnology*. 2018;102:515–38.

Porzi M, Burton-Pimentel KJ, Walther B, Vergères G. Development of personalized nutrition: applications in lactose intolerance diagnosis and management. *Nutrients*. 2021;13(5):1503.

Pushpangadan P, George V, Ijinu TP, Chithra MA. Biodiversity, bioprospecting, traditional knowledge. Sustainable development and value-added products: a review. *Journal of Traditional Medicine & Clinical Naturopathy*. 2018;7(1):1–7.

Rani A, Saini KC, Bast F, Mehariya S, Bhatia SK, Lavecchia R, Zuorro A. Microorganisms: a potential source of bioactive molecules for antioxidant applications. *Molecules*. 2021;26(4):1142.

Satam H, Joshi K, Mangrolia U, Waghoo S, Zaidi G, Rawool S, Thakare RP, Banday S, Mishra AK, Das G, Malonia SK. Next-generation sequencing technology: current trends and advancements. *Biology*. 2023;12(7):997.

Shah AM, Tarfeen N, Mohamed H, Song Y. Fermented foods: their health-promoting components and potential effects on gut microbiota. *Fermentation*. 2023;9(2):118.

Shahbazi R, Sharifzad F, Bagheri R, Alsadi N, Yasavoli-Sharahi H, Matar C. Anti-inflammatory and immunomodulatory properties of fermented plant foods. *Nutrients*. 2021;13(5):1516.

Sharma H, Ozogul F, Bartkiene E, Rocha JM. Impact of lactic acid bacteria and their metabolites on the techno-functional properties and health benefits of fermented dairy products. *Critical Reviews in Food Science and Nutrition*. 2023;63(21):4819–41.

Singh RP, Bhardwaj A. β-glucans: a potential source for maintaining gut microbiota and the immune system. *Frontiers in Nutrition*. 2023;10:1143682.

Singhania RR, Patel AK, Tseng YS, Kumar V, Chen CW, Haldar D, Saini JK, Dong CD. Developments in bioprocess for bacterial cellulose production. *Bioresource Technology*. 2022;344:126343.

Srinivas M, O'Sullivan O, Cotter PD, Sinderen DV, Kenny JG. The application of metagenomics to study microbial communities and develop desirable traits in fermented foods. *Foods*. 2022;11(20):3297.

Suman A, Govindasamy V, Ramakrishnan B, Aswini K, SaiPrasad J, Sharma P, Pathak D, Annapurna K. Microbial community and function-based synthetic bioinoculants: a perspective for sustainable agriculture. *Frontiers in Microbiology*. 2022;12:805498.

Sun W, Shahrajabian MH, Lin M. Research progress of fermented functional foods and protein factory-microbial fermentation technology. *Fermentation*. 2022;8(12):688.

Takao I T, Wang H, Hwang SH, Wang B, Wang L, Somasundaram G. Risk assessment for biopharmaceutical single-use manufacturing: a case study of upstream continuous processing. *Biologicals*. 2023;84:101713.

Tofalo R, Fusco V, Böhnlein C, Kabisch J, Logrieco AF, Habermann D, Cho GS, Benomar N, Abriouel H, Schmidt-Heydt M, Neve H. The life and times of yeasts in traditional food fermentations. *Critical Reviews in Food Science and Nutrition*. 2020;60(18):3103–32.

Tripathi N K, Shrivastava A. Recent developments in bioprocessing of recombinant proteins: expression hosts and process development. *Frontiers in Bioengineering and Biotechnology*. 2019;7:420.

Verma P, Shah MP, editors. *Bioprospecting of Microbial Diversity: Challenges and Applications in Biochemical Industry, Agriculture, and Environment Protection* (2022). Elsevier.

Yao S, Zhao Z, Wang W, Liu X. Bifidobacterium longum: protection against inflammatory bowel disease. *Journal of Immunology Research*. 2021;2021(1):8030297.

Ye L, Bogicevic B, Bolten CJ, Wittmann C. Single-cell protein: overcoming technological and biological challenges towards improved industrialization. *Current Opinion in Biotechnology*. 2024;88:103171.

Yotova M, Utsunomiya Y, Park KH. Acceptance of milk and dairy products in rice-centered food cultures: a comparative study of consumer practices in South Korea and Japan. *Journal of Ritsumeikan Gastronomic Arts and Sciences*. 2022;7:319–30.

Zawistowska-Rojek A, Tyski S. How to improve health with biological agents—narrative review. *Nutrients*. 2022;14(9):1700.

Zhang J, Yang H, Sun Y, Yan B, Chen W, Fan D. The potential use of microalgae for nutrient supply and health enhancement in isolated and confined environments. *Comprehensive Reviews in Food Science and Food Safety*. 2024;23(4):e13418.

14 Marine Microbes
A Treasure Trove of Novel Medicines

Danyal Reyaz, Shriya Pattibaram, Raksa Arun, Srisri Satish Karthik, Vijay Murali RM, Aswini Suresh Kumar, Sayantani Chattopadhyay, Kruthika Prakash, Vasanth Kanth TL, Vanshikaa Karthikeyan, Satish Ramalingam, and Kandasamy Nagarajan Arul Jothi
Department of Genetic Engineering, SRM Institute of Science and Technology, Kattankulathur, Chengalpattu, India

14.1 INTRODUCTION

Ocean, sea, coral reefs, and estuaries are some of the largest and, possibly, the most biodiverse habitats on the planet. These ecosystems occupy over 70% of the earth's surface and are paramount in functioning of global eco-processes such as carbon, nutrient cycling, and climate moderation [1]. Marine ecosystems have the significance that is not only biological but also social and economic since billions of people depend on seas and oceans as their source of food, energy, and recreation.

Seas are large stores of carbon; they take in approximately one third of the carbon dioxide resulting from human activities. This carbon sequestration plays a significant role in reversing some of the effects of global warming [2]. Also, oceans have diverse biological resources; for instance, the number of marine species that are presently unknown is about 80% [3]. These species extend from fish and other large marine mammals to tiny organisms, which remain critical to the ecosystem.

Marine ecosystems are undeniably integral components of the earth's system and hence their health is directly proportional to the earth's health. For instance, coral reefs, which are counted among the most biodiverse ecosystems and are called the 'sea rainforests', occupy not more than 0.1% of the ocean bottom while home to approximately one-quarter of all the marine species. Such energetic systems offer many benefits such as coastal defense, the provision of places for recreation, and discovery of new drugs for medicine. Nonetheless, coral reefs are very vulnerable ecosystems and the loss of reefs due to climate change, pollutants, and overfishing threatens the ocean's bio-diversity and well-being of humans [4].

14.1.1 Overview of Marine Microbial Diversity

The marine organisms that venture around within the ocean include bacteria, archaea, viruses, fungi, and microalgae that are major players of marine systems. These microorganisms are found in all sea habitats, whether they are photic layer or aphotic zone of the ocean. They are among the organisms that are crucial in nutrient cycling, primary production, and contribution toward the sustenance of the ecology [5]. For instance, marine phytoplankton, which are particles of photosynthesizing microorganisms, participate in nearly half of the photosynthetic reach of the world and are the base of the marine food chain and the controllers of global biogeochemical cycles [6].

The diversity of marine microbes is extraordinary. In surface ocean waters alone, there are an estimated 3.6×10^{29} viruses, 3.1×10^{28} prokaryotes, and 5.8×10^{27} eukaryotic microbes [7]. This vast diversity is also coupled with a diverse metabolic flexibility whereby marine microbes can inhabit environments ranging from highly productive coastal regions to the low nutrient oligotrophic waters and other harsh conditions such as deep-sea hydrothermal vents and polar ice caps [8].

Some of the key marine microbes include those that play important roles as decomposers, recycling nutrients like nitrogen and sulfur. In this way, nitrogen-fixing bacteria, for example, work for the creation of the forms of nitrogen which can be appointed by other organisms, thus encouraging primary production in the regions of the ocean where materials are scarce [9]. Sulfur-oxidizing bacteria should be defined in a similar context as presently they are primarily involved in the conversion of sulfide into sulfate, which can then be used by other organisms [10].

14.1.2 Significance of Marine Microbes in Biotechnology and Medicine

Marine microbes are a treasure trove of novel bioactive compounds with immense potential in biotechnology and medicine. Over the past few decades, marine microorganisms have been increasingly recognized for their ability to produce a wide range of secondary metabolites with unique chemical structures and biological activities [11]. These compounds have shown promising applications in various fields, including pharmaceuticals, agriculture, and industrial processes.

DOI: 10.1201/9781003509530-14

14.1.2.1 Pharmaceuticals and Medicine

In the medical field marine microbial metabolites are the antibiotics, antiviral, anticancer, anti-inflammatory, immunosuppressive agents. For example, *Streptomyces* species derived from marine environment has been a rich source of antibiotics and among the compounds Salinosporamide A has shown tremendous anticancer properties [12]. Marizomib, which is known by its natural name Salinosporamide A, has been offered in clinical trials in multiple myeloma and glioblastoma, thus the possibility that marine compounds could offer solutions where none previously existed [13].

Marine fungi also have produced a lot of bioactive compounds. The anticancer agent discodermolide obtained from the marine sponge fungus *Mycobacterium tuberculosis* is another rich source. Thus, it was found that discodermolide possessed high cytotoxic effects against cancer cells by stability of microtubules that makes the compound as a candidate for anticancer therapy. Another example is the polyketide macrolide bryostatin derived from the marine bryozoan *Bugula neritina*, which possesses the hope for the treatment of Alzheimer's disease and cancer due to its effect on protein kinase C activity [14].

Due to the specific conditions that occur within the marine environment, like high pressure, low temperature, and the effects of salinity, marine microbes have developed new metabolic pathways and biosynthetic potential. This has paved way to the formation of structurally complex and pharmacologically active compounds that exists in marine organisms and not in the terrestrial ones [15]. For instance, hamycin is an antifungal compound being isolated from the marine actinomycete *Streptomyces* sp. This drug demonstrates a specific mechanism of action owing to which it is useful in combating new species of fungi that are hardly responsive to any other antifungal drugs [16].

14.1.2.2 Biotechnology and Industrial Applications

Beyond pharmaceuticals, marine microbes offer numerous applications in biotechnology and industrial processes. Enzymes derived from marine microorganisms have unique properties that make them valuable for industrial applications. For instance, marine-derived proteases, lipases, and amylases are used in the food industry for processing and preserving seafood, dairy products, and baked goods [17]. These enzymes are often more stable and active under harsh conditions, such as high salinity and extreme temperatures, compared to their terrestrial counterparts.

Marine microbes are also a promising source of biofuels and bioplastics. Microalgae, for example, can produce high yields of lipids that can be converted into biodiesel, offering a sustainable alternative to fossil fuels [18]. Additionally, certain marine bacteria can produce polyhydroxyalkanoates (PHAs), biodegradable plastics that can be used to reduce the environmental impact of conventional plastics [19].

In the field of agriculture, marine-derived biopesticides and biofertilizers are being developed to promote sustainable farming practices. For example, compounds produced by marine actinomycetes have shown potent insecticidal and fungicidal activities, providing an eco-friendly alternative to synthetic pesticides. Similarly, biofertilizers derived from marine cyanobacteria can enhance soil fertility and promote plant growth by fixing atmospheric nitrogen and producing growth-promoting substances [20].

14.2 DIVERSITY AND ECOLOGY OF MARINE MICROBES

Marine microbes are a vast group which comprises bacteria, archaea, fungi, algae, and viruses. All these groups have their distinct roles in supporting marine ecosystems varying from the biogeochemical processes to the food chain on the marine environment.

1. **Bacteria:** Marine bacteria constitute the largest and perhaps the most diverse microbial group in the seas. They participate in several factors of the ecosystem like breaking down of organic matter, cycle nutrients, and the production of new plant material. Cyanobacteria (blue-green algae) is another example of primary producers that comprise this group of organisms which fix carbon dioxide into organic compounds with the help of photosynthesis [21]. Such heterotrophic bacteria break down organic matter recycling nutrients back to the ecosystem.
2. **Archaea:** Earlier, they were only believed to inhabit extreme environments but archaea are actually dispersed in marine habitats. Its most common habitats are deep-sea and hydrothermal vent systems. Marine archaea are involved in nitrogen and carbon cycling where some groups such as Thaumarchaeota are involved in ammonia oxidation which is central in the nitrification process [22].
3. **Fungi:** Compared with terrestrial fungi, marine fungi have been paid less attention though they are involved in the decomposition of organic matter and the nutrient cycling process. They live in different habitats, these include shallow waters and shores, mud, and the deep-sea environment. Marine fungi can also be mutualistic partners with other marine organisms like algae and sponges, hence helping their hosts in aspects of health and growth [23].
4. **Algae:** Marine algae are the photosynthetic organisms in the marine environment and cover the field of phytoplankton and other large plants known as seaweed. Various algae including diatoms and dinoflagellates are the key components of the marine food chain and contribute to half of the world's primary production [6]. It supports or houses and feeds a host of organisms in the sea commonly known as marine lives.

5. **Viruses**: Viruses in the marine environment are the most abundant organisms within the ocean estimated to be 10^{31} particles in the sea [7]. It also infects marine bacteria, archaea, and eukaryotic microorganisms that affect the microbial community structure and biogeochemical processes. Viruses contribute to the marine carbon cycle especially by breaking up microbial cells and putting out dissolved organic compounds for other microorganisms to use [24].

14.2.1 Habitats and Ecological Niches

Marine microbes inhabit a wide range of environments, from the surface ocean to the deepest ocean trenches. They occupy diverse ecological niches, each with unique physical and chemical characteristics that influence microbial community composition and function. In the sunlit surface waters, or euphotic zone, phytoplankton thrive due to the availability of light and nutrients, driving primary production and forming the base of the marine food web. Coastal ecosystems, including estuaries, mangroves, and coral reefs, are nutrient-rich environments with high microbial activity, supporting diverse communities involved in organic matter decomposition, nutrient cycling, and symbiotic relationships with higher organisms. The deep sea, characterized by high pressure, low temperature, and low nutrient availability, hosts unique microbial communities adapted to these extreme conditions, playing crucial roles in biogeochemical cycles such as the degradation of organic matter that sinks from the surface [25]. Hydrothermal vents, found along mid-ocean ridges, are extreme environments characterized by high temperatures, high pressure, and the presence of toxic chemicals such as hydrogen sulfide. Despite these harsh conditions, hydrothermal vents support diverse microbial communities, including thermophilic and chemolithoautotrophic bacteria and archaea, which derive energy from the oxidation of inorganic compounds, supporting unique ecosystems independent of sunlight [26]. The Arctic and Antarctic oceans are cold, nutrient-rich environments with unique microbial communities adapted to low temperatures. Psychrophilic (cold-loving) bacteria and algae dominate these regions, playing essential roles in primary production and nutrient cycling. Sea ice also harbors diverse microbial communities, including bacteria, algae, and viruses, contributing to polar ecosystem productivity [27].

14.2.2 Adaptations to Extreme Environments

Life in and around the oceans also shows phenomenal features to cope with hostile conditions such as deep-sea vents and the extreme polar seas. These are physical and other changes that enable microbes to survive in harsh conditions such as high and low temperatures, high pressure, and toxic compounds. The microbes that live in the deep ocean are acclimatized to hydrostatic pressures higher than 1,000 atmosphere. Such adaptations are modification of the composition of cell membranes to exhibit fluid nature and functionality when under high pressure and stable proteins and enzymes. For instance, piezophilic or pressure loving bacterium contains membrane lipids with more unsaturated fatty acids to ensure proper fluidity despite the high pressures [28]. Some of hydrothermal vent microbes are thermophilic bacteria and archaea that can live at several temperature above 80°C. Such microbes have enzymes and proteins that are heat resistant and the organisms thrive well at high temperatures. For instance, the DNA polymerase isolated from Thermus aquaticus – Taq polymerase is used in PCR because it is heat stable [29]. Also, they have different glycerol ester lipids, for instance, ether-linked lipids in archaea to enable them withstand high temperatures [30]. Most of the vent microbes are chemolithoautotrophic, meaning, they obtain energy through various chemical reactions with inorganic compounds such as hydrogen sulfide, methane, and hydrogen. Most of these microbes have intricate enzymes and pathways for metabolizing these compounds for energy. For instance, SOB such as those that use the Sox pathway to oxidize sulfide to sulfate to support primary production in lightless conditions [31]. These polar marine microbes survive in low temperatures with the help of ecofriendly enzymes and antifreeze proteins. Thus a metabolism to take place requires the body to undergo some changes for it to operate in subzero temperatures. A range of adaptations was found such as flexible proteins that contain a higher proportion of surface hydrophilicity and less core hydrophobicity that does not allow freezing and preserves the enzyme activity in low temperatures. Also, some algae contain proteins that prevent the nucleation of ice crystals hence preventing the freezing of cells [32]. Marine microbes in different salinity conditions such as estuaries and salt marshes are subjected to osmoregulatory stress. The halo-tolerant organisms experience high external saline stress and internal compatible solutes, which include glycine, betaine, proline, and ectoine to counter the effects of high-salinity stress [33].

14.3 METHODS FOR ISOLATION AND CULTIVATION

14.3.1 Sampling Techniques from Marine Environments

There are a large number of sampling techniques available for examining the sea surface microlayer. Sea surface microlayer is defined as the top 1 mm of the ocean and it acts as a boundary between the surface of the ocean and the lower atmosphere layer.

Some of the sampling techniques are:

a. **Mesh Screen Sampler:**

   It consists of a metal or plastic frame supported by a metal or plastic mesh. When the screen is dipped horizontally in water and withdrawn, the surface layer gets caught within the spaces in the mesh. Depending on the type of mesh used, the screen sampler traverses a depth of 150–400 um.

b. **Glass Plate Sampler:**

This captures the surface layer by dipping the glass plate (approximately 20×20 cm) into the water, and then it is slowly withdrawn. Then a wiper blade is used to remove the surface layer. The depths traversed by such samplers are 20–100 um [34].

c. **Niskin Bottle Sample Collection:**

Particles that settle at approximately a meter in 1–2 h are sampled using Niskin Bottles. However, some of the particles, which are 4 um in diameter, get lost as they sink in the void between the spout and the bottom of the bottle [35].

d. **Autonomous Underwater Vehicles (AUVs):**

AUVs are used to capture peak samples from phytoplankton patches. Traditionally, these samples were captured by using Niskin bottles lowered from a ship. For this human intervention is needed to locate the phytoplankton and inspect it. But the Dorado AUV contains 20 syringe like water samplers called 'gulpers'. Each of these gulpers can collect 1.5 L water samples in 1–2 s [36].

e. **Autonomous Microbial Sampler (AMS):**

AMS helps in obtaining an exogenous DNA-free, uncontaminated microbial sample from the marine environment. AMS sampling can be done through AUVs, Remotely Operated Vehicles (ROVs), etc. These devices are made of a titanium nozzle, which is used to sample in hot environments (>350°C). The fluid components present help in sampling of six samples that are unfiltered. A microcomputer is present, which helps in controlling of all the samples by internal programming or external devices such as AUV computers. Volume pumped, temperature, and other parameters are also recorded during sampling [37].

14.3.2 Isolation and Cultivation Methods

The isolation and culturing techniques of microbes date back to the nineteenth century, when solid agar media, streak plate techniques, and Petri dishes were developed in Robet Koch's laboratory. Thereafter many novel techniques were introduced to culture these microbes [38].

One of the major culturing techniques is 'Microencapsulation'. Here, gel microdroplets are prepared and are dispersed in agarose. These are then mixed with water, containing bacteria, taken from the natural environment. These are then mixed into oil, a non-aqueous environment, to form an emulsion. Then these emulsions are cooled down and the agarose is solidified. Due to constant stirring single bacterial cells are seen in proportion of these microdroplets.

These microdroplets are larger than the bacterial cells, so they can easily be manipulated, like centrifugation, incubation, pipetting, etc. [39].

Another important cultivation technique is 'Agar plate culture' technique. Strains such as *Pseudoalteromonas luteoviolacea* 2ta16 and *Pseudoalteromonas piscicida* JCM 20779 were cultured using this method. These strains were streaked radially on plates having the composition of 0.5% (w/v) peptone, 2% (w/v) agar in commercial seawater substitute, 0.3% (w/v) yeast extract. These cultures are then incubated at 30°C for 18 h in a shaking incubator of 180 RPM [40].

Diffusion chamber (DC) is considered to be one of the most important in situ cultivation techniques. In this, the microbial cells are inoculated and grown in an agar matrix that is sandwiched between the porous membranes. When these chambers are incubated in the natural environment, the microbes acquire all the growth factors required by the environment, while being physically separated from the rest of the microbes in the environment. Through this technique, many microbes from marine sediments have been cultivated.

Dilution to extinction (DTE) cultivation is an excellent method for isolating marine microbes. Here, a physically separated well is inoculated with 1–10 cells, which separates these cells from mixed populations of bacteria, thus reducing their competition with other bacterial colonies. It was through this technique that SAR11, an elusive marine bacteria, was cultivated along with other uncultivated bacterial groups in the marine habitat [41].

It is difficult to cultivate all microbes from the marine environment, so it is important to select, isolate, and cultivate the most important, key players of these marine microbial environments. Through the statements of Paul Carini, these key players are identified to be those microbes that (i) help in bioremediation and biogeochemistry; (ii) have a greater relative abundance; (iii) can produce natural products; and (iv) can diverge from cultured taxa. One of the major methods considered to culture these key players is the 'Reverse Genomics' method. Here microbial cells from a specific taxonomic group can be cultivated by using antibodies that act against the predicted membrane proteins and target these microbes [38].

14.3.3 Challenges in Cultivating Marine Microbes

Cultivation of these marine microbes is very low and severe in the marine environment. This was considered to be the 'great plate count anomaly' due to the high magnitude difference between the number of colonies cultured in the laboratory and the ones seen using epifluorescence microscopy [39].

Though there are abundant bacteria present in the oceans, only a few of them can be cultivated (0.1%). This is due to the distribution of the isolated piezophiles, whose pressure maxima lie among the psychrophiles present in the hadal zones and the vent hyperthermophiles. Though moderate pressure and temperature conditions (20–60°C and ~40–80 MPa) are present in the deep sea sediments, it is still quite difficult to isolate the piezophiles [42].

It is difficult to cultivate these marine microbes due to the following reasons:

i. There are many interactions occurring in the natural environment that may be destroyed in the laboratory conditions. For example, the fast growing bacteria outgrow the slower ones, creating a cell–cell communication imbalance. Similarly, microbes may get inactivated due to the inhibitory compounds produced.
ii. It is generally difficult to cultivate marine bacteria on a substrate or combination of substrates, as there is very little knowledge of the organic substrates present in the sea.
iii. The growth of the marine bacteria may be affected due to the presence of viruses that infect the bacteria, called bacteriophages.
iv. The increased concentrations of substrates that are used for the growth of marine bacteria may become toxic and inhibit their growth as these bacteria grow under oligotrophic conditions [39].

Rarity is considered to be one of the major hurdles while culturing low abundant marine bacteria. Failure in isolation occurs due to the dilution separation method, which misses most of the microbes. The rarity of such species is mediated by external abiotic factors. For example, a species with a narrow environmental niche may be abundant in certain habitats and rarely available in others, all of which are controlled by the abiotic factors of such habitats.

Similarly, many marine bacteria that are dormant are found to be inactive most of the time, and they become active when favorable conditions are present. This is also known as viable but non culturable (VBNC) condition. Dormant bacteria might grow well in certain habitats, and even if the essential factors of such habitats are mimicked in laboratory conditions, they might not grow well, and it is also quite difficult to isolate them [38].

14.4 MARINE MICROBIAL METABOLITES

Secondary metabolites, with their diverse chemical structures and ubiquity in the natural environment around humanity, have been significant sources of medicines and cosmetics throughout historical contexts. With the discovery of Penicillin, a variety of products have been isolated from microorganisms derived from different habitats to supplement human needs. The ocean is home to the largest, unexplored niches of organisms that have arisen with adaptations to combat stressful environments like cold, high pressure, and low light. These organisms give rise to secondary metabolites with unique properties to deal with the same, featuring a vast potential for the pharmaceutical industry to explore.

14.4.1 Types of Secondary Metabolites

14.4.1.1 Polyketides

Polyketides are a large class of natural products derived from a precursor chain containing alternating ketone and methylene groups [43]. They possess many properties like antibacterial, anticancer, and antiviral activity [44]. They form the major category of compounds produced by *Streptomyces*. *Streptomyces pratensis* strain NA-ZhouS1, obtained off marine sediments in Zhousan, East China Sea, produced angucycline-type glycosides called stremycins A/B. Both compounds were found to have antibacterial activity against various bacterial strains, including *Pseudomonas aeruginosa*, methicillin-resistant *Staphylococcus aureus* (MRSA), *Bacillus subtilis*, and *Escherichia coli* [45]. *Streptomyces* sp. bacterium off the coast of Satna Barbara, California, produce Anthramycin having significant activity against MRSA, and *S. aureus* due to DNA/RNA synthesis inhibition [46]. Deep-sea actinobacterium *Nocardiopsis* sp. strain HB-J378 provided a class of angucycline-type aromatic polyketides called nocardiopsistins A-C, which were found to be active against MRSA [47]. The polyketide abyssomicin C, isolated from a marine strain of *Verrucosispora* AB-18032 off the Sea of Japan, showed strong inhibitory activity against MRSA and vancomycin-resistant *S. aureus* while inhibiting the production of folic acid. The absence of this biosynthetic pathway in humans makes it a viable candidate as an antibacterial, antiprotozoal agent [48].

14.4.1.2 Terpenoids

Terpenoids belong to a class of compounds that includes terpenes, diterpenes, and sesquiterpenes. All of them derive their structure from the 5-carbon isoprene [49]. Terpenoids are highly influential in the realm of anticancer and antimalarial drugs [50]. Sesquiterpenoids derived from *Pestalotiopsis* sp., a fungus isolated from the brown alga *Sargassum horneri*, show tyrosinase inhibitory properties. *Streptomyces* sp. SCSIO 10355 produced a sesquiterpene named strepsesquitriol, suggested for use in anti-inflammatory drugs. This compound inhibits the production of TNF-alpha, a known inflammatory cytokine [51]. Altemicidin, produced by *Streptomyces sioyaensis* SA-1758 from the Miyagi prefecture of Japan, was found to exhibit antiproliferative activity on IMC carcinoma and murine lymphoid leukemia L1210 cells [52]. Cyclomarin A, one of the best anti-inflammatory compounds, is derived from the *Streptomyces* strain CNB-982, collected in Mission Bay, California [53]. This compound exhibits strong anticancerous activity in several cancerous cell lines and antiviral activity [54].

14.4.1.3 Alkaloids

Alkaloids are a class of organic compounds that contain at least one nitrogen atom. Alkaloids derived from a marine sponge called *Petrosia similis*, showing HIV inhibition activity [55]. Another sponge of Amphimedon sp. produce bis-pyridine alkaloids, which are potent cytotoxic agents [56]. Streptoindole B is a type of indole alkaloid extracted from the gut of a marine shrimp (*Penaeus* sp.) found in the Zhoushan archipelago in China. It exhibits strong antibacterial activity against MRSA [57]. Alkaloids like Ammosamides A and B and lodopyridine isolated from *Streptomyces* from the Bahamas deep-sea sediments and marine *Saccharomonospora* sp. in the La Jolla submarine canyon respectively showed cytotoxicity to HCT-116 colon cancer cells [58,59]. Tambjamine-like alkaloid, a yellow pigment derived from *Pseudoalteromonas tunicata*, found on the surface of marine invertebrates, shows antifungal activity [60].

14.4.2 Biosynthetic Pathways

14.4.2.1 Polyketide Synthesis

Polyketides are wholly or partly biosynthesized by nonribosomal peptide synthetases (NRPSs). These synthetases contain three core domains, namely the adenylation, thiolation (or peptidyl carrier protein), condensation domains, and one ancillary thioesterase domain to catalyze the assembly of a nonribosomal peptide [61]. A-domains activate the carboxylate group of the initiating substrate (usually an amino acid) to form an aminoacyl-antimicrobial peptide (AMP) intermediate and transfer it to the neighboring thiolation domain via an ATP-dependent reaction. This addition activates the T domain, which attacks the intermediate to form an aminoacyl thioester [62]. C domain models propose two distinct substrate-binding sites that help catalyze the peptide bond formation. One site binds the amino acid to the T domain, accepting the growing peptide. In contrast, the other binds the aminoacyl thioester to the preceding module, being donated down the NRPS [63]. The thioesterase domains catalyze the peptide release from the NRPS via hydrolysis [64] (Figure 14.1).

14.4.2.2 Terpenoid Synthesis

Despite being ubiquitous and diverse, bacteria produce terpenoids rather occasionally. Bioinformatic studies reveal that over 120 genes are involved in bacterial terpene synthases [65]. The preliminary step is the condensation of isopentenyl diphosphate (IPP) and dimethylallyl diphosphate (DMAPP). These structures are building blocks in the mevalonate (MEV) pathway to produce terpenoids [66]. Acetyl-CoA produces MEV after being constantly reduced or condensed. This MEV gets converted into IPP, which is then isomerized to form DMAPP that produces terpenoids [67]. Marine microbial terpenoids differ from the general categorization of terpenoids due to a characteristic called halogenation. This tendency is much higher in marine-derived terpenoids than in terrestrial microorganisms, mainly due to the relatively higher availability of bromine and chlorine in seawater. Marine microorganisms adopt halogenation to amplify the bioactivity of the terpenoids [68].

14.4.2.3 Alkaloid Synthesis

Alkaloids are heterocyclic compounds that are soluble in organic solvents. The biosynthesis of alkaloids follows four crucial steps. In the first step, amine accumulation, primary amines are accumulated beyond average concentrations to be redirected into alkaloid metabolism pathways. These primary amines can either be polyamines or aromatic amines [69]. The next step involves the accumulation of an aldehyde precursor, especially one in tandem with the amine precursor. In the case of a polyamine-derived alkaloid, oxidation deaminates a terminal amine on the precursor molecule. A secondary metabolite is a precursor for aromatic-derived alkaloids [70]. An amine and an aldehyde condense to form an electrophilic iminium cation. This iminium reacts irreversibly with a nucleophile in a Mannich-like reaction to form an alkaloid [71]. The different alkaloid sub-types are a result of nucleophiles with diverse origins.

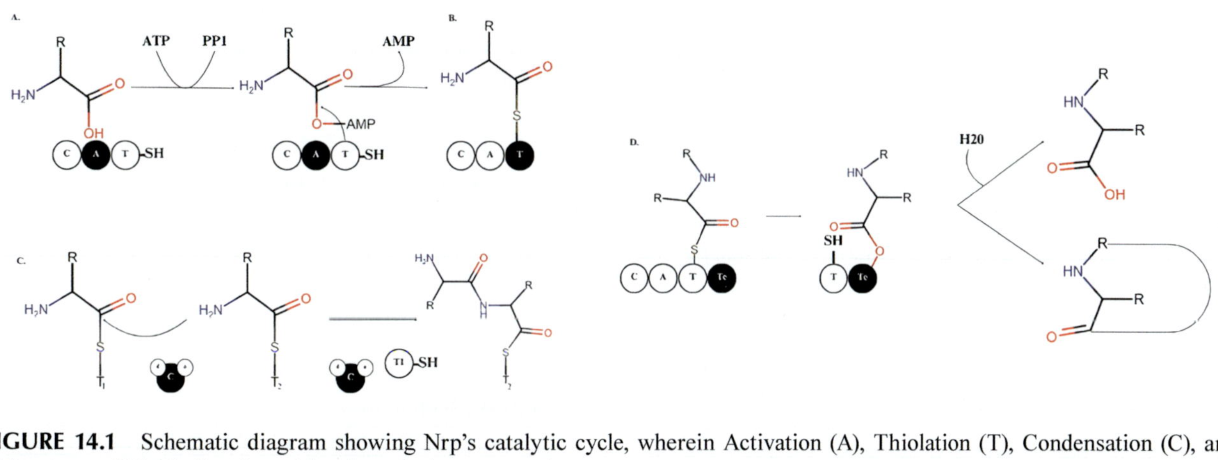

FIGURE 14.1 Schematic diagram showing Nrp's catalytic cycle, wherein Activation (A), Thiolation (T), Condensation (C), and Thioesterase (Te)-Mediated peptide release is demonstrated.

14.5 MARINE MICROBES IN DRUG DISCOVERY

14.5.1 Historical Perspective and Landmark Discoveries

During the mid-20th century, marine organisms were recognized as one of the valuable sources of natural products, from which marine organisms such as invertebrates, sponges, and corals were found to produce a variety of biological metabolites [72]. Over the past years, marine microbes have gained significant amount of attention in the field of drug discovery due to their immense potential in delivering novel therapeutic agents for various diseases such as cancer, antimicrobial resistance, and autoimmune conditions [73–75]. The rich and vast biodiversity of marine microorganisms provide a wide range of bioactive components that can fight against drug-resistant pathogens, thereby making them a valuable source for new antibiotics [76]. The recent shift toward marine drug discovery has led to the identification of various bioactive metabolites from marine organisms, which exhibits promising pharmacological activities and pave the way to overcome the challenges faced in therapy [77]. Landmark discoveries in marine microbes for drug discovery include a variety of marine organisms such as bacteria, fungi, and cyanobacteria which are found to produce a variety of bioactive compounds [76]. These serve as an abundant source of compounds with unique chemical components present in the marine microbes. Salinosporamide A and Tunicamycin are examples of such discoveries that have been identified from marine organisms and have been found to demonstrate therapeutic benefits [78]. Another notable discovery was conducted in the Nagasaki Prefecture, Japan, where over 20,000 marine microbes have been collected to create an original marine microbial extract library containing small and mid-size biomolecules with potential therapeutic applications for autoimmune diseases [73]. Furthermore, another discovery was the development of original marine microbial library constituting the various bioactive compounds produced by a variety of marine organisms including bacteria, fungi, and actinomycetes and the development of shark nanobodies derived from indoor shark farming systems. These shark nanobodies were found to inhibit intracellular protein–protein interaction and can prove to be beneficiary for autoimmune disorders such as familial Mediterranean fever. The above-mentioned discoveries underscore the significance of marine microbes in providing new avenues for drug development and combating emerging challenges like antibiotic resistance [79].

14.5.2 Screening Strategies for Bioactive Compounds

Screening marine bioactive compounds for their antimicrobial properties involves several methodologies and strategies.

14.5.2.1 Culture of Marine Microorganisms

The initial step in harnessing the antimicrobial potential of marine microorganisms involves cultivating these organisms under controlled laboratory conditions. This process is critical for optimizing the production of bioactive metabolites that can be used against plant pathogens.

a. **Media Optimization:**
   The selection and optimization of culture media is critical for promoting the growth of marine microbes. Different media provides varying nutrients to meet the unique needs of these species. Carbon sources, nitrogen sources, and trace elements are altered to establish an optimal environment for growth. By altering these components, researchers can dramatically improve the quantity of microorganism-produced compounds with antibacterial capabilities [80].
b. **Co-Cultivation:**
   Co-cultivation is the process of growing two or more different species of microorganisms together. This approach can stimulate the creation of secondary metabolites that would otherwise not be created if the organisms were cultivated individually. The interaction of diverse species can result in the exchange of signaling molecules, which promotes growth and metabolite production. This technique is especially beneficial for identifying novel compounds with potential antibacterial action since the combined metabolic pathways can produce unique bioactive molecules [81].
c. **In-situ Cultivation:**
   In-situ culturing mimics the natural environment of marine microorganisms, allowing them to grow in conditions similar to their original environments. This method frequently employs specialized devices, such as the I-chip, which isolates microbes while allowing them to interact with their natural environment. Providing a more genuine growing environment, in-situ culturing can activate dormant bacteria and increase the generation of various metabolites that may have antibiotic properties [82].

14.5.2.2 Isolation and Extraction of Natural Compounds

Once the marine microorganisms are cultured successfully, the subsequent step involves isolating and extracting the bioactive compounds that they produce. This process is crucial for identifying and purifying substances that may have antimicrobial properties.

a. **Solvent Extraction:**
   This method employs a solvent to dissolve the desired bioactive chemicals from the biomass of cultivated microorganisms. The choice of solvent

is crucial and is determined by the polarity of the molecules being targeted. Common solvents include ethanol, methanol, and acetone. The procedure normally consists of mixing the solvent with the microbial culture, allowing the compounds to dissolve, and then separating the solvent containing the dissolved compounds from the solid biomass. This approach is extensively used because it is simple and effective for extracting a wide range of chemicals [83].

b. **Liquid–Liquid Extraction:**

Liquid–liquid extraction is a more advanced process for separating chemicals based on their solubility in two immiscible liquids, typically water and an organic solvent. After the first extraction, the mixture is allowed to settle before the two phases are separated. This approach is very beneficial for separating chemicals that cannot be extracted efficiently using a single solvent. Researchers can selectively extract specific bioactive components by changing the pH and employing different solvents, resulting in a purer end product [84].

c. **Solid-Phase Extraction (SPE):**

SPE is a technique for extracting and separating bioactive chemicals from liquid samples using solid adsorbent material. The sample is passed through an adsorbent-containing column, which holds the target molecules while allowing contaminants to pass. After washing the column to eliminate undesirable particles, the desired compounds are eluted using an appropriate solvent. This approach is helpful because it can concentrate and purify chemicals, making it easier to collect high-purity samples for subsequent examination [85].

d. **Purification of Extracted Compounds:**

After extraction, the bioactive components are purified to yield individual chemicals. This purification procedure may include chromatography (high-performance liquid chromatography, HPLC) or recrystallization. These approaches help in separating molecules depending on their chemical properties, such as size, charge, or polarity. The objective is to isolate pure molecules that can be further tested for their antibacterial action against plant diseases [86,87].

14.5.2.3 Activity Screening

Post extraction process, the isolated compounds undergo activity screening to evaluate their activity against specific pathogens. This is typically done using bioassays, where the compounds are tested against bacterial or fungal strains. Common methods include disk diffusion, broth microdilution, and agar well diffusion assays. These results help in identifying which compounds exhibit significant antimicrobial activity [88].

14.5.2.4 Chemical Screening

This method involves the utilization of chemical properties of the isolated compounds. Techniques such as UV spectroscopy, Mass Spectrometry (MS) and Nuclear Magnetic Resonance (NMR) are employed to determine the molecular structure, molecular mass, and other characteristics of the compounds. This information is crucial for understanding the potential mechanisms of action and for further development [88].

14.5.2.5 Mechanism of Action Studies

It is important to understand the way through which the bioactive compounds exert their activity. Hence, the study involves the interaction of compounds with the microbial cells and other pathogens and further study of their effect on their membrane permeability and metabolic pathway [89].

14.5.3 High-Throughput Screening and Omics Approaches

High-throughput screening of marine microbes has become an important strategy in searching for new bioactive compounds for drug discovery. The marine environments, unique in their physicochemical conditions, harbor a wide array of microorganisms, such as bacteria, fungi, and microalgae, that produce structurally diversified and biologically potent secondary metabolites [90]. This may typically begin with sampling different marine habitats, such as deep-sea sediments, hydrothermal vents, and coral reefs, using techniques that allow one to reach normally extreme and inaccessible environments. After isolation, culture follows under controlled laboratory conditions, which may be difficult due to the specific conditions required by these organisms that are sometimes unknown [91]. Extraction methods of metabolites from cultivated microorganisms are applied. Extracts are subjected to high-throughput screening against a broad range of biological targets, including assays that will detect activity across multiple pathways that span from antimicrobial to anticancer and anti-inflammatory activities. Active compounds are defined in terms of their structure by the integration of advanced analytical techniques such as MS and NMR spectroscopy [88]. It is the distinct chemical diversity offered by marine microorganisms, coupled with the efficiency of high-throughput screening, that underlines the potential role of the marine biosphere as a rich reservoir for novel drug leads and therapeutic agents [90] (Figure 14.2).

14.6 ANTIBACTERIAL COMPOUNDS FROM MARINE MICROBES

14.6.1 Overview of Antibacterial Agents Discovered

It is now becoming apparent that marine microbes are a rich source of novel antibacterial substances where firms embrace an outstanding naturally diverse array of bioactive

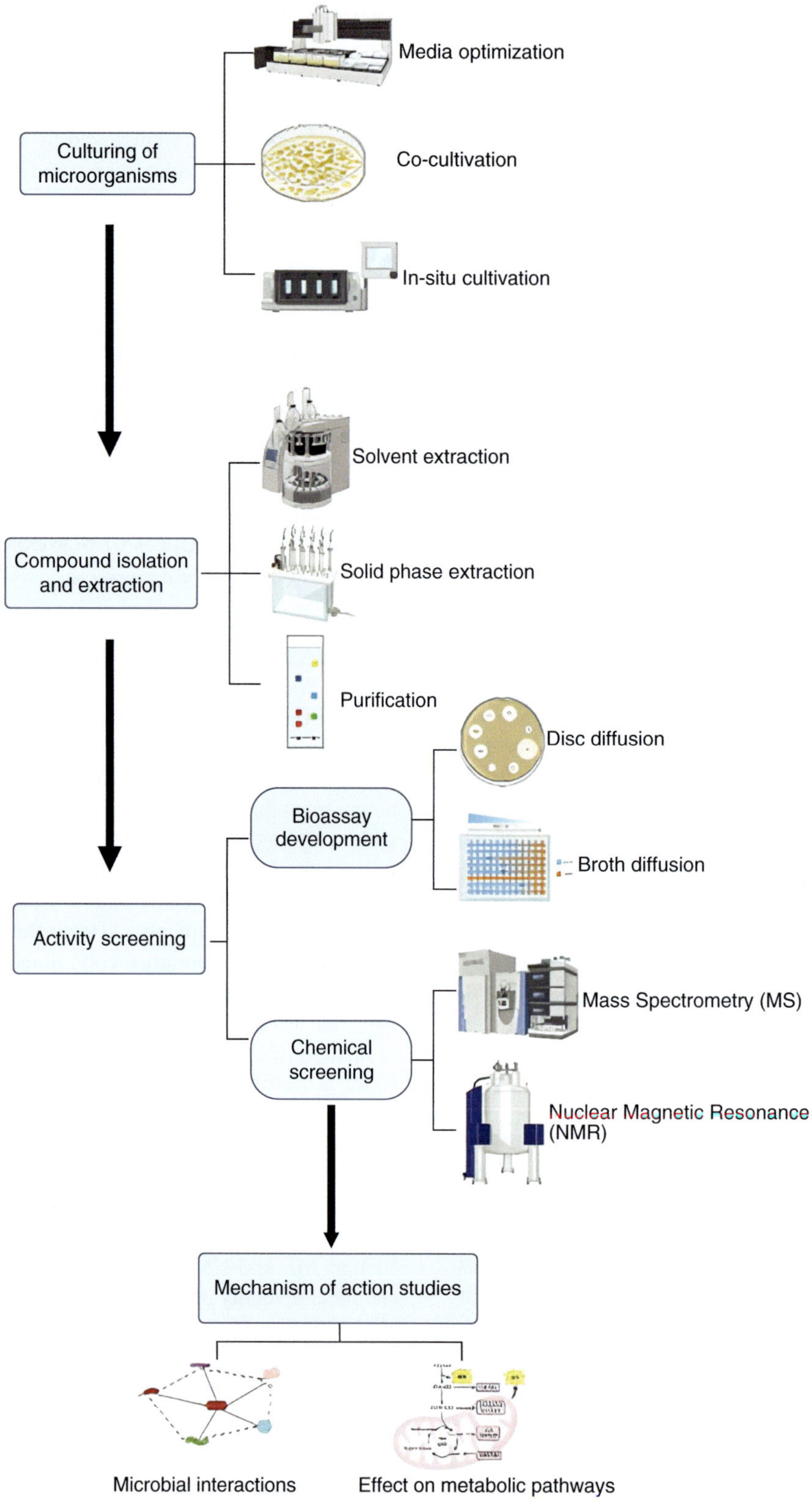

FIGURE 14.2 Workflow of high-throughput screening of marine microorganisms for novel bioactive compounds, from culturing and extraction to bioassay, chemical analysis, and mechanism of action studies.

agents for containing drug-resistant bacteria. These microorganisms such as bacteria, fungi, and algae are located in various and sometimes rather extreme conditions of the sea and as a result synthesize secondary metabolites having rich structural variety and high efficiency in antibacterial effect [90]. Out of all the classes of compounds extracted from marine microbes, polyketides, nitrogenous compounds, terpenoids, and steroids are some of the most recognized offering different modes of action. For example, polyketides are a huge class of secondary metabolites containing a lot of ring structures; they include well-known antibiotics like erythromycin and tetracycline significant in the battle with bacterial infections. Again, nitrogenous composites such as alkaloids and peptides have shown appreciable antibacterial activities; examples are streptomycin and rifampicin used to control tuberculosis and other bacterial diseases, respectively.

Another group of compounds possess antibacterial characteristics that are obtained from isoprene units termed as terpenoids. These compounds extend from monoterpenes to triterpenes and even steroids, and many of them possess splendid antibacterial characteristics [92]. For instance, sesquiterpenes have been effective against MRSA to demonstrate how they could combat antibiotic resistance [93]. Some forecasters affect the synthesis of steroids, the general construction of which can be described as four fused rings and are also considered part of the strategies for combating pathogenic bacteria among marine organisms [94]. Besides these classes, marine natural products including pyrones, sterols, 5α-type steroidal derivatives exhibit high antimycobacterial activity, as well as quinones exhibit great potentiality against anti-tubercular activity especially against *Mycobacterium tuberculosis,* the bacteria causing tuberculosis [95]. The emergence of new drug-resistant tubercular strains is contributing to the extended period of treatment and calls for the development of new anti-tubercular agents where marine-derived compounds are potential sources of such discoveries.

The marine environment has also been of interest in searching for compounds active against *Vibrio* species that cause cholera and seafood diseases [96]. Up till now, 214 anti-*Vibrio* compounds have been identified, where the majority are isolated from marine fungi and bacteria, therefore confirming that they are potent antibacterial compounds [97,98]. The versatility of the structures and significant biological activities of the marine-sourced natural products underscore their prospective role in accessing new antibiotics against the world's current drug resistance problem.

14.6.2 Significant Antibacterial Compounds

Marine microbes have emerged as a prolific source of antibacterial compounds, offering a vast array of bioactive molecules with significant therapeutic potential. Some of these include marine-derived alkaloids, peptides, terpenoids, and polyketides and all of them possess strong antibacterial activity. These natural products have proved efficient against pathogenic bacteria, especially multidrug-resistant ones, which makes the products of immense benefit in the war against antibiotic-resistant strains [90]. For example, marine sponges have given compounds such as bengamide and manzamine based on the note that they have great effects on both Gram-positive and Gram-negative bacteria [99,100]. Likewise, marine actinomycetes, a kind of bacteria famous for their production of secondary metabolites, have given examples like salinosporamide A and marinomycin, which have conspicuous antibacterial activity [101].

These marine-derived antibacterial compounds thus hold many more advantages that go beyond the pure antimicrobial activity. They usually have different modes of action from classical antibiotics, and hence, they do not easily develop cross resistance. For instance, the marine alkaloid manzamine A belongs to a new class of antibiotics that attacks the bacterial cell membrane contrary to other antibiotics. Also, these compounds possess anti-biofilm action, which is very important as it allows combating a chronic bacterial infection, the main goal of which is biofilm formation [102]. The intervention capability in biofilms can help to increase the efficiency of antibiotics that are used, thereby making it possible to achieve synergistic benefits that raise the general efficacy of treatment.

Moreover, these compounds are involved in the marine ecosystem, meaning that compounds have evolved to be effective defenses, pointing to their potency and specificity [103]. The elucidation of the structure of these molecules serves not only as potential therapeutic leads but also spurs the design of synthetic versions and congeners of these classes of antibacterial compounds that can be used to address the problem of resistant pathogens. Marine natural products have also played a key role in the development of the core disciplines of microbial ecology and chemical biology depicting complex host microbial associations of marine organisms.

14.6.3 Mechanisms of Action

The mechanism of action of marine-derived antibacterial compounds which was revealed showed that the compounds affected certain bacterial structures together with some metabolic pathways that are crucial to the existence and growth of bacteria. These compounds are frequently modified to the bacterial cell wall and act in different ways to either retard bacterial growth or kill the bacterial cells. Among the cellular structures that these compounds have been pointed to, one is the bacterial cell membrane. For instance, marine alkaloid manzamine A clings onto the bacterial membrane lipids, which leads to increased bacterial membrane permeability, and ultimately results in cell rupture. This disruption affects the homeostasis of the bacterium and this leads to death of the affected cell [102,104].

For instance, peptides that are extracted from marine organisms include AMPs present in sponges which have channels in the bacterial cell wall. These peptides insert into the membrane and cluster to produce ion/molecule-permeable transmembrane channels, all of which are overactive [105]. This results in phenomena such as depolarization and loss of essential cellular constituents and ends with cell demise. This pore-forming mechanism is even effective on Gram-negative bacteria, which are the bacteria whose outer membrane is very hard for most drugs including antibiotics [106].

TABLE 14.1
Categorization of Antibacterial Compounds Sourced from Marine Microbes

Sr No	Category	Target Bacteria	Mechanism of Action	Benefits	Applications
1	Alkaloids (e.g., Manzamine)	Gram-positive and Gram-negative bacteria	Targets bacterial cell membrane	New mode of bacterial eradication	Biomedical applications, dental care
2	Peptides	Multidrug-resistant strains	Various unique mechanisms	Reduced likelihood of cross-resistance	Wound dressings, coatings for devices
3	Terpenoids	Pathogenic bacteria	Various unique mechanisms	Potent antibacterial properties	Drug development
4	Polyketides (e.g., Salinosporamide A, Marinomycin)	Gram-positive and Gram-negative bacteria	Various unique mechanisms	Strong antibacterial efficacy	Clinical therapy, public health
5	Marine sponges (e.g., Bengamide)	Gram-positive and Gram-negative bacteria	Various unique mechanisms	High potency and specificity	Biomedical applications
6	Marine *Actinomycetes* (e.g., Salinosporamide A, Marinomycin)	Multidrug-resistant strains	Various unique mechanisms	Effective against resistant pathogens	Medical research, drug discovery

For example, polyketides' class compound, namely salinosporamide A, is known to have inhibition effect on bacterial proteasome which is responsible for protein degradation and cell cycle regulation. These compounds lock the proteasome so that it cannot degrade damaged or misfolded proteins any further and as a result, the unwanted proteins accumulate in the bacterial cell. This load interferes with protein balance and induces a stress response that in turn may result in cell cycle arrest and cell death. The proteasome is also involved in the degradation and recycling of proteins where this process is also affected by the action of the bacterium [107].

Marine-derived antibacterial compounds have also been recognized as anti-biofilm agents. Biofilms are complex, ordered societies of bacteria surrounded by a matrix that offers shield against medicines and the body's defenses. Materials like the marine-derived brominated furanone make bacteria unable to form biofilm as it inhibits quorum sensing which is the communication system used by bacteria in forming biofilm. These compounds suppress the quorum sensing, therefore ensuring that bacteria can ill afford to activate genes which ensure the maturation of a biofilm hence ensuring that bacteria are easily dealt with by the immune system or even antibiotics [108].

Some marine compounds are capable of reducing the extracellular matrices of the biofilms and consequently eradicate the shielding barrier to the bacteria and allow the antibacterial agents access to them. For example, the bacteria isolated from marine environment with the capacity to produce enzymes that destroy the components of the biofilm such as polysaccharides, proteins, and DNA within the biofilm's matrix may result in biofilm dispersion and increased antibiotic sensitivity [109].

Marine-derived antibacterial compounds work through various ways to act on bacterial cell membranes, proteasomes, and biofilm structures. Not only do they kill bacteria but they do it in a way that bacteria are less likely to build up a resistance against, which makes their uses in new antibiotics highly promising.

14.7 ANTICANCER COMPOUNDS FROM MARINE MICROBES

14.7.1 Overview of Anticancer Agents

Marine populations represent reservoirs of novel bioactive compounds with a diverse range of chemical structures. Marine organisms like plants, algae, bacteria, fungi, corals, and other organisms produce compounds whose roles in the prevention of tumor formation and the related compound-induced apoptosis and cytotoxicity were analyzed by scientists. The cure for cancer is now viewed as a feasible goal with the advancements in modern drug design and manufacturing. Over the past 50 years, evidence has shown that many marine products derived from plants and microbes exhibit beneficial effects in the prevention and treatment of cancer. Marine drugs like cytarabine, eribulin mesylate, brentuximab vedotin, and trabectedin have proven to be effective against leukemia, metastatic breast cancer, soft tissue sarcoma, and ovarian cancer [110]. Approximately more than 80% of all the approved drugs for chemotherapy are based on bioactive natural compounds [111]. Natural bioactive compounds work by attacking macromolecules expressed by cancer cells, for instance, those molecules released during signal transduction [112]. These natural bioactive compounds in general have been the prime source of compounds for the treatment of many types of cancer. Given this great potential of the marine natural product scaffolds, there has been an increasing interest for exploiting this complex diversity for drug discovery. A focused approach has been suggested to accelerate the development of anticancer drugs from marine organisms with increased efficiency and fewer side effects.

A number of computational methodologies are assisting in the search for marine therapeutic metabolites on the basis of genome analysis to comprehend the mode of action of these compounds and to optimize them for treatments [113]. The Caribbean sponge (*Cryptotethya crypta*) was the first marine organism to be chemically studied in detail [114]. The current progress in anticancer drugs offers a rise in isolation and clinical validation of numerous innovative developments and advances in anticancer therapy. Recent studies have shown that the use of chitosan in developing marine-derived drugs can be effective in anticancer drug development system (DDS). Some of the marine anticancer drugs approved by the Food and Drug Administration (FDA) include: cytarabine (Cytosar U®) which is used for malignant acute myeloid, lymphatic, and myelogenous leukemia and non-Hodgkin's Lymphoma (nHL) therapy; trabectedin (Yondelis®) used in soft tissue sarcoma therapy for breast, prostate, and pediatric sarcomas; Eribulin mesylate (Halaven®) used for breast cancer and liposarcoma treatments among the other drugs which will be discussed in detail later in this chapter.

14.7.2 Anticancer Agents and Their Mechanism of Action

Innate immune pathways like the JAK/STAT signaling and cGAS-STING-mediated immune responses are crucial in the anticancer therapies. Table 14.2 describes the different anticancer agents that are derived from various marine organisms, their chemical constituents, and the mechanism of action.

TABLE 14.2
Tabulation of Some of the Marine Drugs Used in Cancer Treatment Alongside Its Source, Mechanism of Action, and Chemical Composition [115]

Sr No	Drug	Chemical Components	Marine Source	Type of Cancer	Mechanism of Action
1	Eribulin	Halichondrin B, polyether macrolide	Marine sponge (*Halichondria okadai*)	Metastatic breast cancer, liposarcoma	Inhibits microtubule dynamics, leading to G2/M cell cycle arrest and apoptosis of cancer cells [116]
2	Ecol, Fucoidan	Polysaccharides	Alginates, agar, carrageenan, *Pseudomonas* spp. (B-1)		Cytotoxic effect by activation of the innate immune system. Inhibiting angiogenesis, metastasis, and modulating immune responses [117]
3	Cytarabine	Nucleoside analogue	Sponge (*Cryptotethya crypta*)	Malignant acute myeloid, lymphatic, and myelogenous leukemia, nHL	Inhibits DNA synthesis by incorporating into DNA during replication, leading to chain termination.
4	Trabectedin	Alkaloids	Caribbean Sea squirt (*Ecteinascidia turbinata*)	Soft tissue sarcoma and ovarian cancer	Binds to the minor groove of DNA, interfering with the transcription-coupled nucleotide excision repair pathway. This leads to DNA strand breaks and apoptosis of cancer cells.
5	Brentuximab vedotin	Peptide (Dolastatin 10 derivative)	Sea hare (*Dolabella auricularia*)	Relapsed or refractory HL and anaplastic large cell lymphoma	Targets CD30 on cancer cells, delivering MMAE which disrupts microtubule networks, progressing to cell cycle arrest and apoptosis.
6	Midostaurin			AML, myeloid dysplastic syndrome (MDS) and mastocytosis	Inhibits the activities of different protein kinases
7	Plitidepsin	Depsipeptide	Marine tunicate (*Aplidium albicans*)	Multiple Myeloma	Induces apoptosis through inhibition of eukaryotic elongation factor 1A2 (eEF1A2) leading to the disruption of cell cycle and induction of oxidative stress
8	Salinosporamide A	Beta-lactone gamma-lactam	Marine bacterium (*Salinispora tropica*)	Multiple myeloma and glioblastoma	Disrupts protein degradation pathways
9	Bryostatin-1	Macrolide lactone	*Bugula neritina* (a type of bryozoan)	Chronic lymphocytic leukemia, nHL	Modulates PKC activity affecting the cell growth and differentiation. It has also shown potential in enhancing the effects of other chemotherapeutic agents and inducing apoptosis.

(Continued)

TABLE 14.2 (*Continued*)
Tabulation of Some of the Marine Drugs Used in Cancer Treatment Alongside Its Source, Mechanism of Action, and Chemical Composition [115]

Sr No	Drug	Chemical Components	Marine Source	Type of Cancer	Mechanism of Action
10	Lurbinectedin	Alkaloid	Caribbean sea squirt (*Ecteinascidia turbinata*)	Small cell lung cancer	Works almost similar to Trabectedin
11	Didemnin B	Cyclic depsipeptide	Marine tunicate (*Trididemnum solidum*)	Various cancers in clinical trials (predominantly multiple myeloma and leukemia)	Inhibits protein synthesis by blocking elongation factor-1 alpha (EF-103B1), leading to cell cycle arrest and apoptosis.
12	Spisulosine	Sphingolipid	Marine clam (*Spisula polynyma*)	Under investigation for solid tumors	Disruption of actin cytoskeleton and inhibition of sphingolipid metabolic pathway.

14.8 ANTIVIRAL COMPOUNDS FROM MARINE MICROBES

14.8.1 Overview of Antiviral Agents Discovered

Viral infections are a leading cause of mortality worldwide, with viruses like influenza, HSV, HIV, RSV, EV71, DENV, Ebola, and SARS-CoV-2 posing significant health threats. The rapid mutation of viral genomes complicates infection control and often leads to pandemics, exacerbated by global travel and urbanization [118]. Despite advancements in antiviral therapies since the introduction of vidarabine in the 1970s, drug resistance remains a formidable challenge, underscoring the urgent need for novel antiviral agents. Marine-derived secondary metabolites have garnered attention for their potent antiviral properties [119,120]. Each milliliter of seawater harbors a vast number of microorganisms, including cyanobacteria, actinomyces, bacteria, and fungi, which produce a plethora of bioactive compounds. Notable examples include spongouridine from marine sponges, leading to vidarabine for HSV treatment, and cyanovirin-N (CV-N) from cyanobacteria, showing promise against HIV [121,122]. Despite the promising antiviral potential of these marine-derived compounds, their clinical application remains limited, indicating a need for further exploration and development of marine microbial resources for pharmaceutical use [123] (Table 14.3).

14.8.2 Mechanism of Action

a. **Cyanovirin-N:**

CV-N, a lectin from the marine cyanobacterium *Nostoc ellipsosporum*, shows potent activity against HIV [121]. In HIV infection, the virus's gp120 glycoprotein binds to the CD4 receptor on helper T cells, which triggers changes that expose coreceptor sites (CCR5 or CXCR4). This interaction facilitates the fusion of the viral envelope with the host cell membrane, allowing viral RNA and enzymes to enter the host cell. CV-N binds to gp120, blocking its attachment to CD4 receptors, and preventing fusion with the host cell membrane. This inhibition stops viral RNA from entering the host cell, halting reverse transcription into DNA and subsequent viral replication. Thus, CV-N effectively disrupts HIV infection and propagation [129,130].

b. **Griffithsin:**

Griffithsin (GRFT) is isolated from marine red algae *Griffithsia* sp. by the researchers of National Cancer Institute. It is highly potent against HIV in picomolar concentration [131]. It has similar function as CV-N, it binds to gp120 protein and prevents the fusion with the host cell membrane [132]. GRFT binds to the E1E2 glycoprotein on the surface of HCV (hepatitis C virus). By binding it to E1E2 it prevents attachment to CD81 receptor in hepatocytes, thus preventing the spread of infection to healthy cells. It significantly reduces HCV infectivity and shows efficacy in chimeric mice models [133]. GRFT binds to high mannose glycans on HPV's L1 and L2 capsid proteins. This binding prevents the virus from interacting with host cell receptors such as heparan sulfate proteoglycans and integrins. By blocking these interactions, it disrupts HPV attachment to host cell membrane [134].

c. **Fucoidan:**

A sulfated polysaccharide from a brown algae, *Kjellmaniella crassifolia* has a potential in inhibiting the multiplication of Influenza A virus (IAV). Fucoidan KW inhibits viral neuraminidase (NA), blocking new viral particle release and initial

TABLE 14.3
Significant Antiviral Compounds Derived from Marine Microbes

Sr No	Compound	Chemical Components	Organism	Activity Against	Mechanism of Action
1	Cyanovirin-N	Lectin protein	*Nostoc ellipsosporum*	HIV [121]	Binds to glycoprotein in virus envelope
2	EPS-1	Exopolysaccharide	*Bacillus licheniformis*	HSV-2 [124]	Prevents viral attachment to host cells
3	A1	Aromatic	*Microalgae, Cochlodinium polykrikoides*	Influenza virus A and B; RSV A and B; HSV [125]	Interferes with viral replication by inhibiting neuraminidase
4	Griffithsin	Lectin protein with carbohydrate binding domains	*Red algae, Griffithsia*	HIV, HCV, HPV	Binds to glycoproteins and prevents attachment
5	Calcium spirulan	Sulfated polysaccharide	*Arthrospira platensis*	HSV-1, HIV-1, measles	Inhibits viral absorption
6	Naviculan	Sulfated polysaccharide	Diatom, *Navicula directa*	HIV [126]	Prevents viral attachment to host cells
7)	Halovirs A-E	Alkaloid	*Scytalidium* sp.	HSV-1 and HSV-2 [127]	Inhibits viral replication by interfering with viral polymerase
8	Fucoidan	Sulfated polysaccharide	*Kjellmaniella crassifolia*	Influenza A virus [128]	Blocks the entry to host cell by binding to surface glycoprotein
9	Stachyflin	Peptide	*Stachybotrys* sp. *RF-7260*	Influenza A virus	Inhibits viral replication by targeting viral enzymes.
10	Caulerpin	Alkaloid with diketopiperazine ring	Algae *Caulerpa Lamouroux*	HSV-1	Inhibits viral DNA polymerase enzyme
11	Phomasetin	*Polyketide*	*Phoma* sp.	HIV-1	Inhibits reverse transcriptase enzyme
12	Cladosin C	*Terpenoid*	*Cladosporium sphaerospermum*	Influenza A virus (H1N1)	Inhibits viral protease
13	Cytosporin L	Polyketide	*Eutypella sp.*	RSV virus	Inhibits the replication by targeting viral RNA polymerase.
14	ZSU-H85 A	Alkaloid	*Trichoderma* sp.	EV71 virus	Inhibits viral protease
15	Antimycin A10a	Polyketide with a dilactone ring	*Streptomyces kaviengensis*	WEEV virus	Targets viral integrase enzyme
16	Scequinadoline A	Alkaloid	*Dichotomomyces cejpii* F31-1	DENV virus	Blocks replication by targeting reverse transcriptase
17	Xiamycins	Indole alkaloid	*Streptomyces xiamenensis*	PEDV	Inhibits viral protease
18	Ulvan	Sulfated polysaccharide	Green algae, *Ulva* sp.	Influenza A	Prevents the entry to host cells
19	Rubrumlines A	Isoquinoline alkaloid	Fungus, *Eurotium rubrum*	HIV	Inhibits the replication of virus by targeting viral RNA synthesis.
20	Raistrickindole A	Indole alkaloid	Fungus *Penicillium raistrickii*	HCV	Inhibits replication by targeting viral enzymes

infection stages. It also disrupts the EGFR pathway, reducing IAV internalization and replication. Additionally, fucoidan enhances immune response by increasing interferon γ (IFN-γ) and interleukin 2 (IL-2) production. In vitro studies show fucoidan suppresses replication of IAV strains, including H1N1 and H3N2 [128].

d. **Caulerpin:**

Caulerpin, derived from the marine algae *Caulerpa Lamouroux*, shows potential antiviral activity against herpes simplex virus type 1 (HSV-1). Initially, it is phosphorylated by the viral enzyme thymidine kinase, activating it for further modifications. Cellular enzymes then convert it into di- and tri-phosphate forms, essential for its antiviral efficacy. Once fully phosphorylated, caulerpin interacts with the viral DNA polymerase complex (UL30/UL42), disrupting viral DNA strand formation. This disruption occurs due to the absence of the critical 3′-OH group necessary for DNA strand elongation, effectively inhibiting viral replication and preventing the spread of HSV-1 [135].

e. **Stachyflin:**

Stachyflin, produced by *Stachybotrys* sp. RF-7260 is an antiviral compound targeting human influenza viruses by inhibiting the hemagglutinin (HA) protein's conformational change crucial for viral fusion with host cell membranes. It directly interacts with specific sites on HA, preventing the necessary changes for fusion. Stachyflin exhibits a virucidal effect, inactivating the virus outside host cells and remains effective even post-virus adsorption. It primarily targets H1 and H2 subtypes, showing limited activity against H3 and influenza B viruses. Resistance is linked to amino acid substitutions in the HA2 subunit, which may alter HA binding or conformational dynamics [136,137].

f. **Calcium Spirulan:**

Calcium spirulan (Ca-SP), produced from *Arthrospira platensis*, exhibits antiviral properties against enveloped viruses like HSV-1, human cytomegalovirus, and HIV-1. Ca-SP, an acidic polysaccharide isolated from *Spirulina platensis*, inhibits virus replication by blocking viral penetration into host cells, especially during the adsorption phase. At concentrations above 40 μg/mL, it nearly completely blocks HSV-1 penetration. Ca-SP has a high selectivity index (~8,600), indicating potent antiviral activity with low cytotoxicity. Its efficacy is linked to the structural integrity maintained by calcium ions chelated with sulfate groups [138,139].

g. **Halovirs A-E:**

Halovirs A–E that are produced during the saline fermentation of a marine-derived fungus of the genus *Scytalidium*, exhibit specific antiviral mechanisms against HSV-1 and HSV-2. They directly inactivate HSV by binding and neutralizing its infectivity, preventing viral entry into host cells. Halovirs destabilize the viral envelope, disrupting its integrity. Structurally, they contain an Aib-Hyp dipeptide segment and are acylated by fatty acids like myristic or lauric acid, enhancing their interaction with lipid membranes. They operate through a 'carpet-like' mechanism, aligning along the viral membrane to cause disruption. Importantly, halovirs are effective at non-cytotoxic concentrations, making them promising antiviral agents [140].

h. **Xiamycins:**

Xiamycins from marine fungi *Streptomyces xiamenensis* inhibit Porcine Epidemic Diarrhea Virus (PEDV) through multiple mechanisms. They potentially block viral entry by interfering with the interaction between PEDV's spike protein and the aminopeptidase N receptor on host cells. Additionally, xiamycins might disrupt viral fusion processes and modulate cellular signaling pathways such as NF-κB and MAPK, which are crucial for inflammatory responses and viral replication. They could also inhibit viral RNA polymerase or other replication enzymes, thereby reducing viral load. By targeting these pathways, xiamycins impair PEDV's ability to infect and replicate within host cells [127,141].

14.9 OTHER THERAPEUTIC POTENTIALS

Marine organisms have been a great source of natural compounds with unique features resulting from the huge variability of marine habitats and the species residing in them. Most of the marine compounds in use or in clinical trials are drugs for cancer therapy and are also conjugated to antibody to form antibody-drug conjugates (ADCs). These marine agents can target a number of other conditions apart from cancer, viral, and bacterial conditions like fungal, hypertriglyceridemia, obesity, Alzheimer's, and other neural systems diseases.

14.9.1 Antifungal and Anti-parasitic Agents

This class of drugs includes Aplidin, Halicyclamine A, and Manzamine A, among others. Aplidin also known as Plitidepsin is derived from the marine tunicate *Aplidium albicans*. This compound is a depsipeptide that inhibits protein synthesis by targeting the eukaryotic elongation factor 1A2 (eEF1A2). This compound has demonstrated antifungal properties in that it has inhibited the growth of various pathogenic fungi. There has also been evidence that Plitidepsin is being explored for its anti-parasitic potential, particularly in combating malaria by the induction of apoptosis in the parasitic cells. This unique mechanism of action makes this a valuable drug in the development of new antifungal and anti-parasitic therapies.

Halicyclamine A from the marine sponge *Haliclona* sp., is an aminosteroid that works by disrupting membrane integrity and inhibiting sterol synthesis in fungi leading to potent antifungal activity against *Candida* species and other pathogenic fungi. Its use against parasitic infections has also been investigated for its potential against diseases like leishmaniasis and trypanosomiasis [142].

Manzamine A, an alkaloid obtained from the sponge *Haliclona* sp., exhibits both antifungal and anti-parasitic properties. It inhibits cell proliferation and induces apoptosis in the parasitic cells, demonstrating efficacy against

Plasmodium falciparum and *Leishmania donovani*. This ability to target a wide range of parasitic organisms underscores the scope of this drug to develop broad-spectrum anti-parasitic therapies. The above-mentioned compounds and their dual functionality as both antifungal and anti-parasitic highlight their therapeutic versatility.

14.9.2 Anti-inflammatory and Immunomodulatory Compounds

This class of compounds includes the drugs that are used for protection against severe inflammation or immune modulation. These include Pseudopterosin, Scytonemin, Fucoidan etc. Pseudopterosin, a diterpene glycoside extracted from the sea whip *Pseudopterogorgia elisabethae*, is mostly known for its anti-inflammatory properties. This inhibits the production of inflammatory mediators like prostaglandins, leukotrienes, and others, making the treatment of conditions like arthritis and inflammatory skin disease easier. It also enhances wound healing by promoting tissue repair, showcasing its suitability to be used as a therapeutic option for chronic inflammatory conditions [143].

Scytonemin, an indole-alkaloid derived from cyanobacteria *Scytonema* sp., acts as both an anti-inflammatory and immunomodulatory compound. It acts by inhibiting the NF-κB pathway, reducing the production of pro-inflammatory cytokines. Fucoidan is a polysaccharide derived from brown seaweed *Fucus vesiculosus* and has demonstrated both anti-inflammatory and immunomodulatory properties. It is mostly composed of sulfated fucose and modulates immune responses by enhancing the activity of natural killer cells and macrophages. It has shown significant potential in the treatment of inflammatory bowel disease, rheumatoid arthritis, and other autoimmune diseases [144].

14.9.3 Neuroprotective and Cardiovascular Agents

Omega-3 fatty acids (EPA and DHA), primarily found in fish oil, isolated from species such as salmon and mackerel are polyunsaturated fatty acids with extensive neuroprotective [145] and cardiovascular benefits. They modulate the immune processes, improve membrane fluidity, and enhance cell signaling pathways. These have shown to reduce neuroinflammation and oxidative stress which are the critical factors in neurodegenerative disorders like Alzheimer's and Parkinson diseases. Apart from this, these compounds are beneficial in treating cardiovascular disease by lowering the triglyceride levels, reducing blood pressure and preventing arrhythmias. The role of these fatty acids in promoting heart and neural system health makes them essential compounds in the dietary and therapeutic strategies for preventing and managing chronic diseases [146].

Squalamine is another marine agent used in both neural and cardiovascular treatments. It is an amino-sterol derived from the dogfish shark (*Squalus acanthias*). It inhibits angiogenesis and modulates ion channels which contributes to its therapeutic effects. Squalamine has been studied for its potential to prevent neurodegeneration and promote neural survival, making it a valuable drug to treat Alzheimer's disease. Its ability to inhibit abnormal blood vessel formation and reduce cholesterol levels suggests its potential applications in cardiovascular disease management [147].

Ziconotide also known by Prialt is obtained from the venom of the cone snail *Conus magus*, which is a peptide known as omega-conotoxin MVIIA. It works by inhibiting or blocking the N-type calcium channels, which reduces neurotransmitter release and neuronal excitability. Ziconotide has been approved for managing severe chronic pain, particularly in patients with neuropathic pain who are resistant to other treatments. The unique mechanism of action of this drug coupled with its potential analgesic effects makes it suitable in the treatment for chronic pain and neurodegenerative conditions [148].

These marine compounds demonstrate a wide range of therapeutic potentials in different fields of human diseases and conditions beyond their traditional uses, offering promising avenues for the development of new treatment options for the above conditions.

14.10 BIOTECHNOLOGICAL AND PHARMACEUTICAL APPLICATIONS

14.10.1 Marine Enzymes and Their Industrial Applications

Marine enzymes are proteins present in the marine organisms. The marine enzymes catalyze (speeding up the reactions without being consumed in the process) biochemical reactions. The marine enzymes are further categorized into various groups based on functions, namely, hydrolase, isomerase [149]. Hydrolases are further classified into esterase and glycosyl hydrolase. Esterase is a diverse group of hydrolase, which can catalyze the hydrolysis of esters and break them into two main components. The first component is an alcohol (compound containing hydroxyl group (-OH)), this component is formed as a result of the ester bond cleavage by the esterase. The second component is carboxylic acid (compound containing carboxyl group (-COOH)) during hydrolysis process. The esterase is classified based on substrate specificity – lipase, and based on mechanism of action, serine hydrolase, and metal-dependent esterase. Esterases are crucial for industrial applications due to their unique properties and stability under extreme conditions. It is essential in various biological processes such as digestion, cellular signaling, and metabolism. The industrial applications of esterases include synthesis of optically pure substances (compounds containing one enantiomer, enantiomers are pair of molecules that exist in two forms that are mirror image, examples are lactic acid, carbohydrates etc.), preparation of enantiomerically pure drugs, and degradation of human made plastics and organophosphorus compounds such as pesticides like parathion, diazinon etc. Esterase is also used in the degradation of lipophilic components of ink and production of short-chain flavor esters like ethyl acetate, ethyl butyrate etc. [150]. Glycosyl hydrolase also known as glucosidase are found in marine environments like ocean, sea, and coastal areas that provide habitat for marine organisms including fungi, algae, bacteria, and invertebrates. These organisms produce glycosyl hydrolase to adapt to their surroundings. The role of glycosyl hydrolase is breaking down complex substances into simple sugars like glucose, fructose, and galactose. It is also involved in biological processes. The properties of glycosyl hydrolase include salt tolerance, to function effectively in high-salinity conditions (seawater), hyper thermostability, to remain active and stable in high temperatures, barophilicity, ability to withstand high pressure. The industrial applications of glycosyl hydrolase ranges from biorefinery to food industry, in biorefinery, it is involved in conversion of biomass into biofuel such as ethanol (through breakdown of cellulose, and hemicellulose into fermentable sugars). In the food industry, production of sweeteners (examples include aspartame, sucralose etc.), dietary fibers, and prebiotics is by hydrolyzing polysaccharides present in plant materials [151]. Isomerases are enzymes found in marine environments. These enzymes are capable of converting biomolecules into isomers. The industrial application of isomerase is the synthesis of pharmaceuticals and pharmaceutical intermediates for their potential to transform inexpensive biomolecules like l-ribose, d-psicose, lactulose, and d-phenylalanine to valuable isomers d-ribose, d-fructose, lactose, and l-phenylalanine, respectively, with suitable stereochemistry for synthetic medical chemistry. The properties of isomerase are high selectivity, ability to target specific molecules, increased stability, maintaining catalytic activity over extended periods, enhanced activity, catalyzed at faster rate, broader availability, and more accessible for use in different biocatalytic processes [152].

14.10.2 Use of Marine Microbes in Bioremediation

Marine microbes contribute to degradation of pollutants in marine environments. The capabilities of these microbes is to biotransform, decompose, or biodegrade pollutants such as hydrocarbons, heavy metals, and other contaminants. The marine microbes help in the sustainable management of contaminated marine environments to ensure restoration of ecosystem health and balance. It can be utilized in various techniques of bioremediation, namely biostimulation (modification of the marine environment to stimulate existing bacteria capable of bioremediation), bioaugmentation (increasing the marine microbes' ability on the contaminated sites to degrade pollutants), and bioaccumulation (accumulation of chemicals in an organism that takes place if rate of intake exceeds rate of excretion). There are factors to be considered for bioremediation including type of pollutant, marine environment conditions, and microbial community present in the ecosystem [153,154]. Marine oil pollution is caused by oil extraction, processing, transportation, and refining which leads to significant environmental damage. This can impact fisheries, aquatic habitats, aquaculture, and human health. The traditional methods for controlling marine oil pollution include physical method (chemical barriers), chemical methods (using oil dispersants and condensing agents), and biological method (bioremediation), which has low impact and cost-effectiveness. Bioremediation involves complete mineralization of low concentrated biodegradable residual oil, which cannot be removed by physical and chemical methods. Marine microbes produce biosurfactants, which increase contact area between oil and microorganisms to degrade petroleum hydrocarbons in water. Examples of some marine microbes are *Alcanivorax borkumensis* which produce biosurfactant glycolipid, *Pseudomonas* spp. produce rhamnolipids, *Bacillus* spp. produce surfactin, and *Rhodococcus* spp. produce trehalolipids [155]. The types of marine microbes used in bioremediation include bacteria such as *Pseudomonas* spp. (ability to degrade wide range of hydrocarbons and heavy metals), *Alcanivorax* spp. (degrading alkanes found in crude oil like methane, ethane, isopentane etc.), fungi like *Aspergillus sydowii*, *Penicillium chrysogenum* etc., and algae such as *Ulva lactuca* (sea lettuce), *Caulerpa taxifolia*, *Macrocystis pyrifera* (giant kelp) etc. The fungi can degrade complex organic pollutants, particularly utilized in breaking down lignocellulosic material such as wood (cellulose, hemicellulose, and lignin), straw (barley straw, rice straw, and wheat straw), corn stover (leaves, cobs and stalks), bagasse (fibrous residue left after extraction of juice from sugarcane), grasses (switch grass, and miscanthus), and forestry residues (branches, leaves, and saw dust from logging operations). The algae can absorb heavy metals (lead, mercury, cadmium, arsenic etc.) found in contaminated water. The mechanism of bioremediation include bioleaching (using bacteria to extract metals from mine tailings, and contaminated sites), biosorption (utilizing microbial biomass to absorb and concentrate heavy metals from waste water), and biodegradation [156].

14.11 CHALLENGES AND FUTURE PERSPECTIVES

Over the past two decades, microbial extracts have been instrumental in extending life expectancies, enhancing therapeutic outcomes, and transforming medicine. However, this has also led to the emergence of antimicrobial resistance. There is an endless desire to investigate new sources, including marine microorganisms, due to the growing world population and new illnesses. In relation to terrestrial sources, marine microbes as drug sources have not yet been fully investigated, despite their enormous potential. Numerous potentially groundbreaking antibacterial, antifungal, anticancer, and analgesic chemicals are prevalent in the marine environment.

The preservation of marine ecosystems in the face of global warming, invasions of biological species, over-exploitation, pollutants, and habitat loss are recurring challenges that arise at times and locations and have an accumulative impact. These environmental changes dramatically affected the diversity and abundance of species, maritime biodiversity, and the variable number and type of marine animals, rendering it difficult to uncover enough sources of these compounds for therapeutic development [157].

Moreover, a number of circumstances might make the process of isolating and separating bioactive chemicals from marine creatures quite challenging. Any bioactive chemical, whether synthesized or semi-synthesized, needs to be identified in relation to the biological source compound. Due to the structural intricacy and often poor yield of marine compounds, errors may occur in the specification of the chemical formula, actual structure, intramolecular bond arrangement, general configuration, and several stereocenters [158]. This emphasizes the necessity of accurately characterizing these compounds, which is crucial yet difficult to do. In addition, there is a need to use sophisticated methods and procedures for their identification and synthesis due to the complex chemical structures of these natural chemicals, further adding to the challenges of marine drug discovery.

The course of investigation is hampered by the difficulty of accessing deep-sea settings and efficiently surveying marine biodiversity, i.e., the creation of many highly potent new marine compounds has occasionally been impeded by the lack of reliable supplies of an ideal chemical. Further, there is a toxicity risk throughout the therapeutic development process for marine-derived medications [159]. A comprehensive assessment of the safety profiles of many potential marine chemicals is necessary before they are introduced to the market, as they were not successful in clinical trials. Licensing is another barrier to the commercialization of chemicals produced from marine sources, and the tedious task of bringing novel medications to market can be slowed considerably by navigating the complicated web of patent rules and regulations that are unique to marine biotechnology [160].

Rising antimicrobial resistance constitutes an increasing worldwide problem, and marine bacteria may offer crucial new compounds for creating potent medicines. To tackle resistant types and meet the urgent demand for novel antibiotics, further studies are needed to be done on the identification and characterization of these antimicrobial compounds. Innovations in sampling methods currently access these extreme ecosystems that may be a harbor for microbes generating novel compounds with therapeutic uses [161]. Numerous metabolites obtained from deep-sea species have shown medicinal potential that can target a range of ailments, making this investigation crucial.

Recent advances in synthetic biology are essential for augmenting marine microorganisms' potential for medication discovery. Using the application of genetic engineering approaches, it is feasible to initiate metabolic processes in marine microorganisms that were formerly inactive, hence facilitating the generation of innovative bioactive chemicals. This strategy has the potential to increase the production of therapeutic compounds that have remained restricted owing to the difficulties with conventional procedures. It is likewise essential to use cutting-edge screening methods like metabolomics and high-throughput screening to effectively identify bioactive chemicals from marine microorganisms [162].

We anticipate that effective medication research and development will be fueled by major breakthroughs across the field of maritime natural products. This development increases our hope that in the near future, medications produced from marine sources may emerge as a new wave, flooding the pharmaceutical sector and eventually assisting patients.

14.12 ETHICAL, LEGAL, AND ENVIRONMENTAL CONSIDERATIONS

14.12.1 Ethical Issues in Bioprospecting and Marine Conservation

Biopiracy is a significant ethical issue that occurs when corporations use marine resources without obtaining sufficient authorization from the governments or communities that have historically relied on these resources. It is crucial to acknowledge and remunerate local and indigenous groups for their knowledge and resources. This entails conducting open and honest discussions and establishing fair agreements for the distribution of benefits. This calls for a set of rules and regulations that are applicable internationally and are strictly enforced by a judiciary body.

Bioprospecting activities have the potential to disrupt fragile maritime ecosystems due to human greed. In order to maintain ethical standards, it is imperative to do comprehensive evaluations of the environmental impact when collecting marine microorganisms, so as to prevent any negative effects on biodiversity or ecological equilibrium. Adopting sustainable harvesting procedures and implementing monitoring measures are necessary in order to mitigate ecological harm. It also becomes necessary to train individuals in the process of collection of samples from maritime ecosystems so as to prevent any potential damage that could be inadvertently caused to the flora and fauna.

Indigenous peoples frequently hold rich traditional knowledge regarding aquatic habitat and is also a hallmark of some ethnic groups. Ethical bioprospecting should prioritize the respect and incorporation of this knowledge, by actively seeking informed permission and fostering partnership with indigenous groups, in order to ensure that their contributions and rights are duly acknowledged and respected. Failure to do so might result in the extinction of said indigenous groups and loss of culture.

14.12.2 Legal Frameworks and Benefit-Sharing

The legal frameworks that regulate bioprospecting in maritime settings are primarily established by international accords, including the Convention on Biological Diversity (CBD) and the Nagoya Protocol. These frameworks prioritize the requirement of obtaining prior informed permission and ensuring fair distribution of benefits with countries that provide genetic resources. Adhering to these agreements guarantees a just and equal distribution of benefits obtained from marine microorganisms. It is also important to further analyze whether the procurement of said marine microorganisms can harm the local biodiversity.

Different countries have diverse legislative prerequisites for bioprospecting and the safeguarding of maritime resources. Researchers and enterprises must have knowledge of and comply with national regulations that control the collection and utilization of marine biological resources. This encompasses acquiring essential licenses and adhering to legislation pertaining to preservation and intellectual property.

Comprehensive legal frameworks should incorporate explicit benefit-sharing agreements that specify the allocation of earnings derived from marine-based medicines. This guarantees that countries where resources originate from and populations that contribute to their extraction obtain equitable remuneration, and that the advantages are utilized to back conservation efforts and local development projects.

14.12.3 Sustainable Practices in Marine Biotechnology

Sustainable marine biotechnology practices entail the use of measures to prevent the collection of marine microorganisms from posing a hazard to the conservation of species or habitats. Enforcing conservation measures, such as implementing restricted collection zones and quotas, is crucial for the preservation of marine biodiversity and the maintenance of ecosystem health.

Marine biotechnology operations ought to implement strategies that limit the disturbance of the ecosystem. This entails employing non-intrusive sampling methods, refraining from disturbing vulnerable ecosystems, and undertaking environmental impact assessments prior to commencing collection operations. It is essential to continuously monitor marine ecosystems and the effects of biotechnological activity. Conducting study over a long period of time helps to identify any harmful effects on marine environments and enables the creation of adaptive management measures to reduce unwanted impacts. The results of such studies must be implemented into the legal framework, thereby informing and educating decision-makers around the world. Marine biotechnology needs to be in line with sustainable development goals by advocating for techniques that are advantageous to both the environment and human populations. This entails promoting creativity and advancement while guaranteeing that economic benefits are not achieved at the detriment of environmental or societal welfare.

14.13 CONCLUSION

Studies on marine microbes have been found to contain numerous bioactive molecules and hold great research prospects in biotechnology and medicine. Marine microorganisms, being adapted in various kinds of environments, have developed new modes of metabolic functions and biochemical potential, especially found in the tropical areas and deep-sea ecosystems, such as shallow water seas and Hot Springs. It has therefore led to identification of many compounds with antimicrobial, antiviral, antitumor, and other clinical uses.

Bacteria and archaea, fungi and algae, and viruses are essential members of marine ecosystems as they are involved in nutrient cycling, degradation of organic matter, and sometimes in the integrity of the food chains. Due to living in frequently harsh conditions like high pressure, low temperatures, and high salt concentrations, the organisms that are present in these areas possess biochemical characteristics that are useful in industry and pharmacy.

More to the point, chemotherapeutic agents derived from marine microbes have been affirmed to possess therapeutic importance. This work is the beginning toward the discovery of other cytotoxic molecules like cytarabine obtained from marine sponges that has been very useful in the treatment of leukemia and other cancer. Likewise, eribulin mesylate discovered from marine sponge has proved effective in cases of metastatic breast cancer. These examples substantiate that the marine natural products can solve some of the most urgent health problems and open new directions in the pharmacology.

Apart from the use of marine microbes as drugs, these tiny organisms present several advantages for biotechnological and industrial procedures. Marine microorganisms are also involved in the food industry for enzymatic processing and preservation while marine bioplastics and biofuels are the good replacement for the commonly used plastics and fuel. The possibilities of marine microorganisms in agriculture, including biopesticides and biofertilizers, show that these organisms contribute to sustaining agriculture.

The marine microbial research has a bright future with Synthetic Biology and Genetic Engineering facilitating the research for new chemical entity and their biosynthetic pathways. Nevertheless, there are potential problems like supplies, environmental consequences, and the question of biopiracy, which need to be solved for the correct and just usage of marine resources.

Marine microbes are also relatively unexplored though they are perhaps one of the greatest sources of potential biotechnological and medical value. Further studies and investigations are needed to understand the extent of possibilities of such systems and apply them for the advancement of people. With further elucidation of the national and global marine microbial ecosystems and their complex roles, the potential for detailed development of pharmaceutical and technological inventions from the oceans needs to be enhanced.

REFERENCES

[1] J.-P. Gattuso *et al.*, "Contrasting futures for ocean and society from different anthropogenic CO_2 emissions scenarios," *Science*, vol. 349, no. 6243, Jul. 2015, doi: 10.1126/science.aac4722.

[2] C. L. Sabine *et al.*, "The oceanic sink for anthropogenic CO_2," *Science*, vol. 305, no. 5682, pp. 367–371, Jul. 2004, doi: 10.1126/science.1097403.

[3] C. Mora, D. P. Tittensor, S. Adl, A. G. B. Simpson, and B. Worm, "How many species are there on Earth and in the ocean?," *PLoS Biol*, vol. 9, no. 8, p. e1001127, Aug. 2011, doi: 10.1371/journal.pbio.1001127.

[4] O. Hoegh-Guldberg *et al.*, "Coral reefs under rapid climate change and ocean acidification," *Science*, vol. 318, no. 5857, pp. 1737–1742, Dec. 2007, doi: 10.1126/science.1152509.

[5] S. J. Giovannoni and U. Stingl, "Molecular diversity and ecology of microbial plankton," *Nature*, vol. 437, no. 7057, pp. 343–348, Sep. 2005, doi: 10.1038/nature04158.

[6] C. B. Field, M. J. Behrenfeld, J. T. Randerson, and P. Falkowski, "Primary production of the biosphere: integrating terrestrial and oceanic components," *Science*, vol. 281, no. 5374, pp. 237–240, Jul. 1998, doi: 10.1126/science.281.5374.237.

[7] C. A. Suttle, "Viruses in the sea," *Nature*, vol. 437, no. 7057, pp. 356–361, Sep. 2005, doi: 10.1038/nature04160.

[8] E. F. DeLong, "The microbial ocean from genomes to biomes," *Nature*, vol. 459, no. 7244, pp. 200–206, May 2009, doi: 10.1038/nature08059.

[9] J. P. Zehr and R. M. Kudela, "Nitrogen cycle of the open ocean: from genes to ecosystems," *Ann Rev Mar Sci*, vol. 3, no. 1, pp. 197–225, Jan. 2011, doi: 10.1146/annurev-marine-120709-142819.

[10] S. Sievert, R. Kiene, and H. Schulz-Vogt, "The sulfur cycle," Oceanography, vol. 20, no. 2, pp. 117–123, Jun. 2007, doi: 10.5670/oceanog.2007.55.

[11] J. W. Blunt, B. R. Copp, R. A. Keyzers, M. H. G. Munro, and M. R. Prinsep, "Marine natural products," Nat Prod Rep, vol. 33, no. 3, pp. 382–431, 2016, doi: 10.1039/C5NP00156K.

[12] W. Fenical, P. R. Jensen, M. A. Palladino, K. S. Lam, G. K. Lloyd, and B. C. Potts, "Discovery and development of the anticancer agent salinosporamide A (NPI-0052)," *Bioorg Med Chem*, vol. 17, no. 6, pp. 2175–2180, Mar. 2009, doi: 10.1016/j.bmc.2008.10.075.

[13] R. H. Feling, G. O. Buchanan, T. J. Mincer, C. A. Kauffman, P. R. Jensen, and W. Fenical, "Salinosporamide A: a highly cytotoxic proteasome inhibitor from a novel microbial source, a marine bacterium of the new genus *Salinospora*," *Angew Chem Int Ed*, vol. 42, no. 3, pp. 355–357, Jan. 2003, doi: 10.1002/anie.200390115.

[14] D. E. Schaufelberger *et al.*, "The large-scale isolation of Bryostatin 1 from *Bugula neritina* following current good manufacturing practices," *J Nat Prod*, vol. 54, no. 5, pp. 1265–1270, Sep. 1991, doi: 10.1021/np50077a004.

[15] J. F. Imhoff, A. Labes, and J. Wiese, "Bio-mining the microbial treasures of the ocean: new natural products," *Biotechnol Adv*, vol. 29, no. 5, pp. 468–482, Sep. 2011, doi: 10.1016/j.biotechadv.2011.03.001.

[16] C. Imada, "Enzyme inhibitors and other bioactive compounds from marine actinomycetes," *Antonie Van Leeuwenhoek*, vol. 87, no. 1, pp. 59–63, Jan. 2005, doi: 10.1007/s10482-004-6544-x.

[17] A. Trincone, "Marine biocatalysts: enzymatic features and applications," *Mar Drugs*, vol. 9, no. 4, pp. 478–499, Mar. 2011, doi: 10.3390/md9040478.

[18] Y. Chisti, "Biodiesel from microalgae," *Biotechnol Adv*, vol. 25, no. 3, pp. 294–306, May 2007, doi: 10.1016/j.biotechadv.2007.02.001.

[19] G.-Q. Chen and M. K. Patel, "Plastics derived from biological sources: present and future: a technical and environmental review," *Chem Rev*, vol. 112, no. 4, pp. 2082–2099, Apr. 2012, doi: 10.1021/cr200162d.

[20] S. Singh, B. N. Kate, and U. C. Banerjee, "Bioactive compounds from cyanobacteria and microalgae: an overview," *Crit Rev Biotechnol*, vol. 25, no. 3, pp. 73–95, Jan. 2005, doi: 10.1080/07388550500248498.

[21] B. A. Whitton and M. Potts, Eds., *The Ecology of Cyanobacteria*. Dordrecht: Kluwer Academic Publishers, 2002. doi: 10.1007/0-306-46855-7.

[22] M. Könneke, A. E. Bernhard, J. R. de la Torre, C. B. Walker, J. B. Waterbury, and D. A. Stahl, "Isolation of an autotrophic ammonia-oxidizing marine archaeon," *Nature*, vol. 437, no. 7058, pp. 543–546, Sep. 2005, doi: 10.1038/nature03911.

[23] A. S. Amend, D. J. Barshis, and T. A. Oliver, "Coral-associated marine fungi form novel lineages and heterogeneous assemblages," *ISME J*, vol. 6, no. 7, pp. 1291–1301, Jul. 2012, doi: 10.1038/ismej.2011.193.

[24] C. A. Suttle, "Marine viruses — major players in the global ecosystem," *Nat Rev Microbiol*, vol. 5, no. 10, pp. 801–812, Oct. 2007, doi: 10.1038/nrmicro1750.

[25] B. B. Jørgensen and A. Boetius, "Feast and famine — microbial life in the deep-sea bed," *Nat Rev Microbiol*, vol. 5, no. 10, pp. 770–781, Oct. 2007, doi: 10.1038/nrmicro1745.

[26] N. Dubilier, C. Bergin, and C. Lott, "Symbiotic diversity in marine animals: the art of harnessing chemosynthesis," *Nat Rev Microbiol*, vol. 6, no. 10, pp. 725–740, Oct. 2008, doi: 10.1038/nrmicro1992.

[27] J. W. Deming, "Psychrophiles and polar regions," *Curr Opin Microbiol*, vol. 5, no. 3, pp. 301–309, Jun. 2002, doi: 10.1016/S1369-5274(02)00329-6.

[28] D. H. Bartlett, "Pressure effects on in vivo microbial processes," *Biochim Biophys Acta Protein Struct Mol Enzymol*, vol. 1595, no. 1–2, pp. 367–381, Mar. 2002, doi: 10.1016/S0167-4838(01)00357-0.

[29] R. K. Saiki *et al.*, "Primer-directed enzymatic amplification of DNA with a thermostable DNA polymerase," *Science*, vol. 239, no. 4839, pp. 487–491, Jan. 1988, doi: 10.1126/science.2448875.

[30] M. De Rosa and A. Gambacorta, "The lipids of archaebacteria," Prog Lipid Res, vol. 27, no. 3, pp. 153–175, Jan. 1988, doi: 10.1016/0163-7827(88)90011-2.

[31] C. G. Friedrich, F. Bardischewsky, D. Rother, A. Quentmeier, and J. Fischer, "Prokaryotic sulfur oxidation," *Curr Opin Microbiol*, vol. 8, no. 3, pp. 253–259, Jun. 2005, doi: 10.1016/j.mib.2005.04.005.

[32] J. A. Raymond and H. J. Kim, "Possible role of horizontal gene transfer in the colonization of sea ice by algae," *PLoS One*, vol. 7, no. 5, p. e35968, May 2012, doi: 10.1371/journal.pone.0035968.

[33] A. Oren, "Diversity of halophilic microorganisms: environments, phylogeny, physiology, and applications," *J Ind Microbiol Biotechnol*, vol. 28, no. 1, pp. 56–63, Jan. 2002, doi: 10.1038/sj/jim/7000176.

[34] M. Cunliffe, E. Harrison, M. Salter, H. Schäfer, R. Upstill-Goddard, and J. Murrell, "Comparison and validation of sampling strategies for the molecular microbial analysis of surface microlayers," *Aquatic Microbial Ecol*, vol. 57, pp. 69–77, Sep. 2009, doi: 10.3354/ame01330.
[35] E. A. Suter, M. I. Scranton, S. Chow, D. Stinton, L. Medina Faull, and G. T. Taylor, "Niskin bottle sample collection aliases microbial community composition and biogeochemical interpretation," *Limnol Oceanogr*, vol. 62, no. 2, pp. 606–617, Mar. 2017, doi: 10.1002/lno.10447.
[36] Y. Zhang *et al.*, "Targeted sampling by autonomous underwater vehicles," *Front Mar Sci*, vol. 6, Aug. 2019, doi: 10.3389/fmars.2019.00415.
[37] C. D. Taylor *et al.*, "Autonomous Microbial Sampler (AMS), a device for the uncontaminated collection of multiple microbial samples from submarine hydrothermal vents and other aquatic environments," *Deep Sea Res I: Oceanogr Res Papers*, vol. 53, no. 5, pp. 894–916, May 2006, doi: 10.1016/j.dsr.2006.01.009.
[38] D.-S. Mu, Y. Ouyang, G.-J. Chen, and Z.-J. Du, "Strategies for culturing active/dormant marine microbes," *Mar Life Sci Technol*, vol. 3, no. 2, pp. 121–131, May 2021, doi: 10.1007/s42995-020-00053-z.
[39] I. Joint, M. Mühling, and J. Querellou, "Culturing marine bacteria – an essential prerequisite for biodiscovery," *Microb Biotechnol*, vol. 3, no. 5, pp. 564–575, Sep. 2010, doi: 10.1111/j.1751-7915.2010.00188.x.
[40] M. L. Timmermans, K. J. Picott, L. Ucciferri, and A. C. Ross, "Culturing marine bacteria from the genus *Pseudoalteromonas* on a cotton scaffold alters secondary metabolite production," *Microbiologyopen*, vol. 8, no. 5, May 2019, doi: 10.1002/mbo3.724.
[41] D. Jung *et al.*, "Accessing previously uncultured marine microbial resources by a combination of alternative cultivation methods," *Microb Biotechnol*, vol. 14, no. 3, pp. 1148–1158, May 2021, doi: 10.1111/1751-7915.13782.
[42] A. Cario, G. C. Oliver, and K. L. Rogers, "Exploring the deep marine biosphere: challenges, innovations, and opportunities," *Front Earth Sci*, vol. 7, Sep. 2019, doi: 10.3389/feart.2019.00225.
[43] "polyketides," in *The IUPAC Compendium of Chemical Terminology*, Research Triangle Park, NC: International Union of Pure and Applied Chemistry (IUPAC), 2014. doi: 10.1351/goldbook.P04734.
[44] Y. Katsuyama, N. Funa, I. Miyahisa, and S. Horinouchi, "Synthesis of unnatural flavonoids and stilbenes by exploiting the plant biosynthetic pathway in *Escherichia coli*," *Chem Biol*, vol. 14, no. 6, pp. 613–621, Jun. 2007, doi: 10.1016/j.chembiol.2007.05.004.
[45] N. Akhter *et al.*, "Stress-driven discovery of new angucycline-type antibiotics from a marine streptomyces pratensis NA-ZhouS1," *Mar Drugs*, vol. 16, no. 9, p. 331, Sep. 2018, doi: 10.3390/md16090331.
[46] H.-N. Huang, Y.-L. Chan, C.-F. Hui, J.-L. Wu, C.-J. Wu, and J.-Y. Chen, "Use of tilapia piscidin 3 (TP3) to protect against MRSA infection in mice with skin injuries," *Oncotarget*, vol. 6, no. 15, pp. 12955–12969, May 2015, doi: 10.18632/oncotarget.4102.
[47] D. Xu *et al.*, "Nocardiopsistins A-C: new angucyclines with anti-MRSA activity isolated from a marine sponge-derived Nocardiopsis sp. HB-J378," *Synth Syst Biotechnol*, vol. 3, no. 4, pp. 246–251, Dec. 2018, doi: 10.1016/j.synbio.2018.10.008.
[48] Y. Li *et al.*, "Five New amicoumacins isolated from a marine-derived bacterium *Bacillus subtilis*," *Mar Drugs*, vol. 10, no. 2, pp. 319–328, Feb. 2012, doi: 10.3390/md10020319.
[49] "terpenoids," in *The IUPAC Compendium of Chemical Terminology*, Research Triangle Park, NC: International Union of Pure and Applied Chemistry (IUPAC), 2014. doi: 10.1351/goldbook.T06279.
[50] N. J. White, "Qinghaosu (artemisinin): the price of success," *Science*, vol. 320, no. 5874, pp. 330–334, Apr. 2008, doi: 10.1126/science.1155165.
[51] X.-W. Yang *et al.*, "Strepsesquitriol, a rearranged zizaane-type sesquiterpenoid from the deep-sea-derived actinomycete *Streptomyces* sp. SCSIO 10355," *J Nat Prod*, vol. 76, no. 12, pp. 2360–2363, Dec. 2013, doi: 10.1021/np400923c.
[52] A. Takahashi, S. Kurasawa, D. Ikeda, Y. Okami, and T. Takeuchi, "Altemicidin, a new acaricidal and antitumor substance. I. Taxonomy, fermentation, isolation and physico-chemical and biological properties," *J Antibiot*, vol. 42, no. 11, pp. 1556–1561, 1989, doi: 10.7164/antibiotics.42.1556.
[53] M. K. Renner *et al.*, "Cyclomarins A–C, new antiinflammatory cyclic peptides produced by a marine bacterium (*Streptomyces* sp.)," *J Am Chem Soc*, vol. 121, no. 49, pp. 11273–11276, Dec. 1999, doi: 10.1021/ja992482o.
[54] S. Genovese, M. Curini, and F. Epifano, "Chemistry and biological activity of azoprenylated secondary metabolites," *Phytochemistry*, vol. 70, no. 9, pp. 1082–1091, Jun. 2009, doi: 10.1016/j.phytochem.2009.06.016.
[55] T. Venkateshwar Goud, N. Srinivasa Reddy, N. Raghavendra Swamy, T. Siva Ram, and Y. Venkateswarlu, "Anti-HIV active petrosins from the marine sponge *Petrosia similis*," *Biol Pharm Bull*, vol. 26, no. 10, pp. 1498–1501, 2003, doi: 10.1248/bpb.26.1498.
[56] A. G. Atanasov *et al.*, "Discovery and resupply of pharmacologically active plant-derived natural products: a review," *Biotechnol Adv*, vol. 33, no. 8, pp. 1582–1614, Dec. 2015, doi: 10.1016/j.biotechadv.2015.08.001.
[57] A. W. Newaz, K. Yong, X.-Y. Lian, and Z. Zhang, "Streptoindoles A–D, novel antimicrobial indole alkaloids from the marine-associated actinomycete *Streptomyces* sp. ZZ1118," *Tetrahedron*, vol. 104, p. 132598, Jan. 2022, doi: 10.1016/j.tet.2021.132598.
[58] K. N. Maloney *et al.*, "Lodopyridone, a structurally unprecedented alkaloid from a marine actinomycete," *Org Lett*, vol. 11, no. 23, pp. 5422–5424, Dec. 2009, doi: 10.1021/ol901997k.
[59] C. C. Hughes, J. B. MacMillan, S. P. Gaudêncio, P. R. Jensen, and W. Fenical, "The ammosamides: structures of cell cycle modulators from a marine-derived *Streptomyces* species," *Angew Chem Int Ed*, vol. 48, no. 4, pp. 725–727, Jan. 2009, doi: 10.1002/anie.200804890.
[60] S. Egan, S. James, C. Holmström, and S. Kjelleberg, "Correlation between pigmentation and antifouling compounds produced by *Pseudoalteromonas tunicata*," *Environ Microbiol*, vol. 4, no. 8, pp. 433–442, Aug. 2002, doi: 10.1046/j.1462-2920.2002.00322.x.
[61] M. Strieker, A. Tanović, and M. A. Marahiel, "Nonribosomal peptide synthetases: structures and dynamics," *Curr Opin Struct Biol*, vol. 20, no. 2, pp. 234–240, Apr. 2010, doi: 10.1016/j.sbi.2010.01.009.

[123] C. Hao, Z. Xu, C. Xu, and R. Yao, "Anti-herpes simplex virus activities and mechanisms of marine derived compounds," *Front Cell Infect Microbiol*, vol. 13, Jan. 2024, doi: 10.3389/fcimb.2023.1302096.

[124] A. Arena, T. L. Maugeri, B. Pavone, D. Iannello, C. Gugliandolo, and G. Bisignano, "Antiviral and immunoregulatory effect of a novel exopolysaccharide from a marine thermotolerant Bacillus licheniformis," *Int Immunopharmacol*, vol. 6, no. 1, pp. 8–13, Jan. 2006, doi: 10.1016/j.intimp.2005.07.004.

[125] M. Hasui, M. Matsuda, K. Okutani, and S. Shigeta, "In vitro antiviral activities of sulfated polysaccharides from a marine microalga (*Cochlodinium polykrikoides*) against human immunodeficiency virus and other enveloped viruses," *Int J Biol Macromol*, vol. 17, no. 5, pp. 293–297, Jan. 1995, doi: 10.1016/0141-8130(95)98157-T.

[126] J.-B. Lee *et al.*, "Antiviral sulfated polysaccharide from *Navicula directa*, a diatom collected from deep-sea water in Toyama bay," *Biol Pharm Bull*, vol. 29, no. 10, pp. 2135–2139, 2006, doi: 10.1248/bpb.29.2135.

[127] D. Rowley, "Halovirs A–E, new antiviral agents from a marine-derived fungus of the genus *Scytalidium*," *Bioorg Med Chem*, vol. 11, no. 19, pp. 4263–4274, Sep. 2003, doi: 10.1016/S0968-0896(03)00395-X.

[128] W. Wang *et al.*, "Inhibition of influenza A virus infection by fucoidan targeting viral neuraminidase and cellular EGFR pathway," *Sci Rep*, vol. 7, no. 1, p. 40760, Jan. 2017, doi: 10.1038/srep40760.

[129] S. Moir, T.-W. Chun, and A. S. Fauci, "Pathogenic mechanisms of HIV disease," *Ann Rev Pathol: Mech Dis*, vol. 6, no. 1, pp. 223–248, Feb. 2011, doi: 10.1146/annurev-pathol-011110-130254.

[130] S. K. Masenga, B. C. Mweene, E. Luwaya, L. Muchaili, M. Chona, and A. Kirabo, "HIV–host cell interactions," *Cells*, vol. 12, no. 10, p. 1351, May 2023, doi: 10.3390/cells12101351.

[131] T. Mori *et al.*, "Isolation and characterization of griffithsin, a novel HIV-inactivating protein, from the red alga *Griffithsia* sp.," *J Biol Chem*, vol. 280, no. 10, pp. 9345–9353, Mar. 2005, doi: 10.1074/jbc.M411122200.

[132] P. Emau *et al.*, "Griffithsin, a potent HIV entry inhibitor, is an excellent candidate for anti-HIV microbicide," *J Med Primatol*, vol. 36, no. 4–5, pp. 244–253, Aug. 2007, doi: 10.1111/j.1600-0684.2007.00242.x.

[133] P. Meuleman *et al.*, "Griffithsin has antiviral activity against hepatitis C virus," *Antimicrob Agents Chemother*, vol. 55, no. 11, pp. 5159–5167, Nov. 2011, doi: 10.1128/AAC.00633-11.

[134] K. Levendosky *et al.*, "Griffithsin and carrageenan combination to target herpes simplex virus 2 and human papillomavirus," *Antimicrob Agents Chemother*, vol. 59, no. 12, pp. 7290–7298, Dec. 2015, doi: 10.1128/AAC.01816-15.

[135] N. R. P. V. Macedo *et al.*, "Caulerpin as a potential antiviral drug against herpes simplex virus type 1," *Rev Bras Farmacog*, vol. 22, no. 4, pp. 861–867, Aug. 2012, doi: 10.1590/S0102-695X2012005000072.

[136] J. Yoshimoto, M. Kakui, H. Iwasaki, T. Fujiwara, H. Sugimoto, and N. Hattori, "Identification of a novel HA conformational change inhibitor of human influenza virus," *Arch Virol*, vol. 144, no. 5, pp. 865–878, May 1999, doi: 10.1007/s007050050552.

[137] K. Minagawa, S. Kouzuki, and T. Kamigauchi, "Stachyflin and acetylstachyflin, novel anti-influenza a virus substances, produced by *Stachybotrys* sp. RF-7260. II. Synthesis and preliminary structure-activity relationships of stachyflin derivatives," *J Antibiot*, vol. 55, no. 2, pp. 165–171, 2002, doi: 10.7164/antibiotics.55.165.

[138] T. Hayashi, K. Hayashi, M. Maeda, and I. Kojima, "Calcium spirulan, an inhibitor of enveloped virus replication, from a blue-green alga *Spirulina platensis*," *J Nat Prod*, vol. 59, no. 1, pp. 83–87, Jan. 1996, doi: 10.1021/np960017o.

[139] J. Mader *et al.*, "Calcium spirulan derived from *Spirulina platensis* inhibits herpes simplex virus 1 attachment to human keratinocytes and protects against herpes labialis," *J Allergy Clin Immunol*, vol. 137, no. 1, pp. 197-203.e3, Jan. 2016, doi: 10.1016/j.jaci.2015.07.027.

[140] D. Rowley, "Halovirs A–E, new antiviral agents from a marine-derived fungus of the genus *Scytalidium*," *Bioorg Med Chem*, vol. 11, no. 19, pp. 4263–4274, Sep. 2003, doi: 10.1016/S0968-0896(03)00395-X.

[141] D. H. Dethe and M. Shukla, "Enantioselective first total syntheses of the antiviral natural products xiamycins D and E," *Chem Commun*, vol. 57, no. 81, pp. 10644–10646, 2021, doi: 10.1039/D1CC04739F.

[142] F. M. Hauser and R. A. Tommasi, "Total synthesis of (-)-(9R)-7,11-dideoxy-13-deoxodaunomycinone," *J Org Chem*, vol. 56, no. 20, pp. 5758–5759, Sep. 1991, doi: 10.1021/jo00020a010.

[143] R. M. Welch, W. A. Norvell, S. C. Schaefer, J. E. Shaff, and L. V. Kochian, "Induction of iron(III) and copper(II) reduction in pea (*Pisum sativum* L.) roots by Fe and Cu status: does the root-cell plasmalemma Fe(III)-chelate reductase perform a general role in regulating cation uptake?," *Planta*, vol. 190, no. 4, Jun. 1993, doi: 10.1007/BF00224795.

[144] J. H. Fitton, "Therapies from fucoidan; multifunctional marine polymers," *Mar Drugs*, vol. 9, no. 10, pp. 1731–1760, Sep. 2011, doi: 10.3390/md9101731.

[145] T.-K.-O. Nguyen *et al.*, "Phenylpropanoid profiling reveals a class of hydroxycinnamoyl glucaric acid conjugates in Isatis tinctoria leaves," *Phytochemistry*, vol. 144, pp. 127–140, Dec. 2017, doi: 10.1016/j.phytochem.2017.09.007.

[146] P. C. Calder, "Omega-3 polyunsaturated fatty acids and inflammatory processes: nutrition or pharmacology?" *Br J Clin Pharmacol*, vol. 75, no. 3, pp. 645–662, Mar. 2013, doi: 10.1111/j.1365-2125.2012.04374.x.

[147] K. S. Moore *et al.*, "Squalamine: an aminosterol antibiotic from the shark," *Proc Natl Acad Sci USA*, vol. 90, no. 4, pp. 1354–1358, Feb. 1993, doi: 10.1073/pnas.90.4.1354.

[148] L. Oreland, J. Hallman, and M. Damberg, "Platelet MAO and personality – function and dysfunction," *Curr Med Chem*, vol. 11, no. 15, pp. 2007–2016, Aug. 2004, doi: 10.2174/0929867043364838.

[149] S. Ghattavi and A. Homaei, "Marine enzymes: classification and application in various industries," *Int J Biol Macromol*, vol. 230, p. 123136, Mar. 2023, doi: 10.1016/j.ijbiomac.2023.123136.

[150] N. Barzkar *et al.*, "Marine bacterial esterases: emerging biocatalysts for industrial applications," *Appl Biochem Biotechnol*, vol. 193, no. 4, pp. 1187–1214, Apr. 2021, doi: 10.1007/s12010-020-03483-8.

[151] I. D. Antonio Trincone, "Marine glycosyl hydrolases as tool for industrial application," in *Industrial Applications of Glycoside Hydrolases*, Editor: Smriti Shrivastava. Springer, 2020, pp. 151–166. doi: 10.1007/978-981-15-4767-6_5.

[152] A. Trincone, "Application-oriented marine isomerases in biocatalysis," *Mar Drugs*, vol. 18, no. 11, p. 580, Nov. 2020, doi: 10.3390/md18110580.

[153] D. Ganguly, K. L. V. Prasanna, S. Neelapu, and G. Goswami, "Role of microbes in bioremediation," in *Industrial Microbiology and Biotechnology*, Editor: Pradeep Verma. Springer Nature, 2023, pp. 549–584. doi: 10.1007/978-981-99-2816-3_19.

[154] A. Harshvardhan and P. Saikia, "Microbial bioremediation," in *Microbes in Agri-Forestry Biotechnology*, Editors: Gustavo Molina, Zeba Usmani, Minaxi Sharma, Abdelaziz Yasri, Vijai Kumar Gupta. CRC Press, 2022, pp. 221–240. doi: 10.1201/9781003110477-10.

[155] S. Liu *et al.*, "Petroleum spill bioremediation by an indigenous constructed bacterial consortium in marine environments," *Ecotoxicol Environ Saf*, vol. 241, p. 113769, Aug. 2022, doi: 10.1016/j.ecoenv.2022.113769.

[156] P. A. Mendy, A. Kargbo, and M. E. Entonu, "Bioremediation of heavy metal ions from contaminated soil and water by microbes: a review," *Afr J Biol Sci*, vol. 3, no. 2, p. 1, Apr. 2021, doi: 10.33472/AFJBS.3.2.2021.1-8.

[157] A. Rotter *et al.*, "The essentials of marine biotechnology," *Front Mar Sci*, vol. 8, Mar. 2021, doi: 10.3389/fmars.2021.629629.

[158] M. A. Shredah, H. A. H. Ibrahim, and M. S. M. Abd El Hafez, "Challenges and opportunities in marine ecological system and drug discovery," 2023, pp. 282–315. doi: 10.2174/9789815051995123030011.

[159] K. A. Stuart, K. Welsh, M. C. Walker, and R. Edrada-Ebel, "Metabolomic tools used in marine natural product drug discovery," *Expert Opin Drug Discov*, vol. 15, no. 4, pp. 499–522, Apr. 2020, doi: 10.1080/17460441.2020.1722636.

[160] N. Papon, B. R. Copp, and V. Courdavault, "Marine drugs: biology, pipelines, current and future prospects for production," *Biotechnol Adv*, vol. 54, p. 107871, Jan. 2022, doi: 10.1016/j.biotechadv.2021.107871.

[161] I. Ahmed *et al.*, "Exploring marine as a rich source of bioactive peptides: challenges and opportunities from marine pharmacology," *Mar Drugs*, vol. 20, no. 3, p. 208, Mar. 2022, doi: 10.3390/md20030208.

[162] D. Rani, V. Garg, and R. Dutt, "Anticancer potential of azole containing marine natural products: current and future perspectives," *Anticancer Agents Med Chem*, vol. 21, no. 15, pp. 1957–1976, Sep. 2021, doi: 10.2174/1871520621666210112112422.

15 Cosmeceuticals
The Beauty Elixirs from the Deep

Ashiq Shibili Pillatt, Susmita Roy, Samatha M Jain, Surajit Pathak, and Antara Banerjee
Medical Biotechnology Lab, Faculty of Allied Health Sciences,
Chettinad Academy of Research and Education (CARE),
Chettinad Hospital and Research Institute (CHRI), Chennai, India

15.1 INTRODUCTION

The skin serves as the main physical barrier against environmental influences and covers the whole external surface of the body [1]. Human skin ages by becoming thinner, losing its natural flexibility, and losing its capacity to hold moisture [2]. Due to the continuous interaction of the skin with the outside environment, environmental elements including dust, chemical reagents, and ultraviolet (UV) radiation might decrease the antioxidant capacity of the skin and accelerate the process of ageing [3]. Therefore, appropriate skin care is essential for maintaining its structural integrity and functional homeostasis, as well as its overall aesthetic attributes. A new product category called cosmeceuticals has emerged in recent years from the combination of cosmetics and pharmaceuticals which aim to improve skin health and appearance [4]. Dr. Albert Kligman initially used the word "cosmeceuticals" in 1984 to refer to the products that were applied directly to the skin, much like cosmetics, but included biologically active substances that might have pharmacological effects on the skin [5]. Over the last 10 years, the worldwide cosmeceutical industry has grown exponentially due to rising consumer awareness, technological developments, and an increasing need for organic and natural products. In 2023, the global cosmeceuticals market was estimated to be worth approximately USD 63.59 billion. Throughout the forecast period (2024–2032), the market is expected to develop at a compound annual growth rate (CAGR) of 9.1%, from USD 68.67 billion in 2024 to USD 138.26 billion by 2032 [6]. Marine-derived cosmeceuticals represent one of the most fascinating and rapidly growing industries. Over 70% of the Earth's surface is made up of the ocean, which is a huge and mainly unexplored source of bioactive compounds with enormous potential for skincare uses [7]. From shallow to deep seas, the oceans include a vast variety of ecosystems and climatic conditions that support enormous biodiversity in both plants and wildlife. The distinct features of multiple marine habitats have led to diverse biological adaptations, resulting in the generation of an extensive range of bioactive compounds [7,8]. Strong antioxidant, anti-inflammatory, anti-photoaging, anti-melanogenic, anti-tyrosinase, anti-acne, and anti-wrinkling compounds are produced by marine species, such as algae, seaweeds, bacteria, and fungi (Figure 15.1) [9]. The abundance of bioactive components, including polysaccharides, peptides, enzymes, and vitamins, can enhance hydration, protect the skin from environmental stresses, and stimulate the formation of collagen, which makes them ideal for use in cosmeceutical applications [2]. Marine-derived cosmeceuticals have drawn significant attention because of their effectiveness as well as the fact that they fit in with consumers' increasing desire for natural and sustainable products. Marine bioactive compounds are more compatible with human skin than synthetic chemicals, as they are less likely to cause irritation or adverse reactions, therefore providing a more holistic approach to skincare [10]. Algae and seaweed are the most well-known and often utilized components in cosmeceutical products among all marine sources. Strong anti-inflammatory and antioxidant properties are provided by extracts from different forms of algae, which are high in vitamins, minerals, and antioxidants. These will help to reduce oxidative stress and improve the resilience of the skin [11]. In addition to algae, marine microorganisms such as fungi and bacteria have emerged as important sources of bioactive compounds for cosmetic applications. These microbes create secondary metabolites that have distinct structures and abilities that can either replicate or improve the processes that naturally occur in the skin [12]. Similarly, marine collagen is derived from fish and other marine species and is well-known for its exceptional skin-rejuvenating characteristics, which include improving elasticity and reducing wrinkles [13]. As a result, natural products from marine microbes have gained a significant role in the cosmetics industry. However, only a limited number of active compounds derived from marine microorganisms have been utilized in cosmetics. This book chapter summarizes the role of marine microbes-derived cosmeceuticals, their sources, applications, and mechanism of action for maintaining skin health and appearance. It also explores marine-derived supplements as well as the comparison between natural and synthetic cosmeceutical products.

DOI: 10.1201/9781003509530-15

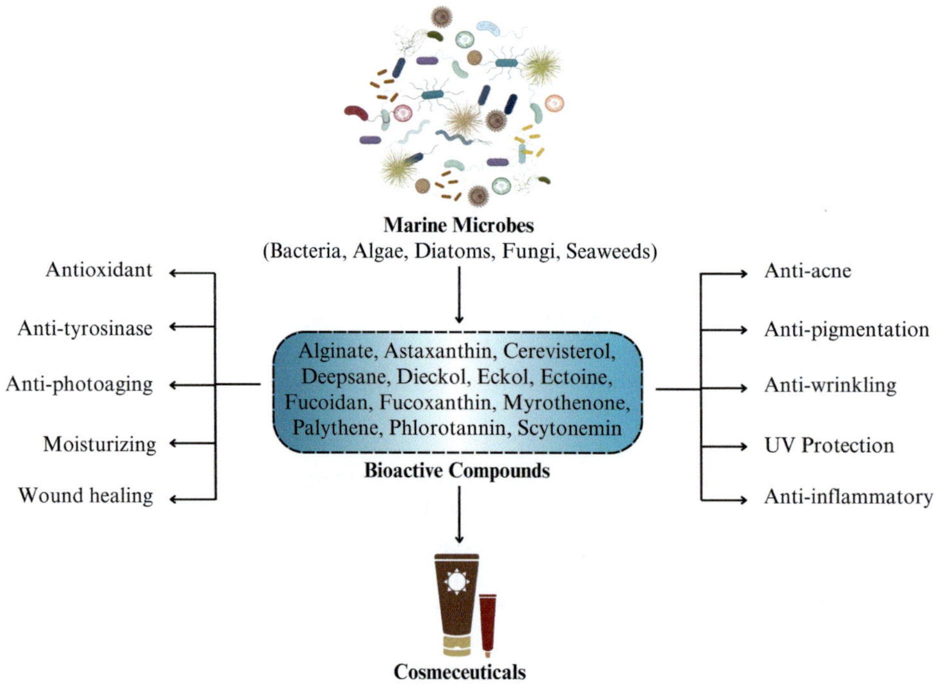

FIGURE 15.1 Pictorial representation of various properties of marine microbe-derived cosmeceuticals.

15.2 HISTORICAL DEVELOPMENTS OF MARINE COSMECEUTICALS

The exploration of marine microorganisms for cosmeceutical use began in the mid-20th century, increased by the understanding that the oceans, which cover more than 70% of the Earth, hold a vast reservoir of unique bioactive compounds. While penicillin was originally discovered from the terrestrial fungus *Penicillium notatum* in the 1920s, later in the 1940s, studies have highlighted that marine microorganisms also possess significant potential for producing novel antibiotic compounds [14]. In the 1990s, marine cosmeceuticals became popularized as researchers revealed that marine microbes generate bioactive compounds with distinctive properties, including antimicrobial, anti-inflammatory, and antioxidant effects [15]. Consequently, skincare treatments and cosmetics derived from marine sources were created to target a range of skin issues. In today's competitive market, researchers are focusing on a new trend in advanced cosmetics that incorporates ingredients derived from the ocean. Bioactive compounds from marine sources like polysaccharides, peptides, enzymes, and vitamins not only serve as effective natural skincare agents but can also be utilized to create nutraceuticals, edible products made from marine organisms and seaweeds as well as innovative cosmetic formulations [9].

15.3 SOURCES OF MARINE MICROBES-DERIVED COSMECEUTICALS

There is a growing trend towards natural bioactive ingredients, driven by the adverse effects and poor absorption associated with synthetic cosmetics. Numerous studies have demonstrated the effectiveness of natural products as viable alternatives to synthetic cosmetics, highlighting their potential benefits even with long-term usage. These products include creams, serums, masks, and supplements designed to target specific skin issues such as ageing, sun protection, and hydration [16]. As marine bioresource technology has advanced, marine-derived natural products have gained considerable interest as cosmeceuticals. With the development of marine bioresource technology, natural products originating from marine sources have gained a lot of attention as cosmeceuticals. Marine organisms such as macroalgae, microalgae, sponges, corals, fungi, and bacteria have become valuable sources of secondary metabolites like fucoidans, galactans, agar, carrageenans, ulvans, porphyrin, alginates, glucans, and others. These compounds are gaining significant application in the field of cosmeceuticals [17]. The ocean supports a wide range of microbial biodiversity across a variety of ecosystems and ecological conditions [18]. The growing popularity of marine-based ingredients has invoked concerns that non-sustainable production methods or large-scale sourcing could derail the already stressed marine ecosystems. Marine microorganisms present a major advantage as sustainable sources for producing valuable compounds, as they can be grown and cultured in fermenters [19,20]. The major sources of cosmeceuticals obtained from marine microorganisms are briefly discussed in Table 15.1.

15.3.1 Bacteria

Bacteria are generally the most prevalent and diverse components of ocean microbial communities, playing a crucial role in biogeochemical cycles and the flow of energy and matter [36].

TABLE 15.1
Cosmeceuticals Derived from Marine Microbes and Their Respective Applications

Marine Microbes	Marine-Derived Compounds	Sources	Cosmetic Applications	Ref:
Bacteria	Deepsane	*Alteromonas macleodii*	It protects sensitive skin from UVB, mechanical, and chemical harm, helps improve skin hydration, and promotes skin regeneration by stimulating collagen production.	[21]
	HE800	*Vibrio diabolicus*	Promotes skin healing and regeneration by stimulating cell proliferation and collagen synthesis and enhances skin hydration.	[22]
	Saproxanthin Myxol	Flavobacteriaceae sp.	It protects the skin from oxidative stress caused by free radicals and UV radiation and reduces pigmentation and dark spots.	[23]
	Thalassotalic acid	*Thalassotalea* sp.	Possesses anti-ageing properties by promoting collagen synthesis and improving skin elasticity, reduces redness or inflammation	[24]
	Ectoine	*Halomonas elongata*	Helps to maintain skin moisture by reducing water loss and possesses anti-inflammatory properties that help to reduce skin irritation and redness.	[25]
Algae	Scytonemin	*Calothrix crustacea, Nostoc commune*	Absorbs UVA and UBA radiation, protects the skin from harmful UV rays, and helps to protect skin cells from oxidative stress.	[26]
	Astaxanthin	*Haematococcus pluvialis*	Possesses antioxidant properties that help the skin from oxidative stress and UV-induced damage, it also reduces hyperpigmentation.	[27]
	β-carotene	*Dunaliella salina*	Reduces fine lines and wrinkles by neutralizing free radicals and supporting skin repair and regeneration, it also reduces hyperpigmentation.	[28]
Diatoms	Docosahexaenoic acid Eicosapentaenoic acid	*Phaeodactylum tricornutum*	Stimulate proteasome activity in skin cells, shield it from the damaging effects of UV radiation	[29]
	Fucoxanthin	*Thalassiosira, Chaetoceros*	Prevents hair loss and melanocyte proliferation, promote keratinocyte differentiation, and the growth of hair and hair follicles.	[30]
Fungi	Cerevisterol Aloesol	*Penicillium* sp.	Possesses antimicrobial, tyrosinase inhibitory, and wound-healing activities. It helps to maintain moisture levels in the skin and prevent dryness.	[31]
	Mycosporine-like amino acids	*Aureobasidium pullulans Hortaea werneckii Cryptococcus liquefaciens*	Absorbs UV light and reduces oxidative damage to skin cells, neutralizes free radicals, reduces oxidative stress, reduces redness and inflammation, enhances skin hydration by improving the skin's moisture-retaining ability.	[32]
	Docosahexaenoic acid	*Schizochytrium* sp. *Aurantiochytrium* sp.	It helps to reduce oxidative stress and inflammation in the skin, improve overall skin moisture levels, and reduce dryness.	[33]
	Oxysporizoline	*Fusarium oxysporum*	Possesses antibacterial activity against methicillin-resistant *Staphylococcus aureus* (MRSA) and MDRSA, antioxidant properties, inhibit melanin production, improve skin tone, and reduce hyperpigmentation	[34]
	Myrothenone	*Myrothecium* sp. *Botrytis* sp.	Possesses anti-tyrosinase activity and helps to prevent and treat acne by inhibiting the growth of acne-causing bacteria on the skin, helping to reduce skin redness and irritation.	[35]

Personal skincare products often incorporate a variety of beneficial compounds, such as those with anti-wrinkle, skin-whitening, moisturizing, and photo-protective effects, many of which are derived from marine microorganisms [19]. The most widely used bioactive substances with anti-wrinkle activity are polysaccharides from microbial origin specifically exopolysaccharides (EPSs). EPSs are mainly extracted from various marine bacterial species including *Pseudoalteromonas* sp. *Alteromonas* sp., and *Vibrio* sp. [37]. Deepsane is an EPS that is extracted from the marine bacteria *Alteromonas macleodii* which is used to protect sensitive skin from UVB, mechanical, and chemical harm [21]. *Vibrio diabolicus*, a marine hydrothermal vent bacteria found in deep waters, creates an EPS called HE800 that resembles hyaluronic acid structurally and has unique characteristics that promote collagen structure [22]. Saproxanthin and Myxol are two rare carotenoids with significant antioxidant activity extracted from marine bacteria that are members of the Flavobacteriaceae family [23]. The marine bacterium *Pseudomonas* sp. creates methylene chloride by inhibiting the formation of tyrosinase, which can reduce the pigmentation of human melanocytes [38]. Similarly, Thalassotalic acids A, B, and C also show anti-tyrosinase activity which is isolated from *Thalassotalea* sp. [24]. Ectoine which increases intermolecular space to enhance the hydration of the cell surface is isolated from halophilic bacteria and actinobacteria [25].

15.3.2 Algae

Marine algae are a rich source of bioactive compounds, offering significant potential for use in cosmeceuticals and nutraceuticals. Marine microalgae are of significant interest in cosmeceuticals due to their ability to absorb UV radiation and protect the skin from roughness, wrinkles, and pigmentation [39]. Scytonemin, a pigment from cyanobacteria (*Chlorogloeopsis* sp., *Calothrix crustacea, Nostoc* sp.), is more efficient than commercial sunscreen formulations in absorbing UVA and UVB rays [26]. The halotolerant microalga *Dunaliella salina* produces UV protective pigment β-carotene more than 10% of its body weight [40]. Astaxanthin extracted from *Haematococcus pluvialis* may decrease different inflammatory mediators and affect the different phases of the oxidative damage cascade, hence it was regarded as a potent antioxidant and superior anti-inflammatory agent [27]. Skin and hair can be softened and smoothened by using oils derived from various microalgae, particularly those from the genus *Chlorella* [41]. Microalgae rich in β-carotene (*Dunaliella salina*, *Spirulina platensis*) and sulphated polysaccharides (*Porphyridium*) can slow down lipid peroxidation, limit oxidative damage, and decrease the production of reaction oxygen species (ROS) [28]. Matsui *et al.* (2003) demonstrated that the red microalga (*Porphyridium*) containing sulphated polysaccharides was found to block leukocyte movement, offering anti inflammatory benefits. Additionally, when applied topically to humans, these polysaccharides helped prevent erythema [42].

15.3.3 Diatoms

The extract of the marine diatom *Phaeodactylum tricornu tum*, rich in ω-3 PUFAs like docosahexaenoic acid (DHA), eicosapentaenoic acid (EPA), and fucoxanthin, boosts proteasome activity in skin cells, particularly fibroblasts, keratinocytes, and melanocytes. This extract enhances skin flexibility and firmness while also protecting it from UV radiation damage [29]. Fucoxanthin and other carotenoid pigments found in diatoms like *Thalassiosira* sp. and *Chaetoceros* sp. are suggested for use in formulations to prevent hair loss because they may influence melanogenesis in skin and hair, promote keratinocyte differentiation, and support the development of hair and hair follicles [30, 43].

15.3.4 Fungi

Several metabolites derived from fungi are known for their various skin cosmeceutical properties, such as photoprotection, anti-acne, skin whitening, and wound healing. Microfungi, which are particularly impacted by global warming and increasing solar radiation, are believed to have evolved adaptive mechanisms to combat UV stress [44]. In a study, it was reported that, Kojic acid isolated from the marine-derived fungus *Altenaria* sp. showed significant tyrosinase inhibitory activity, indicating its potential as a skin-whitening agent, while kojic acid dimethyl ether, kojic acid monomethyl ether, and phomaligol A were inactive [31]. Various fungi derived from the marine environment produce secondary metabolites that may have potential applications in cosmeceuticals. For instance, *Aureobasidium pullulans*, *Hortaea werneckii*, *Cryptococcus liquefaciens*, *Phaeotheca triangularis*, and *Trimmatostroma salinum* are all notable fungal species that are known for their production of mycosporine-like amino acids (MAAs) [32]. The primary producers of DHA are specific species from the Thraustochytriaceae family, namely *Schizochytrium*, *Aurantiochytrium*, and *Ulkenia*. Thraustochytrid oils rich in DHA are presently available as nutraceuticals, but they also have a lot of promise as nutraceuticals and cosmeceuticals [33]. Carotenoids may also be found in marine fungi that belong to the genera *Phaffia*, *Xanthophyllomyces*, and *Rhodotorula*. Significant anti-inflammatory and antioxidant properties of these carotenoids may aid in skin photoprotection. Carotenoids can also prevent harmful processes brought on by or facilitated by UV radiation from the sun [45]. The marine-derived fungus *Fusarium oxysporum* produces a polycyclic quinazoline alkaloid, Oxysporizoline, which exhibits antibacterial activity against methicillin-resistant *Staphylococcus aureus* (MRSA) and multidrug-resistant *Staphylococcus aureus* (MDRSA), along with significant antioxidant properties [34, 46]. Myrothenone and 6-[(E)-hept-1-enyl]-α-pyrone sourced from algicolous fungus *Myrothecium* sp. and *Botrytis sp.* respectively demonstrated significant anti-tyrosinase activity than kojic acid [35].

15.4 MECHANISM OF ACTION OF COSMECEUTICALS

15.4.1 Antioxidant Activity

The generation of ROS is linked to oxidative stress, which is a major factor in the ageing process of the skin. The antioxidant activity is the capability of substances to counteract free radicals and ROS that contribute to oxidative stress. This stress can damage cells, accelerate ageing, and increase the risk of numerous diseases [47, 48]. Cellular components like DNA, proteins, and membrane lipids are among the biological components that are harmed by ROS [11]. Preventing ROS generation is important because the oxidation of membrane lipids is a key factor in skin ageing [49]. Antioxidants are essential components of cosmeceuticals because they shield the skin from oxidative damage, which can be caused by pollutants and UV rays [50]. Antioxidants generated from marine microbes reduce oxidative stress through various methods. They scavenge ROS, directly neutralize free radicals, and strengthen the body's natural antioxidant defences [51]. These compounds can also modulate signalling pathways like MAPK/AP-1/NF-κB/TNF-α/IL-6 involved in oxidative stress responses and inflammation [52]. Astaxanthin is a strong antioxidant with a higher activity than vitamin E derived from the microalga *Haematococcus pluvialis* [53]. By scavenging singlet oxygen and free radicals, it prevents oxidative stress. Astaxanthin also inhibits lipid peroxidation and enhances the cellular antioxidant enzymes catalase and superoxide dismutase (SOD). It reduces the appearance of fine lines and wrinkles by scavenging ROS and shielding DNA, proteins, and cellular lipids from oxidative damage [53, 54]. The cyanobacterium *Spirulina* is well-known for having a high concentration of phycocyanin, a powerful antioxidant. By neutralizing free radicals and preventing lipid peroxidation, phycocyanin shields skin cells from oxidative injury [55]. Similarly, squalene, which is derived from *Aurantiochytrium* sp. and *Halobacterium cutirubrum*, acts as a strong antioxidant. It functions by providing electrons to neutralize free radicals, avoiding cellular damage and promoting the moisture and flexibility of the skin. It is an important component of anti-ageing and moisturizing cosmetics because of its ability to improve skin barrier function and shield against environmental stresses [56].

15.4.2 Anti-inflammatory and Wound-Healing Effects

Wound healing and anti-inflammation are related physiological processes that are essential for preserving tissue integrity and encouraging repair [57]. The anti-inflammatory action is typically mediated through the modulation of signalling pathways involved in inflammation, including the NF-κB, MAPK, and Nrf2 pathways [58]. These cosmeceuticals reduce inflammation by suppressing the activation of inflammatory enzymes such as COX-2 and iNOS, and by inhibiting the production of pro-inflammatory cytokines like TNF-α, IL-1β, and IL-6 [59]. Qi *et al.* (2022) have demonstrated that marine bacteria like *Vibrio* sp., *Pseudoalteromonas* sp., and *Halomonas* sp. produce extracellular polysaccharides (EPS), which modulate the MAPK and NF-κB signalling pathway and have strong anti-inflammatory properties [22]. EPS lowers the production of pro-inflammatory mediators by inhibiting the phosphorylation of p38 MAPK, a crucial protein involved in the inflammatory response [22]. Moreover, by increasing the expression of fibronectin and collagen, EPS accelerates wound healing. A bacterial EPS named EPS-H29, derived from the marine bacterium *Halomonas malpeensis* YU-PRIM-29T, has shown the potential to enhance burn wound healing by promoting dermal fibroblast proliferation and migration, suggesting that it could be developed as a novel topical biomaterial for skin tissue regeneration [60]. Kim et al. (2013) demonstrated that Neoechinulin A, an indolic alkaloid from *Eurotium* sp., exhibits anti-inflammatory effects by inhibiting iNOS, COX-2, IL-1β, and TNF-α expression in LPS-stimulated RAW 264.7 cells. This reduces inflammatory mediators like nitric oxide (NO) and PGE2. Neoechinulin A also prevents MAPK p38 phosphorylation, further alleviating inflammation [61].

15.4.3 Anti-Photoaging Properties

Ageing involves a deterioration in the normal functioning of cells, tissues, and organs, resulting in greater vulnerability to diseases, reduced repair capacity, and ultimately death [62]. When skin ages, the dermis deteriorates, resulting in dryness, larger pores, thinning, wrinkles and fine lines, decreased elasticity, fragility, increased transparency, reduced firmness, and visible blood vessels [63]. This process is driven by both intrinsic and extrinsic factors. Intrinsic factors are mainly related to age via genetics, while extrinsic factors include our lifestyle and environmental parameters like UV radiation, pollutants, smoke, and high temperature [64]. Photoaging is mainly due to exposure to UV radiation from the sun. Being exposed to sunlight triggers several biochemical and physiological effects, including DNA damage, ROS production, signalling pathway activation, the elevation of inflammatory mediators, disruption of the dermal fibre network, and ECM turnover [10]. The skin is naturally shielded from harmful UV radiation by MAAs, which are generated by marine cyanobacteria like *Aphanizomenon flos-aquae*. MAAs absorb UV radiation, which reduces photoaging and stops UV-induced DNA damage [65]. Scytonemin is another algae-derived UV filter extracted from *Nostoc commune*. MAAs absorb UV light in the range of 310–362 nm offering greater protective effectiveness when present outside the cell, whereas scytonemin absorbs UV radiation at its maximum at 386 nm [66]. Additionally, sarmentosamide, produced from *Streptomyces* sp., is a potentially effective anti-ageing compound that inhibits the ERK/JNK pathway and lowers the production of UVB-induced matrix metalloproteinases-1 (MMP-1) in normal human skin fibroblasts [67]. In addition, *Klebsiella aerogenes* produces extracellular cadmium sulfide (CdS) particles under stress, which can absorb UV radiation, providing a protective mechanism against UV-induced damage [68].

15.4.4 Anti-Wrinkling Effects

Wrinkles are caused by the ageing process and are characterized by the formation of folds, creases, and ridges in the skin typically due to the breakdown of collagen and elastin fibres, loss of skin moisture, reduction of elasticity, and shrinking of the epidermis [69, 70]. MMPs play a significant role in photoaging skin by inhibiting the production of collagen [71]. Numerous cells, including keratinocytes, neutrophils, macrophages, fibroblasts, and mast cells, can produce MMPs, which leads to the development of wrinkles [7]. Marine microorganisms have a rich source of bioactive substances with anti-wrinkle characteristics. By focusing on several processes including hydration, antioxidation, and collagen production, these substances provide an integrated approach to skin rejuvenation [72]. Marine collagenase from *Vibrio alginolyticus* facilitates the repair of broken collagen fibres and promotes the growth of new, healthy collagen. This process helps to restore skin elasticity and smoothness [73]. Furthermore, it has been demonstrated that peptides derived from *Chlorella vulgaris* inhibit the production of MMP-1 in human skin cell fibroblasts [74]. The macroscopic appearance of wrinkles and texture in the skin is improved by extracts such as astaxanthin derived from the microalgae *Haematococcus lacustris*, a species of Chlorophyta [57]. Marine microorganism-derived biosurfactants, such as sophorolipids, rhamnolipids, and mannosylerythritol, are now being investigated by researchers for possible application in cosmetics. They help hydrophobic components become more soluble and facilitate efficient passage across the skin barrier. These biosurfactants produce the least amount of skin irritation, which makes them perfect for anti-wrinkling and anti-ageing treatments [75].

15.4.5 Antiacne Activity

Acne vulgaris, usually referred to as acne, is a chronic inflammatory skin illness characterized by the presence of comedones, papules, pustules, and nodules, primarily affecting the face, back, and shoulders [76]. It is a complex condition involving hormonal, immunological, and microbiological factors. The severity and progression of acne are primarily driven by increased sebum production, follicular hyperkeratinization, and bacterial growth [77]. A bacterium called *Propionibacterium acnes* causes inflammation that eventually results in the development of inflammatory lesions in acne [78]. Similarly, *Staphylococcus epidermidis* and *Cutibacterium acnes* can also cause acne [79]. These anaerobic bacteria induce the release of ROS and pro-inflammatory cytokines including TNFα, IL-8, IL-12, and IL-1β which are mediated by Toll-like receptor (TLR). They also produce lipases to break down excess sebum, leading to inflammation and damage to hair follicles [80]. According to Hoz-Romo *et al.* (2022), Actinomycin, Napyradiomycins, and Fridamycin which are derived from *Streptomyces* sp., disrupt the replication of bacterial DNA and the synthesis of proteins in *Cutibacterium acnes* to demonstrate strong antibacterial action. In addition, other secondary metabolites such as phenazines isolated from *Pseudomonas* sp. and *Streptomyces* sp., have shown effectiveness in reducing bacterial load on the skin and preventing the formation of acne lesions [81]. Since marine actinomycetes have antibacterial properties, cosmetics containing their extracts are used more often in acne treatments. Similarly, Phycocyanin, a pigment-protein combination with antibacterial properties, is present in *Spirulina* extracts [82]. By interfering with bacterial metabolism and cell division, phytocyanin suppresses the growth of *Propionibacterium acnes*, hence lowering the number of bacteria on the skin. The anti-inflammatory properties of *Spirulina* are attributed to its increased concentration of gamma-linolenic acid (GLA) and other fatty acids. By preventing the release of pro-inflammatory cytokines and enzymes, which aid in the development of acne, these substances aid in the reduction of inflammation [82].

15.4.6 Anti-Melanogenic and Skin-Whitening Activities

A typical sign of ageing and prolonged UV exposure is hyperpigmentation, which appears as brown spots on the skin, especially in parts of the body that are exposed to the sun frequently [83]. The process of skin whitening involves the use of natural or synthetic substances that reduce the skin's melanin content which results in hypopigmentation [18]. UV light can damage skin cell DNA, increasing Cyclic Adenosine Monophosphate levels, which activates microphthalmia-associated transcription factor protein (MITF) and triggers melanin production [18]. Tyrosinase is a copper-containing membrane enzyme that hydroxylates tyrosine to DOPA and oxidizes it to dopaquinone, affecting early melanogenesis stages. It is crucial for melanin synthesis and regulates skin pigmentation [84]. Melanogenic enzymes associated with tyrosinase, such as TYRP-1, TYRP-2, and peroxidase, are among the depigmenting agents that influence skin pigmentation through transcription and activity [84]. Most widely used skin-whitening cosmetic products contain tyrosinase inhibitors. The application of hydroquinone creams and Kojic acid might help to brighten the dark areas of the skin. Still, prolonged use may lead to irritation, toxicity, reduced skin penetrance, poor stability and harmful effects [85]. This made the researchers develop skin-whitening agents extracted from marine microorganisms and plants. Lim et al. (2022) showed that SNA077, an extract from marine *Streptomyces* sp., inhibits melanogenesis by targeting melanogenic proteins in the cAMP/PKA/CREB signalling pathway in MNT-1 human melanoma and B16 and Melan, a mouse melanocyte cell line [86]. APmarine002 and RoA-050 strains of *Pseudoalteromonas* sp. can generate Pseudoalteromone A, which can suppress tyrosinase activity by reducing melanin-producing gene expression, which results in whitening and brightening effects (Figure 15.2) [87]. According to Kang *et al.* (2011), methylene chloride, a marine bacterial extract from *Pseudomonas* species, decreased the pigmentation of human melanocytes, and Melan-a cells *in vivo* in zebrafish. The inhibition of tyrosinase and the MITF was responsible for the decrease in pigmentation [38].

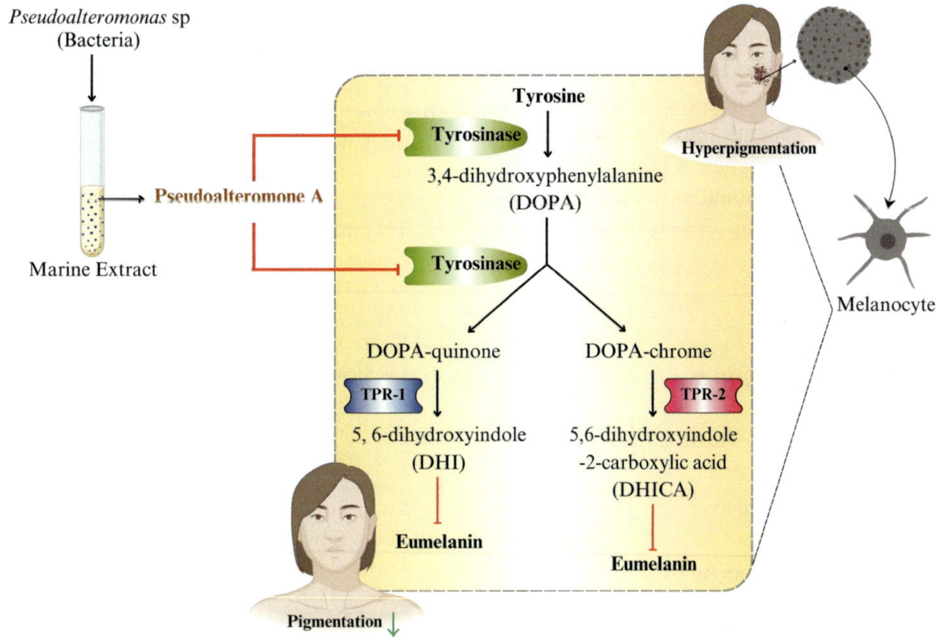

FIGURE 15.2 Diagram illustrating the anti-tyrosinase activity of Pseudoalteromone A derived from *Pseudoalteromonas* sp. in reducing pigmentation.

15.5 NATURAL VS. SYNTHETIC: THE RISE OF MARINE MICROORGANISM-DERIVED COSMETICS

In modern times, the need for cosmetics and cosmeceuticals is swiftly expanding. As customers become increasingly mindful of the possible negative consequences of chemical cosmetics, there is a growing trend of incorporating natural supplements as active components in these products. While terrestrial plants have now grown in popularity as sources of these natural components, the marine environment provides an even wider range of biological elements for cosmetics and cosmeceutical products [2]. Marine microorganism-derived cosmetics are often seen as safer because of their natural origin. They are less prone to irritation, allergic reactions, or other skin disorders and are usually devoid of harsh ingredients, making them safe for delicate skin. In contrast, synthetic chemicals can cause skin irritation, allergies, and other negative effects, particularly when used over an extended period. Certain synthetic substances have been related to hormone imbalances and other health issues [88]. The manufacture and disposal of synthetic cosmetics can pollute the environment. Non-biodegradable pollutants such as microplastics and synthetic flavours may build up in ecosystems, causing harm to biodiversity and the environment. Marine-derived cosmetics are frequently more environmentally friendly since they use marine resources and biodegradable components. The sustainable collection of marine microorganisms may have a lesser environmental effect than the manufacture of synthetic compounds [89,90]. The cosmetics business continues to rely significantly on synthetic goods, owing to their affordability, extensive availability, and rapid effects. As customers grow more health conscious, there is a movement towards natural alternatives [9]. Ultimately, cosmetics generated from marine microbes can be less expensive than synthetic chemicals. While the initial expenditure in research, development, and culture of marine microorganisms may be more extensive, the sustainable and regenerative nature of these resources might result in cheaper production costs over time. Furthermore, these components frequently need less processing and fewer hazardous chemicals, lowering the costs associated with environmental cleaning and regulatory compliance [91].

15.6 CONCLUSION AND FUTURE PERSPECTIVES

Cosmeceuticals have revolutionized the beauty and skincare industry by merging cosmetics and pharmaceuticals. Exploring marine cosmeceuticals has shown potential for improving skin health and appearance via antioxidant, anti-inflammatory, and anti-ageing effects. Marine microbes, including algae, seaweeds, bacteria, and fungi, are abundant sources of bioactive compounds, polysaccharides, peptides, enzymes, and vitamins, which can enhance hydration, protect the skin from environmental stresses, and stimulate the formation of collagen. Looking to the future, several key areas hold promise for advancing the field of marine cosmeceuticals. Further research into the deep-sea environment will produce novel and distinctive bioactive substances with skincare uses. The identification and characterization of these compounds will be made easier by developments in marine biotechnology and genomics.

There is great potential for combining marine microbiology with the latest advances like proteomics, metabolomics, and genomics. With the use of these technologies, research and development procedures can become more effective since they can offer greater insights into the biosynthetic pathways and functional characteristics of compounds produced from marine sources. Additionally, the use of bioinformatics techniques facilitates the discovery of new targets and the creation of innovative cosmeceutical agents.

ABBREVIATIONS

CAGR	Compound Annual Growth Rate
CdS	Cadmium Sulphide
DHA	Docosahexaenoic Acid
EPA	Eicosapentaenoic Acid
EPS	Extracellular Polysaccharides
GLA	Gamma-Linolenic Acid
MAAs	Mycosporine-Like Amino Acids
MITF	Microphthalmia-Associated Transcription Factor Protein
MMP-1	Matrix Metalloproteinases-1
NO	Nitric Oxide
ROS	Reactive Oxygen Species
SOD	Superoxide Dismutase
TLR	Toll-Like Receptor
UV	Ultraviolet

REFERENCES

1. Lopez-Ojeda, W., Pandey, A., Alhajj, M., & Oakley, A. M. (2022). Anatomy, Skin (Integument). In *StatPearls*. StatPearls Publishing.
2. Ding, J., Wu, B., & Chen, L. (2022). Application of Marine Microbial Natural Products in Cosmetics. *Frontiers in Microbiology*, *13*, 892505. https://doi.org/10.3389/fmicb.2022.892505
3. Resende, D. I., Ferreira, M., Magalhães, C., Lobo, J. S., Sousa, E., & Almeida, I. F. (2021). Trends in the Use of Marine Ingredients in Anti-Aging Cosmetics. *Algal Research*, *55*, 102273. https://doi.org/10.1016/j.algal.2021.102273
4. Martin, K. I., & Glaser, D. A. (2011). Cosmeceuticals: the New Medicine of Beauty. *Missouri Medicine*, *108*(1), 60–63.
5. Jain, S., Trivedi, M., Raikwar, M., Lodhi, M., Ali, M. Y., Purohit, A., & Nema, P. (2022). A Review on Herbal Cosmetics and Cosmeceuticals. *Asian Journal of Dental and Health Sciences*, *2*(4), 9–16. https://dx. doi.org/10.22270/ajdhs.v2i4.19
6. Cosmeceuticals Market Size, Share, Growth | Global Report, 2026. (n.d.). *Fortune Business Insights*. Retrieved from https://www.fortunebusinessinsights.com
7. Alves, A., Sousa, E., Kijjoa, A., & Pinto, M. (2020). Marine-Derived Compounds with Potential Use as Cosmeceuticals and Nutricosmetics. *Molecules (Basel, Switzerland)*, *25*(11), 2536. https://doi.org/10.3390/molecules25112536
8. Jesumani, V., Du, H., Aslam, M., Pei, P., & Huang, N. (2019). Potential Use of Seaweed Bioactive Compounds in Skincare-A Review. *Marine Drugs*, *17*(12), 688. https://doi.org/10.3390/md17120688
9. Fonseca, S., Amaral, M. N., Reis, C. P., & Custódio, L. (2023). Marine Natural Products as Innovative Cosmetic Ingredients. *Marine Drugs*, *21*(3), 170. https://doi.org/10.3390/md21030170
10. Costa, J. P., Custódio, L., & Reis, C. P. (2023). Exploring the Potential of Using Marine-Derived Ingredients: From the Extraction to Cutting-Edge Cosmetics. *Marine Drugs*, *21*(12), 620. https://doi.org/10.3390/md21120620
11. Thiyagarasaiyar, K., Goh, B. H., Jeon, Y. J., & Yow, Y. Y. (2020). Algae Metabolites in Cosmeceutical: An Overview of Current Applications and Challenges. *Marine Drugs*, *18*(6), 323. https://doi.org/10.3390/md18060323
12. Rani, A., Saini, K. C., Bast, F., Mehariya, S., Bhatia, S. K., Lavecchia, R., & Zuorro, A. (2021). Microorganisms: A Potential Source of Bioactive Molecules for Antioxidant Applications. *Molecules (Basel, Switzerland)*, *26*(4), 1142. https://doi.org/10.3390/molecules26041142
13. Geahchan, S., Baharlouei, P., & Rahman, A. (2022). Marine Collagen: A Promising Biomaterial for Wound Healing, Skin Anti-Aging, and Bone Regeneration. *Marine Drugs*, *20*(1), 61. https://doi.org/10.3390/md20010061
14. Silber, J., Kramer, A., Labes, A., & Tasdemir, D. (2016). From Discovery to Production: Biotechnology of Marine Fungi for the Production of New Antibiotics. *Marine Drugs*, *14*(7), 137. https://doi.org/10.3390/md14070137
15. Shaikh, H. S. (2022). Cosmeceutical from Marine Origin and Their Collection, Isolation and Extraction: A Review. *Research Journal of Topical and Cosmetic Sciences*, *13*(2), 92–98. https://dx. doi.org/10.52711/2321-5844.2022.00015
16. Nagamani, J. E., Aladakatti, R. H., & Achar, S. S. The Revitalizing Potential of Microalgae in Cosmeceuticals. *International Journal of Innovative Research in Science Engineering and Technology*, 13(6), 11053.
17. Rasmi, Y., Kirboğa, K. K., Rao, T., Ali, M., & Ahmed, M. Z. (2024). Marine Natural Products as a Bioresource for Cosmeceuticals. https://doi.org/10.1039/9781837672288-00223
18. Corinaldesi, C., Barone, G., Marcellini, F., Dell'Anno, A., & Danovaro, R. (2017). Marine Microbial-Derived Molecules and Their Potential Use in Cosmeceutical and Cosmetic Products. *Marine Drugs*, *15*(4), 118. https://doi.org/10.3390/md15040118
19. Blunt, J. W., Carroll, A. R., Copp, B. R., Davis, R. A., Keyzers, R. A., & Prinsep, M. R. (2018). Marine Natural Products. *Natural Product Reports*, *35*(1), 8–53. https://doi.org/10.1039/c7np00052a
20. Romano, G., Costantini, M., Sansone, C., Lauritano, C., Ruocco, N., & Ianora, A. (2017). Marine Microorganisms as a Promising and Sustainable Source of Bioactive Molecules. *Marine Environmental Research*, *128*, 58–69. https://doi.org/10.1016/j.marenvres.2016.05.002
21. Le Costaouëc, T., Cérantola, S., Ropartz, D., Ratiskol, J., Sinquin, C., Colliec-Jouault, S., & Boisset, C. (2012). Structural Data on a Bacterial Exopolysaccharide Produced by a Deep-sea Alteromonas Macleodii Strain. *Carbohydrate Polymers*, *90*(1), 49–59. https://doi.org/10.1016/j.carbpol.2012.04.059
22. Courtois, A., Berthou, C., Guézennec, J., Boisset, C., & Bordron, A. (2014). Exopolysaccharides Isolated from Hydrothermal Vent Bacteria Can Modulate the Complement System. *PloS One*, *9*(4), e94965. https://doi.org/10.1371/journal.pone.0094965
23. Shindo, K., Kikuta, K., Suzuki, A., Katsuta, A., Kasai, H., Yasumoto-Hirose, M., Matsuo, Y., Misawa, N., & Takaichi, S. (2007). Rare Carotenoids, (3R)-Saproxanthin

and (3R,2'S)-Myxol, Isolated from Novel Marine Bacteria (Flavobacteriaceae) and Their Antioxidative Activities. *Applied Microbiology and Biotechnology*, *74*(6), 1350–1357. https://doi.org/10.1007/s00253-006-0774-y

24. Deering, R. W., Chen, J., Sun, J., Ma, H., Dubert, J., Barja, J. L., Seeram, N. P., Wang, H., & Rowley, D. C. (2016). N-Acyl Dehydrotyrosines, Tyrosinase Inhibitors from the Marine Bacterium Thalassotalea sp. PP2–459. *Journal of Natural Products*, *79*(2), 447–450. https://doi.org/10.1021/acs.jnatprod.5b00972
25. Bownik, A., & Stępniewska, Z. (2016). Ectoine as a Promising Protective Agent in Humans and Animals. *Arhiv za higijenu rada i toksikologiju*, *67*(4), 260–265. https://doi.org/10.1515/aiht-2016-67-2837
26. De Jesus Raposo, M. F., de Morais, R. M., & de Morais, A. M. (2013). Health Applications of Bioactive Compounds from Marine Microalgae. *Life Sciences*, *93*(15), 479–486. https://doi.org/10.1016/j.lfs.2013.08.002
27. Singh, K. N., Patil, S., & Barkate, H. (2020). Protective Effects of Astaxanthin on Skin: Recent Scientific Evidence, Possible Mechanisms, and Potential Indications. *Journal of Cosmetic Dermatology*, *19*(1), 22–27. https://doi.org/10.1111/jocd.13019
28. De Jesus Raposo, M. F., de Morais, A. M., & de Morais, R. M. (2015). Marine Polysaccharides from Algae with Potential Biomedical Applications. *Marine Drugs*, *13*(5), 2967–3028. https://doi.org/10.3390/md13052967
29. Pudney, A., Gandini, C., Economou, C. K., Smith, R., Goddard, P., Napier, J. A., Spicer, A., & Sayanova, O. (2019). Multifunctionalizing the Marine Diatom Phaeodactylum Tricornutum for Sustainable Co-production of Omega-3 Long Chain Polyunsaturated Fatty Acids and Recombinant Phytase. *Scientific Reports*, *9*(1), 11444. https://doi.org/10.1038/s41598-019-47875-1
30. Ishika, T., Laird, D. W., Bahri, P. A., & Moheimani, N. R. (2019). Co-cultivation and Stepwise Cultivation of Chaetoceros Muelleri and Amphora sp. for Fucoxanthin Production under Gradual Salinity Increase. *Journal of Applied Phycology*, *31*, 1535–1544. https://doi.org/10.1007/s10811-018-1718-5
31. Li, X., Jeong, J. H., Lee, K. T., Rho, J. R., Choi, H. D., Kang, J. S., & Son, B. W. (2003). γ-Pyrone derivatives, kojic acid methyl ethers from a marine-derived fungus *Altenaria* sp. *Archives of pharmacal research*, *26*(7), 532 534. https://doi.org/10.1007/BF02976876
32. Miyamoto, K. T., Komatsu, M., & Ikeda, H. (2014). Discovery of Gene Cluster for Mycosporine-like Amino Acid Biosynthesis from Actinomycetales Microorganisms and Production of a Novel Mycosporine-like Amino Acid by Heterologous Expression. *Applied and Environmental Microbiology*, *80*(16), 5028–5036. https://doi.org/10.1128/AEM.00727-14
33. Aasen, I. M., Ertesvåg, H., Heggeset, T. M., Liu, B., Brautaset, T., Vadstein, O., & Ellingsen, T. E. (2016). Thraustochytrids as Production Organisms for Docosahexaenoic Acid (DHA), Squalene, and Carotenoids. *Applied Microbiology and Biotechnology*, *100*(10), 4309–4321. https://doi.org/10.1007/s00253-016-7498-4
34. Nenkep, V., Yun, K., & Son, B. W. (2016). Oxysporizoline, an Antibacterial Polycyclic Quinazoline Alkaloid from the Marine-mudflat-derived Fungus Fusarium Oxysporum. *The Journal of Antibiotics*, *69*(9), 709–711. https://doi.org/10.1038/ja.2015.137
35. Mukherjee, A., Das, S., Chakraborty, D., Pal, N., & Das, N. (2024). Fungi's Treasure in Cosmeceuticals-A Comprehensive Chemical Approach. *South African Journal of Botany*, *166*, 311–331. https://doi.org/10.1016/j.sajb.2024.01.036
36. Stal, L. J., & Cretoiu, M. S. (Eds.). (2022). *The Marine Microbiome*. Springer. https://doi.org/10.1007/978-3-030-90383-1.
37. Qi, M., Zheng, C., Wu, W., Yu, G., & Wang, P. (2022). Exopolysaccharides from Marine Microbes: Source, Structure and Application. *Marine Drugs*, *20*(8), 512. https://doi.org/10.3390/md20080512
38. Kang, H. Y., Yoon, T. J., & Lee, G. J. (2011). Whitening Effects of Marine Pseudomonas Extract. *Annals of dermatology*, *23*(2), 144–149. https://doi.org/10.5021/ad.2011.23.2.144
39. Grether-Beck, S., Mühlberg, K., Brenden, H., Felsner, I., Brynjólfsdóttir, A., Einarsson, S., & Krutmann, J. (2008). Bioactive Molecules from the Blue Lagoon: In Vitro and In Vivo Assessment of Silica Mud and Microalgae Extracts for Their Effects on Skin Barrier Function and Prevention of Skin Ageing. *Experimental Dermatology*, *17*(9), 771–779. https://doi.org/10.1111/j.1600-0625.2007.00693.x
40. Gao, X., Jing, X., Liu, X., & Lindblad, P. (2021). Biotechnological Production of the Sunscreen Pigment Scytonemin in Cyanobacteria: Progress and Strategy. *Marine Drugs*, *19*(3), 129. https://doi.org/10.3390/md19030129
41. Mourelle, M. L., Gómez, C. P., & Legido, J. L. (2017). The Potential Use of Marine Microalgae and Cyanobacteria in Cosmetics and Thalassotherapy. *Cosmetics*, *4*(4), 46. https://doi.org/10.3390/cosmetics4040046
42. Matsui, M. S., Muizzuddin, N., Arad, S., & Marenus, K. (2003). Sulfated Polysaccharides from Red Microalgae Have Antiinflammatory Properties In Vitro and In Vivo. *Applied Biochemistry and Biotechnology*, *104*(1), 13–22. https://doi.org/10.1385/abab:104:1:13
43. Peng, J., Yuan, J. P., Wu, C. F., & Wang, J. H. (2011). Fucoxanthin, A Marine Carotenoid Present in Brown Seaweeds and Diatoms: Metabolism and Bioactivities Relevant to Human Health. *Marine Drugs*, *9*(10), 1806–1828. https://doi.org/10.3390/md9101806
44. Wong, H. J., Mohamad-Fauzi, N., Rizman-Idid, M., Convey, P., & Alias, S. A. (2019). Protective Mechanisms and Responses of Micro-fungi Towards Ultraviolet-induced Cellular Damage. *Polar Science*, *20*, 19–34. https://doi.org/10.1016/j.polar.2018.10.001
45. Zhang, D., Yang, X., Kang, J. S., Choi, H. D., & Son, B. W. (2008). Circumdatin I, A New Ultraviolet-A Protecting Benzodiazepine Alkaloid from a Marine Isolate of the Fungus Exophiala. *The Journal of Antibiotics*, *61*(1), 40–42. https://doi.org/10.1038/ja.2008.108
46. Minakshi, Prakash, S., Kumari, H., & Kumar, A. (2024). Natural Bioactive Products from Marine Fungi against Bacterial Infection. In *Fungi Bioactive Metabolites: Integration of Pharmaceutical Applications* (pp. 241–259). Singapore: Springer Nature Singapore. https://doi.org/10.1007/978-981-99-5696-8_8
47. Pizzino, G., Irrera, N., Cucinotta, M., Pallio, G., Mannino, F., Arcoraci, V., Squadrito, F., Altavilla, D., & Bitto, A. (2017). Oxidative Stress: Harms and Benefits for Human Health. *Oxidative Medicine and Cellular Longevity*, *2017*, 8416763. https://doi.org/10.1155/2017/8416763
48. Nakai, K., & Tsuruta, D. (2021). What Are Reactive Oxygen Species, Free Radicals, and Oxidative Stress in Skin Diseases?. *International Journal of Molecular Sciences*, *22*(19), 10799. https://doi.org/10.3390/ijms221910799
49. Rinnerthaler, M., Bischof, J., Streubel, M. K., Trost, A., & Richter, K. (2015). Oxidative Stress in Aging Human Skin. *Biomolecules*, *5*(2), 545–589. https://doi.org/10.3390/biom5020545

50. Chen, J., Liu, Y., Zhao, Z., & Qiu, J. (2021). Oxidative Stress in the Skin: Impact and Related Protection. *International Journal of Cosmetic Science*, *43*(5), 495–509. https://doi.org/10.1111/ics.12728
51. Young, I. S., & Woodside, J. V. (2001). Antioxidants in Health and Disease. *Journal of Clinical Pathology*, *54*(3), 176–186. https://doi.org/10.1136/jcp.54.3.176
52. Xu, H., Zheng, Y. W., Liu, Q., Liu, L. P., Luo, F. L., Zhou, H. C., ... & Li, Y. M. (2018). Reactive Oxygen Species in Skin Repair, Regeneration, Aging, and Inflammation. *Reactive Oxygen Species (ROS) in Living Cells*, *8*, 69–88. https://doi.org/10.5772/intechopen.72747
53. Ambati, R. R., Phang, S. M., Ravi, S., & Aswathanarayana, R. G. (2014). Astaxanthin: Sources, Extraction, Stability, Biological Activities and Its Commercial Applications – A Review. *Marine Drugs*, *12*(1), 128–152. https://doi.org/10.3390/md12010128
54. Davinelli, S., Nielsen, M. E., & Scapagnini, G. (2018). Astaxanthin in Skin Health, Repair, and Disease: A Comprehensive Review. *Nutrients*, *10*(4), 522. https://doi.org/10.3390/nu10040522
55. Fernandes, R., Campos, J., Serra, M., Fidalgo, J., Almeida, H., Casas, A., Toubarro, D., & Barros, A. I. R. N. A. (2023). Exploring the Benefits of Phycocyanin: From Spirulina Cultivation to Its Widespread Applications. *Pharmaceuticals*, *16*(4), 592. https://doi.org/10.3390/ph16040592
56. Gohil, N., Bhattacharjee, G., Khambhati, K., Braddick, D., & Singh, V. (2019). Engineering Strategies in Microorganisms for the Enhanced Production of Squalene: Advances, Challenges and Opportunities. *Frontiers in Bioengineering and Biotechnology*, *7*, 50. https://doi.org/10.3389/fbioe.2019.00050
57. Bhatiya, M., Duttaroy, A. K., Pathak, S., & Banerjee, A. (2023). Cellular Aging: An Introduction, Principle, Hallmarks, and Aging-Associated Diseases. In *Evidence-based Functional Foods for Prevention of Age-related Diseases* (pp. 1–18). Singapore: Springer Nature Singapore. https://doi.org/10.1007/978-981-99-0534-8_1.
58. Saha, S., Buttari, B., Panieri, E., Profumo, E., & Saso, L. (2020). An Overview of Nrf2 Signaling Pathway and Its Role in Inflammation. *Molecules (Basel, Switzerland)*, *25*(22), 5474. https://doi.org/10.3390/molecules25225474
59. Wang, F., Liu, J., An, Q., Wang, Y., Yang, Y., Huo, T., Yang, S., Ju, R., & Quan, Q. (2023). Aloe Extracts Inhibit Skin Inflammatory Responses by Regulating NF-κB, ERK, and JNK Signaling Pathways in an LPS-Induced RAW264.7 Macrophages Model. *Clinical, Cosmetic and Investigational Dermatology*, *16*, 267–278. https://doi.org/10.2147/CCID.S391741
60. Nagaraj, A., Subramaniyan, Y., Surya, S., & Rekha, P. D. (2024). Burn wound healing abilities of a uronic acid containing exopolysaccharide produced by the marine bacterium *Halomonas malpeensis* YU-PRIM-29T. *Applied Biochemistry and Biotechnology*, *196*(11), 8190–8213. https://doi.org/10.1007/s12010-024-04966-8
61. Kim, K. S., Cui, X., Lee, D. S., Sohn, J. H., Yim, J. H., Kim, Y. C., & Oh, H. (2013). Anti-inflammatory Effect of Neoechinulin a from the Marine Fungus Eurotium sp. SF-5989 Through the Suppression of NF-κB and p38 MAPK Pathways in Lipopolysaccharide-Stimulated RAW264.7 Macrophages. *Molecules (Basel, Switzerland)*, *18*(11), 13245–13259. https://doi.org/10.3390/molecules181113245
62. Guo, J., Huang, X., Dou, L., Yan, M., Shen, T., Tang, W., & Li, J. (2022). Aging and Aging-Related Diseases: From Molecular Mechanisms to Interventions and Treatments. *Signal Transduction and Targeted Therapy*, *7*(1), 391. https://doi.org/10.1038/s41392-022-01251-0
63. Suh, S. S., Hwang, J., Park, M., Seo, H. H., Kim, H. S., Lee, J. H., Moh, S. H., & Lee, T. K. (2014). Anti-Inflammation Activities of Mycosporine-Like Amino Acids (MAAs) in Response to UV Radiation Suggest Potential Anti-Skin Aging Activity. *Marine Drugs*, *12*(10), 5174–5187. https://doi.org/10.3390/md12105174
64. Zhang, S., & Duan, E. (2018). Fighting against Skin Aging: The Way from Bench to Bedside. *Cell Transplantation*, *27*(5), 729–738. https://doi.org/10.1177/0963689717725755
65. Zhang, H., Jiang, Y., Zhou, C., Chen, Y., Yu, G., Zheng, L., Guan, H., & Li, R. (2022). Occurrence of Mycosporine-like Amino Acids (MAAs) from the Bloom-Forming Cyanobacteria Aphanizomenon Strains. *Molecules*, *27*(5), 1734. https://doi.org/10.3390/molecules27051734
66. Couteau, C., & Coiffard, L. (2020). Phycocosmetics and Other Marine Cosmetics, Specific Cosmetics Formulated Using Marine Resources. *Marine Drugs*, *18*(6), 322. https://doi.org/10.3390/md18060322
67. Lee, E. S., Lee, E. Y., Yoon, J., Hong, A., Nam, S. J., & Ko, J. (2020). Sarmentosamide, an Anti-Aging Compound from a Marine-Derived Streptomyces sp. APmarine042. *Marine Drugs*, *18*(9), 463. https://doi.org/10.3390/md18090463
68. Pallela, R., Na-Young, Y., & Kim, S. K. (2010). Anti-Photoaging and Photoprotective Compounds Derived from Marine Organisms. *Marine Drugs*, *8*(4), 1189–1202. https://doi.org/10.3390/md8041189
69. McCullough, J. L., & Kelly, K. M. (2006). Prevention and Treatment of Skin Aging. *Annals of the New York Academy of Sciences*, *1067*, 323–331. https://doi.org/10.1196/annals.1354.044
70. Chaudhary, M., Khan, A., & Gupta, M. (2020). Skin Ageing: Pathophysiology and Current Market Treatment Approaches. *Current Aging Science*, *13*(1), 22–30. https://doi.org/10.2174/1567205016666190809161115
71. Panwar, P., Butler, G. S., Jamroz, A., Azizi, P., Overall, C. M., & Brömme, D. (2018). Aging-Associated Modifications of Collagen Affect Its Degradation by Matrix Metalloproteinases. *Matrix Biology: Journal of the International Society for Matrix Biology*, *65*, 30–44. https://doi.org/10.1016/j.matbio.2017.06.004
72. He, X., Gao, X., Guo, Y., & Xie, W. (2024). Research Progress on Bioactive Factors against Skin Aging. *International Journal of Molecular Sciences*, *25*(7), 3797. https://doi.org/10.3390/ijms25073797
73. Bassetto, F., Maschio, N., Abatangelo, G., Zavan, B., Scarpa, C., & Vindigni, V. (2016). Collagenase from Vibrio alginolyticus Cultures: Experimental Study and Clinical Perspectives. *Surgical Innovation*, *23*(6), 557–562. https://doi.org/10.1177/1553350616660630
74. Chen, C. L., Liou, S. F., Chen, S. J., & Shih, M. F. (2011). Protective Effects of Chlorella-derived Peptide on UVB-induced Production of MMP-1 and Degradation of Procollagen Genes in Human Skin Fibroblasts. *Regulatory Toxicology and Pharmacology: RTP*, *60*(1), 112–119. https://doi.org/10.1016/j.yrtph.2011.03.001
75. Kalasariya, H. S., Yadav, V. K., Yadav, K. K., Tirth, V., Algahtani, A., Islam, S., Gupta, N., & Jeon, B. H. (2021). Seaweed-Based Molecules and Their Potential Biological Activities: An Eco-Sustainable Cosmetics. *Molecules*, *26*(17), 5313. https://doi.org/10.3390/molecules26175313

76. Parać, E., Špiljak, B., Lugović-Mihić, L., & Bukvić Mokos, Z. (2023). Acne-like Eruptions: Disease Features and Differential Diagnosis. *Cosmetics*, *10*(3), 89. https://doi.org/10.3390/cosmetics10030089
77. Januário, A. P., Félix, R., Félix, C., Reboleira, J., Valentão, P., & Lemos, M. F. L. (2021). Red Seaweed-Derived Compounds as a Potential New Approach for Acne Vulgaris Care. *Pharmaceutics*, *13*(11), 1930. https://doi.org/10.3390/pharmaceutics13111930
78. Vasam, M., Korutla, S., & Bohara, R. A. (2023). Acne Vulgaris: A Review of the Pathophysiology, Treatment, and Recent Nanotechnology Based Advances. *Biochemistry and Biophysics Reports*, *36*, 101578. https://doi.org/10.1016/j.bbrep.2023.101578
79. Fournière, M., Latire, T., Souak, D., Feuilloley, M. G. J., & Bedoux, G. (2020). Staphylococcus Epidermidis and Cutibacterium Acnes: Two Major Sentinels of Skin Microbiota and the Influence of Cosmetics. *Microorganisms*, *8*(11), 1752. https://doi.org/10.3390/microorganisms8111752
80. Makrantonaki, E., Ganceviciene, R., & Zouboulis, C. (2011). An Update on the Role of the Sebaceous Gland in the Pathogenesis of Acne. *Dermato-endocrinology*, *3*(1), 41–49. https://doi.org/10.4161/derm.3.1.13900
81. De La Hoz-Romo, M. C., Díaz, L., & Villamil, L. (2022). Marine Actinobacteria a New Source of Antibacterial Metabolites to Treat Acne Vulgaris Disease-A Systematic Literature Review. *Antibiotics (Basel, Switzerland)*, *11*(7), 965. https://doi.org/10.3390/antibiotics11070965
82. Dranseikienė, D., Balčiūnaitė-Murzienė, G., Karosienė, J., Morudov, D., Juodžiukynienė, N., Hudz, N., Gerbutavičienė, R. J., & Savickienė, N. (2022). Cyano-Phycocyanin: Mechanisms of Action on Human Skin and Future Perspectives in Medicine. *Plants*, *11*(9), 1249. https://doi.org/10.3390/plants11091249
83. Babitha, S., & Kim, E. K. (2011). Effect of Marine Cosmeceuticals on the Pigmentation of Skin. *Marine Cosmeceuticals: Trends and Prospects*, 63–66.
84. Chang T. S. (2009). An Updated Review of Tyrosinase Inhibitors. *International Journal of Molecular Sciences*, *10*(6), 2440–2475. https://doi.org/10.3390/ijms10062440
85. Owolabi, J. O., Fabiyi, O. S., Adelakin, L. A., & Ekwerike, M. C. (2020). Effects of Skin Lightening Cream Agents - Hydroquinone and Kojic Acid, on the Skin of Adult Female Experimental Rats. *Clinical, Cosmetic and Investigational Dermatology*, *13*, 283–289. https://doi.org/10.2147/CCID.S233185
86. Song, Y. S., Balcos, M. C., Yun, H. Y., Baek, K. J., Kwon, N. S., Kim, M. K., & Kim, D. S. (2015). ERK Activation by Fucoidan Leads to Inhibition of Melanogenesis in Mel-Ab Cells. *The Korean Journal of Physiology & Pharmacology: Official Journal of the Korean Physiological Society and the Korean Society of Pharmacology*, *19*(1), 29–34. https://doi.org/10.4196/kjpp.2015.19.1.29
87. Lim, S. J., Min, D. J., Kim, S., Lee, J., Lee, E. S., Kim, H., Cho, S. Y., Beak, H. S., Lee, C. S., Nam, S. J., & Ko, J. (2021). Pseudoalteromone A, a Ubiquinone Derivative from Marine Pseudoalteromonas spp., Suppresses Melanogenesis. *Marine Drugs*, *19*(11), 612. https://doi.org/10.3390/md19110612
88. Martins, A. M., & Marto, J. M. (2023). A sustainable Life Cycle for Cosmetics: From Design and Development to Post-use Phase. *Sustainable Chemistry and Pharmacy*, *35*, 101178. https://doi.org/10.1016/j.scp.2023.101178
89. Cubas, A. L. V., Bianchet, R. T., Reis, I. M. A. S. dos, & Gouveia, I. C. (2022). Plastics and Microplastic in the Cosmetic Industry: Aggregating Sustainable Actions Aimed at Alignment and Interaction with UN Sustainable Development Goals. *Polymers*, *14*(21), 4576. https://doi.org/10.3390/polym14214576
90. Amberg, N., & Fogarassy, C. (2019). Green Consumer Behavior in the Cosmetics Market. *Resources*, *8*(3), 137. https://doi.org/10.3390/resources8030137
91. Prajaputra, V., Isnaini, N., Maryam, S., Ernawati, E., Deliana, F., Haridhi, H. A., Fadli, N., Karina, S., Agustina, S., Nurfadillah, N., Arisa, I. I., Desiyana, L. S., & Bakri, T. K. (2024). Exploring Marine Collagen: Sustainable Sourcing, Extraction Methods, and Cosmetic Applications. *South African Journal of Chemical Engineering*, *47*, 197–211. https://doi.org/10.1016/j.sajce.2023.11.006

16 Advancements in Marine Biosensing
Applications and Future Trends

Sheena Mariam Thomas, Anu Shibi Anilkumar, and Ramakrishnan Veerabathiran
Human Cytogenetics and Genomics Laboratory, Faculty of Allied Health Sciences, Chettinad Hospital and Research Institute, Chettinad Academy of Research and Education, Chennai, India

16.1 INTRODUCTION

Biosensors are analytical tools that detect changes in biological processes and convert them into electrical signals (Metkar and Girigoswami, 2019). They utilize various biological components, including tissues, enzymes, cells, bacteria, and acids, to assess whether the transducer produces a voltage or current output (Haleem et al., 2021). This technology has seen significant development due to its potential applications, particularly in clinical treatment, pharmaceuticals, biomedical industries, and healthcare. The versatility of biosensors makes them invaluable in managing human health by detecting diseases, preventing them, aiding patient rehabilitation, and monitoring health. They are effective in detecting microorganisms such as bacteria, viruses, and pathogens, thereby contributing to public health management (Choi, 2020). The working mechanism of a basic biosensor is depicted in Figure 16.1.

Microbial biosensors, in particular, are gaining popularity for their ease of use and wide-ranging applications. They are utilized for rapid detection of organic and inorganic toxicants in environments, such as heavy metals in wastewater. In food science, they assess nutritional values and monitor fermentation processes for substances like ethanol. Additionally, microbial biosensors play crucial roles in clinical diagnostics by detecting contaminants in food and identifying substances like hormones, pathogens, and DNA (Nnachi et al., 2022). A microbial biosensor consists of immobilized living or non-living microbial cells integrated with a transducer. Non-living cells, acquired through permeabilization or from whole cells with periplasmic enzymes, have emerged as a cost-effective substitute for enzymes. Microbial biosensors, in recent times, are mainly grouped into two categories based on their sensing technology: electrochemical and optical (D'souza, 2001). Microbial biosensors are utilized in a range of biological systems, making use of the light emitted by luminescent bacteria as a quick, sensitive, and non-intrusive method of testing. Nature is home to bioluminescent bacteria (BLB), which may be found in both terrestrial and marine settings. Additionally, genetically modified microbes (GEM) have been used to construct bioluminescent whole-cell biosensors (WCBs) that detect the presence of heavy metals, pesticides, and organic pollution. Microbial biosensors are mostly used in the environmental sector. Therefore, marine microbial biosensors are analytical tools that combine biological components with physical transducers to identify contaminants in the environment and evaluate the quality of the water (Ali et al., 2020).

Light production in living things, or bioluminescence (BL), is emerging as a viable substitute for the more common fluorescence. BL is very attractive for integrating small and portable devices. With a broad spectrum of applications ranging from molecular imaging to the construction of biosensors, BL offers a powerful manifold system based on several luciferases and luciferin analogs (Calabretta et al., 2022). Innovative methods for marine-centered biomedical research necessitate collaborative efforts among diverse experts including medical practitioners, scientists, biotechnologists, and marine microbiologists. While marine organisms serve as valuable biomedical models for studying human diseases, the potential impact of microbes that inhabit the marine ecosystem in this realm remains largely uncharted. By investigating the precise mechanisms of microbial metabolite functions, improving dialogue between microbial biotechnology and marine sciences, understanding chemical interplays in marine settings, and creating suitable bioassays, novel categories of drugs sourced from marine microbes can be pioneered (Bhatnagar and Kim, 2010). Marine microbial biosensors present substantial benefits for environmental monitoring, marine biology, and oceanography. These sensors, which use marine microbes, can deliver real-time data and analysis within complex marine ecosystems. This chapter provides a brief overview of marine microbial biosensors and explores the emerging trends in their development and application.

16.2 MARINE MICROBIAL BIOSENSORS

Marine biosensors are critical for understanding and monitoring marine environments and resources. These advanced devices detect and measure vital chemical compounds and organisms, including methane, radon, ferrous ions, carbon dioxide, microorganisms, nutrients, and seafood. Human

 DOI: 10.1201/9781003509530-16

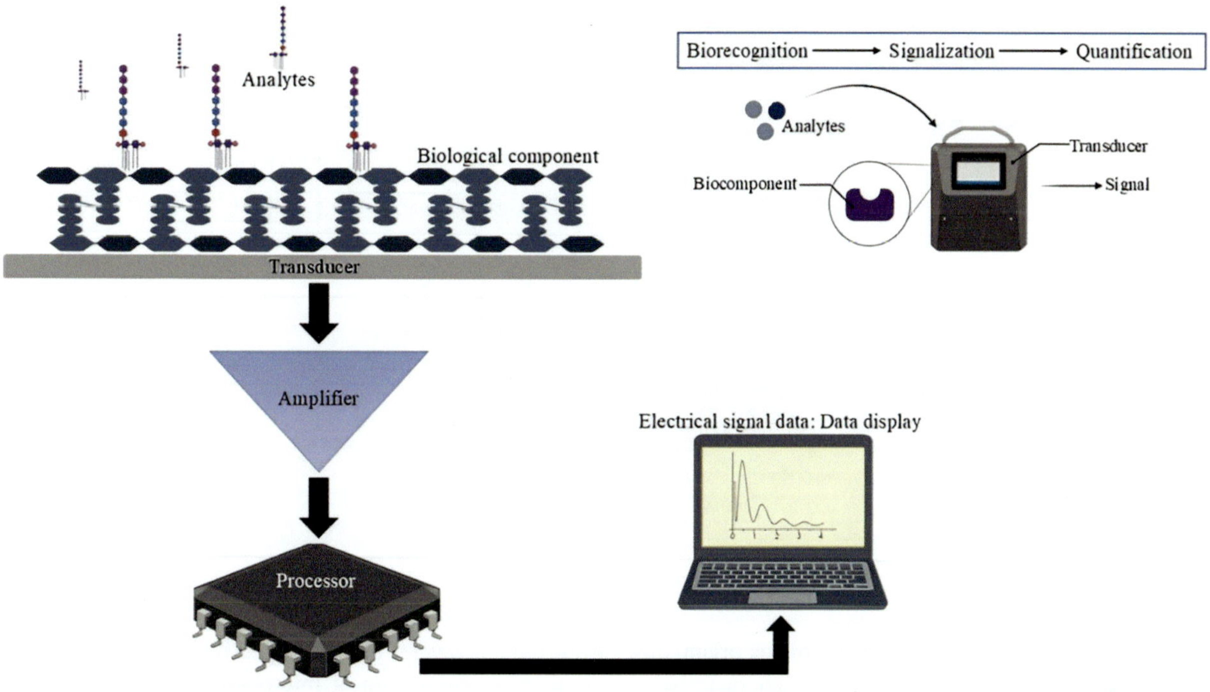

FIGURE 16.1 The working mechanism of a basic biosensor.

activities such as mineral resource exploration, waste dumping, and harbor construction significantly impact marine environments, highlighting the need for accurate monitoring tools. Marine biosensors provide real-time, precise data, enabling comprehensive assessments of marine health and the effects of human activities. This capability supports the sustainable management and conservation of marine resources (Liu et al., 2023). Living cells, including microbes and human cells, perceive and react to a variety of stimuli such as nutrients, attractants, repellents, environmental stressors, and other biological signals. These signals are transmitted through intricate signal transduction pathways within the cell, culminating in specific responses. The natural sensors that microbes utilize to detect these signals can be repurposed to create new gene networks, allowing the generation of microbial biosensors capable of detecting and reporting various substances. Synthetic biology is focused on modifying living organisms, such as microbial biosensors, to carry out novel functions aimed at addressing real-world issues. Diverse methods are being investigated to develop microbial biosensors with the capability to detect a wider range of clinically relevant substances. High-throughput techniques hold the potential to identify new sensing components through processes like genome mining or by integrating metabolomic screens with streamlined biosensor discovery strategies (Inda and Lu, 2020).

16.2.1 Mechanisms of Action of Microbial Biosensors

Numerous enhanced microbial biosensors have been discovered in the last several decades, and they appear promising for a wide range of applications. Microbial biosensors are analytical instruments made up of a microbe that recognizes a target substrate and uses physiology, electrical current, or biochemistry to convert the received signal into a quantifiable reaction. Microbial biosensors' utility is based on a variety of standard optical (e.g., BL or fluorescence), electrochemical, and sensory-regulated sensing and identification techniques (Lim et al., 2015).

In the typical construction of a biosensor, the biological recognition element identifies the target compound, and a transducer is employed to transform the biological reaction into a detectable signal. This signal can be linked to the sample's concentration and determined through electrochemical, optical, acoustic, mechanical, or electronic means. Biosensors have been extensively researched and applied in various fields, including homeland security, food safety, public health, and environmental monitoring; ever since the introduction of the initial glucose biosensor by Clark and Lyons in 1962, the advancement of biosensors has incorporated various biological recognition elements, including cofactors, enzymes, antibodies, bacteria, organelles, tissues, and cells from higher organisms. Enzymes are particularly prevalent as recognition elements in biosensors due to their remarkable sensitivity and specificity (Balootaki and Hassanshahian, 2014). On the other hand, enzyme purification is expensive and time-consuming. Furthermore, there's a chance that the in vitro working environment will cause the enzyme activity to drop. Microorganisms (including algae, yeast, and bacteria) offer a viable choice for biosensor production as they can be easily cultivated in large quantities through cell culture. Additionally, microbial cells exhibit higher vigor and stability in vitro and are simpler

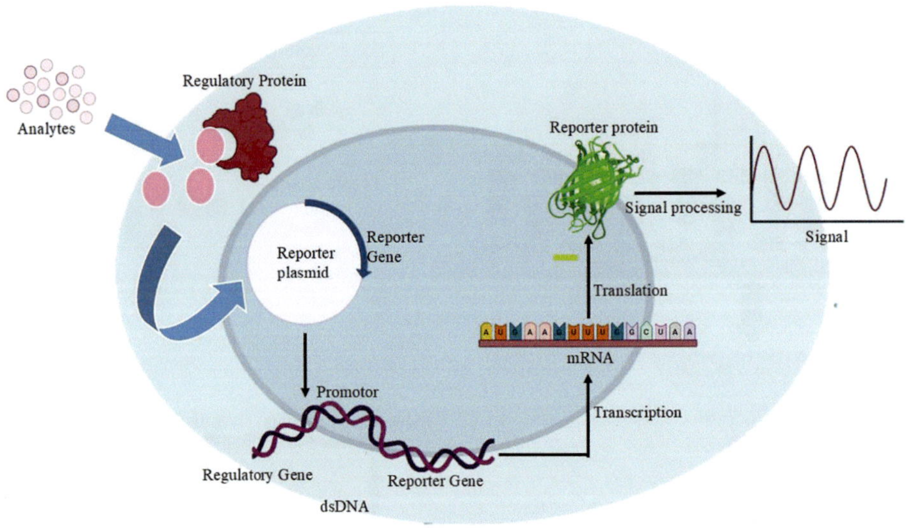

FIGURE 16.2 The action mechanism of a microbial biosensor in a live cell.

to handle compared to cells from more complex organisms such as plants, animals, and humans. These characteristics have the potential to streamline the manufacturing process and improve the efficacy of biosensors (Chandra et al., 2020). The action mechanism of the microbial biosensor is illustrated in Figure 16.2.

The process of developing a biosensor includes selecting the microorganisms to be used in it. Substrate specificity and detection sensitivity are the most important aspects of the issue. Through the use of potentiometric and amperometric transducers, it was possible to anticipate the cells' selectivity to carbohydrates, alcohols, and organic acids through the investigation of many *Gluconobacter* strains. In a microbial biosensor, a membrane or gel strip with immobilized microorganisms, securely attached to the transducer's active area, acts as the bio-receptor. Diffusion rate and sensor stability are two major factors that influence the immobilization decision. Broadly speaking, there are two categories of immobilization techniques: passive, which uses the natural adhesive power of microorganisms, and active, which fixes the cells by chemical or physical means. Next, cells are added to the gel or polymer matrix to ensure that they remain within the polymer's spatial network. The immobilization caused by the simultaneous adherence of viable and non-viable cells to the various materials was ascertained by the adsorption method. Usually, the cells aggregate through a process known as flocculation. When cross-linking agents are exposed to the microbial cell wall, the covalent attachment technique is often used to reveal the functional groups on the carrier or transducer surface (Reshetilov et al., 2010).

16.2.2 Classification of Microbial Biosensors

Due to the rapid growth of this area of research, the application of microbial biosensors has increased widely and, therefore, is mainly classified based on their sensing techniques, even though several other perspectives influence the wide application and design of microbial biosensors. The classification of microbial biosensors is primarily divided into two categories: optical and electrochemical microbial biosensors and is depicted in Table 16.1.

16.2.2.1 Electrochemical Microbial Biosensors

The primary characteristic of an electrochemical biosensor is its capacity to transform the interaction between a biorecognition element and an analyte into electrical signals that reflect the concentration of the analyte (Zhang et al., 2019). The three primary categories of electrochemical biosensors include amperometric, which detects variations in current upon the binding of biorecognition agents and analytes; potentiometric, which measures electric potential; and impedimetric, which gauges impedance (Hameed et al., 2018). The research conducted by Wang et al. (2016) created an electrochemical biosensor capable of detecting *Escherichia coli* O157 in food, water, and environmental samples with a detection limit of 102 CFU/mL. Despite their rapid performance, electrochemical biosensors have certain limitations. Factors such as redox interferences and the saturation energy of enzyme-substrate interactions can impact the electrochemical process, particularly when enzymes are employed as biorecognition agents. Therefore, alternative biorecognition elements, such as nucleic acids and nanoparticles, may be employed. For example, non-functionalized gold nanoparticles have demonstrated significant potential in the detection of *E. coli* O157. Furthermore, because microbial infections are often unevenly distributed in environmental, food, or water samples, detecting them with electrochemical methods can be challenging, especially without sample preparation (Wan et al., 2016).

Amperometric microbial biosensors have numerous applications in the environment. Chemical warfare weapons, petroleum additives, and pesticides have all been made using synthetic organophosphorus (OP) substances, which

TABLE 16.1
Various Types of Microbial Biosensors Based on Their Classification, Sub-type, Description, and Typical Applications

Classification	Sub-type	Description	Applications
Electrochemical Microbial Biosensors	Amperometric Biosensors	Measure the current generated by the oxidation or reduction of analytes. Commonly involve carbon nanotubes or carbon paste electrodes for enhanced sensitivity.	Environmental applications, glucose detection, and detection of other metabolites.
	Conductometric Biosensors	Measure changes in conductivity in response to analyte interaction. These are fast, sensitive, and often used for ion detection.	Water quality monitoring, pesticide detection, and ion detection in various solutions.
	Potentiometric Biosensors	Measure the potential difference between a working electrode and a reference electrode. Often used for detecting ions or specific compounds like antibiotics.	Detection of antibiotics, ions, and other analytes in environmental and clinical settings.
	Impedimetric Biosensors	Measure the impedance or resistance to the flow of electric current, providing information about the analyte concentration.	Monitoring of cell growth, detection of pathogens, and monitoring of environmental pollutants.
Optical Microbial Biosensors	Bioluminescence-Based Sensors	Utilize the light emitted by bioluminescent organisms or genetically modified microbes to detect the presence of specific analytes.	Environmental monitoring, including the detection of pollutants and toxins. Also used in clinical diagnostics and food safety assessments.
	Fluorescence-Based Sensors	Use the fluorescence emitted by certain microbes, often involving GFP (Green Fluorescent Protein) or other fluorophores to detect analytes.	Detection of low-concentration analytes in environmental monitoring, clinical diagnostics, and biochemical research.
Other Types of Microbial Biosensors	Aptamer-Based Biosensors	Use aptamers (short DNA or RNA molecules) that bind specifically to target molecules, with microbial cells used to signal the binding event.	Detection of pathogens, toxins, and small molecules in clinical and environmental settings.
	Enzyme-Based Biosensors	Utilize enzymes produced by microbes to catalyze reactions with specific analytes, converting them into detectable products.	Glucose monitoring, detection of lactose, and other enzymatically detectable substances in food, clinical, and environmental applications.
	Nucleic Acid-Based Biosensors	Involve the detection of specific DNA or RNA sequences using microbial cells. These biosensors often use hybridization or amplification techniques for high sensitivity.	Detection of genetically modified organisms (GMOs), pathogens, and environmental DNA/RNA for biodiversity assessments.
	Whole-Cell Biosensors	Use entire microbial cells to detect changes in environmental conditions or the presence of specific substances.	Environmental monitoring, detection of heavy metals, organic pollutants, and in situ bioremediation assessments.

are known to induce serious neurological and muscle diseases. Molecular biology technology has allowed for the identification of several genes encoding the OP-degrading enzyme, which are then used as the recognition element in WCBs. Moreover, amperometric microbial biosensors offer quick and accurate instruments for fermentation and health applications. Approximately 85% of the biosensor market is accounted for by glucose detection, which is very important for diabetes diagnosis and food and fermentation quality control. Numerous microbial biosensors have been created to detect glucose by utilizing the oxygen consumption linked to respiratory activity in microorganisms. *Pseudomonas fluorescens* and *Gluconobacter oxydans* are two of the microbial strains acting as sensing elements. Carbon nanotubes or carbon paste electrodes were added to all of the transducers in this instance to improve the devices' amperometric responsiveness (Su et al., 2011).

Conductometric Microbial Biosensors: The conductometric microbial biosensor responds to analytes quickly and sensitively, which makes it desirable. In this context, a conductometric biosensor using the *Chlorella vulgaris* microalgae was utilized for the detection of heavy metal ions and pesticides in water samples (Chouteau et al., 2005). A technique involved freezing *C. vulgaris* onto a membrane with bovine serum albumin, followed by placement onto a platinum-interdigitated electrode. The resulting biosensors had a detection limit of 10 ppb for Cd^{2+} and Zn^{2+} after 30 minutes of exposure; however, Pb^{2+} did not cause noticeable interference as it was ideally absorbed by the albumin. Additionally, since organophosphate (OP) chemicals can inhibit acetylcholinesterase activity, the same microbial biosensor was also effective for detecting OP pesticides. An analogous conductometric biosensor that utilized *C. vulgaris* as the sensing component achieved a remarkably low detection limit of 1 ppb for Cd^{2+} (Guedri and Durrieu, 2008). Another conductometric biosensor was developed for the detection of urea, incorporating lyophilized *Brevibacterium ammoniagenes* encased in pH-sensitive polyaniline on a platinum twin-wire electrode. Because of the bacteria's catabolic activity, urea was converted to ammonia, raising the pH in the area. The working electrode sensed the change in the carbon paste electrode resistivity caused by the pH fluctuation (Jha et al., 2009).

Potentiometric Microbial Biosensors: These assess the voltage distinction between the working electrode and the reference electrode, which are segregated by a selective membrane, to measure the concentration of analytes in a sample. A potentiometric biosensor was developed to swiftly and accurately detect cephalosporin antibiotics. This biosensor employs a pH electrode and was modified utilizing permeabilized *Pseudomonas aeruginosa* (Kumar et al., 2008). Proton generation occurred close to the pH electrode as a result of the microbial layer's enzyme activity, which caused the hydrolysis of cephalosporin. The response was induced by the variance in electrical potential between the working and reference electrodes. A study in 2008 reported the use of a different potentiometric biosensor to identify lactam residues in milk (Ferrini et al., 2008).

As the microbial fuel cell (MFC) develops further, it may be used as a biosensor for target chemical monitoring and in situ analysis (Li et al., 2010). A renewable anaerobic consortium was utilized to continuously monitor the glucose levels in the anode electrolyte of a single-chamber MFC incorporating a proton exchange membrane (PEM) (Kumlanghan et al., 2007). Upon glucose oxidation, the anaerobic organisms transferred electrons to the external circuit. Consequently, there could be a correlation between the glucose concentration in the sample solution and the maximum closed-circuit potential, which corresponds to the peak current under a consistent circuit load (Tront et al., 2008). *Geobacter sulfurreducens* served as the external electron acceptor. The system underwent testing with an influent solution containing varying concentrations of acetate, and the resulting current was observed for fluctuations. Furthermore, MFC-based biosensors have been employed for BOD (biochemical oxygen demand) detection. In the development and optimization of an MFC biosensor for BOD, anaerobic sludge and carbon cloth were utilized as the anodes (Di Lorenzo et al., 2009).

16.2.2.2 Optical Microbial Biosensors

Optical biosensors are devices that use optical standards to transduce biochemical interactions into an appropriate output signal (Mudenkattil et al., 2022). A variety of optical characteristics can change to reveal biosensing events, such as refractive index, luminescence, adsorption, or fluorescence (Lechuga, 2005). BL and fluorescence microbial biosensors, in particular, have been the subject of extensive research and are depicted in Figure 16.3. The versatility and resilience to electrical noise are two benefits of optical microbial biosensors that use bacteria cells as biosensing components. Because of their affordability, compact size, and variable shape, optical fibers, the optical waveguides at the core of every optical biosensor, have been widely used.

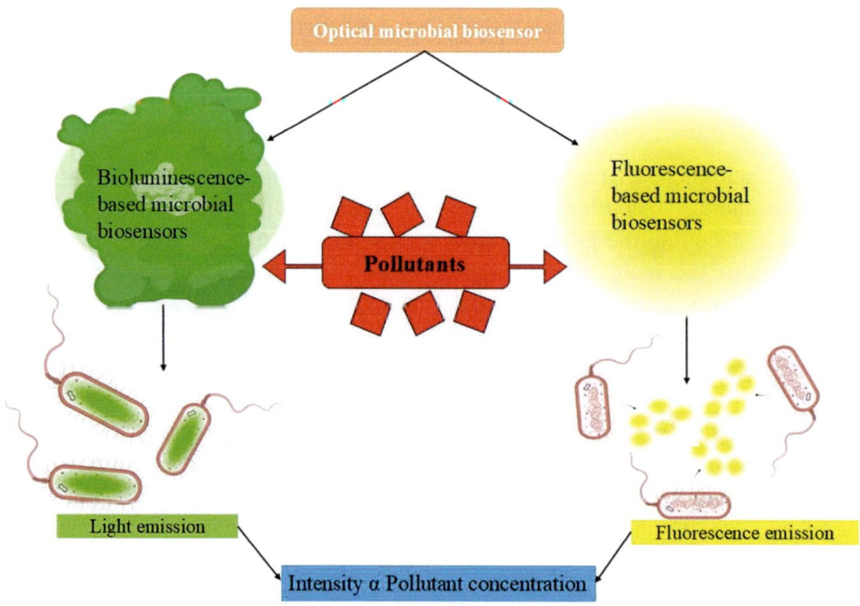

FIGURE 16.3 The working principle of optical microbial biosensors.

Furthermore, microbial biosensor-based optical fibers are portable and simple to employ in the field for online monitoring (Ivask et al., 2007).

BL-based microbial biosensors: The luminescence produced by pertinent microbial communities is chiefly influenced by the bioavailable portion of the identified pollutant, rendering them dependable and efficient instruments for evaluating risks and monitoring the environment, particularly in the assessment of toxicity and heavy metals (Chinalia et al., 2008). Numerous varieties of BLB have been employed in scientific investigations, such as naturally occurring BLB, genetically engineered bacteria that generate light, biosensors, and immobilized bioluminescent systems (Girotti et al., 2008). Thereby, low power consumption, robust design, single-chip production, and cheap cost of ownership of these biosensors allow them to be used in a wide range of non-laboratory applications. Even so, there are inherent issues with the overall light-collecting efficiency of such biosensors. Nevertheless, genetic methodologies, like fixing luminescent substances and entire cells on solid substrates, have the potential to boost the sensitivity, selectivity, and durability of these light-producing tools (Ben-Yoav et al., 2009).

Bacterial-based fluorescence microbial biosensors: In situations with low levels of the analyte, the brightness of fluorescence emission correlates linearly with the analyte concentration, particularly in whole cells that exhibit bacterial-driven fluorescence triggered by an external light stimulus. Bioluminescent biosensors sensitive to specific analytes have been created utilizing green fluorescent protein (GFP) and oxygen-responsive fluorescent substances. GFP and analogous fluorescent proteins provide remarkable durability, whereas BL enables quicker and more sensitive identification of the target analyte in comparison to fluorescence. Limited research has compared these methods, and colorimetric techniques based on color shifts have not been widely explored until recently. For example, a sensitive biosensor (cytosensor) was created using toxin-sensitive live cells and chromatophores to detect interactions between fish cells and microbes with high sensitivity to toxins (Xu and Ying, 2011).

16.2.2.3 Other Types of Microbial Biosensors

Apart from the microbial biosensors discussed above, other varieties rely on transducers, which have been recently reported.

Aptamer-based microbial biosensors: Aptamers are single-stranded oligodeoxynucleotides derived from synthetic RNA or ssDNA libraries that are quandary to recipient molecules with specificity and elevated affinity, typically 40–100 nt in length. Aptamers, often referred to as "molecular antibodies", exhibit binding affinities and selectivities similar to antibodies, with dissociation constants in the nM-pM range. The SELEX technology screens aptamers in vitro, avoiding animal or cell involvement and allowing the screening of toxic and non-immunogenic substances. Aptamers are easily synthesized and duplicated via PCR, offering short cycles, low costs, and consistency across batches. They target a wide range of substances, from small molecules to macromolecules and even complete viruses, bacteria, and cells. Aptamers have excellent thermal and chemical stability, allowing for long-term storage and transport at room temperature, and can be renatured after denaturation. Their use in biosensors enables repeated applications under diverse conditions. Additionally, aptamers can be chemically modified for detection and application purposes, boasting a small molecular weight, good biocompatibility, no immunogenicity, and strong tissue penetration, making them suitable for molecular imaging, drug delivery, and intracellular applications. Their predictable secondary structure and modifiability enhance their affinity and specificity, supporting the development of versatile detection methods (Ye et al., 2019).

Enzyme-based microbial biosensors: Applying immobilized enzymes to a specified reaction that is monitored using a designated transducer is the basic model design of enzyme-based biosensors. The many groups of enzymes that are employed, including oxidoreductases, hydrolases, ligases, isomerases, and lyases, have been linked to specific transducers that enable the development of biosensors for a range of clinical, medical, and pharmaceutical purposes. These biosensors have an easy-to-transport, constantly functional design. The use of immobilized glucose oxidase or glucose dehydrogenase mostly determines blood glucose levels. Invertase is mostly used for measuring sucrose, aside from this. Despite extensive application research, these biosensors have several drawbacks, primarily impacting their ability to detect environmental contamination because they lack substrate-acting enzyme systems (Cherian and Baskar, 2020).

Nucleic acid-based biosensors use genetic material, such as DNA, RNA, or nucleic acid analogs, as the probe. Signals are produced based on the interaction between the materials. These biosensors are primarily utilized in the disciplines of genotyping, gene expression investigations, illness diagnostics, and drug development. The primary issue with this biosensor is that the probe becomes immobile on it, which often compromises the study's performance (Hua et al., 2022).

16.3 MICROBIAL CELLS: BIOSENSOR AGENTS

Microbes are widely distributed and can metabolize a wide variety of chemical compounds. This ability has been used to create biosensors, which are useful as biological sensing agents. Microbes can also be easily genetically modified through recombinant technology or mutation, adapting to harsh environments and developing the capacity to break down new molecules over time. It is also an affordable way to obtain intracellular enzymes (Santos-Beneit, 2024). Bacteria and yeasts are the primary types of microbes used for microbial biosensor development.

Biosensing Elements: The majority of microorganisms utilized in biosensors are single-celled organisms lacking complex multicellular structures. Until recently, humans only perceived bacteria as singular cells (Tecon and Van der Meer, 2006). In addition to whole cells, cellular organelles (such as mitochondria) and particles (such as cell walls) can also be considered for multifunctional biocatalysts. Microorganisms are divided into two categories: viable cells and nonviable cells. Viable, intact cells serve as biological entities and are increasingly important in the advancement of microbial biosensors. While they may not exhibit the same level of specificity as the distinct biorecognition of individual molecules, they offer significant benefits in the development of microbial biosensors and are still valuable for detecting complex sequences of events that are characteristic of healthy, functioning cells (Belkin, 2003). Microbial biosensors based on live cells often trigger specific responses in two different ways. One of the analytes of interest (pollutants in the environment, for example) inhibits microbial respiration. The other approach relies on the metabolism of living cells and measures respiratory metabolic activity by measuring the total substrate assimilation capability of microorganisms (Moraskie et al., 2021). Living cells have two primary functions: measuring biological oxygen demand and uptake of additional nutrients associated with growth or metabolism. Additionally, they are capable of detecting pollutants or hazardous substances in viable cells. The cells either reduce their metabolic activity or perish when exposed to toxic concentrations that are too high. As the level of toxins in the surrounding cells grows, the signal formed generated by the reporter genes decreases proportionally. Nevertheless, this form of biosensor is constrained by its potential to react non-selectively to any substance that poses a threat to the cells (Zhou et al., 2024).

Bacterial Strains: The process of choosing certain bacterial strains is carried out by using microorganisms with specified substrate preferences, from which the strains were identified based on their predicted sensitivity and selectivity. The database that serves as the basis of the data was built using the substrate-specific properties of several classes of microorganisms from various taxa. These strains were typically evaluated based on their substrate sensitivity for organic acids, alcohols, and carbohydrates (Reshetilov, 2005).

Genetically Engineered Strains: A WCB that offers advantages such as being non-invasive, providing real-time results, being cost-effective, and easy to use, is based on transgenically modified bioluminescent bioreporters sensitive to volatile markers. Living cells are used as sensing components in the construction of a WCB. Their simplicity, affordability, and relatively high sensitivity have made them a popular choice for medical diagnostics and environmental monitoring in recent decades. WCBs using transgenically modified bioreporters are particularly appealing because their reporter-containing plasmids can be easily customized to detect specific environmental toxicants. Applications for this kind of WCB include heavy metals, antibiotics, indoor air quality, volatiles identification, and water quality monitoring (Veltman et al., 2023).

16.4 APPLICATIONS OF MICROBIAL BIOSENSORS

Heavy metals represent a major environmental hazard because of their persistence and toxicity. Unable to degrade naturally, these metals accumulate within living organisms, causing a range of issues. Consequently, there is a critical need for low-cost, precise, and rapid methods to monitor heavy metal contamination (Dai and Choi, 2013). While effective in heavy metal quantification, traditional methods like spectroscopy, voltammetry, and chromatography are often unsuitable for on-site analysis and cost-effective screening. This has driven the development of innovative, economical techniques.

One promising approach involves a love-wave bacteria-based sensor platform for real-time heavy metal detection in liquid environments. By immobilizing bacteria or polyelectrolytes on the sensor, changes in mass and viscoelastic properties caused by heavy metal interactions can be monitored in real time, providing a rapid and sensitive analytical tool (Gammoudi et al., 2010). Additionally, "lights-on" biosensors, genetically engineered microorganisms equipped with a metal-responsive unit linked to a promoterless reporter gene, offer another approach to heavy metal detection. These biosensors are frequently employed in toxicity testing for both pure substances and environmental pollutants. When a bioavailable compound, such as a heavy metal or organic pollutant, enters the sensor cell and combines with the regulatory protein, it triggers the countenance of the reporter gene. BL, a commonly used reporter system, is particularly advantageous due to its speed and simplicity. GFP and β-galactosidase are also utilized as reporters; the LuxCDABE gene cassette, which allows for BL, is especially useful for environmental applications because of its simplicity and speed (Ivask et al., 2009).

In addition to heavy metals, pesticide detection in complex environments has become a focal point in environmental monitoring. A range of biosensors have been developed, many of which rely on cholinesterases—enzymes sensitive to pesticide inhibition—as their core biocatalyst. OP and carbamate pesticides, which inhibit cholinesterase, are effective insecticides but present significant toxicity risks to animals and humans (Kröger et al., 2002). Innovative approaches have also explored the use of photosynthetic cyanobacteria. For example, the biosensors developed by Rawson et al. in 1989 used *Synechococcus*, a cyanobacterium highly susceptible to herbicides, as a biocatalyst. This sensitivity makes *Synechococcus* a promising candidate for environmental monitoring applications, further expanding the toolkit available for detecting harmful substances in the environment (Rawson et al., 1989). Türcmiş et al. (2018) developed an optical biosensor incorporating an array of algal bio-mediators designed for real-time monitoring of key marine pollutants. The biosensor's effectiveness was assessed by testing its response to three commonly found pesticides—simazine, atrazine, and diuron. Sima Iravani and colleagues (2024) created a simple electrochemical biosensor that employs a strain of locally isolated *Scenedesmus* species as the complete-cell recognition component to measure the levels of Cd^{2+} cations (Iravani et al., 2024).

Further advancements in biosensor technology have extended their applications in detecting a variety of environmental contaminants. Catterall et al. (2010) developed and optimized an enhanced FM-TOX method, integrating novel features, with *E. coli* serving as the biocatalyst. This method effectively evaluates the toxicity of various contaminants, such as chlorinated phenols, heavy metals, and other inorganic compounds, emphasizing its application in environmental monitoring and toxicology. These examples of biosensor technology demonstrate the diverse and crucial roles that these tools play in environmental monitoring and protection (Catterall et al., 2010).

Challenges such as crude oil spills and their derivatives, resulting from accidents during offshore extraction, transportation, or usage, require sophisticated detection methods in marine environments. In response to this, Sevilla et al. (2015) developed a biosensor system for detecting alkanes in seawater, employing either *E. coli* or marine hydrocarbonoclastic bacteria. The biosensor utilizing *Alcanivorax borkumensis* demonstrated superior performance, particularly at low concentrations of pure alkanes or petrol, compared to other tested strains (Sevilla et al., 2015).

Similarly, the application of electrochemical biosensors in detecting marine contaminants using *Staphylococcus aureus* as described by Liu et al. (2023) provides valuable insights into how salinity influences the biotoxicity of pollutants. This underscores the importance of understanding environmental variables in the context of pollution detection and management (Liu et al., 2022). In freshwater systems, the work by Binnerup and Sorensen (1992) highlights the critical need for sensitive detection of nitrates and nitrites to prevent eutrophication, which can lead to the degradation of water quality. This bioassay, utilizing anaerobically grown *P. aeruginosa*, represents an important tool for monitoring and controlling nutrient levels in natural water bodies (Binnerup and Sorensen, 1992).

Advancements in biosensor technology have also expanded to include the detection of heavy metals, which are still significant environmental pollutants. Moraskie et al. (2021) developed microbial whole-cell biosensors (MWCBs) for antimonite and arsenite detection using the *E. coli* ars operon, demonstrating the expanding applications of biosensors in this area. This technology offers a sensitive and specific approach to monitoring toxic metals in various environments (Moraskie et al., 2021). WCBs, engineered by Calvanese et al. (2024), provide a promising solution for in situ iron analysis, further extending the utility of biosensors. This technology, which involves genetically engineering bacteria to respond specifically to different iron species, provides a more straightforward and cost-effective alternative for real-time environmental monitoring (Calvanese et al., 2024).

Furthermore, incorporating halophilic bacteria into MFCs has gained considerable attention for renewable energy production and self-powered biosensing. These microorganisms are especially well-suited for environments with variable and high-saline concentrations, like certain types of wastewaters. Their distinct evolutionary adaptations enable them to operate efficiently in high-salinity conditions, making them promising candidates for microbial electrochemical sensing (Gaffney et al., 2021).

Maintaining water quality control is critical for efficient fisheries management and sustainable expansion of aquatic farming. According to Cunha et al. (2018), an optical sensor that uses quantum dots (QDs) and DNA aptamers can detect aquatic toxins generated by freshwater and marine microorganisms, such as cyanobacteria, dinoflagellates, and diatoms, in real time. Seafood quality declines rapidly due to high levels of microbial activity following harvest, leading to changes in the sensory attributes and chemical composition of marine products (Cunha et al., 2018). Trimethylamine oxide (TMAO) plays a crucial role in the non-protein nitrogen (NPN) component and serves as an indicator of overall quality in marine fish as well as certain freshwater species (Gram et al., 1990; Anthoni et al., 1990). Various applications of the microbial biosensors are illustrated in Figure 16.4.

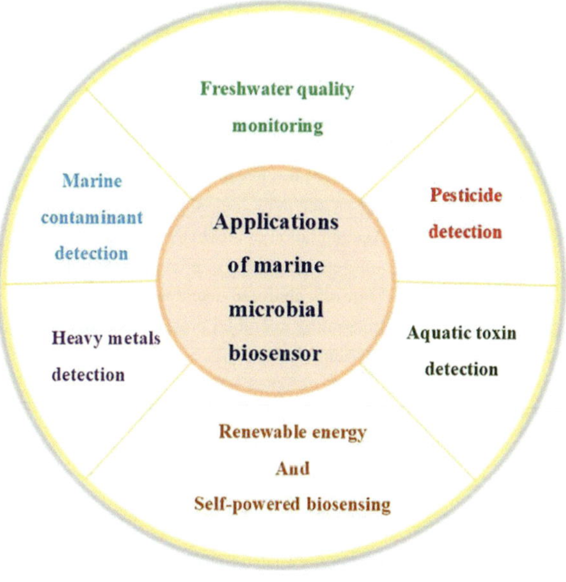

FIGURE 16.4 Applications of marine microbial biosensors.

Collectively, these advancements in biosensor technology highlight the increasing importance of innovative detection methods in addressing environmental challenges. Whether monitoring the health of coral reefs, detecting marine contaminants, controlling nutrient levels in freshwater, or analyzing toxic metals, biosensors are proving to be indispensable tools in the ongoing efforts to protect and sustain both marine and terrestrial ecosystems.

16.5 CHALLENGES OF MICROBIAL BIOSENSORS

The effective monitoring capability of innovative microbial-derived biosensors for environmental pollutants has made them an attractive alternative to conventional sensing technologies in terms of cost. Table 16.2 provides insights on various challenges of microbial biosensors outlining the key obstacles in their development, application, and commercialization.

Comprehensive development of WCBs might be highly helpful in isolated or remote locations where test sample transportation is not feasible, especially against emerging contaminants, chemicals, potentially harmful pollutants, and heavy metals. As a result of many techno-societal obstacles that have limited their larger-scale exploitation, only a small number of the numerous microbe-based biosensors that have been developed over the past few years have been effectively used for in situ monitoring or commercialized. For example, none of the existing microbial biosensors can identify substances at concentrations below 0.1 M to disrupt promoter activity. Additionally, biosensors are frequently non-specific, reacting differently to a collection of compounds than to a single component. Additionally, a major limitation of microbial biosensors for real-time detection applications is their slow response time, which is often due to the period needed for the expression of the reporter gene. By improving host strains, modifying novel strains of synthesizing modifying enzymes, creating more sensitive promoters, and employing surface-expressed proteins, scientists and the scientific community have made great strides in improving the sensitivity, specificity, and response times of microbial biosensors. As an example, surface-expressed proteins may interact directly with substrates without going through the cell, increasing the sensitivity and speed of the biosensors created by microbes (Bilal and Iqbal, 2019).

The inherent challenge of maintaining cell viability and activity under unfavorable environmental circumstances devoid of nutrients or containing compounds that hinder growth is another issue that significantly impacts the widespread use of microbial biosensors (Baumstark-Khan et al., 2005). An ideal biosensor should be able to withstand and sustain its biosensing functions and efficiency in extreme field conditions, such as acidic or alkaline pH levels, high temperatures, or the presence of organic solvents (Lei et al., 2006). Hydrogels are a suitable and practical encapsulation type that was developed in response to the need for microbial biosensors in low-resource and unfavorable situations (Van der Meer and Belkin, 2010). Furthermore, it would be risky to release biosensors created from genetically engineered microbes into the environment. The majority of biosensor cells are conjugated with supporting materials, which eliminates the risk of an unfavorable outcome. More significantly, microbial-derived biosensors may be optimized and regulated to dispel public and community misconceptions about genetically modified microbial biosensors (Johnson et al., 2017).

TABLE 16.2
Various Microbial Challenges Outlining Their Application

Challenge	Description
Stability and Shelf Life	Microbial cells used in biosensors can lose viability over time, leading to reduced sensor performance and shortened shelf life.
Sensitivity and Specificity	Achieving high sensitivity and specificity can be difficult, especially in complex samples with multiple interfering substances.
Response Time	Some microbial biosensors have longer response times due to the time required for microbial cells to interact with analytes and produce a signal.
Environmental Conditions	Variability in environmental conditions (e.g., pH, temperature, humidity) can affect the performance of microbial biosensors.
Reproducibility	Ensuring consistent performance across different batches of biosensors is challenging, often due to biological variability.
Biofouling	Accumulation of unwanted biological material on sensor surfaces can interfere with the sensor's ability to detect target analytes.
Integration with Electronic Systems	Integrating microbial biosensors with electronic systems for data processing and transmission can be complex.
Scale-up and Commercialization	Moving from laboratory prototypes to large-scale production while maintaining functionality and cost-effectiveness is difficult.
Regulatory and Safety Concerns	Ensuring that microbial biosensors meet regulatory standards and addressing safety concerns related to the use of live microbes is essential.

16.6 CONCLUSION

Marine microbial biosensors represent a promising frontier in both environmental monitoring and biomedical applications. A significant advantage of their service is the ability to detect contaminants and analyze marine ecosystems in real time, offering substantial benefits for the management of public health and the conservation of the environment. As interdisciplinary research continues to evolve, integrating marine biology, microbiology, and biotechnology, there is potential for developing innovative biosensors that harness the unique properties of marine microbes.

By further understanding microbial interactions and metabolite mechanisms, researchers can pave the way for novel therapeutic agents and improved environmental monitoring tools. The ongoing advancements in this field highlight the necessity for continued exploration of marine microbial biosensors, which may hold the key to addressing pressing global challenges related to health and environmental sustainability.

ACKNOWLEDGMENT

The authors thank the Chettinad Academy of Research Education for their constant support and encouragement.

REFERENCES

Ali, A. A., Altemimi, A. B., Alhelfi, N., & Ibrahim, S. A. (2020). Application of biosensors for detection of pathogenic food bacteria: A review. *Biosensors*, *10*(6), 58.

Anthoni, U., Børresen, T., Christophersen, C., Gram, L., & Nielsen, P. H. (1990). Is trimethylamine oxide a reliable indicator for the marine origin of fish?. *Comparative Biochemistry and Physiology Part B: Comparative Biochemistry*, *97*(3), 569–571.

Balootaki, P. A., & Hassanshahian, M. (2014). Microbial biosensor for marine environments. *Bulletin of Environment, Pharmacology and Life Sciences*, *3*, 1–3.

Baumstark-Khan, C., Cioara, K., Rettberg, P., & Horneck, G. (2005). Determination of geno-and cytotoxicity of groundwater and sediments using the recombinant SWITCH test. *Journal of Environmental Science and Health*, *40*(2), 245–263.

Belkin, S. (2003). Microbial whole-cell sensing systems of environmental pollutants. *Current Opinion in Microbiology*, *6*(3), 206–212.

Ben-Yoav, H., Elad, T., Shlomovits, O., Belkin, S., & Shacham-Diamand, Y. (2009). Optical modeling of bioluminescence in whole cell biosensors. *Biosensors and Bioelectronics*, *24*(7), 1969–1973.

Bhatnagar, I., & Kim, S. K. (2010). Immense essence of excellence: Marine microbial bioactive compounds. *Marine Drugs*, *8*(10), 2673–2701.

Bilal, M., & Iqbal, H. M. (2019). Microbial-derived biosensors for monitoring environmental contaminants: Recent advances and future outlook. *Process Safety and Environmental Protection*, *124*, 8–17.

Binnerup, S. J., & Sørensen, J. (1992). Nitrate and nitrite microgradients in barley rhizosphere as detected by a highly sensitive denitrification bioassay. *Applied and Environmental Microbiology*, *58*(8), 2375–2380.

Calabretta, M. M., Lopreside, A., Montali, L., Zangheri, M., Evangelisti, L., D'Elia, M., & Michelini, E. (2022). Portable light detectors for bioluminescence biosensing applications: A comprehensive review from the analytical chemist's perspective. *Analytica Chimica Acta*, *1200*, 339583.

Calvanese, M., D'Angelo, C., Tutino, M. L., & Lauro, C. (2024). Whole-cell biosensor for iron monitoring as a potential tool for safeguarding biodiversity in polar marine environments. *Marine Drugs*, *22*(7), 299.

Catterall, K., Robertson, D., Hudson, S., Teasdale, P. R., Welsh, D. T., & John, R. (2010). A sensitive, rapid ferricyanide-mediated toxicity bioassay developed using *Escherichia coli*. *Talanta*, *82*(2), 751–757.

Chandra, P., Enespa, Singh, R., & Arora, P. K. (2020). Microbial lipases and their industrial applications: A comprehensive review. *Microbial Cell Factories*, *19*, 1–42.

Cherian, E., & Baskar, G. (2020). Enzyme biosensors based on enzyme inhibition for pesticide and heavy metal detection. *Enzyme Inhibition – Environmental and Biomedical Applications*, *1*, 59–72.

Chinalia, F. A., Paton, G. I., & Killham, K. S. (2008). Physiological and toxicological characterization of an engineered whole-cell biosensor. *Bioresource Technology*, *99*(4), 714–721.

Choi, J. R. (2020). Development of point-of-care biosensors for COVID-19. *Frontiers in Chemistry*, *8*, 517.

Chouteau, C., Dzyadevych, S., Durrieu, C., & Chovelon, J. M. (2005). A bi-enzymatic whole cell conductometric biosensor for heavy metal ions and pesticides detection in water samples. *Biosensors and Bioelectronics*, *21*(2), 273–281.

Cunha, I., Biltes, R., Sales, M. G. F., & Vasconcelos, V. (2018). Aptamer-based biosensors to detect aquatic phycotoxins and cyanotoxins. Sensors, *18*(7), 2367.

Dai, C., & Choi, S. (2013). Technology and applications of microbial biosensor.

Di Lorenzo, M., Curtis, T. P., Head, I. M., & Scott, K. (2009). A single-chamber microbial fuel cell as a biosensor for wastewaters. *Water Research*, *43*(13), 3145–3154.

D'souza, S. F. (2001). Microbial biosensors. *Biosensors and Bioelectronics*, *16*(6), 337–353.

Ferrini, A. M., Mannoni, V., Carpico, G., & Pellegrini, G. E. (2008). Detection and identification of β-lactam residues in milk using a hybrid biosensor. *Journal of Agricultural and Food Chemistry*, *56*(3), 784–788.

Gaffney, E. M., Simoska, O., & Minteer, S. D. (2021). The use of electroactive halophilic bacteria for improvements and advancements in environmental high saline biosensing. *Biosensors*, *11*(2), 48.

Gammoudi, I., Tarbague, H., Othmane, A., Moynet, D., Rebière, D., Kalfat, R., & Dejous, C. (2010). Love-wave bacteria-based sensor for the detection of heavy metal toxicity in liquid medium. *Biosensors and Bioelectronics*, *26*(4), 1723–1726.

Girotti, S., Ferri, E. N., Fumo, M. G., & Maiolini, E. (2008). Monitoring of environmental pollutants by bioluminescent bacteria. *Analytica Chimica Acta*, *608*(1), 2–29.

Gram, L., Wedell-Neergaard, C., & Huss, H. H. (1990). The bacteriology of fresh and spoiling Lake Victorian Nile perch (Lates niloticus). *International Journal of Food Microbiology*, *10*(3–4), 303–316.

Guedri, H., & Durrieu, C. (2008). A self-assembled monolayers based conductometric algal whole cell biosensor for water monitoring. *Microchimica Acta*, *163*, 179–184.

Haleem, A., Javaid, M., Singh, R. P., Suman, R., & Rab, S. (2021). Biosensors applications in medical field: A brief review. *Sensors International*, *2*, 100100.

Hameed, S., Xie, L., & Ying, Y. (2018). Conventional and emerging detection techniques for pathogenic bacteria in food science: A review. *Trends in Food Science & Technology*, *81*, 61–73.

Hua, Y., Ma, J., Li, D., & Wang, R. (2022). DNA-based biosensors for the biochemical analysis: A review. *Biosensors*, *12*(3), 183.

Inda, M. E., & Lu, T. K. (2020). Microbes as biosensors. *Annual Review of Microbiology*, *74*(1), 337–359.

Iravani, S., Ahmadzadeh, H., Sajjadizadeh, H. S., & Rounaghi, G. (2024). Scenedesmus sp. MN738556 as a whole cell biosensor for sub-part per billion detections of Cd2+ cations in freshwater sources. *Microchemical Journal*, *197*, 109836.

Ivask, A., Green, T., Polyak, B., Mor, A., Kahru, A., Virta, M., & Marks, R. (2007). Fibre-optic bacterial biosensors and their application for the analysis of bioavailable Hg and As in soils and sediments from Aznalcollar mining area in Spain. *Biosensors and Bioelectronics*, *22*(7), 1396–1402.

Ivask, A., Rõlova, T., & Kahru, A. (2009). A suite of recombinant luminescent bacterial strains for the quantification of bioavailable heavy metals and toxicity testing. *BMC Biotechnology*, *9*, 41.

Jha, S. K., Kanungo, M., Nath, A., & D'Souza, S. F. (2009). Entrapment of live microbial cells in electropolymerized polyaniline and their use as urea biosensor. *Biosensors and Bioelectronics*, *24*(8), 2637–2642.

Johnson, A. O., Gonzalez-Villanueva, M., Wong, L., Steinbüchel, A., Tee, K. L., Xu, P., & Wong, T. S. (2017). Design and application of genetically-encoded malonyl-CoA biosensors for metabolic engineering of microbial cell factories. *Metabolic Engineering*, *44*, 253–264.

Kröger, S., Piletsky, S., & Turner, A. P. (2002). Biosensors for marine pollution research, monitoring and control. *Marine Pollution Bulletin*, 45(1–12), 24–34.

Kumar, S., Kundu, S., Pakshirajan, K., & Dasu, V. V. (2008). Cephalosporins determination with a novel microbial biosensor based on permeabilized Pseudomonas aeruginosa whole cells. *Applied Biochemistry and Biotechnology*, *151*, 653–664.

Kumlanghan, A., Liu, J., Thavarungkul, P., Kanatharana, P., & Mattiasson, B. (2007). Microbial fuel cell-based biosensor for fast analysis of biodegradable organic matter. *Biosensors and Bioelectronics*, *22*(12), 2939–2944.

Lechuga, L. M. (2005). Optical biosensors. *Comprehensive Analytical Chemistry*, *44*, 209–250.

Lei, Y., Chen, W., & Mulchandani, A. (2006). Microbial biosensors. *Analytica Chimica Acta*, *568* (1–2), 200–210.

Li, F., Sharma, Y., Lei, Y., Li, B., & Zhou, Q. (2010). Microbial fuel cells: The effects of configurations, electrolyte solutions, and electrode materials on power generation. *Applied Biochemistry and Biotechnology*, *160*, 168–181.

Lim, J. W., Ha, D., Lee, J., Lee, S. K., & Kim, T. (2015). Review of micro/nanotechnologies for microbial biosensors. *Frontiers in Bioengineering and Biotechnology*, *3*, 61.

Liu, C., Yu, H., Zhang, B., Liu, S., Liu, C. G., Li, F., & Song, H. (2022). Engineering whole-cell microbial biosensors: Design principles and applications in monitoring and treatment of heavy metals and organic pollutants. *Biotechnology Advances*, *60*, 108019.

Liu, Y., Lu, H., & Cui, Y. (2023). A review of marine in situ sensors and biosensors. *Journal of Marine Science and Engineering*, *11*(7), 1469.

Metkar, S. K., & Girigoswami, K. (2019). Diagnostic biosensors in medicine–a review. *Biocatalysis and Agricultural Biotechnology*, *17*, 271–283.

Moraskie, M., Roshid, M. H. O., O'Connor, G., Dikici, E., Zingg, J. M., Deo, S., & Daunert, S. (2021). Microbial whole-cell biosensors: Current applications, challenges, and future perspectives. *Biosensors & Bioelectronics*, *191*, 113359.

Mudenkattil, S., Girigoswami, A., Jayaprakash, T., & Girigoswami, K. (2022). Development of nanobiosensors for human immunodeficiency virus detection: a mini review. *Exploration of Medicine*, *3*(5), 479–493.

Nnachi, R. C., Sui, N., Ke, B., Luo, Z., Bhalla, N., He, D., & Yang, Z. (2022). Biosensors for rapid detection of bacterial pathogens in water, food and environment. *Environment International*, *166*, 107357.

Rawson, D. M., Willmer, A. J., & Turner, A. P. (1989). Whole-cell biosensors for environmental monitoring. *Biosensors*, *4*(5), 299–311.

Reshetilov, A. N. (2005). Microbial, enzymatic, and immune biosensors for ecological monitoring and control of biotechnological processes. *Applied Biochemistry and Microbiology*, *41*, 442–449.

Reshetilov, A. N., Iliasov, P. V., & Reshetilova, T. A. (2010). The microbial cell based biosensors. *Intelligent and Biosensors*, 289–322.

Santos-Beneit F. (2024). What is the role of microbial biotechnology and genetic engineering in medicine?. *MicrobiologyOpen*, *13*(2), e1406.

Sevilla, E., Yuste, L., & Rojo, F. (2015). Marine hydrocarbonoclastic bacteria as whole-cell biosensors for n-alkanes. *Microbial Biotechnology*, *8*(4), 693–706.

Su, L., Jia, W., Hou, C., & Lei, Y. (2011). Microbial biosensors: A review. *Biosensors and Bioelectronics*, *26*(5), 1788–1799.

Tecon, R., & Van der Meer, J. R. (2006). Information from single-cell bacterial biosensors: What is it good for?. *Current Opinion in Biotechnology*, *17*(1), 4–10.

Tront, J. M., Fortner, J. D., Plötze, M., Hughes, J. B., & Puzrin, A. M. (2008). Microbial fuel cell biosensor for in situ assessment of microbial activity. *Biosensors and Bioelectronics*, *24*(4), 586–590.

Türemiş, M., Silletti, S., Pezzotti, G., Sanchís, J., Farré, M., & Giardi, M. T. (2018). Optical biosensor based on the microalga-paramecium symbiosis for improved marine monitoring. *Sensors and Actuators B: Chemical*, *270*, 424–432.

Van Der Meer, J. R., & Belkin, S. (2010). Where microbiology meets microengineering: Design and applications of reporter bacteria. *Nature Reviews Microbiology*, *8*(7), 511–522.

Veltman, B., Ma, J., Harpaz, D., Xing, F., & Eltzov, E. (2023). Genetically engineered bacterial strains constructed as a whole-cell biosensor for specific volatiles identification of infected potato tubers with a soft rot disease. *Sensors and Actuators B: Chemical*, *387*, 133788.

Wan, J., Ai, J., Zhang, Y., Geng, X., Gao, Q., & Cheng, Z. (2016). Signal-off impedimetric immunosensor for the detection of Escherichia coli O157: H7. *Scientific Reports*, *6*(1), 19806.

Xu, X., & Ying, Y. (2011). Microbial biosensors for environmental monitoring and food analysis. *Food Reviews International*, *27*(3), 300–329.

Ye, W., Liu, T., Zhang, W., Zhu, M., Liu, Z., Kong, Y., & Liu, S. (2019). Marine toxins detection by biosensors based on aptamers. *Toxins*, *12*(1), 1.

Zhang, Z., Zhou, J., & Du, X. (2019). Electrochemical biosensors for detection of foodborne pathogens. *Micromachines*, *10*(4), 222.

Zhou, Y., Zheng, S., & Qin, W. (2024). Electrochemical biochemical oxygen demand biosensors and their applications in aquatic environmental monitoring. *Sensing and Bio-Sensing Research*, *44*, 100642.

17 The Silent Shift
How Ocean Acidification and Rising Temperature Affect Marine Organisms

Alka Rana[1], *Priya*[1], *Manish Thakur*[2], *and Neeta Raj Sharma*[2]
[1]Department of Microbiology, School of Bioengineering & Biosciences, Lovely Professional University, Punjab, India
[2]School of Bioengineering & Biosciences, Lovely Professional University, Punjab, India

17.1 INTRODUCTION

During the Industrial Revolution in the 18th century, carbon dioxide emissions had drastically increased, primarily due to the burning of fossil fuels. Consequently, the level of carbon dioxide rose considerably from 280 parts per million (ppm) (Padilla-Gamiño et al., 2022). By the end of this century, if carbon dioxide emissions are not controlled, the level of atmospheric carbon dioxide could reach 580–1,000 ppm (Ling et al., 2024). An increase in the concentration of carbon dioxide in marine ecosystems may cause serious health issues because the ocean absorbs a large amount of atmospheric carbon dioxide. The chemical reaction that lowers the sea pH is known as ocean acidification (OA) (Leung et al., 2022). Carbonic acid forms when atmospheric carbon dioxide dissolves in seawater. Therefore, carbonic acid is unstable in seawater and breaks down into carbonate and bicarbonate, releasing hydrogen ions during the process. These extra hydrogen ions make seawater more acidic by lowering its pH. The pH of the ocean has already dropped by 0.1 units since the industrial era started. The balance of the carbonate system in seawater changes as the level of carbon dioxide increases (Sumaila et al., 2011). The level of carbonate ions in the environment is expected to decrease sharply by around half. This reduction in carbonate ions can have various negative effects, particularly on marine life and ecosystems that rely on them (Findlay & Turley, 2021). This decreases the carbonate saturation state of seawater, which measures the amount of carbonate required for marine organisms to build their shells and skeletons from calcium carbonate. When the carbonate saturation state decreases, it becomes harder for organisms, such as corals, shellfish, and other plankton, to form calcium carbonate shells or skeletons (Hill & Hoogenboom, 2022).

17.2 OVERVIEW OF OCEAN ACIDIFICATION AND TEMPERATURE RISE

The pH of the oceans is decreasing continuously owing to the increased absorption of carbon dioxide from the atmosphere, which is known as OA. Human activities, such as the extraction and burning of fossil fuels, deforestation, and increased population, have led to an increased amount of carbon dioxide in the atmosphere, now far surpassing the levels seen before the industrial era (Godbold & Calosi, 2013). The ocean absorbs approximately 25% of the CO_2 released, which reacts with seawater to form carbonic acid. This process results in a decrease in pH, an increase in acidification, and changes in carbonate chemistry. With the beginning of the industrialization era, it has been noticed that the average pH of the ocean has decreased by 0.1% which is a global threat that is accumulating in the world's oceans. These changes occur at a faster rate, disturbing the entire marine life. The chemical reactions are as follows.

$$CO_2 + H_2O \rightarrow H_2CO_3 + HCO_3 \rightarrow H^+CO_3^{2-} + 2H^+$$

The outcome of OA leads to a decreased rate of $CaCO_3$ production in marine calcifiers, while the rates of bioerosion and $CaCO_3$ dissolution have significantly increased (Ross et al., 2022). In addition to OA, an increase in temperature may increase the prevalence and severity of coral bleaching. Thus, the risk of the physical breakdown of reefs also increases. Consequently, concerns have been raised that because of OA and rising temperatures, coral reefs are shifting from growing (accretion) to net erosion (breaking down), which threatens their structural and ecological functions (Noyes et al., 2024). Some studies have reported the relationship between net ecosystem calcification (NEC) and seawater aragonite saturation rate. NEC can be calculated as the total calcification minus the total $CaCO_3$ dissolution in the coral reef ecosystem, while the aragonite saturation state (Ωa) is a thermodynamic index indicating the anticipated net direction of $CaCO_3$ formation or dissolution (Andersson et al., 2007). NEC and Ωa have a positive correlation, but the exact nature may vary across different reef ecosystems. A positive correlation indicates that as the NEC rate decreases, the Ωa also decreases. This means that, as OA progresses and Ωa decreases, the net calcification of coral reef ecosystems tends to decline (Meléndez et al., 2022). This relationship can be influenced by many factors along with carbon chemistry, these factors may

DOI: 10.1201/9781003509530-17

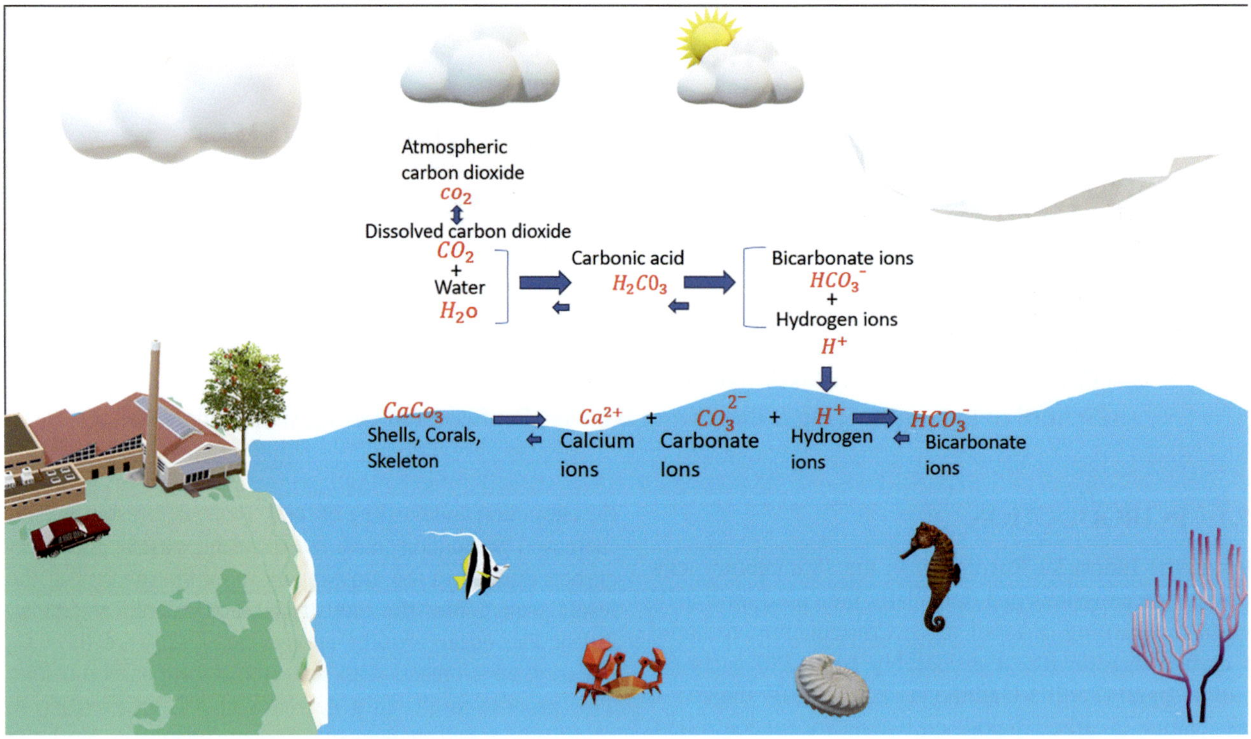

FIGURE 17.1 Mechanism of ocean acidification.

include temperature, light, nutrition, and hydrography. All of these factors directly or indirectly affect NEC. As discussed earlier, as OA progresses, the relationship between NEC and Ωa is also likely to change in several ways, such as a reduced NEC rate (leading to a shift of coral reefs from net calcification to net dissolution) and shifting thresholds, which means the geochemical threshold at which NEC becomes zero (i.e., calcification equals dissolution) may shift (Afdal et al., 2023). The potential for adaptation is also influenced, as calcifying organisms might adapt to lower Ωa levels. All of these changes influence the relationship between NEC and Ωa, making some coral communities more resilient (Figure 17.1).

The excess heat trapped by greenhouse gases (GHGs) is absorbed by the ocean, leading to an increase in the temperature of the water. This phenomenon adversely affects marine life, as many species have tolerance up to a specific temperature, and exceeding these limits can cause reduced growth, increased mortality, and impaired reproduction (Chukwuka et al., 2023). Warming also leads to a shift in species distribution, as these organisms tend to move to places with adequate temperatures to survive, which subsequently disrupts the established food webs and creates new competitive interactions (Samiullah & Khanum, 2024). The existence of humans and other life forms depends on the concentration of GHGs, and any change in their concentration can lead to disruption of different life forms (Huang et al., 2024). An increase in carbon dioxide concentration leads to an increase in the temperature of the ocean, and this increase in temperature accelerates the melting of ice sheets and glaciers, leading to an increase in average sea level.

Marine microorganisms, which play a crucial role in biogeochemical cycles and environmental functioning, have also been severely affected by OA and rising temperatures, as every microorganism requires optimal pH and temperature for growth (Sun, 2023). The increase in oceanic pH only favors limited microbial species while disrupting the metabolic functions of other species, such as photosynthesis and respiration, leading to a change in the microbial community composition, which further affects ecological processes such as nutrient cycling and carbon sequestration (Das & Mangwani, 2015). Microorganisms, such as *Coccolithophores* and *Foraminifera,* are calcifying microorganisms that are adversely affected by acidification, as they have a direct impact on their survival as well as their role in ecological processes (Joint et al., 2011). In addition to acidification, rising temperature also affects species distribution; as the temperature of the water increases, some microbial species may shift their distribution to the cooler region, which can alter the geographical distribution of microbial communities and their associated functions (Nagelkerken & Connell, 2022). The loss of diversity also increased with an increase in temperature. The synergistic effect of OA and warming is far more concerning than their individual effects on species (Yun et al., 2021). The combined effect of these two phenomena can impair the overall growth and survival of important microbial groups such as phytoplankton. They potentially lead to increased stratification, which could alter the nutrient fluxes and light levels experienced by microbes. These changes affect higher trophic levels in fish, seabirds, and marine mammals (Hu et al., 2022). Addressing climate

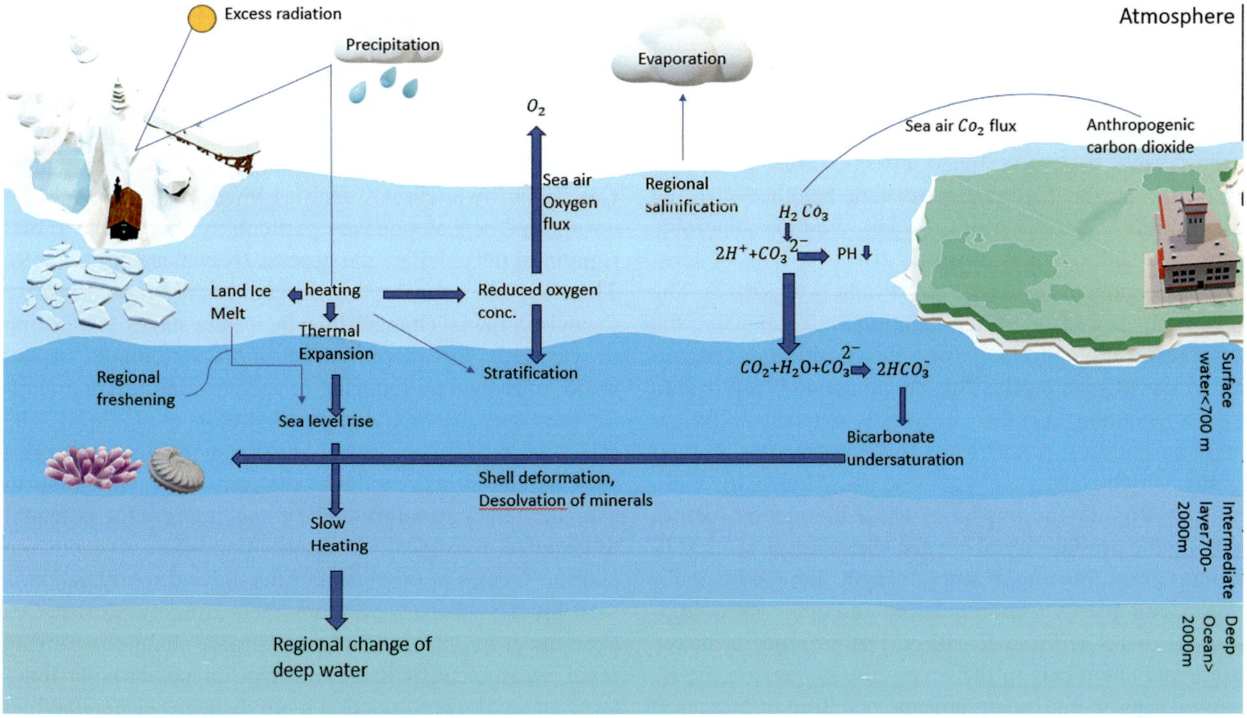

FIGURE 17.2 Effect of global warming on marine life.

change by reducing GHG emissions is crucial for mitigating its impacts and preserving the health and resilience of marine ecosystems (Figure 17.2).

17.3 IMPACT OF OCEAN ACIDIFICATION ON MARINE SHELLED MOLLUSKS

Mollusca is a diverse group of invertebrate animals that includes gastropods (snails and slugs), bivalves (oysters, mussels, etc.), cephalopods (octopuses, squids, etc.), and approximately 76,000 species that have been identified yet. These shelled mollusks play a significant role in maintaining marine ecosystems by fulfilling essential needs, including the formation of structural habitats for other benthic organisms and the purification of water (Baag & Mandal, 2022). Shelled mollusks are important both ecologically and economically, providing essential ecosystem services and supporting a significant global aquaculture industry. OA affects the growth and shell production of juvenile and shelled mollusks, but the effects may vary from species to species, or even among the same species (Gazeau et al., 2013). The early life history stages of shelled mollusks, particularly embryos and larvae, are highly vulnerable to OA stress, with far more negative effects than on juveniles and/or adults (Unzueta-Martínez et al., 2021). Increasing the CO_2 level has a negative impact on shell thickness, development, and survival rate of embryos and larvae, whereas the impact varies for juveniles. Most studies show that OA has no effect or a minor effect on juveniles and adults (Shan et al., 2024).

17.4 DISRUPTION OF THE MARINE FOOD WEB

The marine food web is a network of interconnected food chains within the ocean environment, and acidification of the ocean can disturb these food chains, leading to the disruption of the entire food web (Peña et al., 2021). The food quality of primary producers, such as phytoplankton and algae, can be directly affected by the increasing pH of the ocean and subsequent trophic transfer to secondary and tertiary producers. OA significantly affects the fatty acid and polyunsaturated fatty acid (PUFA) composition of phytoplankton (Gao et al., 2018). Studies have shown that the PUFA content decreased by approximately 3% in the diatom *Cylindrotheca fusiformis* grown under elevated conditions. A decrease in the PUFA content in phytoplankton can constrain the growth and reproduction of zooplankton that feed on them (Du et al., 2021). The accumulation rate of phenolic compounds in diatoms and coccolithophores also increased because of OA. Due to the presence of these phenolic compounds, the nutritional value for higher trophic levels has been decreasing (du Pontavice et al., 2021). It is suggested that OA can influence the transfer of metabolites mainly secondary metabolites from primary to secondary producers (Jin et al., 2020). Examples include increased iodine levels in the kelp *Saccharina japonica* under OA, which were reflected in the tissues of the abalone that fed on it. As the OA progresses, the community composition of phytoplankton shifts, which leads to the shortening of the food chain and promotes algal bloom growth, which reduces the export flux of organic matter and disrupts zooplankton (Fabry, 2008).

17.5 SYNERGISTIC EFFECTS WITH OTHER ENVIRONMENTAL STRESSORS

Acidification and warming synergistically interact with other stressors to intensify their impacts on marine ecosystems, and these combined effects pose significant threats to marine life and ecosystem health. Warming water contains less dissolved oxygen, creating anaerobic or anoxic conditions that adversely affect the lives of aerobic microorganisms in the ocean (Prada et al., 2017). The stress rate is increased by acidification, thus impairing the rate of respiration and increasing the sensitivity of organisms to low oxygen levels. This phenomenon is referred to as deoxygenation. Another stressor is nutrient pollution: excess agricultural runoff leads to the formation of algal blooms, which consume oxygen and cause eutrophication. This can amplify the impact of these blooms by further reducing the availability of oxygen (Schubert et al., 2023). Changes in salinity and freshwater input, along with acidification, can disrupt osmoregulation and other physiological processes (Griffin et al. 2022). The presence of heavy metals and chemicals in the ocean affects marine life by accumulating in marine organisms, resulting in increased physiological stress, making them highly sensitive to other stressors. The combined effects of acidification and warming can increase the toxicity of some pollutants (Wei et al., 2023).

17.6 PHYSIOLOGICAL IMPACTS ON MARINE ORGANISMS

Global warming affects ocean ecosystems by altering temperature, acidity-level circulation patterns, oxygen availability, stratification, and nutrient distribution. Microorganisms are essential for biogeochemical cycles and are susceptible to temperature change. The balance of microbial communities can be disrupted by variations in temperature. Therefore, the rate of evaporation increases across the ocean, leading to a shallow surface mixed layer (Medeiros & Souza, 2023). Due to global warming, the metabolism and diversity of microbes, along with ocean biogeochemistry, face significant threats. Globally, there are some noticeable shifts in the distribution of phytoplankton communities due to rising temperatures, which impact the biosphere and alter the primary production of the oceans (Czaja et al., 2023). Increasing GHGs and global warming severely affect the growth of marine life, including fish, algae, and phytoplankton (Shi & Li, 2024). OA and global warming are the primary causes of these effects. Some fish species are highly sensitive to temperature and are already extinct. Over the past 60 years, tuna and phytoplankton have decreased by 20% in the Indian Ocean coral reef species due to global warming. Scientists estimate that if this trend continues, plankton and fish populations could shrink by 50%–90% in 50 years (Lin, 2023). Tropical Pacific waters, especially plankton, are expected to warm by more than 3°C by the end of this century.

17.7 EFFECTS ON REPRODUCTIVE SUCCESS AND DEVELOPMENT

OA significantly impacts certain marine invertebrates, especially their reproductive stages, cleavage, larval development, settlement, reproduction, and early stage. Calcifying invertebrates such as bivalves, corals, echinoderms, and cephalopods are particularly susceptible to disruption of the calcification process (Kawahata et al., 2019). The early life stages of these organisms are more sensitive to environmental changes than their later stages. According to one study, OA can affect marine invertebrates' capacity to reproduce by reducing oxygen availability and altering how they use their energy (Figuerola et al., 2021). The disturbance of this process poses a major threat to species such as corals and mollusks that rely on calcification to safeguard their structures. These species have the potential to generate ecological imbalances that affect reproduction and the species of other organisms. Sexual reproduction is vital for the survival, evolution, and diversity of a species (Porteus et al., 2021). However, climate change is causing rapid ocean changes that can affect important body functions and make reproduction more difficult. Understanding the weak points and sensitive stages of reproductive cycles can help us predict how these changes will affect marine life, from individual species to entire ecosystems.

In 2016, commercial aquaculture generated approximately 26 million tons of invertebrates, accounting for 77% of marine and coastal aquaculture and 32% of worldwide aquaculture (Zunino et al., 2021). Invertebrate aquaculture is being increasingly used in restoration projects to help natural populations that have difficulty recovering on their own (Tai et al., 2021). A reliable supply of viable gametes or larvae, either produced under controlled hatchery settings or harvested from the wild, is essential for both commercial and restoration aquaculture. However, OA may limit their ability to reproduce, which could impede aquaculture productivity.

In contrast to aquaculture, wild fisheries rely only on marine species that reproduce effectively in their native environments. The detrimental effects of acidification on reproduction would immediately result in fewer wild stocks for fisheries (Duarte et al. 2022). Furthermore, human intervention is becoming increasingly necessary to maintain and restore populations of marine invertebrates listed as threatened or endangered. The ability of populations to survive and procreate successfully once interventions are over determines the success of such programs (Mangi et al., 2018). Therefore, understanding how low pH levels impact reproductive processes and their potential effects on the quantity and distribution of targeted species is crucial for aquaculture, fisheries, and conservation activities.

17.8 ECONOMIC AND SOCIETAL IMPLICATIONS

Climate change directly and indirectly affects the aquaculture industry. Direct impacts include the physiological and physical conditions of shellfish and finfish stocks in aquaculture systems (Tan & Zheng, 2020). Indirect effects include alterations in primary and secondary productivity, input availability, and production costs (Wilson et al. 2020). This may further increase product pricing, the cost of fish oil and fish meal, and other important resources for agricultural producers and fishers. Due to changes in environmental conditions, they directly influence the geographic distribution of invertebrates and marine fish. This is often observed in the ecological responses of marine species. OA affects marine resources, such as fish populations, shellfish, and coral reefs.

17.9 CONSERVATION AND MITIGATION STRATEGIES

OA and global warming are two major threats to marine ecosystems driven by climate change and CO_2 emissions. To address these issues and protect marine biodiversity, both global and local efforts are required. Mitigation strategies include a reduction in the concentration of CO_2 in the environment by transitioning to renewable resources, improving energy efficiency in industries and households, protecting and restoring the blue carbon ecosystem (mangroves) that absorbs CO_2, and minimizing nutrient runoff that worsens OA (Maulu et al., 2021). Conservation strategies include the preservation of marine life by establishing marine protected areas to protect vulnerable species and implementing sustainable fisheries management to reduce stress on marine organisms. Coral reef protection is achieved by developing coral nurseries and breeding heat-resistant and acidification-tolerant coral species (Clements & Chopin, 2017), strengthening ecosystem resilience by promoting diversity and managing local stressors to increase ecosystem resilience to acidification. The major strategy includes public awareness and policy action, which can be achieved by educating communities and advocating stronger international agreements (Billé et al., 2013).

17.10 CONCLUSION

In conclusion, a major effect of rising carbon dioxide emissions brought on by human activity—mainly since the Industrial Revolution—is OA. Because calcifying animals like corals, mollusks, and certain plankton species depend on calcium carbonate for growth and survival, the continuous drop in ocean pH jeopardizes marine ecosystems. Wide-ranging effects could result in disturbances to food webs, marine biodiversity, and the general well-being of oceanic habitats. The consequences for fisheries and aquaculture also emphasize how urgently comprehensive conservation and mitigation measures are needed to prevent further deterioration of these essential resources. Both domestic and international initiatives are necessary to solve the problems caused by OA and global warming, putting policies in place to lower carbon emissions and improve the resilience. In order to protect marine biodiversity and ensure the sustainability of fisheries and aquaculture in the face of climate change, marine ecosystems must be resilient. Therefore, conserving marine life and maintaining the health of ocean ecosystems for future generations depend on comprehending and reducing the effects of OA.

REFERENCES

Afdal, Bengen, D. G., Wahyudi, A. J., Rastina, Prayitno, H. B., & Koropitan, A. F. (2023). Variation in CO_2 fluxes, net ecosystem production, and calcification in the tropical waters of seagrass and coral reefs. *Regional Studies in Marine Science*, *68*, 103290. https://doi.org/10.1016/j.rsma.2023.103290

Andersson, A. J., Bates, N. R., & Mackenzie, F. T. (2007). Dissolution of carbonate sediments under rising pCO_2 and ocean acidification: Observations from Devil's Hole, Bermuda. *Aquatic Geochemistry*, *13*(3), 237–264. https://doi.org/10.1007/s10498-007-9018-8

Baag, S., & Mandal, S. (2022). Combined effects of ocean warming and acidification on marine fish and shellfish: A molecule-to-ecosystem perspective. *Science of the Total Environment*, *802*, 149807. https://doi.org/10.1016/j.scitotenv.2021.149807

Billé, R., Kelly, R., Biastoch, A., Harrould-Kolieb, E., Herr, D., Joos, F., Kroeker, K., Laffoley, D., Oschlies, A., & Gattuso, J.-P. (2013). Acting against ocean acidification: A review of management and policy options. *Environmental Management*, *52*(4), 761–779. https://doi.org/10.1007/s00267-013-0132-7

Chukwuka, A. V., Omogbemi, E. D., Adeogun, A. O. (2023). Habitat sensitivity in the West African coastal area: Inferences and implications for regional adaptations to climate change and ocean acidification. *Environmental Monitoring and Assessment*, *196*(1), 79. https://doi.org/10.1007/s10661-023-12171-z

Clements, J. C., & Chopin, T. (2017). Ocean acidification and marine aquaculture in North America: Potential impacts and mitigation strategies. *Reviews in Aquaculture*, *9*(4), 326–341. https://doi.org/10.1111/raq.12140

Czaja, R., Holmberg, R., Pales Espinosa, E., Hennen, D., Cerrato, R., Lwiza, K., O'Dwyer, J., Beal, B., Root, K., Zuklie, H., & Allam, B. (2023). Behavioral and physiological effects of ocean acidification and warming on larvae of a continental shelf bivalve. *Marine Pollution Bulletin*, *192*, 115048. https://doi.org/10.1016/j.marpolbul.2023.115048

Das, S. & Mangwani, N. (2015). Ocean acidification and marine microorganisms: Responses and consequences. *Oceanologia*, *57*(4), 349–361. https://doi.org/10.1016/j.oceano.2015.07.003

Du, D., Lu, Y., Zhou, Y., Li, Q., Zhang, M., Han, G., Cui, H., & Jeppesen, E. (2021). Bioaccumulation, trophic transfer, and biomagnification of perfluoroalkyl acids (PFAAs) in the marine food web of the South China Sea. *Journal of Hazardous Materials*, *405*, 124681. https://doi.org/10.1016/j.jhazmat.2020.124681

du Pontavice, H., Gascuel, D., Reygondeau, G., Stock, C., & Cheung, W. W. L. (2021). Climate-induced decrease in biomass flow in marine food webs may severely affect predators and ecosystem production. *Global Change Biology*, *27*(11), 2608–2622. https://doi.org/10.1111/gcb.15576

Duarte, J. A., Villanueva, R., Seijo, J. C., & Vela, M. A. (2022). Ocean acidification effects on aquaculture of a high resilient calcifier species: A bioeconomic approach. *Aquaculture*, *559*, 738426. https://doi.org/10.1016/j.aquaculture.2022.738426

Fabry, V. J. (2008). Marine calcifiers in a high-CO_2 ocean. *Science*, *320*(5879), 1020–1022. https://doi.org/10.1126/science.1157130

Figuerola, B., Hancock, A. M., Bax, N., Cummings, V. J., Downey, R., Griffiths, H. J., Smith, J., & Stark, J. S. (2021). A review and meta-analysis of potential impacts of ocean acidification on marine calcifiers from the Southern Ocean. *Frontiers in Marine Science*, *8*, 584445. https://doi.org/10.3389/fmars.2021.584445

Findlay, H. S., & Turley, C. (2021). Chapter 13—Ocean acidification and climate change. In T. M. Letcher (Ed.), *Climate Change (*Third Edition) (pp. 251–279). Elsevier. https://doi.org/10.1016/B978-0-12-821575-3.00013-X

Gao, K., Zhang, Y., & Häder, D.-P. (2018). Individual and interactive effects of ocean acidification, global warming, and UV radiation on phytoplankton. *Journal of Applied Phycology*, *30*(2), 743–759. https://doi.org/10.1007/s10811-017-1329-6

Gazeau, F., Parker, L. M., Comeau, S., Gattuso, J.-P., O'Connor, W. A., Martin, S., Pörtner, H.-O., & Ross, P. M. (2013). Impacts of ocean acidification on marine shelled molluscs. *Marine Biology*, *160*(8), 2207–2245. https://doi.org/10.1007/s00227-013-2219-3

Godbold, J. A., & Calosi, P. (2013). Ocean acidification and climate change: Advances in ecology and evolution. *Philosophical Transactions of the Royal Society B: Biological Sciences*, *368*(1627), 20120448. https://doi.org/10.1098/rstb.2012.0448

Griffin, A. J., Anderson, Z., Ballard, J., Bates, N. R., Garley, R., Johnson, R., Martz, T., Pacheco, F., Takeshita, Y., & Andersson, A. J. (2022). Seasonal changes in seawater calcium and alkalinity in the Sargasso Sea and across the Bermuda carbonate platform. *Marine Chemistry*, *238*, 104064. https://doi.org/10.1016/j.marchem.2021.104064

Hill, T. S., & Hoogenboom, M. O. (2022). The indirect effects of ocean acidification on corals and coral communities. *Coral Reefs*, *41*(6), 1557–1583. https://doi.org/10.1007/s00338-022-02286-z

Hu, N., Bourdeau, P. E., Harlos, C., Liu, Y., & Hollander, J. (2022). Meta-analysis reveals variance in tolerance to climate change across marine trophic levels. *Science of the Total Environment*, *827*, 154244. https://doi.org/10.1016/j.scitotenv.2022.154244

Huang, R., Zhang, P., Zhang, X., Chen, S., Sun, J., Jiang, X., Zhang, D., Li, H., Yi, X., Qu, L., Wang, T., Gao, K., Hall-Spencer, J. M., Adams, J., Gao, G., & Lin, X. (2024). Ocean acidification alters microeukaryotic and bacterial food web interactions in a eutrophic subtropical mesocosm. *Environmental Research*, *257*, 119084. https://doi.org/10.1016/j.envres.2024.119084

Jin, P., Hutchins, D. A., & Gao, K. (2020). The impacts of ocean acidification on marine food quality and its potential food chain consequences. *Frontiers in Marine Science*, *7*, 543979. https://doi.org/10.3389/fmars.2020.543979

Joint, I., Doney, S. C., & Karl, D. M. (2011). Will ocean acidification affect marine microbes? *The ISME Journal*, *5*(1), 1–7. https://doi.org/10.1038/ismej.2010.79

Kawahata, H., Fujita, K., Iguchi, A., Inoue, M., Iwasaki, S., Kuroyanagi, A., Maeda, A., Manaka, T., Moriya, K., Takagi, H., Toyofuku, T., Yoshimura, T., & Suzuki, A. (2019). Perspective on the response of marine calcifiers to global warming and ocean acidification—Behavior of corals and foraminifera in a high CO_2 world "hot house." *Progress in Earth and Planetary Science*, *6*(1), 5. https://doi.org/10.1186/s40645-018-0239-9

Leung, J. Y. S., Zhang, S., & Connell, S. D. (2022). Is ocean acidification really a threat to marine calcifiers? A systematic review and meta-analysis of 980+ studies spanning two decades. *Small*, *18*(35), 2107407. https://doi.org/10.1002/smll.202107407

Lin, S. (2023). Phosphate limitation and ocean acidification co-shape phytoplankton physiology and community structure. *Nature Communications*, *14*(1), 2699. https://doi.org/10.1038/s41467-023-38381-0

Ling, Z., Pan, J., Kontchouo, F. M. B., Liu, S., Lu, X., Guo, X., Penzik, M. V., Kozlov, A. N., Huang, Y., & Zhang, S. (2024). Current situation of marine CO_2 sequestration and analysis of related environmental issues. *Fuel*, *366*, 131288. https://doi.org/10.1016/j.fuel.2024.131288

Mangi, S. C., Lee, J., Pinnegar, J. K., Law, R. J., Tyllianakis, E., & Birchenough, S. N. R. (2018). The economic impacts of ocean acidification on shellfish fisheries and aquaculture in the United Kingdom. *Environmental Science & Policy*, *86*, 95–105. https://doi.org/10.1016/j.envsci.2018.05.008

Martins Medeiros, I. P., & Souza, M. M. (2023). Acid times in physiology: A systematic review of the effects of ocean acidification on calcifying invertebrates. *Environmental Research*, *231*, 116019. https://doi.org/10.1016/j.envres.2023.116019

Maulu, S., Hasimuna, O. J., Haambiya, L. H., Monde, C., Musuka, C. G., Makorwa, T. H., Munganga, B. P., Phiri, K. J., & Nsekanabo, J. D. (2021). Climate change effects on aquaculture production: Sustainability implications, mitigation, and adaptations. *Frontiers in Sustainable Food Systems*, *5*, 609097. https://doi.org/10.3389/fsufs.2021.609097

Meléndez, M., Salisbury, J., Gledhill, D., Langdon, C., Morell, J. M., Manzello, D., & Sutton, A. (2022). Net ecosystem dissolution and respiration dominate metabolic rates at two western Atlantic reef sites. *Limnology and Oceanography*, *67*(3), 527–539. https://doi.org/10.1002/lno.12009

Nagelkerken, I., & Connell, S. D. (2022). Ocean acidification drives global reshuffling of ecological communities. *Global Change Biology*, *28*(23), 7038–7048. https://doi.org/10.1111/gcb.16410

Noyes, T. J., Garley, R., & Bates, N. R. (2024). Calcification and trophic responses of mesophotic reefs to carbonate chemistry variability. *Frontiers in Marine Science*, *10*, 1274915. https://doi.org/10.3389/fmars.2023.1274915

Padilla-Gamiño, J. L., Alma, L., Spencer, L. H., Venkataraman, Y. R., & Wessler, L. (2022). Ocean acidification does not overlook sex: Review of understudied effects and implications of low pH on marine invertebrate sexual reproduction. *Frontiers in Marine Science*, *9*, 977754. https://doi.org/10.3389/fmars.2022.977754

Peña, V., Harvey, B. P., Agostini, S., Porzio, L., Milazzo, M., Horta, P., Le Gall, L., & Hall-Spencer, J. M. (2021). Major loss of coralline algal diversity in response to ocean acidification. *Global Change Biology*, *27*(19), 4785–4798. https://doi.org/10.1111/gcb.15757

Poloczanska, E. S., Burrows, M. T., Brown, C. J., García Molinos, J., Halpern, B. S., Hoegh-Guldberg, O., Kappel, C. V., Moore, P. J., Richardson, A. J., Schoeman, D. S., & Sydeman, W. J. (2016). Responses of marine organisms to climate change across oceans. *Frontiers in Marine Science*, *3*, 62. https://doi.org/10.3389/fmars.2016.00062

Porteus, C. S., Roggatz, C. C., Velez, Z., Hardege, J. D., & Hubbard, P. C. (2021). Acidification can directly affect olfaction in marine organisms. *Journal of Experimental Biology*, *224*(14), jeb237941. https://doi.org/10.1242/jeb.237941

Prada, F., Caroselli, E., Mengoli, S., Brizi, L., Fantazzini, P., Capaccioni, B., Pasquini, L., Fabricius, K. E., Dubinsky, Z., Falini, G., & Goffredo, S. (2017). Ocean warming and acidification synergistically increase coral mortality. *Scientific Reports*, *7*, 40842. https://doi.org/10.1038/srep40842

Ross, C. L., Warnes, A., Comeau, S., Cornwall, C. E., Cuttler, M. V. W., Naugle, M., McCulloch, M. T., & Schoepf, V. (2022). Coral calcification mechanisms in a warming ocean and the interactive effects of temperature and light. *Communications Earth & Environment*, *3*(1), 1–11. https://doi.org/10.1038/s43247-022-00396-8

Samiullah, M., & Khanum, R. (2024). Climate change and ocean acidification: Unraveling the complex interactions and implications for marine ecosystems. *International Journal of Natural Sciences*, *4*(1), 1. https://doi.org/10.47604/ijns.2397

Schubert, N., Alvarez-Filip, L., & Hofmann, L. C. (2023). Systematic review and meta-analysis of ocean acidification effects in Halimeda: Implications for algal carbonate production. *Climate Change Ecology*, *4*, 100059. https://doi.org/10.1016/j.ecochg.2022.100059

Shan, E., Zhang, X., Yu, Z., Hou, C., Pang, L., Guo, S., Liu, Y., Dong, Z., Zhao, J., Wang, Q., & Yuan, X. (2024). Seawater warming rather than acidification profoundly affects coastal geochemical cycling mediated by marine microbiome. *Science of the Total Environment*, *957*, 177365. https://doi.org/10.1016/j.scitotenv.2024.177365

Shi, Y., & Li, Y. (2024). Impacts of ocean acidification on physiology and ecology of marine invertebrates: A comprehensive review. *Aquatic Ecology*, *58*(2), 207–226. https://doi.org/10.1007/s10452-023-10058-2

Sumaila, U. R., Cheung, W. W. L., Lam, V. W. Y., Pauly, D., & Herrick, S. (2011). Climate change impacts on the biophysics and economics of world fisheries. *Nature Climate Change*, *1*(9), 449–456. https://doi.org/10.1038/nclimate1301

Sun, Y. (2023). Effects of ocean acidification on the marine organisms. *Highlights in Science, Engineering and Technology*, *69*, 342–348. https://doi.org/10.54097/hset.v69i.12130

Tai, T. C., Sumaila, U. R., & Cheung, W. W. L. (2021). Ocean acidification amplifies multi-stressor impacts on global marine invertebrate fisheries. *Frontiers in Marine Science*, *8*, 596644. https://doi.org/10.3389/fmars.2021.596644

Tan, K., & Zheng, H. (2020). Ocean acidification and adaptive bivalve farming. *Science of the Total Environment*, *701*, 134794. https://doi.org/10.1016/j.scitotenv.2019.134794

Unzueta-Martínez, A., Downey-Wall, A. M., Cameron, L. P., Ries, J. B., Lotterhos, K. E., & Bowen, J. L. (2021). Ocean acidification alters the diversity and structure of oyster associated microbial communities. *Limnology and Oceanography Letters*, *6*(6), 348–359. https://doi.org/10.1002/lol2.10214

Wei, Y., Ding, D., Gu, T., Xu, Y., Sun, X., Qu, K., Sun, J., & Cui, Z. (2023). Ocean acidification and warming significantly affect coastal eutrophication and organic pollution: A case study in the Bohai Sea. *Marine Pollution Bulletin*, *186*, 114380. https://doi.org/10.1016/j.marpolbul.2022.114380

Wilson, T. J. B., Cooley, S. R., Tai, T. C., Cheung, W. W. L., & Tyedmers, P. H. (2020). Potential socioeconomic impacts from ocean acidification and climate change effects on Atlantic Canadian fisheries. *PLOS One*, *15* (1), e0226544. https://doi.org/10.1371/journal.pone.0226544

Yun, M. S., Sun, J., Lovejoy, C., & Lee, S. H. (2021). Editorial: Microbial response to a rapidly changing marine environment: Global warming and ocean acidification. *Frontiers in Microbiology*, *12*, 731732. https://doi.org/10.3389/fmicb.2021.731732

Zunino, S., Libralato, S., Melaku Canu, D., Prato, G., & Solidoro, C. (2021). Impact of ocean acidification on ecosystem functioning and services in habitat-forming species and marine ecosystems. *Ecosystems*, *24*(7), 1561–1575. https://doi.org/10.1007/s10021-021-00601-3

18 The Heavy Metal Menace
Impact on Marine Microorganisms

Rekha Gaba
Department of Chemistry, DAV University, Jalandhar, India

18.1 INTRODUCTION

The wide open spaces of our seas, which are brimming with life, hide a secret world—a microscopic symphonic orchestra in which innumerable marine microorganisms are essential to preserving the fragile equilibrium of the marine ecosystem. The unseen orchestra members of the marine food chain, bacteria, archaea, and protists, break down organic matter, cycle vital nutrients, and impact global biogeochemical cycles. The vivid tapestry of marine life, from the smallest plankton to the majestic whales, is supported by their persistent work. But the rising flood of heavy metal pollution is causing an increasing amount of dissonance to this delicate balance.

The long-lasting effects of heavy metals including lead, mercury, cadmium, copper, and zinc are caused by a variety of human activities that release them into our waters. These unwanted visitors disturb the well-planned symphony of marine life, in contrast to natural ingredients that are easily integrated into biological processes. It is essential to comprehend how heavy metals mute the essential sounds of marine microbes in order to protect the health of our oceans and the services they offer.

Heavy metal pollution comes from a number of sources, such as mining activities, burning fossil fuels, industrial discharges, and agricultural runoff. These contaminants can dissolve in the water column and build up in marine sediments, where they can easily infiltrate the food chain. At the base of the web, marine microorganisms are the first to respond to this environmental attack. Their high surface area to volume ratio and small stature make them especially vulnerable to the negative effects of heavy metals.

Heavy metals have a wide-ranging and complex effect on marine microorganisms. By imitating necessary components and interfering with the delicate dance of enzymes within the microbial cell, these contaminants behave like cunning intruders. The workhorses of cellular operations, enzymes perform a wide range of functions, from constructing intricate molecules to decomposing nourishment. Heavy metals can throw a wrench in the works and impede essential processes like growth, reproduction, and respiration when they find their way into the enzyme machinery. The entire marine environment may be impacted by this disturbance if microbial populations' total metabolic activity declines.

Beyond interfering with enzymes, heavy metals directly endanger DNA, the building block of life. These impurities have the potential to produce dangerous free radicals, which are extremely reactive chemicals that disrupt genetic material's delicate structure. Mutations in DNA can result in abnormalities in vital biological operations, impeding microorganisms' capacity to proliferate and adjust to shifting environmental circumstances. Moreover, heavy metals have the ability to upset the delicate ion balance in microbial cells, which can result in the loss of vital electrolytes and, eventually, cell death.

The effects of contaminated heavy metals go much beyond the realm of single-celled organisms. A reduction in the variety and quantity of marine microorganisms, the base of the marine food web, has a domino effect. Decreased microbial activity can affect the cycling of nutrients, causing ecological imbalances. From plankton to fish, these imbalances can impair the growth and survival of larger creatures by upsetting the food chain. In the end, the production and health of our seas may suffer significantly if the microbial symphony is silenced.

But throughout this entire depressing story, there is a ray of hope. According to research, some marine microorganisms have amazing adaptations that allow them to survive and even flourish in environments with high concentrations of heavy metals. These hardy microorganisms have developed complex detoxification systems that use techniques like intracellular sequestration, which stores metals securely inside the cell away from critical functions, and cell wall binding, which traps metals outside the cell. The use of living organisms to clean up contaminated surroundings, or bioremediation, is an interesting new application for this innate ability of some bacteria.

Through the utilization of these intriguing microorganisms' metal-resistant characteristics, researchers may be able to create organic remedies for cleaning up marine ecosystems contaminated by heavy metals. Targeted bioremediation techniques can be developed by having a better understanding of the creative detoxifying processes these hardy microorganisms use. Essentially, we can protect the well-being of marine microorganisms and explore the possibility of a healthier and cleaner ocean by researching the effects of heavy metals on the tiny symphony of the ocean.

This introduction explores the vital role that marine microorganisms play in the ecosystem of the ocean, the increasing

DOI: 10.1201/9781003509530-18

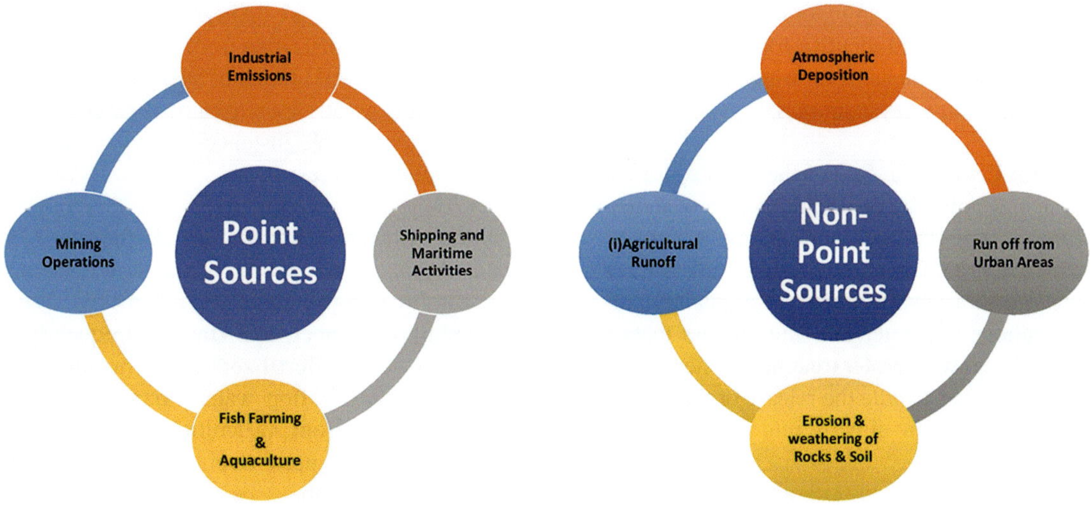

FIGURE 18.1 Different sources of heavy metals in marine ecosystem.

TABLE 18.1
Some Studies at Different Places around the Globe

Sr. No.	Heavy Metals	Area	References
1	Cu, Pb, Zn, Cr, Cd, As, Hg, Ni	NTZK5 core	Hou et al. [1]
2	Cu, Pb, Zn, Cr	Yangtze River estuary	Chen et al. [2]
3	Cu, Zn, Cr, Cd, Hg, Ni, Ba, V	Mediterranean Sea	Robledo Ardila et al. [3]
4	Cu, Pb, Zn, Cr, Cd, As, Hg, Ni	Jiangsu offshore, China	He et al. [4]
5	Cu, Cr, Ni	Karoon water way river, Iran	Diagomanolin et al. [5]
6	Zn, V, Cr, Cu, Pb, Sc, As, Cd	South China Sea	Chen et al. [6]
7	Hg, As, Cu, Pb, Cd, Zn, Cr	Pearl River estuary, Bohai Sea China	Du et al. [7]
8	Cr, Ni, Cu, Zn, Pb, As, Cd, Al, Fe, Mn	South China Sea, Vietnam	Sattarova et al. [8]
9	Cr, Mn, Fe, Co, Ni, Cu, Zn, Hg	Ulhas estuary, Mumbai, India	Thorat et al. [9]
10	Fe, Mn, Cr, Cu, Zn, Cd, Pb	Netravathi-Gurupur estuary, India	Saha et al. [10]
11	Al, As, Cd, Co, Cr, Cu, Fe, Mn, Ni, Pb, Sb, Zn	Musa Estuary, Persian Gulf	Madadi et al. [11]
12	Al, V, Cr, Mn, Fe, Co, Ni, Cu, Zn, As, Se, Sr, Mo, Cd, Hg, Pb	Al-Khobar coastline, Arabian Gulf, Saudi Arabia	Alharbi et al. [12]
13	Cu, Pb, Zn, Cd, Cr	Xincun Bay	Hao et al. [13]
14	Cd, Pd, Cr	Tunisian coasts	Elleuch et al. [14]
15	V, Cr, Ni, Cu, Zn, Cd, Pb	Red Sea, Saudi Arabia	Alnasser et al. [15]
16	As, Ba, Cd, Cr, Cu, Hg, Ni, Pb, Se, V, Zn	Mallorca island (Mediterranean Sea), Spain	Ardila et al. [16]

risk posed by heavy metal pollution, and the various ways in which these pollutants interfere with microbial processes. It also emphasizes how bioremediation can be used as a natural way to lessen the effects of heavy metal contamination.

18.2 SOURCES AND ACCUMULATION OF HEAVY METALS IN THE MARINE ENVIRONMENT

Heavy metal pollution of the environment has become a more pressing problem. According to reports, there could be both planned and unintentional effects on living things from this. Furthermore, the primary activities responsible for environmental degradation are both natural and man-made processes. These actions have the potential to contaminate various environmental domains both directly and indirectly. Moving further, the various sources of heavy metals are well explained in the next sections and shown in Figure 18.1. A lot of studies have been reported on heavy metal accumulation in marine ecosystem. Some of the latest studies are mentioned in Table 18.1.

18.2.1 Point Sources

The recognizable, localized sources from which heavy metals enter the ocean environment are known as point sources of heavy metal marine contamination. Point sources are concentrated and simpler to identify for focused management tactics than non-point sources, which are diffuse and dispersed. Here are some key examples:

i. **Industrial Emissions**: One of the main causes of heavy metal pollution in coastal waters is industrial facilities. Lead, mercury, cadmium, copper, and zinc are just a few of the heavy metals that are released into wastewater as a result of manufacturing operations that often use and process metals [16]. Municipal wastewater treatment systems may be ill-equipped to handle the release of these metals, which can occur directly into rivers or indirectly [17–20].

ii. **Mining Operations**: Onshore and offshore mining activities are major causes of heavy metal pollution. Large amounts of waste rock and tailings are produced during the mining and processing of metals, and these materials frequently include significant concentrations of heavy metals. These have the potential to contaminate rivers, estuaries, and eventually the open ocean by seeping into nearby waterways [21,22]. Furthermore, mishaps such as mine spills may result in abrupt and disastrous discharges of heavy metals, seriously harming the environment.

iii. **Shipping and Maritime Activities:** The worldwide maritime sector contributes to heavy metal pollution in a number of ways. Heavy metals like copper are frequently included in anti-fouling coatings used on ship hulls to stop the growth of barnacles and other organisms. Heavy metals are released into the water column by these paints when they deteriorate, especially in ports and shipping lanes [23–25]. Moreover, unintentional releases of ballast water, which is utilized to keep ships stable, may bring in heavy metals from other areas, endangering local ecosystems.

iv. **Fish Farming and Aquaculture:** Aquaculture adds to the pollution of heavy metals even though it provides a sustainable source of fish and shellfish. Anti-fouling coatings are frequently used on nets and cages in fish farming operations, which can cause heavy metals like copper to leak into the nearby water [21]. Heavy metals in the water and sediments can also be a result of the use of antibiotics and other chemicals in aquaculture.

18.2.2 Non-Point Source

The origins of heavy metal pollution in the water can be easily identified as point sources, such as mining operations and industrial discharges, while non-point sources pose a more subtle threat. These diffuse sources are hard to locate and manage since they come from a larger geographic area. Nonetheless, the total burden of heavy metals in the marine environment is greatly increased by their combined action. Here are some notable instances of heavy metal pollution from non-point sources:

i. **Agricultural Runoff:** One of the main causes of heavy metal pollution that is not from point sources is agriculture. Heavy metals like lead, cadmium, copper, and zinc are frequently found in pesticides, fertilizers, and manure [26,27]. Rain and irrigation water carry these contaminants away from agricultural land, where they eventually end up in rivers and finally make their way to the ocean.

ii. **Atmospheric Deposition:** By air deposition, heavy metals can also find their way into the marine ecosystem. Heavy metals are released into the atmosphere via burning fossil fuels, burning coal, and industrial pollutants. Rain or dry fallout can then carry these airborne pollutants over great distances and deposit them on the ocean's surface [28].

iii. **Runoff from Urban Areas:** An additional important source of non-point source heavy metal pollution is stormwater runoff from urban areas. With their web of streets, parking lots, and rooftops, urban environments gather pollutants such as brake linings, tire dust, and air deposition. Heavy metals are eventually carried into storm drains, rivers, and the ocean by this contaminated runoff during rainy seasons [29–32].

iv. **Erosion and Weathering of Rocks and Soil:** Natural heavy metal discharge into the marine environment can also occur as a result of soil and rock weathering and erosion. On the other hand, human actions like land use changes and deforestation can quicken these processes, which increases the mobilization of heavy metals [33].

18.3 MECHANISMS OF HEAVY METAL CONSUMPTION BY MICROORGANISMS

Microorganisms employ diverse strategies to combat heavy metal toxicity. One primary mechanism involves the uptake of heavy metals through transporters located on the cell membrane. Once inside the cell, heavy metals can bind to cellular components like proteins, DNA, and enzymes. Some microorganisms can sequester heavy metals in specialized compartments or expel them from the cell through efflux pumps. Additionally, certain microorganisms possess enzymatic systems that can convert heavy metals into less toxic or more easily removable forms.

18.4 MULTIFACETED DETRIMENTAL EFFECTS ON MICROORGANISMS

Heavy metals have extensive and diverse negative effects on marine microbes. These contaminants impede the basic basis of microbial existence by interfering with a number of cellular functions. Here, we examine a few of the main processes that heavy metals use to cause harm.

18.4.1 Interference with Enzyme Activity

Enzymes are essential biological catalysts that drive various metabolic processes within microorganisms. Heavy metals can interfere with enzyme activity through multiple mechanisms. One common way is by directly binding to the enzyme's active site, inhibiting its function. Moreover, heavy metals can disrupt enzyme structure and stability, leading to denaturation and loss of activity. Furthermore, heavy metals can induce the generation of reactive oxygen species (ROS) within the cell, causing oxidative damage to enzymes and other cellular components. The workhorses of cellular operations, enzymes perform a wide range of functions, including as waste clearance, nutrition uptake, and energy production. Enzyme activity can be affected by heavy metals in two main ways. First of all, they misplace themselves within the structures of enzymes by imitating necessary elements such as iron or zinc. The form and function of the enzyme are disrupted by this displacement, making it more difficult for it to attach to substrates and catalyze essential activities [34]. Second, heavy metals have the ability to bind directly to crucial functional groups in enzymes, further reducing their activity. The inhibition of growth, reproduction, and general survival of marine microorganisms can result from this disturbance of enzyme function, which can have a domino effect on metabolic processes.

18.4.2 Disruption of Cellular Membrane Integrity

The microbial cell's cell membrane regulates the passage of nutrients and waste materials, serving as the gatekeeper of the cell. There are various ways in which heavy metals might compromise this essential barrier's integrity. They have the ability to change the phospholipid makeup of the membrane, which increases permeability and allows vital cellular components to seep out [35]. Moreover, heavy metals have the ability to interact with membrane proteins, impairing their functionality and the membrane's ability to permeabilize selectively. The ability of microorganisms to maintain cellular homeostasis may be compromised as a result of this disruption, which may also result in the entry of toxic substances and the loss of vital electrolytes [36].

The process by which heavy metal absorption causes cellular damage and dysfunction in bacteria is complex and involves multiple pathways. It is well recognized that heavy metals, including arsenic, cadmium, lead, and mercury, are poisonous to microorganisms. When they enter microbial cells, they can have a significant impact on the viability, metabolism, and structure of the cells. Determining the effects of heavy metal pollution on microbial populations and ecosystem health requires an understanding of these pathways. Below, we look at some of the main ways that microbes' cells are disrupted by heavy metal absorption (Figure 18.2).

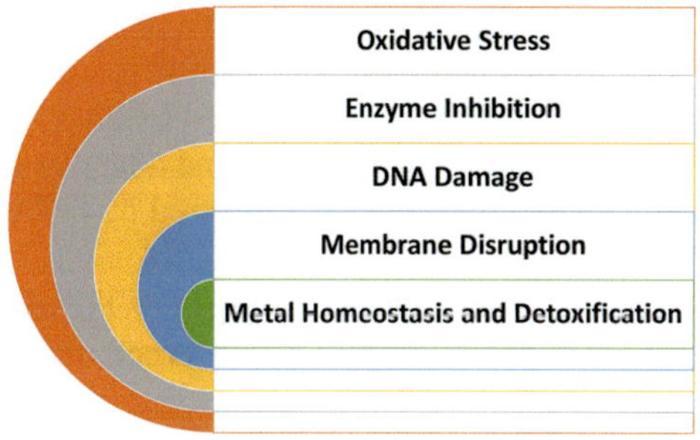

FIGURE 18.2 Different means of disruption of cell of microorganism by heavy metal absorption.

Inducing oxidative stress [37] is one of the main ways that heavy metals damage microbial cells. ROS are produced by heavy metals and can cause oxidative damage to DNA, lipids, and proteins in cells. When ROS interact with cellular constituents, structural changes and functional degradation can result. This oxidative stress has the potential to impair vital biological functions and cause cell death.

Heavy metals can attach to an enzyme's active site [38], preventing it from functioning and interfering with metabolic processes. Heavy metal inhibition can affect a lot of enzymes that are important in biological activities like respiration, photosynthesis, and DNA replication. Heavy metals can hinder cellular metabolism and energy production by interfering with enzyme function, which can stunt microorganism growth and viability.

Cell structure and function depend on the integrity of the cell membrane. By interacting with membrane lipids and proteins, heavy metals can affect the fluidity and permeability of membranes, disrupting their integrity. This disturbance may impair the efficient passage of ions and molecules across the membrane, upset the equilibrium of osmotic pressure, and eventually result in the lysis or leakage of cell contents.

18.4.3 Mutagenesis and Damage to DNA

When heavy metals come into contact with DNA molecules, they can cause damage such as cross-linking and strand breakage. This DNA damage can cause mutagenesis and genomic instability by interfering with transcription and replication mechanisms. In severe situations, DNA damage [39] caused by heavy metals can set off pathways leading to programmed cell death or cause mutations that could change a species' phenotype or decrease its fitness.

Beyond the cellular barrier, heavy metals have deleterious effects that directly affect DNA, the genetic code that codes for all life. Harmful free radicals, which are extremely reactive chemicals that disrupt genetic material's delicate structure, can be produced by heavy metals. Microorganisms' capacity to flourish and adapt to shifting environmental conditions can be hampered by DNA mutations, which can cause problems in vital biological operations. Moreover, several heavy metals have the ability to physically bind with DNA, which prevents cell division and replication.

18.4.4 Undermining Microbial Diversities

Heavy metals harm marine microorganisms not just on an individual basis but also on the level of the entire microbial population. The degree to which different microbial species are sensitive to heavy metals varies. The overall variety of microbes can decrease as a result of sensitive species being selectively eliminated by heavy metal pollution [40]. This loss of diversity may have significant effects on how the marine environment functions. Microbial communities are essential to the decomposition, cycling of nutrients, and biogeochemical processes. These essential processes may be interfered with by a decrease in microbial diversity, which could have an effect on the whole food web.

18.4.5 Effect on Ecosystem Function and Microbial Growth

The basic processes of microbial life, growth, and reproduction are affected by the harmful impacts of heavy metals. Microorganisms' capacity to obtain nutrients, produce energy, and proliferate is directly hampered by disruptions to enzyme activity and cellular membrane integrity. Moreover, heavy metal-induced DNA damage can result in mutations that impair vital cellular functions, further impeding growth and reproduction.

Ecological imbalances can result from heavy metal-induced declines in the variety and abundance of microbial communities, which can upset important biogeochemical cycles including the carbon and nitrogen cycles. Decreased microbial activity in the soil can prevent organic matter from breaking down, which might result in an accumulation of nutrients that are inaccessible to plants. Reduced microbial activity in aquatic habitats can interfere with the water column's nutrient cycle, which will ultimately influence primary productivity and the food chain as a whole.

18.4.6 Biomagnification and Bioaccumulation

The bioaccumulation and biomagnification of heavy metals is a serious concern as well [41]. Heavy metals are absorbed by microorganisms from their surroundings. Heavy metal concentrations in the food chain result from the consumption of these species by higher trophic level organisms, and they may eventually reach dangerous levels in top predators and even humans, who eat seafood for protein.

18.5 BIOREMEDIATION: A MICROBIAL MARVEL

Despite the detrimental effects, some marine microorganisms possess remarkable adaptations to tolerate and even thrive in environments with high heavy metal concentrations. These microbes have evolved sophisticated mechanisms for detoxification, including binding metals to their cell walls or sequestering them within the cell. This inherent ability of certain microbes paves the way for a promising approach called bioremediation. By harnessing the metal-resistant properties of these microorganisms, scientists can potentially develop natural solutions to clean up heavy metal-contaminated marine environments.

18.6 CONCLUSION

Heavy metals pose a significant threat to the health and function of marine microorganisms. However, research into these resilient microbes also reveals a glimmer of hope. Their ability to tolerate and even detoxify heavy metals offers exciting possibilities for bioremediation. Further research on these fascinating organisms and the development of effective bioremediation strategies are crucial for safeguarding the future of our oceans.

18.7 FUTURE DIRECTIONS

Research on the interaction between heavy metals and marine microorganisms is ongoing. Future efforts should focus on:

- Identifying novel heavy metal-resistant microbes with high bioremediation potential.
- Elucidating the detoxification mechanisms employed by these microbes.
- Developing large-scale bioremediation strategies for real-world application.

By combining a deeper understanding of the problem with the power of microbial solutions, we can work toward a future where our oceans are free from the detrimental effects of heavy metal contamination.

REFERENCES

1. Hou, Z., Li, J., Cheng, H., Sun, Y., Li, B., Zhang, Y., & Li, H. (2024, March). Temporal variation and pollution status of heavy metals in core sediments from Jiangsu intertidal area, China. *Marine Pollution Bulletin*, *200*, 116111. https://doi.org/10.1016/j.marpolbul.2024.116111.
2. Chen, B., Liu, J., Qiu, J., Zhang, X., Wang, S., & Liu, J. (2017, March). Spatio-temporal distribution and environmental risk of sedimentary heavy metals in the Yangtze River Estuary and its adjacent areas. *Marine Pollution Bulletin*, *116*(1–2), 469–478. https://doi.org/10.1016/j.marpolbul.2016.11.055.

3. Robledo Ardila, P. A., Álvarez-Alonso, R., Árcega-Cabrera, F., Durán Valsero, J. J., Morales García, R., Lamas-Cosío, E., Oceguera-Vargas, I., & DelValls, A. (2024, February). Assessment and review of heavy metals pollution in sediments of the Mediterranean Sea. *Applied Sciences, 14*(4), 1435. https://doi.org/10.3390/app14041435
4. He, L., Jiang, S., Chen, X., Cheng, H., Ye, P., Chen, L., Liu, Y., Yin, Y., & Leng, X. (2023, December). Historical trends and pollution assessment of heavy metals in core sediments from the Jiangsu offshore area, China. *Regional Studies in Marine Science, 65*, 103112. https://doi.org/10.1016/j.rsma.2023.103112.
5. Diagomanolin, V., Farhang, M., Ghazi-Khansari, M., & Jafarzadeh, N. (2004, June). Heavy metals (Ni, Cr, Cu) in the Karoon waterway river, Iran. *Toxicology Letters, 151*(1), 63–67. https://doi.org/10.1016/j.toxlet.2004.02.018.
6. Chen, F., Mao, S., Li, G., Tian, Y., Miao, L., Xu, W., Zhu, X., & Yan, W. (2024, March). Anthropogenic multipollutant input to the offshore South China Sea. *Science of the Total Environment, 916*, 170228. https://doi.org/10.1016/j.scitotenv.2024.170228.
7. Du, M., Zheng, M., Liu, A., Wang, L., Pan, X., Liu, J., & Ran, X. (2022, August). Effects of emerging contaminants and heavy metals on variation in bacterial communities in estuarine sediments. *Science of the Total Environment, 832*, 155118. https://doi.org/10.1016/j.scitotenv.2022.155118.
8. Sattarova, V., Aksentov, K., Alatortsev, A., Duc Luong, L., Shakirov, R., Ivanov, M., & Legkodimov, A. (2021, December). Distribution and contamination assessment of trace metals in surface sediments of the South China Sea, Vietnam. *Marine Pollution Bulletin, 173*, 113045. https://doi.org/10.1016/j.marpolbul.2021.113045.
9. Thorat, B. R., Prasad, P., & Ram, A. (2023, June). Heavy metal accumulation in a moderately polluted Ulhas estuary, Western India. *Regional Studies in Marine Science, 60*, 102818. https://doi.org/10.1016/j.rsma.2023.102818.
10. Saha, A., Das, B., Sarkar, D., Samanta, S., Vijaykumar, M., Khan, M. F., Kayal, T., Jana, C., Kumar, V., Gogoi, P., & Chowdhury, A. R. (2024, February). Trace metals and pesticides in water-sediment and associated pollution load indicators of Netravathi-Gurupur estuary, India: implications on coastal pollution. *Marine Pollution Bulletin, 199*, 115950. https://doi.org/10.1016/j.marpolbul.2023.115950.
11. Madadi, R., Mejjad, N., & De la Torre, G. E. (2023, May). Geochemical speciation, ecological risk, and source identification of heavy metal(loid)s in sediments and waters from Musa Estuary, Persian Gulf. *Marine Pollution Bulletin, 190*, 114836. https://doi.org/10.1016/j.marpolbul.2023.114836.
12. Alharbi, T., Alfaifi, H., & El-Sorogy, A. (2017, June). Metal pollution in Al-Khobar seawater, Arabian Gulf, Saudi Arabia. *Marine Pollution Bulletin, 119*(1), 407–415. https://doi.org/10.1016/j.marpolbul.2017.03.011.
13. Hao, Z., Qian, J., Zheng, F., Lin, B., Xu, M., Feng, W., & Zou, X. (2024, February). Human-influenced changes in pollution status and potential risk of sediment heavy metals in Xincun Bay, a typical lagoon of Hainan, China. *Marine Pollution Bulletin, 199*, 115949. https://doi.org/10.1016/j.marpolbul.2023.115949.
14. Elleuch, J., Hmani, R., Drira, M., Michaud, P., Fendri, I., & Abdelkafi, S. (2021, December). Potential of three local marine microalgae from Tunisian coasts for cadmium, lead and chromium removals. *Science of the Total Environment, 799*, 149464. https://doi.org/10.1016/j.scitotenv.2021.149464.
15. Alnasser, A. S., Halawani, R. F., & Quicksall, A. N. (2024, January). Assessment of heavy metal distribution, risk, and sourcing in mangrove sediments from three Saudi Arabian Red Sea locations. *Marine Pollution Bulletin, 198*, 115821. https://doi.org/10.1016/j.marpolbul.2023.115821.
16. Ardila, P. A. R., Alonso, R. L., Valsero, J. J. D., García, R. M., Cabrera, F. R., Cosío, E. L., & Laforet, S. D. (2023, January). Assessment of heavy metal pollution in marine sediments from southwest of Mallorca island, Spain. *Environmental Science and Pollution Research, 30*(7), 16852–16866. https://doi.org/10.1007/s11356-022-25014-0.
17. Gworek, B., Bemowska-Kałabun, O., Kijeńska, M., & Wrzosek-Jakubowska, J. (2016, September). Mercury in Marine and oceanic waters—a review. *Water, Air, & Soil Pollution, 227*(10), 371. https://doi.org/10.1007/s11270-016-3060-3.
18. Lv, J., Hu, R., Wang, N., Zhu, L., Zhang, X., Yuan, X., & Liu, B. (2021, September). Distribution and movement of heavy metals in sediments around the coastal areas under the influence of multiple factors: a case study from the junction of the Bohai Sea and the Yellow Sea. *Chemosphere, 278*, 130352. https://doi.org/10.1016/j.chemosphere.2021.130352.
19. Liu, G., Yu, Y., Hou, J., Xue, W., Liu, X., Liu, Y., Wang, W., Alsaedi, A., Hayat, T., & Liu, Z. (2014, December). An ecological risk assessment of heavy metal pollution of the agricultural ecosystem near a lead-acid battery factory. *Ecological Indicators, 47*, 210–218. https://doi.org/10.1016/j.ecolind.2014.04.040.
20. Niu, L., Li, J., Luo, X., Fu, T., Chen, O., & Yang, Q. (2021, December). Identification of heavy metal pollution in estuarine sediments under long-term reclamation: ecological toxicity, sources and implications for estuary management. *Environmental Pollution, 290*, 118126. https://doi.org/10.1016/j.envpol.2021.118126.
21. Turner, A. (2010, February). Marine pollution from antifouling paint particles. *Marine Pollution Bulletin, 60*(2), 159–171. https://doi.org/10.1016/j.marpolbul.2009.12.004.
22. Fernando, W. A. M., Ilankoon, I., Syed, T. H., & Yellishetty, M. (2018, March). Challenges and opportunities in the removal of sulphate ions in contaminated mine water: a review. *Minerals Engineering, 117*, 74–90. https://doi.org/10.1016/j.mineng.2017.12.004.
23. Ali, H., Khan, E., & Sajad, M. A. (2013, May). Phytoremediation of heavy metals—concepts and applications. *Chemosphere, 91*(7), 869–881. https://doi.org/10.1016/j.chemosphere.2013.01.075.
24. Singh, R., Singh, S., Parihar, P., Singh, V. P., & Prasad, S. M. (2015, February). Arsenic contamination, consequences and remediation techniques: a review. *Ecotoxicology and Environmental Safety, 112*, 247–270. https://doi.org/10.1016/j.ecoenv.2014.10.009.
25. Pozo-Antonio, S., Puente-Luna, I., Lagüela-López, S., & Veiga-Ríos, M. (2014, August). Techniques to correct and prevent acid mine drainage: a review. *DYNA, 81*(186), 73. https://doi.org/10.15446/dyna.v81n186.38436.26.
26. Goncalves, A. C., Nacke, H., Schwantes, D., & Coelho, G. F. (2014, March). Heavy metal contamination in Brazilian Agricultural soils due to application of fertilizers. *Environmental Risk Assessment of Soil Contamination*. https://doi.org/10.5772/57268.
27. Kumar, S., Islam, R., Akash, P. B., Khan, M. H. R., Proshad, R., Karmoker, J., & MacFarlane, G. R. (2022, June). Lead (Pb) contamination in agricultural products

and human health risk assessment in Bangladesh. *Water, Air, & Soil Pollution*, *233*(7), 257. https://doi.org/10.1007/s11270-022-05711-9.
28. Nriagu, J. O., & Pacyna, J. M. (1988, May). Quantitative assessment of worldwide contamination of air, water and soils by trace metals. *Nature*, *333*(6169), 134–139. https://doi.org/10.1038/333134a0.
29. Vithanage, M., Bandara, P. C., Novo, L. A., Kumar, A., Ambade, B., Naveendrakumar, G., Ranagalage, M., & Magana-Arachchi, D. N. (2022, September). Deposition of trace metals associated with atmospheric particulate matter: environmental fate and health risk assessment. *Chemosphere*, *303*, 135051. https://doi.org/10.1016/j.chemosphere.2022.135051.
30. Zwolak, A., Sarzyńska, M., Szpyrka, E., & Stawarczyk, K. (2019, July). Sources of soil pollution by heavy metals and their accumulation in vegetables: a review. *Water, Air, & Soil Pollution*, *230*(7), 164. https://doi.org/10.1007/s11270-019-4221-y.
31. Aonghusa, C. N., & Gray, N. F. (2002, January). Laundry detergents as a source of heavy metals in Irish domestic wastewater. *Journal of Environmental Science and Health, Part A*, *37*(1), 1–6. https://doi.org/10.1081/ese-100108477.
32. Zhang, L., Zhu, G., Ge, X., Xu, G., & Guan, Y. (2018, October). Novel insights into heavy metal pollution of farmland based on reactive heavy metals (RHMs): pollution characteristics, predictive models, and quantitative source apportionment. *Journal of Hazardous Materials*, *360*, 32–42. https://doi.org/10.1016/j.jhazmat.2018.07.075.
33. Vareda, J. P., Valente, A. J., & Durães, L. (2019, September). Assessment of heavy metal pollution from anthropogenic activities and remediation strategies: a review. *Journal of Environmental Management*, *246*, 101–118. https://doi.org/10.1016/j.jenvman.2019.05.126.
34. Balali-Mood, M., Naseri, K., Tahergorabi, Z., Khazdair, M. R., & Sadeghi, M. (2021, April). Toxic mechanisms of five heavy metals: mercury, lead, chromium, cadmium, and arsenic. *Frontiers in Pharmacology*, *12*. https://doi.org/10.3389/fphar.2021.643972.
35. Yasutake, A., Sawada, M., Shimada, A., Satoh, M., & Tohyama, C. (2004, May). Mercury accumulation and its distribution to metallothionein in mouse brain after sub-chronic pulse exposure to mercury vapor. *Archives of Toxicology*, *78*(9), 489–495. https://doi.org/10.1007/s00204-004-0572-1.
36. Assefa, S., & Köhler, G. (2020, February). Intestinal microbiome and metal toxicity. *Current Opinion in Toxicology*, *19*, 21–27. https://doi.org/10.1016/j.cotox.2019.09.009
37. Khalid, M., Hassani, S., & Abdollahi, M. (2020, April). Metal-induced oxidative stress: an evidence-based update of advantages and disadvantages. *Current Opinion in Toxicology*, *20–21*, 55–68. https://doi.org/10.1016/j.cotox.2020.05.006.
38. Ospina, E., Alonso, B., Casado, C. M., & Armada, M. P. G. (2023, November). Electrocatalytic self-assembled multilayer structures based on thiolated Fc-DAB dendrimers: determination of heavy metal ions by enzymatic inhibition. *Catalysis Today*, *423*, 114293. https://doi.org/10.1016/j.cattod.2023.114293.
39. Zhou, S., Wei, C., Liao, C., & Wu, H. (2008, January). Damage to DNA of effective microorganisms by heavy metals: impact on wastewater treatment. *Journal of Environmental Sciences*, *20*(12), 1514–1518. https://doi.org/10.1016/s1001-0742(08)62558-9.
40. Xie, Y., Fan, J., Zhu, W., Amombo, E., Lou, Y., Chen, L., & Fu, J. (2016, May). Effect of heavy metals pollution on soil microbial diversity and bermudagrass genetic variation. *Frontiers in Plant Science*, *7*, 755. https://doi.org/10.3389/fpls.2016.00755.
41. Danovaro, R., Cocozza di Montanara, A., Corinaldesi, C., Dell'Anno, A., Illuminati, S., Willis, T. J., & Gambi, C. (2023, November). Bioaccumulation and biomagnification of heavy metals in marine micro-predators. *Communications Biology*, *6*(1), 1206. https://doi.org/10.1038/s42003-023-05539-x.

19 Delving Deep into the Microplastics Crisis in the Indian Aquatic Environment
An Overview

Anantpreet Kaur[1], Harpreet Walia[1], and Samriti Khosla[2]
[1]Department of Botany and Environment Studies, DAV University, Jalandhar, India
[2]Department of Physics, DAV University, Jalandhar, India

19.1 INTRODUCTION

The Earth's oceans cover over 70% of its surface and are vital in supporting life on the planet. However, in recent decades marine ecosystems have faced unusual challenges due to human activities with the rise of microplastics (MPs) emerging as a particularly alarming issue. MPs are defined as plastic particles less than 5 mm which have become pervasive in marine environments globally posing significant risks to aquatic life, ecosystems, as well as human health (Gasperi *et al*., 2018; Ebere *et al*., 2019; Veerasingam *et al*., 2020).

India is experiencing a rapid growth in its plastic industry, especially in western regions. These regions account for 47% of consumption and generate approximately 26 million metric tons annually, solidifying India's important role in the global plastic waste generation community (Vaid *et al*., 2021). The widespread utilization of diverse synthetic polymers results in a significant amount of non-biodegradable waste, ultimately forming MPs in aquatic ecosystems (Gupta *et al*., 2021).

Nonetheless, the roots of this issue trace back several decades. The first observations of plastic pollution in the ocean date back to 1972, when researchers made unexpected findings in the western North Atlantic. These early observations revealed the presence of preproduction pellets and degraded fragments in plankton tows, marking the foundation of the understanding of plastic's impact on marine ecosystems. Later studies in the North Pacific and South Atlantic regions further highlighted the occurrence of plastic debris in the oceans (Eriksen *et al*., 2018). Since then, several investigations into oceanic plastic debris have been conducted (Veerasingam *et al*., 2020). MPs are formed through the breakdown of larger primary MPs and consist of various forms such as fragments, fibers, pellets, films, and spheres (Hale *et al*., 2020; Selvam *et al*., 2021). The pharmaceutical and cosmetics industries are contributors to the presence of spheres in aquatic environments. On the contrary, secondary MPs emerge from the gradual degradation or fragmentation of larger plastic items due to UV radiation, mechanical wear from waves, or biological breakdown by microorganisms (Selvam *et al*., 2021).

Recent research has revealed the presence of MPs trapped in sea ice, deposited along shorelines globally and scattered across the sea floor. Synthetic fibers have been observed precipitating from the skies. These findings collectively indicate large-scale contamination across diverse ecosystems (Eriksen *et al*., 2018). Thus, MPs are prevalent in all environmental matrices ranging from the Arctic to Antarctica, polluting seawater, sediments, rivers, soil, and even the air we breathe (Prata *et al*., 2019). They serve as an intermediate stage between macroscopic plastic debris and nanoscale materials less than 1 μm, with smaller particles exhibiting a 'Trojan Horse' effect, posing greater threats due to their potential to introduce significant amounts of metals and metalloids into the food chain (Hale *et al*., 2020; Hildebrandt *et al*., 2021).

India's vast coastline, stretching over 7,500 km, serves as a center for activities such as fishing, shipping, tourism, and recreation. However, the combination of a dense population, rising plastic consumption, and an extensive river network has significantly heightened the threat of MP pollution in aquatic environments. While India contributes considerably to global plastic waste, research specifically addressing MP pollution within its aquatic ecosystems remains limited. This highlights the urgent need to review existing studies and identify critical knowledge gaps to guide future research efforts (Vanapalli *et al*., 2021).

In this chapter, we will explore the composition, sources, distribution, and impacts of MPs, as well as the strategies to mitigate MP pollution, with a particular focus on their implications for the Indian aquatic ecosystem.

19.2 COMPOSITION OF MPs

The composition of MPs varies significantly, reflecting their origin, manufacturing processes, and environmental degradation. Analytical methods like Fourier Transform Infrared

DOI: 10.1201/9781003509530-19

Spectroscopy (FTIR) have proven effective in identifying the primary polymer types present in Indian aquatic ecosystems. Research has shown that the dominant MP materials on Indian coastlines include polyethylene, polystyrene, and polypropylene (Karthik *et al.*, 2018). In contrast, sediments from the Ganga River predominantly contain polyethylene and polyethylene terephthalate (PET) (Sarkar *et al.*, 2019).

Studies on groundwater pollution in Chennai and Tamil Nadu revealed the presence of PET alongside polyamide (PA), highlighting the penetration of MPs into deeper water sources (Ganesan *et al.*, 2019). Using advanced Micro-Fourier Transform Infrared Spectroscopy (μ-FTIR), Gupta *et al.* (2021) identified an extensive range of polymers in estuarine MP samples. This analysis detected approximately 33 polymer types, including polyacrylamide (PAM), polyacetylene, PA, polyvinyl pyrrolidone (PVP), polyvinyl chloride (PVC), and polyimide (PI).

A thorough understanding of MP composition is vital for determining their origins, evaluating their behavior in aquatic environments, and predicting their ecological consequences. Also, identifying the most common plastic polymers is a key step toward formulating effective strategies to mitigate MP contamination in India's water bodies.

19.3 SOURCES OF MPs

The degradation of plastics through natural (biotic) and environmental (abiotic) processes generates plastic particles at microscopic scales, ranging from MPs (≤5 mm) to nanoplastics (≤1 μm) (Trevisan *et al.*, 2020). MPs originate from diverse sources, each contributing to their prevalence in aquatic environments. One of the primary contributors to MP pollution in estuarine ecosystems is the breakdown of larger plastic debris, often increased by mismanaged effluents from sewage treatment plants and washing machines (Gupta *et al.*, 2021). These effluents release significant quantities of MPs into estuarine systems, facilitating their accumulation and dispersion.

Other sources of MPs include the degradation of synthetic fibers from textiles, direct emissions from industrial activities, and the fragmentation of plastic waste in landfills and marine environments. Microbeads from personal care products and plastic litter from terrestrial activities, including runoff and improper waste disposal, further intensify MP pollution in aquatic ecosystems. Simultaneously, industrial effluents contribute to the release of heavy metals such as chromium, arsenic, copper, aluminum, and iron into aquatic ecosystems compounding environmental contamination (Khalid *et al.*, 2021). Coastal and marine ecosystems face additional pressures from rivers, stormwater, agricultural runoff, industrial and sewage discharges, shipping operations, fishing activities, and littering, all of which significantly escalate plastic pollution (Sivadas *et al.*, 2022).

The Ganges River, one of India's major water bodies, is the second largest contributor to plastic pollution globally, releasing approximately 105,000 tons of plastic into the Indian Ocean annually (Lechthaler *et al.*, 2021). Supporting this, global models indicate that rivers like the Ganges play a pivotal role in transporting MPs into the ocean (Veerasingam *et al.*, 2020). While the Ganges-Brahmaputra River system ranks sixth in global plastic waste contributions, recent studies emphasize that factors such as land use, proximity to waste generation points, and coastline length have a more pronounced influence on plastic pollution than river basin size. Interestingly, smaller rivers with higher pollution loads often play a disproportionate role in discharging plastic waste into the ocean, positioning Indian rivers as the second largest contributors to global marine plastic pollution (Sivadas *et al.*, 2022).

An estimated 1–3 billion MP particles are discharged daily into the Bay of Bengal, with concentrations progressively increasing from river sources to the sea. However, the variation in MP concentrations along the course of rivers remains poorly understood. Addressing this knowledge gap is essential for identifying key sources and pathways of MP pollution and implementing effective management strategies (Napper *et al.*, 2021).

19.4 DISTRIBUTION AND FATE OF MPs

MPs are distributed globally through fragmentation and transportation processes, facilitated by wind and water currents. These processes transfer plastics from inland freshwater systems, such as rivers and lakes, to the open ocean, where they ultimately accumulate along coastlines or settle on the seafloor (Eriksen *et al.*, 2018). The levels of MPs in aquatic environments vary widely, ranging from a few particles to several thousand particles per cubic meter. Modeling studies indicate that MPs measuring 0.333–4.75 mm in size represent over 90% of the plastic particles present in surface ocean waters, highlighting the predominance of this size range in marine ecosystems (Trevisan *et al.*, 2020).

Once introduced into marine environments, MPs undergo a dynamic journey involving transport, distribution, and degradation. This complex interplay is illustrated in Figure 19.1, which depicts the pathways linking terrestrial sources to marine ecosystems, emphasizing the interconnectedness of land-based and aquatic environments in the dispersion of MPs. Global studies have documented the presence and characteristics of MPs across diverse environments, including Rameswaram Coral Island in Tamil Nadu; coastal waters near Kuala Nerus and Kuantan Port in Malaysia; drinking water treatment facilities in Germany; freshwater systems like Vembanad Lake in Kerala and Lake Huron in Canada; and urban estuaries in China (Ganesan *et al.*, 2019).

Recent advances in research suggest that the prevalence of MPs, particularly smaller particles below 10 μm, has been significantly underestimated. Concentrations of these tiny particles can range from millions to tens of millions per cubic meter in marine waters, both offshore and nearshore. Such findings mark 5–7 orders of magnitude increase in previously reported MP levels, underscoring the growing concern about nanoplastics as a critical yet often overlooked component of plastic pollution (Trevisan *et al.*, 2020).

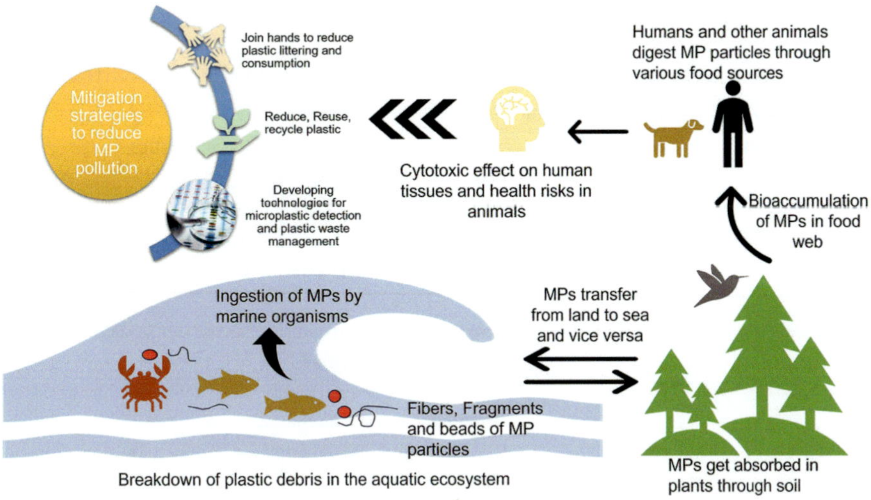

FIGURE 19.1 Microplastic pathways – land to sea interactions and mitigation strategies to reduce microplastic pollution.

Nanoplastics owing to their minuscule size are particularly challenging to detect and remove. Their small size facilitates ingestion by a wide range of organisms, penetration into tissues, trophic transfer through food webs, and interactions with organic matter and other pollutants. These factors make nanoplastics a potentially hazardous component of marine litter (Trevisan *et al.*, 2020). Despite their likely significance, nanoplastics remain one of the least studied aspects of plastic pollution, demanding further investigation into their accumulation in organisms and associated ecological and health impacts.

The persistence of MPs in the marine environment lasting hundreds to thousands of years amplifies their long-term threat to ecosystems. From zooplankton to apex predators, numerous marine species ingest MPs, which can bioaccumulate and biomagnify through the food web.

India with its extensive network of rivers, lakes, and coastlines faces significant challenges from MP pollution. Densely populated regions and industrial activities exacerbate the problem, leading to the widespread presence of MPs in aquatic systems. Research has highlighted notable concentrations of MPs in Indian waters. For instance, studies on the Ganga River sediment revealed that river discharge serves as a key pathway for MP transport, effectively carrying them from terrestrial sources to marine environments (Sarkar *et al.*, 2019; Amrutha and Warrier, 2020). Table 19.1 provides an overview of MP distributions in various regions of India, shedding light on their prevalence in marine ecosystems. These findings call for urgent measures to address the issue of MPs in the country and beyond.

19.5 IMPACT OF MPs

The widespread presence of MPs in marine ecosystems poses significant ecological, biological, and socio-economic impacts. Research by Saha *et al.* (2021) identified MPs in shellfish and finfish species from the Sal Estuary, Goa, raising concerns about their potential impacts on human health and ecological stability. The study highlighted that smaller-sized MPs (10–30 μm) were predominantly found in marine organisms, with key polymers such as PAM, polyacetylene, and ethylene vinyl alcohol being the most common. This contamination, particularly in seafood consumed locally, enhances the risks posed to human health and marine biodiversity. MPs are known to cause physical injuries, digestive tract blockages, and disruptions in feeding behaviors in marine life (Veerasingam *et al.*, 2020). These findings stress the critical need for monitoring and reducing MP pollution to protect marine ecosystems and safeguard the seafood industry (Saha *et al.*, 2021).

MPs also act as vectors for harmful pollutants in aquatic environments, exhibiting a strong affinity for hydrophobic organic contaminants like polycyclic aromatic hydrocarbons (PAHs) (Trevisan *et al.*, 2020). Fibrous MPs are particularly concerning as they adsorb PAHs, which are associated with genotoxic effects (Gasperi *et al.*, 2018). Moreover, MPs have been found to transport heavy metals and persistent organic pollutants (POPs) through food webs, amplifying their toxicological effects (Veerasingam *et al.*, 2020). This phenomenon, often referred to as the 'Trojan Horse' effect, is especially problematic for filter-feeding organisms, which are at greater risk of ingesting MPs and their associated contaminants (Trevisan *et al.*, 2020). Studies also indicate that exposure to certain polymers, such as PVC and PE, can impair the immune system of fish (Iri *et al.*, 2021).

MPs not only harm marine organisms physically but also serve as platforms for pollutant accumulation. For example, heavy metals have been detected on MP surfaces using Scanning Electron Microscopy with Energy-Dispersive X-ray Spectroscopy (SEM-EDX) (Ganesan *et al.*, 2019). Recent advancements in Inductively Coupled Plasma Mass spectrometry (ICP-MS) have further facilitated the analysis of metal distribution in MPs, providing valuable insights into how metals may enter and move through marine food webs (Hildebrandt *et al.*, 2021).

TABLE 19.1
Microplastic Concentrations in Various Aquatic Environments across India

Location/Region	Type of water body	Average microplastic concentration	Notable characteristics	References
Veeranam Lake, Tamil Nadu	Lake	28 items/km^2	Freshwater lake, relatively low concentration; influenced by local fishing and agricultural activities.	Bharath *et al.* (2021)
Vembanad Lake, Kerala		252.80±25.76 particles/m^2	High tourism and fisheries activity; microfibers from domestic waste predominant.	Sruthy and Ramasamy (2017)
Kosasthalaiyer River, Tamil Nadu	River	0.67 particle/L	River near urban and industrial areas; significant input from local runoff and urbanization.	Lechthaler *et al.* (2021)
Adyar River, Tamil Nadu	River	0.33 particle/L	Located near Chennai, highly urbanized area with considerable runoff and sewage discharge.	Lechthaler *et al.* (2021)
Muthirappuzhayar River, Kerala	River	0.20 particle/L	Rural river with limited human settlement; low contamination levels.	Lechthaler *et al.* (2021)
The Brahmaputra River a) MP size 20-150 μm	River	531–3,485 MP/kg dry weight	High levels of microplastics due to heavy industrial activity and significant population density in catchment areas.	Tsering *et al.* (2021)
b) MP size 150-5,000 μm		20–240 MP/kg dry weight	Larger particles present in river sediment; mostly fragments from construction and agricultural runoff.	
The Indus River a) MP size 20-150 μm	River	525–1,752 MP/kg dry weight	River traverses through industrial regions; significant presence of small plastic particles from textile industries and urban waste.	Tsering *et al.* (2021)
b) MP size 150-5,000 μm		60–340 MP/kg dry weight	Presence of both microfibers and plastic fragments.	
The Ganges, India/ Bangladesh	River	0.038 MP/L	Concentration linked to high population density and waste disposal practices.	Napper *et al.* (2021)
Netravathi River, Karnataka	River	288 pieces/m^3	River near coastal areas; influenced by industrial and domestic effluents.	Amrutha and Warrier (2020)
Ganga River	River	210.25±124.65 items/kg	High concentration due to significant urban and industrial pollution, particularly near urban centers like Varanasi.	Sarkar *et al.* (2019)
Yamuna River, Delhi	River	1.78×10^3 particles/m^3	High urban pollution.	Vaid *et al.* (2022)
Sal Estuary, Goa	Coastal	48±19 particles/L	Estuarine environment with significant contributions from domestic and tourism activities.	Saha *et al.* (2021)
Girgaon Beach, Mumbai (Arabian Sea)	Coastal	3.54±0.01 mg/kg	Beaches heavily impacted by waste from local fishing, tourism, and industrial activity.	Tiwari *et al.* (2019)
Tuticorin Beach, Tamil Nadu (Bay of Bengal)	Coastal	2.75±0.03 mg/kg	High industrial activity and shipping-related plastic waste.	Tiwari *et al.* (2019)
Dhanushkodi Beach, Tamil Nadu (Bay of Bengal)	Coastal	1.05±0.01 mg/kg	Beach near a remote town; lower concentrations due to minimal industrial impact but influenced by nearby fishing communities.	Tiwari *et al.* (2019)
Tamil Nadu Coast	Coastal	46.6±37.2 particles/m^2	A stretch of coastline impacted by tourism, agriculture, and urban runoff.	Karthik *et al.* (2018)
Rameswaram Island, the Gulf of Mannar	Coastal	403 pieces	Coastal area with significant tourist activity; plastic fragments predominantly.	Vidyasakar *et al.* (2018)
Chennai coast	Coastal	304 pellets (March 2015)	High contamination from urban and industrial runoff, with plastic pellets being the dominant type.	Veerasingam *et al.* (2016)
		896 pellets (November 2015)	Seasonal variation in pollution; greater concentration during the monsoon.	
Mumbai Beaches, Maharashtra	Coastal	7.49 g/m^2	High microplastic concentrations due to proximity to large urban centers and industrial zones.	Jayasiri *et al.* (2013)
Alang-Sosiya Ship-breaking Yard, Gulf of Cambay	Coastal/ sediment	81 mg/kg of sediment	Area impacted by ship-breaking industry, contributing to microplastic contamination from disused ships.	Reddy *et al.* (2006)

The impact of MPs extends beyond marine systems, affecting terrestrial environments as well. Their accumulation in soil has been shown to disrupt soil functions, altering physical properties such as porosity, bulk density, and water retention, as well as impacting chemical and biological processes (Meng *et al.*, 2022).

Human exposure to MPs, primarily through contaminated food, is an emerging health concern. These particles can bind with heavy metals, increasing their bioavailability in the human gastrointestinal tract. Furthermore, MPs and metal nanoparticles have been linked to cytotoxic effects on human tissues, presenting additional health risks (Khalid *et al.*, 2021). The economic repercussions of MP pollution, including harm to fisheries, tourism, and coastal infrastructure, further aggravate the challenges faced by coastal communities.

19.6 MITIGATING MP POLLUTION

India has initiated several measures to address the growing problem of MP pollution through regulatory reforms and policy advancements. The Plastic Waste Management (Amendment) Rules 2021 introduced important changes, including the Central Pollution Control Board's (CPCB) enforcement of a ban on single-use plastic (SUP) items as of July 1, 2022. These prohibited items include plastic ear buds, balloon sticks, flags, candy sticks, ice cream sticks, and polystyrene (thermocol) decorations. Other items such as plastic plates, cups, glasses, cutlery (forks, spoons, knives), straws, trays, and packing films for sweet boxes, invitation cards, and cigarette packets, along with plastic banners under 100 μm in thickness were also banned. By the end of 2022, the mandated minimum thickness for carry bags made of virgin or recycled plastic increased to 120 μm, aiming to promote reuse and enhance recyclability. Thicker plastic bags, being costlier, also act as a deterrent to casual use, and littering, presenting a progressive step compared to similar global measures (Singh and Biswas, 2023).

Despite these regulatory strides, stricter enforcement and robust implementation mechanisms are critical to achieving meaningful results. Collaboration with international organizations is equally vital to mitigate cross-border impacts of MPs. For example, Japan and the United Nations Environment Programme (UNEP) launched a joint initiative in 2019 to address marine plastic litter in Southeast Asia and India. This partnership focused on supporting local and provincial governments, particularly in cities like Mumbai, Agra, and along the Ganges River, to combat plastic waste (UNEP, 2019). These collaborative efforts demonstrate the importance of global partnerships in tackling the pervasive issue of plastic pollution.

To understand the broader implications of MP pollution, extensive research and sampling are necessary to examine their interaction with pollutants such as polychlorinated biphenyls (PCBs), POPs, and heavy metals. Investigating the combined toxicological impacts of MPs and these pollutants on coastal ecosystems and food webs is crucial for developing effective mitigation strategies (Karthik *et al.*, 2018).

Advocating for the adoption of sustainable practices, particularly the principles of the 5Rs – reduce, reuse, recycle, refuse, and repair – can further strengthen efforts to combat MP pollution. By focusing on reducing the demand for plastic, promoting reusability, enhancing recycling infrastructure, refusing unnecessary plastic products, and repairing items to extend their lifespan, this approach addresses the issue at multiple levels. Implementing these principles through public awareness campaigns and collaborative initiatives can lead to a significant reduction in plastic waste and MP contamination in aquatic environments.

Projections suggest that if current trends persist, the volume of plastic in oceans could exceed the biomass of fish by 2050 (Begai and Henam, 2021). Therefore, efforts to mitigate MP contamination in India's aquatic ecosystems should include investments in innovative technologies for plastic waste management, intensified public education campaigns, and targeted research to bridge knowledge gaps and inform policy decisions (Figure 19.1). These strategies, when combined, can pave the way for effective solutions to combat the escalating issue of plastic pollution.

19.7 CONCLUSION

MP pollution in India's aquatic ecosystems is a growing environmental concern, with rivers, lakes, and coastal areas showing significant contamination. Key sources of MPs include industrial activities, urban runoff, and tourism, with fibrous particles and plastic fragments being the most common types. Despite advances in research, challenges remain in quantifying and tracing MP sources due to methodological limitations and data gaps. Standardized protocols and improved analytical techniques are crucial for better understanding the full extent of pollution and its impact on ecosystems. Effective mitigation strategies will require coordinated efforts in plastic waste management, regulatory frameworks, and public awareness. Addressing these challenges is essential for reducing MP pollution and safeguarding India's aquatic environments.

REFERENCES

Amrutha K, Warrier AK (2020). The first report on the source-to-sink characterization of microplastic pollution from a riverine environment in tropical India. *Science of the Total Environment*, 739:140377. https://doi.org/10.1016/j.scitotenv.2020.140377

Begai A, Henam SD (2021). Beating plastic pollution: UNEP's priorities and partnership in India. *Environmental Policy and Law*, 51:265–269. https://doi.org/10.3233/EPL-201069

Bharath K, Srinivasalu S, Natesan U, Ayyamperumal R, Kalam N, Anbalagan S, Sujatha K, Alagarasan C (2021). Microplastics as an emerging threat to the freshwater ecosystems of Veeranam lake in south India: a multidimensional approach. *Chemosphere*, 264(2):128502. https://doi.org/10.1016/j.chemosphere.2020.128502

Ebere EC, Wirnkor VA, Ngozi VE (2019). Uptake of microplastics by plant: a reason to worry or to be happy? *World Scientific News*, 131:256–267. https://www.researchgate.net/publication/334083970_Uptake_of_Microplastics_by_Plant_a_Reason_to_Worry_or_to_be_Happy

Eriksen M, Thiel M, Prindiville M, Kiessling T (2018). Microplastic: what are the solutions? In: Wagner M, Lambert S (eds) *Freshwater Microplastics. The Handbook of Environmental Chemistry*, 58. Springer, Cham. https://doi.org/10.1007/978-3-319-61615-5_13

Ganesan M, Nallathambi G, Srinivasalu S (2019). Fate and transport of microplastics from water sources. *Current Science*, 117(11):1879–1885. https://doi.org/10.18520/cs/v117/i11/1879-1885

Gasperi J, Wright SL, Dris R, Collard F, Mandin C, Guerrouache M, Langlois V, Kelly FJ, Tassin B (2018). Microplastics in air: are we breathing it in? *Current Opinion in Environmental Science & Health*, 1:1–5. https://doi.org/10.1016/j.coesh.2017.10.002

Gupta P, Saha M, Rathore C, Suneel V, Ray D, Naik A, Unnikrishnan K, Dhivya M, Daga K (2021). Spatial and seasonal variation of microplastics and possible sources in the estuarine system from central west coast of India. *Environmental Pollution*, 288: 117665. https://doi.org/10.1016/j.envpol.2021.117665

Hale RC, Seeley ME, La Guardia MJ, Mai L, Zeng EY (2020). A global perspective on microplastics. *Journal of Geophysical Research: Oceans*, 125:e2018JC014719. https://doi.org/10.1029/2018JC014719

Hildebrandt L, Nack FL, Zimmermann T, Profrock D (2021). Microplastics as a Trojan horse for trace metals. *Journal of Hazardous Materials Letters*, 2:100035. https://doi.org/10.1016/j.hazl.2021.100035

Iri AH, Shahrah MHA, Ali AM, Qadri SA, Erdem T, Ozdur IT, Icoz K (2021). Optical detection of microplastics in water. *Environmental Science and Pollution Research*, 28: 63860–63866. https://doi.org/10.1007/s11356-021-12358-2

Jayasiri HB, Purushothaman CS, Vennila A (2013). Quantitative analysis of plastic debris on recreational beaches in Mumbai, India. *Marine Pollution Bulletin*, 77:107–112. https://dx.doi.org/10.1016/j.marpolbul.2013.10.024

Karthik R, Robin RS, Purvaja R, Ganguly D, Anandavelu I, Raghuraman R, Hariharan G, Ramakrishna A, Ramesha R (2018). Microplastics along the beaches of southeast coast of India. *Science of the Total Environment*, 645:1388–1399. https://doi.org/10.1016/j.scitotenv.2018.07.242

Khalid N, Aqeel M, Noman A, Khan SM, Akhter N (2021). Interactions and effects of microplastics with heavy metals in aquatic and terrestrial environments. *Environmental Pollution*, 290:118104. https://doi.org/10.1016/j.envpol.2021.118104

Lechthaler S, Waldschläger K, Sandhani CG, Sannasiraj SA, Sundar V, Schwarzbauer J, Schüttrumpf H (2021). Baseline study on microplastics in Indian rivers under different anthropogenic influences. *Water*, 13:1648. https://doi.org/10.3390/w13121648

Meng F, Yang X, Riksen M, Geissen V (2022). Effect of different polymers of microplastics on soil organic carbon and nitrogen – a mesocosm experiment. *Environmental Research*, 204:111938. https://doi.org/10.1016/j.envres.2021.111938

Napper IE, Baroth A, Barrett AC, Bhola S, Chowdhury GW, Davies BF, Duncan EM, Kumar S, Nelms SE, Niloy NH, et al. (2021). The abundance and characteristics of microplastics in surface water in the transboundary Ganges River. *Environmental Pollution*, 274:116348. https://doi.org/10.1016/j.envpol.2020.116348

Prata JC, da Costa JP, Duarte AC, Rocha-Santos T (2019). Methods for sampling and detection of microplastics in water and sediment: a critical review. *Trends in Analytical Chemistry*, 110:150–159. https://doi.org/10.1016/j.trac.2018.10.029

Reddy MS, Basha S, Adimurthy S, Ramachandraiah G (2006). Description of the small plastics fragments in marine sediments along the Alang-Sosiya ship-breaking yard, India. *Estuarine, Coastal and Shelf Science*, 68:656–660. doi:10.1016/j.ecss.2006.03.018

Saha M, Naik A, Desai A, Nanajkar M, Rathore C, Kumar M, Gupta P (2021). Microplastics in seafood as an emerging threat to marine environment: a case study in Goa, west coast of India. *Chemosphere*, 270:129359. https://doi.org/10.1016/j.chemosphere.2020.129359

Sarkar DJ, Sarkar SD, Das BK, Manna RK, Behera BK, Samanta S (2019). Spatial distribution of meso and microplastics in the sediments of river Ganga at eastern India. *Science of the Total Environment*, 694:133712. https://doi.org/10.1016/j.scitotenv.2019.133712

Selvam S, Manisha A, Roy PD, Venkatramanan S, Chung SY, Muthukumar P, Jesuraja K, Elgorban AM, Ahmed B (2021). Microplastics and trace metals in fish species of the Gulf of Mannar (Indian Ocean) and evaluation of human health. *Environmental Pollution*, 291:118089. https://doi.org/10.1016/j.envpol.2021.118089

Singh S, Biswas MK (2023). Management strategies for single-use plastics: lessons to learn from Indian approach of minimizing microplastic waste. *Environmental Science Advances*, 2:1680–1695. https://doi.org/10.1039/d3va00222e

Sivadas SK, Mishra P, Kaviarasan T, Sambandam M, Dhineka K, Murthy MVR, Nayak S, Sivyer D, Hoehn D (2022). Litter and plastic monitoring in the Indian marine environment: a review of current research, policies, waste management, and a roadmap for multidisciplinary action. *Marine Pollution Bulletin*, 176:113424. https://doi.org/10.1016/j.marpolbul.2022.113424

Sruthy S, Ramasamy EV (2017). Microplastic pollution in Vembanad Lake, Kerala, India: The first report of microplastics in lake and estuarine sediments in India. *Environmental Pollution*, 222:315–322. https://doi.org/10.1016/j.envpol.2016.12.038

Tiwari M, Rathod TD, Ajmal PY, Bhangare RC, Sahu SK (2019). Distribution and characterization of microplastics in beach sand from three different Indian coastal environments. *Marine Pollution Bulletin*, 140:262–273. https://doi.org/10.1016/j.marpolbul.2019.01.055

Trevisan R, Uzochukwu D, Di Giulio RT (2020). PAH sorption to nanoplastics and the Trojan horse effect as drivers of mitochondrial toxicity and PAH localization in Zebrafish. *Frontiers in Environmental Science*, 8:78. doi: 10.3389/fenvs.2020.00078

Tsering T, Sillanpää M, Sillanpää M, Viitala M, Reinikainen SP (2021). Microplastics pollution in the Brahmaputra River and the Indus River of the Indian Himalaya. *Science of the Total Environment*, 789:147968. https://doi.org/10.1016/j.scitotenv.2021.147968

UNEP (2019). Japan and UN Environment announce new cooperation to boost knowledge on marine litter in Southeast Asia and India, UNEP, Bangkok. https://www.unep.org/news-and-stories/press-release/japan-and-un-environment-announce-new-cooperation-boost-knowledge accessed on 26 March 2024.

Vaid M, Mehra K, Gupta A (2021). Microplastics as contaminants in Indian environment: a review. *Environmental Science and Pollution Research*, 28:68025–68052. https://doi.org/10.1007/s11356-021-16827-6

Vaid M, Mehra K, Sarma K, Gupta A (2022). Investigations on the co-occurrence of microplastics and other pollutants in the River Yamuna, Delhi. *Water Supply*, 22(12):8767. https://doi.org/10.2166/ws.2022.408

Vanapalli KR, Dubey BK, Sarmah AK, Bhattacharya J (2021). Assessment of microplastic pollution in the aquatic ecosystems – an Indian perspective. *Case Studies in Chemical and Environmental Engineering*, 3:100071. https://doi.org/10.1016/j.cscee.2020.100071

Veerasingam S, Ranjani M, Venkatachalapthy R, Bagaev A, Mukhanov V, Litvinyuk D, Verzhevskaia L, Guganathan L, Vethamony P (2020). Microplastics in different environmental compartments in India: analytical methods, distribution, associated contaminants and research needs. *Trends in Analytical Chemistry,* 133:116071. https://doi.org/10.1016/j.trac.2020.116071

Veerasingam S, Mugilarasan M, Venkatachalapathy R, Vethamony P (2016). Influence of 2015 flood on the distribution and occurrence of microplastic pellets along the Chennai coast, India. *Marine Pollution Bulletin*, 109:196–204. https://doi.org/10.1016/j.marpolbul.2016.05.082

Vidyasakar A, Neelavannan K, Krishnakumar S, Prabaharan G, Priyanka TSA, Magesh NS, Godson PS, Srinivasalu S (2018). Macrodebris and microplastic distribution in the beaches of Rameswaram Coral Island, Gulf of Mannar, Southeast coast of India: a first report. *Marine Pollution Bulletin*, 137:610–616. https://doi.org/10.1016/j.marpolbul.2018.11.007

20 Human Pathogen Spillover in Oceans

Pallavi Narania and Tejinder Kaur
Department of Zoology, DAV University, Jalandhar, India

20.1 INTRODUCTION

According to the National Ocean and Atmospheric Administration (NOAA), the oceans make up around 71% of the planet's surface and hold 97% of all the water on earth. As such, the marine ecosystem that covers an extensive portion of the planet gives us access to a variety of marine foods and creatures as well as other chemicals that have practical uses in both industry and medicine. Despite having a wealth of important resources, they also contain some life-threatening species for both marine life and humans. There exist a variety of ways in which these pathogens will enter the marine environment. Usually, a disease that accidently enters the aquatic food chain and infects one organism will find a simple route to propagate from that organism to a healthy one via the marine trophic system without the help of any external agents. However, a number of anthropogenic activities such as water contamination, water pollution, climate change, etc., carried out in or in close proximity to the ocean will facilitate the spread of these pathogens. As a result, there is an oceanic spillover of possible human illness. The precise definition of spillover is the introduction of a parasite into a host population that has never been infected by another species (Wells and Clark, 2019). The term "zoonotic spillover" refers to the possible infection that may spread from wild animals to people (Ellwanger and Cheis, 2021). In 2021, Ellwanger and Cheis asserted that spillover is a significant factor in the transmission of infectious diseases from non-human species to humans either directly or indirectly through a food web. These spillovers can often lead to diseases posing a serious concern toward public health. A number of pandemics have been linked to zoonotic spillover such as from Plague of Athens to COVID-19 (Weiss and Sankaran, 2022). The ocean is not an exception to spillover, just like on land. Owing to a variety of natural and man-made circumstances, the oceans have also and continue to witness disease spillover through zoonosis. Aquaculture, which ought to be a newly developed and well-established food industry in many countries, faces contamination from human disease spillover into the oceans. This is because possible infections could wander into the ocean and contaminate the industry.

The absence of comprehensive and methodical research on the interactions between pathogens and various physical and biological entities has contributed to the spillover events within the oceans which will continue to be a major global concern for both human as well as the ocean health.

20.2 HOW ZOONOTIC SPILLOVER OCCURS?

The occurrence of pathogen spillover in any ecosystem depends on the host-pathogen interaction. In general, zoonotic spillover is caused by interaction of host, pathogen, and vector. Figure 20.1 illustrates the key components and their interaction responsible for zoonotic spillover into the environment, which involves (a) the host species responsible for the pathogen's release; (b) the infected species; and (c) the vector that helps spread the infection.

It was also observed that in some cases, through the process of genetic recombination, a novel strain of the pathogen may emerge in the host when two more vectors, each carrying a different virus, infect the same host (Wasik et al., 2019). It will eventually boost the pathogen's survival and persistence in the host, allowing it to become available for the spillover process (Perez et al., 2023). The likelihood of a spillover incident will mostly depend on how common and severe the illness is in reservoir hosts, who are often non-human animals and synthetic materials, like plastics (Plowright et al., 2017). There is a higher chance of spillover incidents when several anthropogenic activities are involved, including handling animals, eating meat from wild animals, poaching, and marine pollution. Wet markets which trade in live animals have contributed significantly

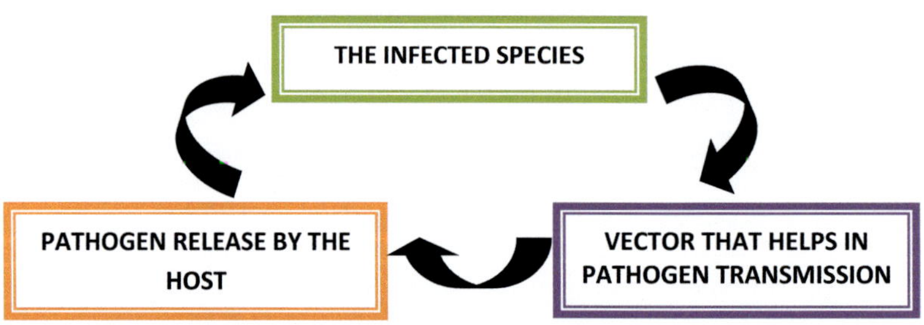

FIGURE 20.1 Key components responsible for the zoonotic spillover of pathogens.

DOI: 10.1201/9781003509530-20

to the spread of human infections into the ocean and vice versa through these animals as they encourage human contact with a variety of pathogenic species (Ellwanger and Cheis, 2021). Some other ways through which pathogens can spread include biofluids, meat, blood, infected surfaces used for sales, and aerosols released into living spaces Wißmann et al., 2021).

20.3 CHARACTERISTICS FEATURES OF POTENTIAL PATHOGEN TO CAUSE ZOONOSIS IN THE OCEANS

There are several characteristics that facilitate the phenomena of pathogenic spillover into the oceans because they help the pathogens get beyond the host and geographic obstacles that stand in their way of spreading illness. Below is a discussion of a few of them.

20.3.1 Presence of RNA Genome

As compare to DNA, there are high chances of mutations in RNA molecule because of their less stable structure. Thus, pathogens having RNA as their genetic material have high chances of recombination in its genome, enhancing their adaptability toward different kinds of barriers present on their way (Pauciullo et al., 2024).

For instance, some RNA viruses such as Alphavirus, papilloma virus, adenovirus have more potential for zoonosis (Ellwanger and Cheis, 2021).

20.3.2 Association of Pathogen with Specific Reservoir Host

In order to spread throughout the environment, the diseases favor some host animals over others. This may be due to some preferable host factors such as long life span, immune system, long-term co-existence with humans, interaction with multiple ecosystem for food which allow their rapid growth and easy transmission. It was observed that wild animals like bats are the most effective host for the transmission of pathogen as bats are known to carry numerous pathogens with them as their unique immune system properties allows the virus particles to co-exist inside them without causing any illness (Letko et al., 2020). Order Rodentia is followed by the Order Chiroptera, and primates are most likely to carry a number of zoonotic viruses causing cross-contamination leading to zoonosis (Keesing and Ostfeld, 2021).

20.3.3 Presence of Biodiversity

Dilution effects play an important role in zoonotic spillover of the pathogens. The presence of biodiversity is inversely proportional to the availability of the pathogen to get introduced into the new ecosystem as it dilutes the number of pathogens, hence avoiding its spillage into the nearby environment (Strauss et al., 2024). One such example was that of Lyme disease where restoration of vertebrate diversity had reduced the chances of spillover from wildlife to humans (Sokolow et al., 2019). With the increased use of land for various anthropogenic activities, there is a substantial loss of biodiversity, thereby increasing the chances of infectious diseases among humans.

20.4 SOME MAJOR FACTORS CONTRIBUTING TO THE PATHOGEN SPILLOVER

Every zoonotic spillover event is a multifactorial occurrence, meaning that a number of causes contribute to its occurrence (Plowright et al., 2017). The frequency and intensity of interactions between the pathogen, vector, and host are determined by a number of direct and indirect parameters. Some of them are covered below:

1. **Human-Reservoir Interaction:**
   Generally speaking, one of the most important parts of a spillover event is the interaction of diverse human and animal species. It was observed that human-animal interaction contributed significantly in the process of zoonotic spillover into the oceans (Plowright et al., 2017). The reservoir hosts gets infected majorly through interaction with pathogen from the meat industry, livestock farmers, domestic and wild animals (Ellwanger and Cheis, 2021). Thus, more the interaction, higher the chances of pathogen intensity in an area.
2. **The Infected Host's Population and Distribution within a Specific Environment:**
   There is a direct positive correlation between the chances of spillover event with respect to the number and prevalence of the infected host species in a specific region at a specific period of time (Ellwanger and Cheis, 2021; Escudero-Pérez et al., 2023).
3. Frequency of host-pathogen interaction intensity and ability of pathogen to transmit from one host to another (Borremans et al., 2019). Zoonosis is enhanced by geographical range overlap (Albery et al., 2020).
4. The pathogen's interaction with the vector surface, also referred to as horizontal gene transfer (Bowley et al., 2021).
5. The pathogen's transit length and mobility across the sea surface is an important factor in deciding the severity of the spillover into the sea water (Bowley et al., 2021).
6. **Evolutionary Relation and Genetic Distance between the Host Species:**
   It is believed that a pathogen has a higher probability of spreading among phylogenetically related species than it does among phylogenetically distant hosts (Ellwanger and Cheis, 2021).

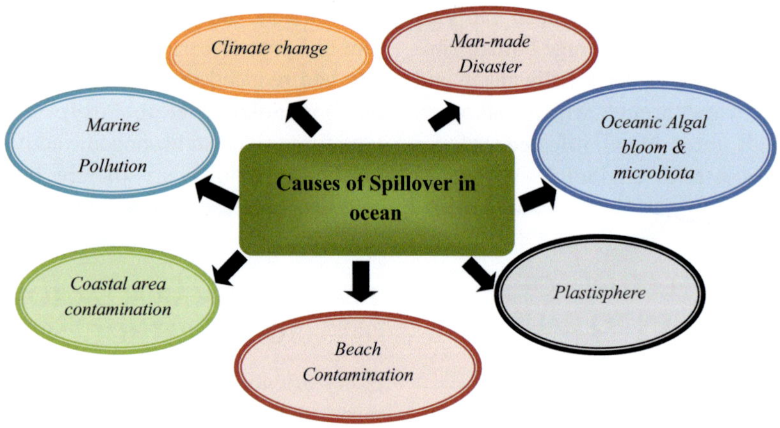

FIGURE 20.2 Multiple variables influencing the spread of human pathogens in oceans.

Very similar host and reservoir would encourage the vector transfer from one host to the new healthy host as they share a highly similar genetic makeup, therefore making it easier for the disease to enter the body cells without any need of adaptability in their strategy of infection.

Every single one of these factors is constantly connected to the several reasons that lead to the spread of human infections in the sea. Some of the major contributing factors responsible for the pathogen spillover into the ocean through human interventions are shown in Figure 20.2.

20.4.1 Microplastics Leading to Formation to Plastisphere

Over time, microplastics have become a major cause of the extensive human-caused ocean pollution. They serve as a distinctive adherent surface on which pathogens can adhere. Masó and his colleagues investigated the relationship between human pathogens and plastic for the first time in 2003. In oceanic water, when any existing hydrophobic biochemical substance such as nutrient, infochemicals (toxin, pheromones), polysaccharide, and microbe gets attached to the residing microplastics in the sea, it forms a coating or a film over the surface, then they are referred to as ecocorona or plastisphere (Bowley et al., 2021; Athulya and Chandrasekaran, 2023). The formation of plastisphere was first demonstrated by Zettler and his colleagues in 2013. Ecocorona is formed due to the process of sorption. It resembles the natural substances which were utilized by the organism of lower trophic level like bivalves, crustaceans, phytoplanktons, and marine vertebrates, thus paving their way into the marine food web together with the hidden human pathogen (Zhang et al., 2022). In 2021, according to Bowley and his colleagues, a biofilm structure typically consisted of pathogens belonging to many species as well as a single species that adhered to its surface. Microplastics are referred to any plastic particle with size less than 5 mm in diameter (Ziani et al., 2023). On the basis of the source of its emergence, microplastics are categorized further into two types, i.e., primary and secondary microplastics. Primary microplastics are the ones made for the cosmetics and textile industries whereas secondary plastics refer to the tiny plastic particle formed due to the chemical and mechanical breakdown of the large plastic through the action of oceanic waves and solar radiations (Periyasamy and Tehrani-Bagha, 2022).

Without any control, surveillance, and proper regulation, plastic waste is being discarded into the oceans (Avio et al., 2017). Over time, this accumulates as microplastics, which marine animals can readily consume. Another significant finding regarding microplastics was made by experts who discovered that these particle surfaces are associated with human pathogens, primarily protozoan parasites (Beans, 2023). This laid a base for the recognition that human activity can cause human pathogens to leak into the ocean. Consequently, it is now evident that microplastics have the ability to act as a vector for the spread of disease-causing viruses into new regions of the ocean as well as a reservoir host for diseases that affect humans (Emenike et al., 2023).

Many tests were conducted to ascertain the interaction between potential pathogens and microplastics in order to understand how these microplastics were associated with the pathogenic bacteria inside the ocean to induce pathogen spillover in ocean water. For example, Zhang and colleagues' (2022) study examined parasites that the World Health Organization (WHO) lists as parasites that are often disregarded as a cause of illnesses associated with eating shellfish and their undesirable persistence in the marine ecosystem. They discovered the association between microfibers found in polyester and polyethylene microbeads and the three zoonotic protozoan parasites, *Giardia enterica*, *Toxoplasma gondii*, and *Cryptosporidium parvum* (Kerlin, 2022). They also came to the conclusion that the nature and dimensions of the microplastic had an impact on the kind of pathogen that was linked to the surface of the microplastic. For example, *C. parvum* preferably gets attached with larger pore-sized microplastics. Another study found that the irregularities and the pores existing over the microplastic surface provide an ideal surface for the attachment of *Aspergillus flavus* and various other fungal and bacterial

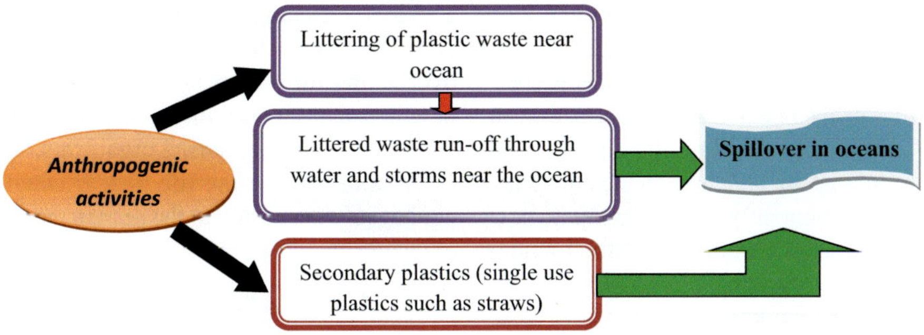

FIGURE 20.3 A potential explanation for how human activity spreads microplastics that carry harmful pathogens into the oceans.

strains. Some other bacterial forms found to be associated with the microplastics are *Paenibacillus sp.*, *Bacillus sp.* (Park and Kim, 2019). In the open ocean, *Vibrio* appears to be widely distributed together with plastisphere (Zettler et al., 2013). It was suggested that the sticky biofilm that accumulated on plastic surfaces over time seems essential to the plastic-pathogen interaction as well as for their spillover into the ocean (Zhang et al., 2022). Together with certain alterations in the host's immune responses against these prospective pathogens, microplastics were also thought to be responsible for the microplastic-associated pathogen's development of a more sophisticated mechanism of resistance against antibiotics (Beans, 2023). According to research, microplastics contain 100–5,000 times more antimicrobial-resistant bacteria than the surrounding ocean water, which serves as a hotspot for the growth of bacteria with greater resistance to antibiotics (Bowley et al., 2021).

In addition to chemical pollutants, it was discovered that microplastics can also enter and get mixed well in ocean due to ocean dynamics such as turbulence, leading to vertical mixing of sea water, sedimentation of solid particulate into the bottom of the ocean, biofilm formation around microplastics, movement of surface water due to wind, and beaching (Guo et al., 2024). However, human activity in nearby areas of the ocean is the major source of microplastic pollution. These include industrial effluent, irrigation, waste water treatment plants, and fishing activities that release microplastics into the ocean (Amelia et al., 2021). Some of the other anthropogenic activities are represented in Figure 20.3.

When this pollution enters the sea, it is easily consumed by a variety of aquatic creatures, including zooplanktons and filter feeders (bivalves). As a result, the infectious agent adhering to the plastic becomes biomagnified, severely impacting and contaminating both terrestrial and marine food chains. Thus, the introduction of human pathogens into the oceans impacts both human health and the marine ecology. In turn, zoonotic illnesses related to marine pollution are caused by humans themselves.

20.4.2 Algal Bloom and Microbiota of the Ocean

As per the NOAA fact sheet, the oceanic ecology is negatively impacted by harmful algal blooms (HABs), also referred to as red tides. HAB is characterized as the occurrence of an excess of toxin-producing algae in the water. Even though algal blooms frequently benefit the marine ecology, excessive production of them can result in the release of harmful chemicals and potent toxins into the sea water (Figgatt, 2017). In US Commission on Ocean Policy (2004), it was stated that these toxins may have neurotoxic, respiratory disease and diarrhetic effects along with marine food poisoning to aquatic creatures as well as terrestrial organism linked with the infected individuals (Fleming et al., 2006). Some other deteriorating effects must be unfavorable climatic conditions, such as a rise in sea water temperature that is suitable for disease growth (Dai et al., 2024). One important cause of HAB is human interference with water sources, such as fertilizer run-off from agriculture and the release of industrial effluents containing nitrogen and phosphorus (Newton and Melaram, 2023). Because of it, the accumulation of phosphates and nitrates in the water will encourage the rapid growth of algae to the point where they begin to pose a major hazard to the surrounding ecosystem. The most recent study found that the microbial contact with algae cells is controlled by the release of infochemicals, biochemicals, and other dissolved nutrients into the ocean water (Kuhlisch et al., 2024). These compounds will act as a mediator in the pathogen's cellular communication with marine creatures and other pathogens.

20.4.3 Climate Change and Sea Water Warming

The host-pathogen interaction is always susceptible to change or should alter if change in global climate takes place (Cohen et al., 2018). NOAA defined global warming as an increase in air temperature and is considered as one of the most significant changes in the global climate that have been recorded till January 2024. According to evaluations published by the Intergovernmental Panel on Climate Change (IPPC), the primary reason of the unusual warming of ocean water is human-caused carbon emissions into the atmosphere. According to recent research, the excessive discharge of various nutrients and pollutants into the ocean has increased the surface water's temperature, which has led to the growth of infectious organisms (viruses, bacteria, and algae) on them. Because of the impact of global warming, pathogens that were previously limited to tropical waters can now travel poleward due to an increase in sea surface temperature (Cohen et al., 2018). *Vibrio cholerae* is one such example of a pathogen. Climate change not only is responsible for the human pathogen spillover into the ocean but it also alters the dynamics of the ocean and the host-pathogen relationship inside it, ultimately affecting the aquatic food web and concerning ecosystem and human health (Liao et al., 2024). This shift in dynamics will help anticipate how altered host-pathogen interactions would affect the foundation and functioning of the ecosystem.

In 2018, Cohen and his co-workers examined how these interactions have changed as a result of climate change and have identified certain effects of them on zoonosis. Some of these effects are listed below:

i. Elevation in stress level inside host
ii. Diversification of pathogen's range
iii. Rise in virulence of the pathogen
iv. Host range modification

It was observed that due to increase in the ocean temperature, there is an elevation in the pathogenicity of pathogen due to quick metabolism (Sun et al., 2022).

20.4.4 Coastal Water and Beach Contamination

In 2017, the United Nations Environment Programme claimed that 37% of the world's population resides within 100 km of a shoreline, with a population density that is two times greater than the global average. Of this 37%, three-quarters of coast area is more likely to degrade as a result of human activity. The coast areas of approximately 86% of Europe and 69% of Asia are among the most susceptible regions for this degradation. According to Griffin and his colleagues in 2003, the neglected release of partly (only primary treatment) or untreated waste water and some recreational water exposure will eventually contribute to the spread of viral pathogens into the ocean especially enteric viruses. This contaminated and untreated water is directly discharged through outfalls into the oceans. Previous research has found a minimal amount of viruses in the ocean absorbing outfall water. It was noteworthy, nevertheless, that swimmers who go near contaminated beaches run a higher chance of getting sick (Griffin et al., 2003). These pathogenic viruses will lead to an array of illness such as gastrointestinal problems and respiratory illness to the infected person through zoonotic infection through oceanic water.

20.4.5 Marine Pollution

The pollution of oceanic water is also another major concern for spillover events. It can be caused to a variety of reasons all indicating toward the negligence of the human toward the marine ecosystem. As per the NOAA reports, some of the major pollutants are fertilizers and pesticide water run-off from agricultural land into the ocean, industrial discharge, habitat destruction due to usual oil spillage accidents affecting the immune system of marine organisms, coral bleaching, disturbing marine food web due to overfishing for own benefit and sometimes introduction of new invasive species, increasing the stress and competition for survival among marine organisms.

As a result, marine life gets affected and the infection again gets transmitted to the human world through zoonosis.

20.5 DETECTION MEASURES

Scientists have claimed that the major portion of ocean containing organism or other synthetic particles is still unexplored. Thus, there is a need for very sensitive and robust detection technology for the effective quantitative and qualitative analysis of sea water. For the detection of the pathogen, detection techniques are divided into different types based on the component being tested to check the presence of the bacteria. Some of them are listed below:

A. **Direct Methods:**
   1. **Cell Counting:**
      The viable cell counting is done for the water sample using bacterial agar plates by measuring the colony-forming units (CFU) per millimeter (Meyer et al., 2023).
   2. **Most Probable Number (MPN) Method:**
      It is generally used for the detection of coliform bacteria in the water through traditional as well as automated version of MPN system (Rompré et al., 2002: Aziz et al., 2024).
   3. **ATP, Gases, and Turbidity Measurements:**
      ATP and gases generated by the pathogens are quantified in the sample using different detection techniques such as Biotrace (Arroyo et al., 2017), Celsis system (Hansen et al., 2019), Bactec and BacT microbial detection systems (Li et al., 2019). Similarly, turbidity of the water is measured using bioactivity monitor system to detect the water turbidity caused by microbial activity (Farrell et al., 2018).

B. **Indirect Methods:**
   1. **Direct Fluorescent Antibody (DFA) Technique Using Alcian Blue Staining Dye:**
      For visualization of the biofilm matrix exopolymer is deposited over the surface of the microplastic due to the different body secretions made by the pathogen (Zhang et al., 2022).
   2. **Confocal Microscopy:**
      Because of great portability, advancements in microscopy have made it possible to employ the technique in field studies for quantification and characterization of the biofilm and associated pathogen (Sommerfeld et al., 2014).
   3. **Immunological Assays:**
      Various immunological assays can be employed for the detection as well as quantification of the pathogenic antigens or antibodies

present in the water sample, such as different types of Enzyme Linked Immunosorbent Assay (ELISA) and immunodiffusion techniques (Oon et al., 2023).

4. **Sequencing Technology:**

   The complex interaction between disease-causing pathogens and microplastics may now be studied more quickly and robustly by means of recent improvements in next-generation sequencing techniques (Wani et al., 2023). Other techniques for the detection of nucleic acids includes microarray (Aw and Rose, 2012), mass spectrometry (Ho and Reddy, 2010), pulse field gel electrophoresis (Sharma-Kuinkel et al., 2016), and some others.

5. **Polymerase Chain Reaction (PCR):**

   Generally, quantitative real time PCR is extensively used together with viability assay for the early and quick detection of pathogen in the water body as it tells us about the real-time conditions of the water sample of the area under study (Botes et al., 2013).

6. **Filtration Techniques:**

   Membrane-based filters are the most commonly used filters for the primary detection process of the water pathogen in the oceanic water that may prove potential pathogens for any kind of zoonotic infections. They are made available in various filtration systems such as in reverse osmosis filters, activated carbon filters commonly used in homes to get rid of pathogen-containing water. Membrane filters are able to detect enterococci and coliform pathogens such as *Escherichia coli, Salmonella, Shigella, Yersinia,* and others (Aziz et al., 2024).

20.6 CONTROL AND PREVENTION

Among the typical precautions that must be taken are better sanitation practices for wild animals, increased monitoring at the site of the pathogen-host interaction, control over the loss of biodiversity, and identification of the ecological and social variables that are causing spillover episodes (Ellwanger and Cheis, 2021). Therefore, there is a need of a holistic approach system that must be employed combining the technological, social, and political actions system for successful control and prevention of the pathogen spillover events into the oceans. A few of these are discussed below:

20.6.1 Microplastic Surveillance

It is crucial to ensure that the production process for plastics used in personal care and cosmetic products (PCCP) should be safe and incapable of producing microplastic to prevent its association with the pathogenic surfaces causing zoonotic diseases (Kakade et al., 2024). The biodegradation of microplastics can be accelerated with the employment of microplastic-degrading bacteria. Bacteria like *Bacillus* and *Pseudomonas* have proven acceptable in microplastic degradation (Zhai et al., 2023). Certain microplastic-related bacteria, such as *Bacillus* and *Paenibacillus*, have also been shown to aid in the decomposition of the plastic they are associated with during overflow (Park & Kim, 2019).

Technology constantly helps civilization, especially in the era of sophisticated machine learning and artificial intelligence. As an example, in 2012, the North Sea Foundation and Plastic Soup created the smartphone application "Beat the Microbead," which scans a product's barcode to determine whether or not it includes microbead. In order to make the app a worldwide success for international users, the United Nations Environment Programme and a non-governmental organization (NGO) with headquarters in the United Kingdom teamed forces in 2013. As a result, consumers can avoid utilizing microplastic-containing products.

Governments of various nations have taken initiatives in controlling the microplastics disposal in the oceans. In this regard, per reports from the NOAA, the former US President Barack Obama prohibited the use of microplastic beads in personal care products and other cosmetic items under the Microbead-Free Waters Act of 2015 in an effort to lessen the amount of secondary microplastics that end up in the environment. In 2012, as per the UN environment agency reports, a number of nations, including Sweden, Belgium, Austria, and the Netherlands, have jointly banned PCCP microplastics to save human health from secondary health issues resulting from disrupted food chains and to protect the marine ecology.

To lessen the pollution caused by plastic trash, the public can always rely on the 3Rs (Reduce, Reuse, and Recycle) philosophy in addition to government initiatives.

20.6.2 For Algal Bloom

According to the well-known phrase, prevention is always preferable to treatment. Therefore, the most effective method of avoiding HAB growth is to stop the influx of dissolved nutrients into saltwater and coastline. Experts are currently working to create satellite images and remote sensing systems to closely monitor the occurrence and spread of HAB into the ocean, with support from NOAA, despite certain limitations in using such system such as dynamic movement of algae covers, effect of climate change over algal growth, etc. (Liu et al, 2022).

20.6.3 Early Detection Systems

Novel tools and techniques are being made to efficiently detect the health indicators of the marine ecosystem. Modern imaging techniques should be able to work on field along with statistical analysis of the area distribution (Aguzzi et al., 2024). With the aid of advanced biochemical and molecular assays, one can monitor the water quality in real time for the early detection of the pathogens, thereby preventing their zoonotic transmission in the surrounding environment

(Zhang et al., 2021). Specific pathways concerned with the pathogen under study can be traced using epidemiological studies of the most probable route of infection for zoonotic spillover into the oceans (Plowright et al., 2017).

20.6.4 Public Awareness and Engagement

Awareness campaigns for the local coastal communities and for general public can help in educating the right sea shore practices and risks of pathogen contamination through polluted waters, safeguarding the local ecosystem (Thushari and Senevirathna, 2020). Government should take a step forward in training the medical professionals as well as general public in diagnosing the pathogen so that early control and prevention should take place at the individual level, helping in predicting the possible outcomes.

20.6.5 Maintenance of Coastal Water Quality

In addition to improved sea food industries, recreational activities near the ocean such as swimming, boating, sailing, sunbathing, fishing, etc., need to be monitored carefully. This can be done by issuing advisories by the tourism departments to the coastal managers as well as to the general public.

20.7 CONCLUSION

Human pathogen spillover into the oceans is an emerging concern that poses risks to both marine ecosystems and human health. Spillover events have the potential to cause harm, including epidemics and, in the worst situations, pandemics. Therefore, it is crucial to understand the factors responsible for the zoonotic spillover events into the oceans carrying human pathogens into the oceanic life. However, no species ought to be eradicated or demonized because of possible threats to human safety. Respect is owed to each species' ecological significance in the natural world. In addition to the natural causes of zoonosis, a number of human-caused activities play a significant role in the spillover episodes that occur in the ocean. Apart from human activities, there are several others factors associated with the spillover into the oceans such as pathogenic preference toward the host organisms, the genetic makeup of the different hosts for cross transmission, etc. In order to effectively manage the infectious disease through oceans, we should have good and precise detection measures for the early detection of the probable infection in the ecosystem, saving time, cost, and the lives of hundreds before the occurrence of actual pandemic or endemic disease. Detection can be done directly on sea water sample or indirectly to obtain a better understanding about the pathogen that needs to be dealt with. Even though zoonosis through spillover incidents returns, the only way to stay safe from them is to effectively manage the steps taken to lower the risks involved and improve management. Recent public emergencies due to zoonotic spillovers had led to recent endemics and pandemics such as COVID-19 and monkey pox disease. Thus, it is not advisable to completely neglect these threats and employ strategies to minimize the risks of occurrence or recurrence of such public health emergencies. Preventive measures, including improved wastewater management, regulation of human activities, sustainable agriculture, and public education, are critical to mitigating the risks of zoonotic spillover. The preventive measures can be employed from the community level by maintaining hygiene at work station dealing with animal handling. The government can help in monitoring the safety and handling of animals, thereby reducing the oceanic contamination and minimizing the risk of human pathogens entering the oceans from various routes such as beach pollution, climate change effects, and microplastics emission. By addressing these challenges through a combination of technological, regulatory, and community-based strategies, we can protect both the oceans and human populations from the impacts of zoonotic diseases.

REFERENCES

Aguzzi, J., Thomsen, L., Flögel, S., Robinson, N. J., Picardi, G., Chatzievangelou, D., Bahamon, N., Stefanni, S., Grinyó, J., Fanelli, E., Corinaldesi, C., Del Rio Fernandez, J., Calisti, M., Mienis, F., Chatzidouros, E., Costa, C., Violino, S., Tangherlini, M., & Danovaro, R. (2024). New technologies for monitoring and upscaling marine ecosystem restoration in deep-sea environments. *Engineering*, *34*, 195–211. https://doi.org/10.1016/j.eng.2023.10.012

Albery, G. F., Eskew, E. A., Ross, N., & Olival, K. J. (2020). Predicting the global mammalian viral sharing network using phylogeography. *Nature Communications*, *11*(1), 2260. https://doi.org/10.1038/s41467-020-16153-4

Amelia, T. S. M., Khalik, W. M. A. W. M., Ong, M. C., Shao, Y. T., Pan, H.-J., & Bhubalan, K. (2021). Marine microplastics as vectors of major ocean pollutants and its hazards to the marine ecosystem and humans. *Progress in Earth and Planetary Science*, *8*(1), 12. https://doi.org/10.1186/s40645-020-00405-4

Arroyo, M. G., Ferreira, A. M., Frota, O. P., Rigotti, M. A., De Andrade, D., Brizzotti, N. S., Peresi, J. T. M., Castilho, E. M., & De Almeida, M. T. G. (2017). Effectiveness of ATP bioluminescence assay for presumptive identification of microorganisms in hospital water sources. *BMC Infectious Diseases*, *17*(1), 458. https://doi.org/10.1186/s12879-017-2562-y

Athulya, P. A., & Chandrasekaran, N. (2023). Interactions of natural colloids with microplastics in aquatic environment and its impact on FTIR characterization of polyethylene and polystyrene microplastics. *Journal of Molecular Liquids*, *369*, 120950. https://doi.org/10.1016/j.molliq.2022.120950

Avio, C. G., Gorbi, S., & Regoli, F. (2017). Plastics and microplastics in the oceans: from emerging pollutants to emerged threat. *Marine Environmental Research*, *128*, 2–11. https://doi.org/10.1016/j.marenvres.2016.05.012

Aw, T. G., & Rose, J. B. (2012). Detection of pathogens in water: from phylochips to qPCR to pyrosequencing. *Current Opinion in Biotechnology*, *23*(3), 422–430. https://doi.org/10.1016/j.copbio.2011.11.016

Aziz, S., Mazhar, A. R., Ubaid, A., Shah, S. M. H., Riaz, Y., Talha, T., & Jung, D.-W. (2024). A comprehensive review of membrane-based water filtration techniques. *Applied Water Science*, *14*(8), 169. https://doi.org/10.1007/s13201-024-02226-y

Beans, C. (2023). Are microplastics spreading infectious disease? *Proceedings of the National Academy of Sciences of the United States of America*, *120*(31), e2311253120. https://doi.org/10.1073/pnas.2311253120

Borremans, B., Faust, C., Manlove, K. R., Sokolow, S. H., & Lloyd-Smith, J. O. (2019). Cross-species pathogen spillover across ecosystem boundaries: mechanisms and theory. *Philosophical Transactions of the Royal Society B: Biological Sciences*, *374* (1782), 20180344. https://doi.org/10.1098/rstb.2018.0344

Botes, M., de Kwaadsteniet, M., & Cloete, T. E. (2013). Application of quantitative PCR for the detection of microorganisms in water. *Analytical and Bioanalytical Chemistry*, *405*(1), 91–108. https://doi.org/10.1007/s00216-012-6399-3

Bowley, J., Baker-Austin, C., Porter, A., Hartnell, R., & Lewis, C. (2021). Oceanic hitchhikers – assessing pathogen risks from marine microplastic. *Trends in Microbiology*, *29*(2), 107–116. https://doi.org/10.1016/j.tim.2020.06.011

Climate Change: Global Temperature | NOAA Climate.Gov. (2024, January 18). https://www.climate.gov/news-features/understanding-climate/climate-change-global-temperature

Cohen, R. E., James, C. C., Lee, A., Martinelli, M. M., Muraoka, W. T., Ortega, M., Sadowski, R., Starkey, L., Szesciorka, A. R., Timko, S. E., Weiss, E. L., & Franks, P. J. S. (2018). Marine host-pathogen dynamics: influences of global climate change. *Oceanography*, *31*(2), 182–193. https://www.jstor.org/stable/26542664

Dai, C., Yuan, F., Wang, D., Yang, X., Du, J., Yu, W., & Zhang, C. (2024). Settling velocity of submillimeter microplastic fibers in still water. *Science of the Total Environment*, *907*, 168054. https://doi.org/10.1016/j.scitotenv.2023.168054

Ellwanger, J. H., & Chies, J. A. B. (2021). Zoonotic spillover: understanding basic aspects for better prevention. *Genetics and Molecular Biology*, *44*(Suppl 1), e20200355. https://doi.org/10.1590/1678-4685-GMB-2020-0355

Emenike, E. C., Okorie, C. J., Ojeyemi, T., Egbemhenghe, A., Iwuozor, K. O., Sahu, O. D., Okoro, H. K., & Adeniyi, A. G. (2023). From oceans to dinner plates: the impact of microplastics on human health. *Heliyon*, *9*(10), e20440. https://doi.org/10.1016/j.heliyon.2023.e20440

Environment, U. N. (2017, August 18). *Coastal Zone Management | UNEP—UN Environment Programme*. https://www.unep.org/topics/ocean-seas-and-coasts/regional-seas-programme/coastal-zone-management

Escudero-Pérez, B., Lalande, A., Mathieu, C., & Lawrence, P. (2023). Host–pathogen interactions influencing zoonotic spillover potential and transmission in humans. *Viruses*, *15*(3), 599. https://doi.org/10.3390/v15030599

Farrell, C., Hassard, F., Jefferson, B., Leziart, T., Nocker, A., & Jarvis, P. (2018). Turbidity composition and the relationship with microbial attachment and UV inactivation efficacy. *Science of the Total Environment*, *624*, 638–647. https://doi.org/10.1016/j.scitotenv.2017.12.173

Figgatt, M. (2017). Harmful algal bloom–associated illnesses in humans and dogs identified through a pilot surveillance system—New York, 2015. *MMWR. Morbidity and Mortality Weekly Report*, *66*, 1182–1184. https://doi.org/10.15585/mmwr.mm6643a5

Fleming, L. E., Broad, K., Clement, A., Dewailly, E., Elmir, S., Knap, A., Pomponi, S. A., Smith, S., Solo Gabriele, H., & Walsh, P. (2006). Oceans and human health: emerging public health risks in the marine environment. *Marine Pollution Bulletin*, *53*(10–12), 545–560. https://doi.org/10.1016/j.marpolbul.2006.08.012

Griffin, D. W., Donaldson, K. A., Paul, J. H., & Rose, J. B. (2003) Pathogenic human viruses in coastal waters. *Clinical Microbiology Reviews*, *16*(1), 129–143. https://doi.org/10.1128/CMR.16.1.129-143.2003

Guo, M., Noori, R., & Abolfathi, S. (2024). Microplastics in freshwater systems: dynamic behaviour and transport processes. *Resources, Conservation and Recycling*, *205*, 107578. https://doi.org/10.1016/j.resconrec.2024.107578

Hansen, C. B., Kerrouche, A., Tatari, K., Rasmussen, A., Ryan, T., Summersgill, P., Desmulliez, M. P. Y., Bridle, H., & Albrechtsen, H. J. (2019). Monitoring of drinking water quality using automated ATP quantification. *Journal of Microbiological Methods*, *165*, 105713. https://doi.org/10.1016/j.mimet.2019.105713

Ho, Y.-P., & Reddy, P. M. (2010). Identification of pathogens by mass spectrometry. *Clinical Chemistry*, *56*(4), 525–536. https://doi.org/10.1373/clinchem.2009.138867

Kakade, A., Mi, J., & Long, R. (2024). Microplastics in the world oceans and strategies for their control. *Reviews of Environmental Contamination and Toxicology*, *262*(1), 20. https://doi.org/10.1007/s44169-024-00069-x

Kerlin, K. E. (2022, April 26). *Pathogens Can Hitch a Ride on Plastic to Reach the Sea*. UC Davis. https://www.ucdavis.edu/health/news/pathogens-can-hitch-ride-plastic-reach-sea

Keesing, F., & Ostfeld, R. S. (2021). Impacts of biodiversity and biodiversity loss on zoonotic diseases. *Proceedings of the National Academy of Sciences*, *118*(17), e2023540118. https://doi.org/10.1073/pnas.2023540118

Kuhlisch, C., Shemi, A., Barak-Gavish, N., Schatz, D., & Vardi, A. (2024). Algal blooms in the ocean: hot spots for chemically mediated microbial interactions. *Nature Reviews Microbiology*, *22*(3), 138–154. https://doi.org/10.1038/s41579-023-00975-2

Letko, M., Seifert, S. N., Olival, K. J., Plowright, R. K., & Munster, V. J. (2020). Bat-borne virus diversity, spillover and emergence. *Nature Reviews Microbiology*, *18*(8), 461–471. https://doi.org/10.1038/s41579-020-0394-z

Li, G., Sun, J., Pan, S., Li, W., Zhang, S., Wang, Y., Sun, X., Xu, H., & Ming, L. (2019). Comparison of the performance of three blood culture systems in a Chinese tertiary-care hospital. *Frontiers in Cellular and Infection Microbiology*, *9*, 285. https://doi.org/10.3389/fcimb.2019.00285

Liao, H., Lyon, C. J., Ying, B., & Hu, T. (2024). Climate change, its impact on emerging infectious diseases and new technologies to combat the challenge. *Emerging Microbes & Infections*, *13*(1), 2356143. https://doi.org/10.1080/22221751.2024.2356143

Liu, S., Glamore, W., Tamburic, B., Morrow, A., & Johnson, F. (2022). Remote sensing to detect harmful algal blooms in inland waterbodies. *Science of the Total Environment*, *851*, 158096. https://doi.org/10.1016/j.scitotenv.2022.158096

Masó, M., Garcés, E., Pagès, F., & Camp, J. (2003). Drifting plastic debris as a potential vector for dispersing Harmful Algal Bloom (HAB) species. *Scientia Marina*, *67*(1), 107–111. https://scientiamarina.revistas.csic.es/index.php/scientiamarina/article/view/459/view/459

Meyer, C. T., Lynch, G. K., Stamo, D. F., Miller, E. J., Chatterjee, A., & Kralj, J. M. (2023). A high-throughput and low-waste viability assay for microbes. *Nature Microbiology*, *8*(12), 2304–2314. https://doi.org/10.1038/s41564-023-01513-9

Newton, A. R., & Melaram, R. (2023). Harmful algal blooms in agricultural irrigation: risks, benefits, and management. *Frontiers in Water*, *5*, 1325300. https://doi.org/10.3389/frwa.2023.1325300

Oon, Y.-L., Oon, Y.-S., Ayaz, M., Deng, M., Li, L., & Song, K. (2023). Waterborne pathogens detection technologies: advances, challenges, and future perspectives. *Frontiers in Microbiology*, *14*, 1286923. https://doi.org/10.3389/fmicb.2023.1286923

Our impact on the cosmetics industry. Beat the Microbead. Retrieved December 7, 2024, from https://www.beatthemicrobead.org/impact/global-impact/

Park, S. Y., & Kim, C. G. (2019). Biodegradation of micro-polyethylene particles by bacterial colonization of a mixed microbial consortium isolated from a landfill site. *Chemosphere*, *222*, 527–533. https://doi.org/10.1016/j.chemosphere.2019.01.159

Pauciullo, S., Zulian, V., La Frazia, S., Paci, P., & Garbuglia, A. R. (2024). Spillover: mechanisms, genetic barriers, and the role of reservoirs in emerging pathogens. *Microorganisms*, *12*(11), 2191. https://doi.org/10.3390/microorganisms12112191

Perez, L. J., Forberg, K., Cloherty, G. A., & Berg, M. G. (2023). Temporal and coevolutionary analyses reveal the events driving the emergence and circulation of human mamastroviruses. https://doi.org/10.1080/22221751.2023.2217942

Periyasamy, A. P., & Tehrani-Bagha, A. (2022). A review on microplastic emission from textile materials and its reduction techniques. *Polymer Degradation and Stability*, *199*, 109901. https://doi.org/10.1016/j.polymdegradstab.2022.109901

Plowright, R. K., Parrish, C. R., McCallum, H., Hudson, P. J., Ko, A. I., Graham, A. L., & Lloyd-Smith, J. O. (2017). Pathways to zoonotic spillover. *Nature Reviews Microbiology*, *15*(8), 502–510. https://doi.org/10.1038/nrmicro.2017.45

Rompré, A., Servais, P., Baudart, J., de-Roubin, M.-R., & Laurent, P. (2002). Detection and enumeration of coliforms in drinking water: current methods and emerging approaches. *Journal of Microbiological Methods*, *49*(1), 31–54. https://doi.org/10.1016/S0167-7012(01)00351-7

Sharma-Kuinkel, B. K., Rude, T. H., & Fowler, V. G. (2016). Pulse field gel electrophoresis. *Methods in Molecular Biology*, *1373*, 117–130. https://doi.org/10.1007/7651_2014_191

Sommerfeld Ross, S., Tu, M. H., Falsetta, M. L., Ketterer, M. R., Kiedrowski, M. R., Horswill, A. R., Apicella, M. A., Reinhardt, J. M., & Fiegel, J. (2014). Quantification of confocal images of biofilms grown on irregular surfaces. *Journal of Microbiological Methods*, *100*, 111–120. https://doi.org/10.1016/j.mimet.2014.02.020

Sokolow, S. H., Nova, N., Pepin, K. M., Peel, A. J., Pulliam, J. R. C., Manlove, K., Cross, P. C., Becker, D. J., Plowright, R. K., McCallum, H., & De Leo, G. A. (2019). Ecological interventions to prevent and manage zoonotic pathogen spillover. *Philosophical Transactions of the Royal Society B: Biological Sciences*, *374* (1782), 20180342. https://doi.org/10.1098/rstb.2018.0342

Strauss, A. T., Hobbie, S. E., Reich, P. B., Seabloom, E. W., & Borer, E. T. (2024). The effect of diversity on disease reverses from dilution to amplification in a 22-year biodiversity × N × CO_2 experiment. *Scientific Reports*, *14*(1), 10938. https://doi.org/10.1038/s41598-024-60725-z

Sun, B., Sun, B., Zhang, B., & Sun, L. (2022). Temperature induces metabolic reprogramming in fish during bacterial infection. *Frontiers in Immunology*, *13*, 1010948. https://doi.org/10.3389/fimmu.2022.1010948

Thushari, G. G. N., & Senevirathna, J. D. M. (2020). Plastic pollution in the marine environment. *Heliyon*, *6*(8), e04709. https://doi.org/10.1016/j.heliyon.2020.e04709

Wani, A. K., Akhtar, N., Naqash, N., Rahayu, F., Djajadi, D., Chopra, C., Singh, R., Mulla, S. I., Sher, F., & Américo-Pinheiro, J. H. P. (2023). Discovering untapped microbial communities through metagenomics for microplastic remediation: recent advances, challenges, and way forward. *Environmental Science and Pollution Research*, *30*(34), 81450–81473. https://doi.org/10.1007/s11356-023-25192-5

Wasik, B. R., De Wit, E., Munster, V., Lloyd-Smith, J. O., Martinez-Sobrido, L., & Parrish, C. R. (2019). Onward transmission of viruses: how do viruses emerge to cause epidemics after spillover? *Philosophical Transactions of the Royal Society B: Biological Sciences*, *374*(1782), 20190017. https://doi.org/10.1098/rstb.2019.0017

Weiss, R. A., & Sankaran, N. (2022). Emergence of epidemic diseases: zoonoses and other origins. *Faculty Reviews*, *11*, 2. https://doi.org/10.12703/r/11-2

Wells, K., & Clark, N. J. (2019). Host specificity in variable environments. *Trends in Parasitology*, *35*(6), 452–465. https://doi.org/10.1016/j.pt.2019.04.001

Wißmann, J. E., Kirchhoff, L., Brüggemann, Y., Todt, D., Steinmann, J., & Steinmann, E. (2021). Persistence of pathogens on inanimate surfaces: a narrative review. *Microorganisms*, *9*(2), 343. https://doi.org/10.3390/microorganisms9020343

Zettler, E. R., Mincer, T. J., & Amaral-Zettler, L. A. (2013). Life in the "plastisphere": microbial communities on plastic marine debris. *Environmental Science & Technology*, *47*(13), 7137–7146. https://doi.org/10.1021/es401288x

Zhai, X., Zhang, X.-H., & Yu, M. (2023). Microbial colonization and degradation of marine microplastics in the plastisphere: a review. *Frontiers in Microbiology*, *14*, 1127308. https://doi.org/10.3389/fmicb.2023.1127308

Zhang, E., Kim, M., Rueda, L., Rochman, C., VanWormer, E., Moore, J., & Shapiro, K. (2022). Association of zoonotic protozoan parasites with microplastics in seawater and implications for human and wildlife health. *Scientific Reports*, *12*(1), 6532. https://doi.org/10.1038/s41598-022-10485-5

Zhang, S., Li, X., Wu, J., Coin, L., O'Brien, J., Hai, F., & Jiang, G. (2021). Molecular methods for pathogenic bacteria detection and recent advances in wastewater analysis. *Water*, *13*(24), 3551. https://doi.org/10.3390/w13243551

Ziani, K., Ioniță-Mîndrican, C.-B., Mititelu, M., Neacșu, S. M., Negrei, C., Moroșan, E., Drăgănescu, D., & Preda, O.-T. (2023). Microplastics: a real global threat for environment and food safety: a state of the art review. *Nutrients*, *15*(3), 617. https://doi.org/10.3390/nu15030617

Index

For Product Safety Concerns and Information please contact our EU representative GPSR@taylorandfrancis.com Taylor & Francis Verlag GmbH, Kaufingerstraße 24, 80331 München, Germany